AF846698

Anthropocene – Humanities and Social Sciences

Series Editors
Nathanaël Wallenhorst, Université Catholique de l'Ouest (UCO), Angers, France
Christoph Wulf, Anthropologie und Erziehung, Freie Universität Berlin, Berlin, Germany

This series is the first collection to address the subject of education within the Anthropocene. The inspiration behind the series is a dialogue between Earth system science (ESS) and human and social sciences (which, when their object of study is the socio-political implications of our entering the Anthropocene are now officially called the environmental humanities). The series offers perspectives from a variety of disciplines. The raison d'être of this new collection is to draw together high quality studies dealing with the gravity of today's environmental situation whilst also viewing the concept of the Anthropocene from a critical perspective by expanding on work engaged in in human and social sciences over recent years.

The main objectives of this series are to:

- bring together studies from Earth system science and the environmental humanities in the same collection, from an educational angle
- engage with the question of the future and how to prepare for it in a time of uncertainty

This series welcomes topics including, but not limited to:

- pedagogy
- didactics in the Anthropocene
- health education
- education for governance and democracy
- addressing the challenges of contemporary societies and ecological changes
- transforming agricultural production methods
- preparing for the future

The collection would make a valuable contribution to the academic world and research, which is currently seeking new editorial spaces for us to publish scientific works on the implications of entering into a new geological epoch.

Works published in this series include monographs, edited volumes, and handbooks and textbooks.

The series is intended for researcher working on the Anthropocene with all its socio-political implications and also students in different disciplines following courses on environmental issues. It may, however, also be of interest to all social stakeholders interested in environmental questions, who are becoming increasingly numerous including, but not limited to, policy makers, activists, entrepreneurs.

We aim to make a first decision within 1 month of submission.

In case of a positive first decision the work will be provisionally contracted: the final decision about publication will depend upon the result of the anonymous peer review of the complete manuscript. We aim to have the complete work peer-reviewed within 3 months of submission.

Proposals should include:

- A short synopsis of the work or the introduction chapter
- The proposed Table of Contents • CV of the lead author(s)
- List of courses for possible course adoption

The series discourages the submission of manuscripts that are below 75,000 words in length.

For more information, please contact publishing editor, christi.lue@springer.com

Nathanaël Wallenhorst

A Critical Theory for the Anthropocene

Springer

Nathanaël Wallenhorst
Dean of the Faculty of Education
Université catholique de l'Ouest (UCO)
Angers, France

ISSN 2731-6343 ISSN 2731-6351 (electronic)
Anthropocene – Humanities and Social Sciences
ISBN 978-3-031-37737-2 ISBN 978-3-031-37738-9 (eBook)
https://doi.org/10.1007/978-3-031-37738-9

Translation of the academic work written in French: "Une théorie critique pour l'Anthropocène" (unpublished) funded by the "Chaire interculturalité" (University of Haute Alsace, France). The translation is by Ben Engel.

This work was supported by Chaire interculturalité de l'Université de Haute Alsace (France)

© The Editor(s) (if applicable) and The Author(s), under exclusive license to Springer Nature Switzerland AG 2023
This work is subject to copyright. All rights are solely and exclusively licensed by the Publisher, whether the whole or part of the material is concerned, specifically the rights of reprinting, reuse of illustrations, recitation, broadcasting, reproduction on microfilms or in any other physical way, and transmission or information storage and retrieval, electronic adaptation, computer software, or by similar or dissimilar methodology now known or hereafter developed.
The use of general descriptive names, registered names, trademarks, service marks, etc. in this publication does not imply, even in the absence of a specific statement, that such names are exempt from the relevant protective laws and regulations and therefore free for general use.
The publisher, the authors, and the editors are safe to assume that the advice and information in this book are believed to be true and accurate at the date of publication. Neither the publisher nor the authors or the editors give a warranty, expressed or implied, with respect to the material contained herein or for any errors or omissions that may have been made. The publisher remains neutral with regard to jurisdictional claims in published maps and institutional affiliations.

This Springer imprint is published by the registered company Springer Nature Switzerland AG
The registered company address is: Gewerbestrasse 11, 6330 Cham, Switzerland

Paper in this product is recyclable.

Foreword

The Chair in Interculturalities at the University of Haute-Alsace is supporting the publication of *A critical theory for the Anthropocene* for at least two reasons:

- To provide the scientific community with a reference tool in the field, useful for both experts and young researchers;
- To resonate with its own line of research, Interculturality for Transitions.

Over the last few years, the concept of transition has played an increasingly important role in thinking and acting towards a more resilient society in the face of the crises inherent to the Anthropocene era (climatic, energetic, biologic, social, economic, etc.). Historically, it is closely linked to the concept of sustainable development. Yet this concept is being challenged by some of those promoting the transition concept, who are claiming that it is more operational, more concrete and more effective than the "empty concept" of sustainable development (Bourg, 2012), whose economic, social and environmental goals (Morand-Deviller, 2019) do not seem compatible in a long-term systemic approach.

While government officials regularly refer to it in their speeches, the notion of transition mainly emphasises concrete investment, based on citizens' empowerment, through local initiatives and approaches that are considered to be more tangible. The diversity and wealth of in situ experiments are seen as guarantees of resilience and contextual adaptation, making it possible to accelerate system change while directly involving citizens. However, some of the pitfalls specific to this approach seem similar to those encountered by sustainable development, in particular the issue of transforming the macro-economic and societal model. The transition must demonstrate the capacity of local initiatives to form a system, by changing scale. Another issue lies in the inclusion of such an approach in a predominantly English-speaking international framework, which is still very much focused on sustainable development. Ultimately, the aim remains to collectively build new ways of living together that are fair, respectful of the environment and adapted to local contexts.

- The main objective of the Chair in Interculturalities is to explore and discuss the opportunities and challenges of transition from a dual intercultural perspective: The "diverse diversities" (Dervin, 2012) of the participants will enable us to engage in an inter- and cross-disciplinary dialogue in order to identify issues, avenues of research and potential joint projects. Through our shared reflections, we hope gradually to build a common framework around the notion of transition;
- The operational dimension of the concept of transition invites us to reflect on how and at what level dialogues can be imagined in order to initiate the many transitions that seem necessary today. The profound systemic changes that these perspectives entail require a dialectic that will gradually lead to the establishment of new societal systems. In this context, intercultural dynamics need to be observed in order to gradually increase the effectiveness of the dialogues leading to transition.

The Chair's contribution to this book is thus a continuation of our work, of intercultural encounters and interdisciplinary dialogues, in a perspective of reflective, critical and forward-looking questioning, particularly in the Anthropocene.

Université de Haute-Alsace
Mulhouse, France
Loïc Chalmel
loic.chalmel@uha.fr

Université de Haute-Alsace
Mulhouse, France
Johann Chalmel
johann.chalmel@uha.fr

Bibliography

Bourg, D., & Papaux A. (dir.). (2015). « Article *Transition* ». *Dictionnaire de la pensée écologique*, Ed. PUF.

Bourg, D. (2012). « Transition écologique, plutôt que développement durable ». In *Penser le développement durable*, Victoires Éditions.

Dervin, F. (2011). *Impostures interculturelles*. L'Harmattan.

Acknowledgment

Special thanks go to my editor Christi Lue, who believed in this project from the start and honoured me with her trust, and to Professor Christoph Wulf with whom we are working to develop living ways of becoming human in the twenty-first century.

I would like to thank Professor Loïc Chalmel, who chairs the "Interculturality Chair" (Université de Haute Alsace, France) for the grant to supervise the translation.

Finally, I would like to thank Ben Engel for his extremely valuable and thorough help with the translations.

Warning to the reader

This work was produced in French before being translated into English. The bibliography is made up of the books and articles with which I have worked (whose references can be found in the bibliography), in French, English and German. When a book originally written in French or German has been translated into English, I indicate this to the reader in brackets, but when I quote a book I indicate the page from which it comes in the version I worked with (in French or in German), the translation is then mine.

Contents

Chapter 1
Foundations of a Critical Theory for the Anthropocene

Abstract This first chapter is a general introduction. It lays the foundations of a critical theory for the Anthropocene. It is organised into four parts: (1) Between Prometheism and post-Prometheism (polemical function); (2) Linking Earth, politics and education to prepare for the future (inventive function); (3) Conviviality as a paradigm of political education (creative function); (4) The proposed way forward.

Keywords Critical theory · Anthropocene · Prometheism · Postprometheism · Earth system

A lifeless planet undergoes changes over time: it evolves in response to changes in the level of energy received from the star around which it orbits; it may be impacted by objects from outer space; and it is subject to its own geological forces. Our Earth, estimated to be 4.54 billion years old (Dalrymple, 2001), is not lifeless: it has been influenced by lifeforms over the course of billions of years. Thus, cyanobacteria[1] have gradually facilitated the emergence of an atmosphere including CO_2. Over time, plants appeared, and decreased the Earth's albedo,[2] thus increasing the temperature (Wolff, 2014, p. 255). 541 million years ago, the so-called "Cambrian explosion" began, characterised by the complexification and diversification of life, where all the major animal phyla appeared. This transformed the biosphere. From that point on, life and the Earth have been continuously interacting. On five occasions, in less than two million years, climate changes wped out more than 75% of species (both animal and plant), on land and in the oceans: 445 million years ago; 380 million years ago; 252 million years ago; 200 million years ago; and 66 million years ago, this last mass extinction event marking the end of the Cretaceous era, during which the dinosaurs died out (Barnosky et al., 2011). Around 2.6 million years ago, the genus *Homo habilis* – the first member of the genus *Homo*, with the

---

[1] Cyanobacteria are algae that absorb CO_2 from the atmosphere and release O_2.

[2] The Earth's albedo is the capacity of the Earth's surface to reflect solar radiation (albedo is the reflectivity of a surface).

© The Author(s), under exclusive license to Springer Nature Switzerland AG 2023

N. Wallenhorst, *A Critical Theory for the Anthropocene*, Anthropocene – Humanities and Social Sciences, https://doi.org/10.1007/978-3-031-37738-9_1

ability to craft tools out of stone – appeared in Africa. This is considered a geological marker, as it marks the beginning of the Quaternary geological period. 1.6 million years later, the genus *Homo erectus* appeared, representing a new departure in the species, with their ability to manipulate fire. The mastery of this technique, in addition to being useful for hunting and protection from nocturnal predators, allowed the first hominids access to a protein-rich diet, leading to a leap forward in their physical and mental capacities. The change in hominids' diet made possible by fire led to the tripling of brain volume (Steffen et al., 2007, p. 614). As a result, humans now have the highest brain-to-body ratio of all animals, with brains measuring about 1300 cm^3. This is what allowed the complexification of language, and subsequently the development of civilisations and writing (Steffen et al., 2011a, p. 846). Hominids very quickly learned to use tools to alter their environment and increase their power to act. The earliest carved stones of which we are currently aware date from 2.6 million years ago (Semaw et al., 2003) and are attributed to *Homo habilis*.[3]

Having a transformative effect on your environment is no bad thing. Indeed, it is the very mark of life. Like all living things, humans need to interact with their environment and transform it to their advantage. Humans, like many other living things, carve out niches within their ecosystems. However, one of the crucial singularities of humanity is that we are an ultra-social species, who build very significant socio-cultural niches in ecosystems (Ellis, 2015, p. 287). The appearance of anatomically modern humans in Africa is dated to about 350,000 years ago (Hubelin et al., 2017).[4] It is through their impact on the environment that palaeontologists can track the movements of *Homo sapiens* across the globe. Indeed, in each place explored by *Homo sapiens*, an extinction of the megafauna (animals larger than 44 kg) is observed. *Homo sapiens* is known to have arrived in Australia about 46,000 years ago, and in the Americas about 15,000 or 14,000 years ago (Williams et al., 2016, p. 8). Another important element in *Homo sapiens*' historical interaction with their environment is the emergence of agriculture. Around 10,000 years ago, simultaneously in four different locations around the globe, *Homo sapiens* developed agriculture, made possible by climate stabilization (Steffen et al., 2011b, p. 741). While 95% of *Homo sapiens*' habitation of the planet was marked by a nomadic, hunter-gatherer existence, the other 5% is defined by settled agriculture and animal husbandry, enabled by the favorable climate of the Holocene (Zalasiewicz et al., 2008).

Agriculture marks a turning point in humans' relationship with their environment and with each other. Humans learned to control ecosystems and to produce surplus crops and livestock for their own benefit. Under their influence, ecosystems underwent a progressive transformation that was eventually to become radical and

[3] We have recently found traces of the use of cut stones dating back to about 3.39 million years ago (where stone puncture marks were identified on bones, showing that hominids cutting the meat and extracted the marrow) (McPherron et al., 2010), calling pre-existing knowledge into question.

[4] This is a point where thinking has evolved. While it was previously thought that *Homo sapiens was* 195,000 years old (McDougall et al., 2005), a 2017 paper by French paleo-anthropologist Jean-Jacques Hubelin et al. seems to have proven that *Homo sapiens* is about 350,000 years old.

irreversible. This transformation was concomitant with the development of urban civilisations (in Mesopotamia, Egypt, India, China, etc.), with monotheistic religions and hierarchical societies. After a few thousand years of sedentary existence, humans discovered the energy of fossil coal, which was first used during the Song dynasty in China, about a thousand years ago, in the iron industry. Fossil coal mining then developed in Europe during the thirteenth century in England, reaching 360,000 tonnes per year by 1600. Elsewhere, other energy sources were being used, such as wood and charcoal (Steffen et al. 2011a, p. 846). A little over a century later, the British learned how to transform this energy, thanks to the steam engine, patented by the Scotsman James Watt. During the nineteenth century, there was a turning point in the history of humans' interaction with their environment. This ability to transform energy made it possible to industrialise production in various parts of the world. This caught the attention of a group of authors who were the forefathers of ecological thought. Examples of their work can be read in the anthology of ecological thought by Bourg and Fragnière (2014). Their number includes the French lawyer Eugène Huzar who published *The End of the World through Science* (1855/2008); the American diplomat and philologist George Perkins Marsh, author of *The earth as modified by human action* (1874/1970); the French geographer and activist Elisée Reclus author of *La Terre: description des phénomènes de la vie du globe* (*The Earth: description of the phenomena of the life of the globe*, 1881); the French ecologist and naturalist Maurice de Tribolet, author of a study on 'Les animaux disparus depuis l'apparition de l'homme' ("Animals that have become extinct since the appearance of man", 1886); the Russian geographer and meteorologist Alexander Woeikof, author of 'The Influence of Man on the Earth' (1901); the German geologist Ernst Fischer who published 'Man as a geological factor' ('*Der Mensch als geologischer Faktor*') in 1915; or the Russian geochemist Vladimir Vernadsky with, among other works, 'The autotrophy of mankind' in 1925.[5]

During the 19th and 20th centuries, the use of tools to work with materials was refined and developed, with the emergence of a series of techno-sciences, culminating in the discovery of thermonuclear fusion and the invention of the atomic bomb. From the 1950s onwards, everything gathered pace. As the result of prodigious mastery of matter marked by belief in the boundlessness of technoscience, of nature's resources, of emancipation through consumption, we are now entering a new stage in our relationship with the Earth. This new stage is marked by uncertainty about the future, and the need to choose the direction that human beings wish to take in their continuing 'adventure' of life. For some time now, one species, *Homo sapiens*, has been having an increasing impact on its environment. Today, this impact is becoming global, and is considered to be unsustainable by researchers in Earth System Sciences. Indeed, the impact of the anthropogenic alterations to the chemical make-up of the atmosphere and oceans will be felt for thousands of years to come. The Earth is now entering a new geological epoch: the Anthropocene. We, humans, have permanently altered the habitable conditions of the biosphere, and humanity's very

[5] This list is not exhaustive; other scientists have alerted their contemporaries (Bourg & Fragnière, 2014).

survival is at risk. Human beings have acquired geological power, and have tipped the Earth into a new geological epoch that is far less favourable to human life. *Anthropos* is at the origin and centre of the Anthropocene.

The scale of ecosystem transformation was increased tenfold by the thermo-industrial and productivist revolution, then by the generalization of infinite trade based on planned obsolescence, and finally by the combination of nuclear fusion techno-science and military policies during WWII. There is clearly strong causality whereby humans are responsible for the planet's entry into the Anthropocene. Yet is there not a causal origin common to all human civilisations in their efforts to master nature in order to generate surpluses for human benefit, bringing about radical and irreversible ecosystemic transformations? A number of questions arise. *Homo sapiens* is a social and political species, with consistent differentiations between its different members. Are all human beings responsible for these global environmental changes and their impact on the Earth system as a whole? How have we transformed our environment to such a degree without a political choice? Thus, is man truly "by nature" a political animal (Aristotle, *Politics, I.*, 2)? Is it possible to heighten the political component of the human adventure? What type of anthropological evolution would be necessary for such a consolidation? In the face of the current situation, do we not need a profound shift in human behaviour in order to ensure the survival of the species?

The major risk of the contemporary period is the degradation of the conditions of human life on Earth. Given the imminence of this threat, it is important to think about ways in which humanity can survive, both from the point of view of the very conditions of physical existence, and from the point of view of the idea of humanity (Bourg, 2009, p. 71). It is thus necessary to think afresh about human nature and the human condition, in the context of existential reflection (Arnsperger, 2010, p. 40) to lay the groundwork for other ways of thinking and acting (Flahault, 2005, p. 382), placing emphasis on the long term. The dawn of the Anthropocene brings this political question back to the forefront, in a new light. The question has been addressed by authors from Kant to Rawls, Plato and Marx. It has ethics at its core: "How can we envisage a viable collective future?" (Braud, 2002, p. 10). The French philosopher Maurice Bellet (Braud, 2002, p. 10) puts the question thus: "How can we 'give ourselves a world and a life to live'?" (1993, p. 164). We appear to be insufficiently prepared for the future, and have failed to truly politically invest in it (Villalba in Badré et al., 2011, pp. 208–211). When we look to the future, in addition to the disquieting prospect of our individual deaths (Arnsperger, 2008, p. 95), we now see the looming spectre of the demise of our species (our adventure will come to an end). The question is not only that of the wellbeing or comfortable life of humanity to come; for some authors, it is the radical question of the very survival of the species (Perret, 2010, p. 106). Clearly, we must think about what unites us and what we, collectively, have to do. In *Discourse on Method* (1637, Part IV), Réné Descartes declared: "I think, therefore I am", setting the tone for the next several centuries of thinking, in which emphasis has been placed on that very 'I' (the individual). However, in today's and tomorrow's world, it is important instead to think in terms of a "we". To whom, though, does this "we" refer? Is it we, who view ourselves as being differentiated from non-humans, whose goal is to become "masters and possessors of nature" (*Discourse on Method*, 1637, Part VI)? Or does the 'we' actually include those non-humans from whom we usually distance ourselves?

Thus, this research works on a set of normative, prospective and speculative questions: what kind of anthropological shift is necessary in the Anthropocene? What lessons must be learned, given the entry into the Anthropocene, to allow the human adventure to endure? What kind of education is needed? What might be some of the principles of educating for politics in the Anthropocene? The object of this work is neither the Anthropocene, nor the concept of the Anthropocene, nor indeed the development of a political sociology in the Anthropocene, but the political anthropology revealed both by the dawn of the Anthropocene and by the work of the Earth System Sciences on the Anthropocene. In light of these reflections, we will then be able to think of the fundamentals of education for politics, within the framework of education *in* the Anthropocene.

There are two driving directions behind these different questions. The first is a questioning of the processes of alienation typical of the modern world, revealed through the materialisation of the Anthropocene, both in the deteriation of the conditions of coexistence and in our representations of the human adventure within the bounds of the Earth. The second is the proposal of a direction or a path (conceptual and to be implemented) by which to combat this alienation, which we perceive as having the potential to annihilate the human adventure. This twofold movement is characteristic of the critical theory of the Frankfurt School. This is a particularly fruitful intellectual tradition of German origin, which has reflected on many of the major issues of the twentieth century and those of the early twenty-first century.[6] The Frankfurt School has the resources to address this almost unthinkable fact: we have altered the climatic and environmental stability of the Holocene, which facilitated the emergence of the world's great civilisations, through agriculture and the production of agricultural surpluses. What anthropological impacts will result from this corruption of the Earth system? What will be the civic implications of the very foundations of human civilisation? To date, no contemporary critical theorist has dealt with the subject of the Anthropocene, and a critical theory for the Anthropocene has yet to be developed.[7] Such a theory would usefully be informed by a critical dialogue with the German philosopher and sociologist Hartmut Rosa, whose two

---

[6] Transhumanism or 9/11 with Jürgen Habermas, the acceleration inherent to capitalism with Hartmut Rosa, or the growth of inequalities and social conflicts with Alex Honneth.

[7] What needs to be developed? A critical theory *in the* Anthropocene? *Of* the Anthropocene or *for* the Anthropocene? Clearly, all three are interesting. It is indeed important, *in* the Anthropocene, to pursue the enterprise of critical theory in order to identify what can help reify humanity, and rid the human adventure of the yoke of capitalism. However, we also need to delve into an increasingly critical theorisation *of* the Anthropocene. Why and how has it come to pass that we have entered the Anthropocene? What are the anthropological and civilisational forces behind it? Finally, it is also necessary, on the basis of these two critical points, to develop a critical theory *for* the Anthropocene: that is, one that allows us to continue to live, think and act in the Anthropocene. It is a matter of learning to weather the storm of the Anthropocene, and emerge intact (if not unchanged). Thus, a critical theory *for* the Anthropocene actually means a critical theory *for* this

major books, *Acceleration* (2013) and *Resonance* (2016), are enjoying worldwide success at the time of writing. These two books reflect the dual nature of the critical theory, in the framework of which Rosa constantly reminds us that his work sits. Rosa identifies *acceleration* as the main alienating force in the contemporary period (which he describes as 'late modernity'): the hegemonic growth imperative of capitalism forces us into a dehumanising spiral of perpetual and unsustainable acceleration (with the pace of life constantly increasing).[8] This destructive alienation of humanity, he proposes, could be countered by what he terms *resonance*. Such resonance operates outside the mechanisms of acceleration, and would allow the emergence of a post-capitalist world. Rosa proposes to haul humanity out of the rut into which we have fallen as the result of unbounded neoliberalism. The ambition seems almost unrealistically far-fetched. Although, in *Resonance*, Rosa displays a real sensitivity to nature, and is aware of the entry into the Anthropocene, this is neither the starting point of his work, nor even the main threat which, in his view, we currently face. Indeed, he sometimes gives the impression of downplaying the importance of the facts of global warming. On the other hand, one element is particularly interesting: another name by which the Anthropocene is referred to is 'the Great Acceleration' (Steffen et al., 2004, 2015; Waters et al., 2016). Could it be that resonance (and its learning) in opposition to acceleration offer the tools we need to chart a course through the tempestuous waters of 'the Great Acceleration'?

* *
*

In Gilles Deleuze's posthumous publication *Lettres et autres textes*, (*Letters and Other Texts*) published in 2015, the French philosopher points to three reasons for a book to exist (pp. 86–87): to correct an error, to remedy an oversight, and to establish a new concept. Thus, a new book or new research has three main functions: a polemic function, an inventive function and a creative function[9] – each of which is only relevant if it is integrated into the scientific debate, and in dialogue with contemporary research.

What error does this research address? Analysis of the scientific literature of the Anthropocene reveals a fundamental anthropological error, which represents a threat to humanity itself. This error is the modern idea that Promethean power-seeking behaviour of *homo oeconomicus* – aiming to maximise their own individual gains, by pushing past boundaries – can work as a social system. This research aims to counter what the Ancient Greeks called *hybris* – destructive madness or excess.

What kind of oversight does this work aim to rectify? This research highlights an oversight: the failure to take account of the Earth, of its bio-geochemical flows of matter, and of the resilient and interdependent fabric of life that is the basis of

---

purpose. The term does not mean 'in favour of' the Anthropocene, or 'in order to favour' it, but on the contrary, 'in order to restore the balance of civilisation in the Anthropocene'.

[8] We conducted a lengthy interview in German with Hartmut Rosa that was translated and rewritten and published twice (Rosa & Wallenhorst, 2017a, b).

[9] This plan, which takes up the three functions noted by Deleuze, is inspired by the organisation of French philosopher Etienne Tassin's preface written to the reissue of his book *Le trésor perdu – Hannah Arendt, l'intelligence de l'action politique*, published by Klincksieck in 2017.

politics, in our thinking. In extending the scope of the debate, this work proposes that society, in the broadest sense, must also include non-human entities.

What novel concepts does this research propose? In the course of this work, three interrelated concepts are developed: that of an anthropological shift allowing for the consolidation of politics, that of the human adventure, and that of convivial citizenship. Thus, this work proposes that convivialism be assimilated as an educational paradigm to accompany an anthropological shift, allowing for the consolidation of politics, enabling the human adventure to survive in the Anthropocene. This constitutes the critical theory for the Anthropocene advanced by this work.

1.1 Between Prometheism and Post-prometheism (Polemic Function)

1.1.1 The Anthropocene at the Heart of the War of Ideas at the Beginning of the Twenty-First Century

The contemporary period is marked by two strong features which have an impact on our globalised societies: (1) the techno-scientific and digital revolution and the convergence into NBIC,[10] with all its implications and commercial applications; and (2) the advent of the Anthropocene, marking an irreversible alteration of the conditions which make Planet Earth habitable, threatening humanity's very survival. Because these major changes impact all spheres of daily life (health, industry, education, military, leisure, transport, food, etc.), there has been a resurgence in political thinking in recent years. Several manifestos have been published, each of which represents a new voice in the public debate, and a new political approach.

First, there is the *Manifesto for an Accelerationist Politics* by Nick Srnicek and Alex Williams. Their 20-page manifesto, published on a website, has been widely commented on and critiqued, and has aroused a great deal of enthusiasm among intellectuals around the world. The idea is simple: it is necessary to accelerate techno-scientific progress in order to bring about a global shift from a capitalist system to a post-capitalist one. The objective is the emergence of workless societies, in which everyone can live in dignity. The underlying anthropology is resolutely Promethean. In the same techno-scientific and Promethean vein but with a capitalist and liberal ethos, the *Ecomodernist Manifesto* was published in 2015, inspired by the eco-pragmatism of the *Breakthrough Institute*. The call for ecological Prometheism is one of this thinktank's defining features. In the eyes of the Breakthrough Institute, the Anthropocene is indicative of humanity's power, not of our *hybris*. The nineteen authors of this manifesto are optimistic about the future,[11]

[10] NBIC stands for nano-bio-info-cogno technology.

[11] One of the foremost proponents of a 'good Anthropocene' is the American geographer Erle C. Ellis, whom we interviewed on the question of a good Anthropocene and the underlying political anthropology. That interview is published in an issue of *Political Reasons* (Ellis et al., 2019).

because of the power humanity has gained through our knowledge of the complex workings of the Earth system, and our technical mastery. Thus, they speak of a '*Great* Anthropocene'. The first was British in origin; the second emerged from the west coast of the United States. In 2013, a third manifesto (of French origin, this time) was published, and made waves around the world: the *Convivialist Manifesto*, which followed in the footsteps of the MAUSS (*Mouvement anti-utilitariste en Sciences sociales* [Anti-Utilitarian Movement in Social Sciences]). Whilst the first two manifestos have opposing political stances (one advocating a regulated, post-capitalist economy; the other a liberalised capitalist economy), what they have in common is their hyper-modernity, with the proposal to see the Promethean activities of the modern era through to their conclusion.[12] The main feature of the *Convivialist Manifesto* which sets it apart from its two predecessors is its post-Promethean focus. Convivialism is defined, in this manifesto, as 'a quest for […] a mode of living together (*con-vivere*) that values human relationships and cooperation and enables us to challenge one another without resorting to mutual slaughter and in a way that ensures consideration for others and for nature' (Srnicek & Williams, 2013, p. 14). The *Convivialist Manifesto* is set against the boundless *hybris* of *homo oeconomicus*. This short text marks the beginning of a post-Promethean political anthropology project. The authors propose to break with the Promethean pursuit of modern individuals' goals, and instead propose cleaving to the conviviality that comes from commonality between individuals. Here, the sharing of human existence is an expression of the post-Prometheism that we define, in this book, as the abandonment of a demiurgic search for power based on transgressions (of the planet's limits, of our own bio/physiological/cognitive bodily limits, of the rules of distribution of goods and wealth between individuals, and so forth).[13] It is necessary to rediscover the art of living together, the convivialists tell us, and it is this rediscovery and renewal, based on interdependence, that we identify as post-Promethean. The *Convivialist Manifesto* would be only a small and unimportant text, were it not for that fact that it represents the coming together of a set of alternative ideas in this

---

This interview is a counterpoint to another interview we conducted with Dominique Bourg, published in the same issue of *Political Reasons* (Bourg et al., 2019).

[12] It should be noted here that the common use of the term 'Promethean' may refer to the Faustian figure of technological man in the twentieth century, and to western use of technology. In line with the analyses of French intellectual Jacques Ellul, the adjective 'Promethean' can be associated with techno-scientific enterprise. The use of the adjective 'Promethean' here signifies an additional dimension to the use of technology. Referring to the myths of Prometheus, this adjective allows us to think of human power in the context of transgression. 'Promethean' here refers to an anthropological characteristic that cannot be reduced to the West. Prometheism is defined in this work as the combination of techno-scientific mastery with a demiurgic quest for power, based on transgression, exemplifying *hybris*.

[13] The idea of transgressing limits is ambivalent here, since it is not so much a question of transgressing laws enacted by a society, but of going beyond (or even far beyond) what might be thought of as a limit by common sense.

period of postmodernity[14] – a period of disenchantment to which the Anthropocene sounds a death knell. Indeed, not only do the 50 authors (since joined by nearly 4000 signatories) mention in their introduction that the common denominator between is this focus on the mode of sharing existence[15]; in recent years, we have seen the publication of numerous manifestos with the same convivialist and post-Promethean orientation. A notable example is the *Animal Manifesto* by the French philosopher Corine Pelluchon, published in 2017, proposing to consider animals, not as citizens, but as political subjects, and thus extending to animals the conviviality at the foundation of a life worth living. Faced with the bleak reality of the Anthropocene, the *Manifest für das Anthropozän* (Manifesto for the Anthropocene)[16] by German philosophers Andreas Weber and Hildegard Kurt, published in 2015, develops a romanticised, poetic approach to coexistence (the authors cite the German translation of the *Convivialist Manifesto* in the conclusion). In this direct convivialist extension, three researchers, Michel Bauwens, Vasilis Kostakis and Alex Pazaitis wrote *The Commons Manifesto – Peer to Peer*. In the light of the development of *Peer-to-Peer* technologies (which have allowed things like Wikipedia, freeware, Facebook and Uber to emerge), these authors identify the emergence of a new relational style allowing the advent of a post-capitalist world, provided, of course, that they are not appropriated by capitalism, as is precisely what happened with Facebook and Uber. In a less academic version, two manifestos were published in early 2019, acknowledging the entry into the Anthropocene and expounding on the need to transform the way in which we live together. The first of these, the *Manifesto for Climate Justice,*[17] was published in March 2019. It is signed by '*Notre Affaire À Tous*' (Our Common Cause) – a movement that originated in 2015 in the Netherlands, where citizens are committed to reducing their greenhouse-gas emissions. The movement supports a collection of citizens and families who are

---

[14] It is important to clarify the meaning of the adjective 'postmodern', which will be used regularly in this work. It first appeared in Lyotard's *The Postmodern Condition*, and has been used in the social sciences to describe the emerging post-industrial society (Baudouin, 2002, p. 50). This postmodernity is mainly characterised by the fall of Grand Narratives allowing us to 'decipher the meaning of the human condition' (Baudouin, 2002, p. 51). For Maurice Bellet, whose work we shall study extensively in this book, postmodernity refers to the period when 'we no longer know where we are going' (Bellet, 2008, p. 642), and Man is confronted with a 'destructive power [...] which is constantly chipping away at the support mechanisms' (Bellet, 1993, p. 139). Postmodernism can be understood as an 'idea of delegitimisation [which] denounces the imperialism of scientific reason that aims to optimise the efficiency of global systems, through technological performance' (Lambert, 2001, p. 73). The end of the twentieth century marks postmodernity, where faith in humanity has been eroded, generating doubt about the future and an inability to anticipate it. Thus, it is important in postmodernity to revisit fundamental questions about the meaning of the human condition and the conditions for a human life. The work of Boutinet (1990, 1998, 2004, 2010) is the main basis for using the adjective 'postmodern' to describe the contemporary period in this work.

[15] The authors also point out that a set of differences remain.

[16] The full title is *Lebendigkeit sei! Für eine Politik des Lebens. Ein Manifest für das Anthropozän*, i.e. *Let there be vitality! For a politics of life. A manifesto for the Anthropocene.*

[17] The full title is *How We Will Save the World – Manifesto for Climate Justice*.

suing lawmakers or large corporations. *Notre Affaire À Tous* has joined forces with Greenpeace France, Oxfam France and the *Fondation pour la nature et l'Homme* (Foundation for Nature and Man) in the 'Case of the Century', bringing a lawsuit against the French State for failing to act upon its stated objectives of reducing greenhouse-gas emissions. How can we change everything together, right now? These activists seem to be writing a solid text based on scientific data.[18] The text opens with particularly strong words: 'It is becoming impossible not to think about it. Not to know. The Earth is hunting us. It cannot take it anymore. Our presence is too painful. In a century of industrial acceleration, we have captured so much energy, and used so many resources, that we have altered the balance of the planet to our sole advantage. So it is heating up, rumbling, sliding into the unknown, towards another version of itself, another ecosystem that is already shaking the foundations of our world' (Notre Affaire À Tous, 2019, p. 3). The tone is in line with another manifesto published at exactly the same time, at the beginning of 2019, by the French MP Delphine Batho[19]: *Ecologie intégrale – Le manifeste*. Written in the same tone as the *Manifesto for Climate Justice*, Delphine Batho's manifesto mentions that all aspects of our existence are going to experience radical upheaval. She states that the sustainability of the human adventure is compromised. It will become increasingly difficult to live together and act in concert, and it is important to rethink what underpins our existence. This is clearly the first time that we have seen agreement between the tones of political and scientific discourse.[20] These two manifestos published in 2019 mark a break in style, and the matters raised are increasingly being examined. Civic engagement is gathering momentum. Changing lifestyles against a backdrop of anthropological mutation as the very condition for the sustainability of the human adventure, are beginning to enter the political arena.[21]

---

[18] A group of researchers were involved in the drafting of this manifesto, including French historian Christophe Bonneuil, French philosopher Catherine Larrère, and French urbanist Guillaume Faburel.

[19] Delphine Batho, born in 1973, served as Minister of Ecology, Sustainable Development and Energy from 21 June 2012 to 2 July 2013 under President François Hollande, and has been a Member of Parliament for Deux-Sèvres since August 2013.

[20] The book is postfaced by Dominique Bourg, and we can see how much his philosophical work inspired this manifesto that brings together all his driving ideas.

[21] It is also worth noting the *Manifesto of the Appalled Economists* (2010) and the *New Manifesto of the Appalled Economists* (2015), initially signed by the French economists Philippe Askenazy, Thomas Coutrot, André Orléan and Henri Sterdyniak, and today signed by more than 10,000 people. In particular, the authors propose abandoning the foundations of financial neoliberalism and emancipation of States from the alienation of the financial markets. What is astonishing is that at no point does the adjective 'atterrés' (Appalled, in the original French) represent wordplay with the 'landing' that the earth could represent for the economy. The thinking is in no way ecological and we do not see a break with the growth paradigm (only with the ways in which wealth is distributed). On the other hand, in *The New Manifesto of the Appalled Economists*, published in 2015, ecology is a thread running through the economic thinking of the collective of authors who signed it.

1.1.2 *The Error of Promethean Anthropology in Terms of Modernity*

The work on the Anthropocene in Earth System Sciences is an interesting indicator of political anthropology issues. The concept of the Anthropocene is now increasingly mobilised in academic literature, mainly in the field of Earth System Sciences, and since 2010, in social sciences or the emerging environmental humanities. As the American environmentalist Erle C. Ellis (2017, p. 526), identified, in 2017, Google Scholar found 42,800 articles have been written containing the term 'Anthropocene', 3000 of which had it in their title. In 2019, the same search engine identified 80,000 articles; then 160,000 in 2020; then 324,000 in May 2022! Rarely in the course of scientific history has there been so rapid a surge in interest in a notion. The notion of the Anthropocene influences researchers, whether they are in Earth System Sciences or the Social Sciences in general. This notion has a power of attraction, fascination, passion or competition. Many researchers change the direction of their work when they encounter the notion of the Anthropocene. The academic content relating to the Anthropocene highlights the way in which this notion from the Earth System Sciences is linked to politics and reveals anthropological issues. Anthropology is predominantly Promethean, marked by acceptance that any kind of technological progress which is widely adopted will inevitably and ineluctably continue.

Is not the Promethean anthropology of modernity – and the acceleration (Rosa, 2013) typical of hypermodernity (Aubert, 2005) – a mistake? This can be seen in the *Accelerationist Manifesto*, and in the majority of related texts (de Sutter, 2016a, b; Cuboniks, 2016; Brassier, 2016). Indeed, Srnicek and Williams (2013, section 03.21) write, 'We declare that only the Promethean politics of maximum control over society and its environment can deal with global problems or achieve win out over capitalism'. They aspire to 'an alternative modernity that neoliberalism is inherently incapable of bringing about' (section 01.1) and believe it is necessary to 'accelerate the process of technological evolution' in order to win social conflicts (paragraph 03.7). This same Promethean anthropology can be found at the heart of the *Ecomodernist Manifesto*: 'A good Anthropocene requires that humans use their ever-increasing technical, economic and social capacities to improve the human condition, stabilise the climate, and protect nature. (...) These technological and socio-economic processes are central to economic modernisation and environmental protection' (Part 1). The tone is decidedly technophilic: 'modern technologies, by making more efficient use of the flows and services of natural ecosystems, offer a real chance to reduce the overall human impact on the biosphere. To adopt these new technologies is to find paths to a good Anthropocene' (Part 3). Here, too, the point is made that the acceleration of Promethean mastery through technology is needed: 'Full access to modern energies is an essential prerequisite for human development and for the decoupling of development from nature' (Part 4);

'Accelerating technological progress will require the active participation of the private sector, entrepreneurs, civil society, and the State' (Part 6). This Promethean anthropological approach, pursuing our quest for power and mastery over nature, is also discernible, underlying a range of scientific articles relating to the Anthropocene (Crutzen, 2002; Steffen et al., 2007; Zalasiewicz et al., 2010; Ellis, 2011a, b; Crutzen & Schwägerl, 2011; Steffen et al., 2011a). This Promethean anthropology is driven by the ambition to develop freedom through increasing human power, collective self-mastery and mastery of the environment. Yet is this development of freedom not an illusion? In fact, this Prometheism is based on an anthropology built on the inevitability of the techno-scientific, capitalist and neo-liberal race to the top,[22] as can be identified on the basis of a number of scientific articles on the Anthropocene. According to this viewpoint, ultimately, humans have no alternative choice to this endless quest for power – even if it means transgressing all anthropological and environmental limits.

Yet today, the idea that we can overcome our limitations through progress is seems to ring hollow, in light of the environmental emergency we now face (Villalba, 2015, p. 59; Arnsperger & Bourg, 2017). Since the work of German philosopher Günther Anders, *Die Antiquiertheit des Menschen – Über die Seele im Zeitalter der zweiten industriellen Revolution*[23] (*The Obsolescence of Man – On the Soul in the Age of the Second Industrial Revolution*), published in Germany in 1956[24] and in which there has been a resurgence of interest of late, some authors have decried the Promethean anthropology of modernity as unsound. This includes the French philosopher François Flahault, in his history of human excess through the analysis of the Promethean imaginary developed during modernity (Flahault, 2008a, b); and French philosopher and environmentalist Hicham-Stéphane Afeissa in his genealogy of ecological discourse (2014); and French philosopher Jean-Philippe Pierron in his analysis of the different modalities of apprehension of water[25] (2013) and his proposal of a poetics of water, aiming to create a new non-utilitarian and non-Promethean imaginary (2018).

[22] This will be demonstrated in the first part of this book.

[23] Günther Anders published a second volume in 1980 entitled *Die Antiquiertheit des Menschen – Über die Zerstörung des Lebens im Zeitalter der dritten industriellen Revolution* (*The Obsolescence of Man – On the Destruction of Life in the Age of the Third Industrial Revolution*).

[24] In order to identify the date of contribution to the scientific and philosophical debate, we refer in this book to the date of the initial publication, working with a translated version published later, or a reprint of the original work. The way the bibliography is referenced identifies the version of the publication we have worked with; for example: 1972 (original edition 1961).

[25] Between the waters of Orpheus, Prometheus and Hermes, corresponding to the premodern, modern and postmodern eras.

1.1.3 Is Man Really 'by Nature a Political Animal'?

Swiss Anthropocene philosopher Alexander Federau ends his book *For a Philosophy of the Anthropocene* (2017) with a reflection on Hannah Arendt's (the German emigree to the USA) prologue in *The human condition,*[26] claiming that it is time for a better answer to the question of the human condition than the revolution that she wishes to lead, leaving planet Earth and rebuilding from scratch with her own creation. How can we think about the human condition in the Anthropocene, and allow everyone an existence less dominated by destructive Promethean *hubris*? Indeed, the Anthropocene reveals the ongoing and future evolution of the human condition. This identification of the novel aspect of the current human condition is one of the most urgent tasks of the contemporary period (Palsson et al., 2013, p. 8). As the Icelandic anthropologist Gísli Pálsson and his co-authors (2013, p. 8) note, *The Human Condition* is a particularly valuable resource for this.[27] Indeed, Arendt clearly indicates the way in which different forms of human activity (labour, work and action) connect human beings to one another and to the material world. Arendt understands the Modern Era as having modified the way in which human beings exist in the world, causing alienation from the artificial world and from nature. Thus, 'A major challenge for the social sciences and humanities is to explore the extent to which the human condition, as analysed by Arendt, has changed in the Anthropocene epoch, and the nature of this change'[28] (Palsson et al., 2013, p. 8). The Promethean anthropology, characterised by inescapability, needs to be criticised, and in criticizing it, it may be necessary to employ an Arendtian reading of the political, with the concept of action being central, characterised by the possibility of human freedom – and the corollary of educability, which is not the same as natality.

[26] Arendt wrote this book in English, entitled *The Human Condition*. This title highlights an important feature of this fundamental anthropology book, which is an analysis of man's relationship with the world, with the aim being to resist totalitarianism. Thus, it has an historical component. The German title is *Vita activa* and articulates Arendt's reflection with its counterpart, to which she would respond many years later with her unfinished book *The Life of the Mind*. It signifies the complementarity of the two human activities – active life and contemplative life – and refers, here, to an anthropological and historical component. The French version is entitled *Condition de l'homme moderne*. The publisher decided not to call it *La condition humaine* to avoid confusion with André Malraux's novel *La condition humaine* published in 1933, which had a major impact on intellectualism in the first half of the twentieth century. The French title emphasises another dimension with a critique of modernity and a historical analysis. All three titles of this book correspond well to the subject matter (which is not always the case with French translations of Arendt's work) but each one highlights an aspect of her work. In this work we will refer to this book by its original title, *The Human Condition*, which is the title of the work that marked the intellectual life of the time. The excerpts quoted, on the other hand, come from the 1961 edition of *Condition de l'homme moderne*, translated by Georges Fradier.

[27] It is also noteworthy that the French climatologist Michel Magny, in his book *Aux racines de l'Anthropocène – Une crise écologique reflet d'une crise de l'homme* (2019) develops a critique of modernity based on Arendtian thinking.

[28] This work was written in French, with quotations originally written in English or German. The original German quotations by Andreas Weber appear as footnotes.

Given that some of the anthropological views in Earth System Sciences are marked by the Promethean ineluctability of a consumerist race to the top, due to distrust of humans' capacity to act in concert, Arendt's conceptions of the political appear to be particularly valuable. Indeed, for Arendt, the political refers precisely to humans' ability to act together. On reading the work of various Earth System researchers on the notion of the Anthropocene, one question that seems to need to be addressed is that of the *anthropos* responsible for the Anthropocene. To which *anthropos* does the Anthropocene refer? In the light of that question, the broader question arises: is man really 'by nature a political animal'? (Aristotle, *Politics, I.* 2). Certainly, from the work of Christophe Bonneuil, the Swedish historian Andréas Malm or Alexander Federau, we understand that it is not all humans, but just a few, who are responsible for tipping the planet into the Anthropocene. What is the *anthropos* of those handful of thermo-industrial capitalists of the nineteenth century whose techno-scientific and economic model spawned the globalisation so typical of the twentieth century? Can this *anthropos* be 'by nature a political animal'? On the other hand, which *anthropos* is *not* responsible for the entry into the Anthropocene? It would seem that the Anthropocene undermines the Aristotelian conception of the political animal and, on the contrary, highlights the extent to which, for the last two centuries, man has been an economic animal. Prior to *homo oeconomicus*, the debate on the dating of the entry into the Anthropocene reveals a cause in humans' mastery of nature, in order to generate agricultural surpluses, leading to a radical and irreversible transformation of ecosystems. *Homo oeconomicus* is the latest representation of this anthropological aspect, common to all civilisations. While humans in the past two centuries have been characterized by the maximisation of individual interests, it is clear that humans are capable of acting together sometimes,[29] and it is now important to allow humans to become more political. A break with the individualism of *Homo oeconomicus* is necessary. Indeed, it seems that here, acting together is absolutely necessary to make the politics between humans emerge. If the Anthropocene is, as we regularly hear, indicative of humanity's geological power, it is also (and above all) the manifestation of our political powerlessness (Bonneuil, 2014). Not only do we appear politically impotent, but the important reference to the political in Aristotle's definition of humanity no longer seems entirely appropriate in the Anthropocene. How can Man be a political animal if the political is defined as acting together? Though *humans* are indeed animals, only humans are *political* animals when they act together (even if Aristotle's understanding of the political nature of humans also means that humans are fundamentally relational and connected to others).

[29] In recent years, we have seen examples of the 'miracle' of political action, with all the popular protests of the Arab Spring from December 2010; the *Maillots Jaunes* (Yellow Vests) movement from October 2018; the climate strike by schoolchildren and students, ongoing since the winter of 2018; the protests in Algeria in February 2019 against President Abdelaziz Bouteflika's standing for re-election for a fifth term; or the Hong Kong mobilisations that gained momentum in June 2019.

1.1.4 A Problematic Political Anthropology

The Anthropocene serves as a powerful conceptual tool because it is a material consequence of the western way of life, and the resulting environmental transformation, and it brings political questions to the forefront. The re-examination of what we want and can become is a direct extension of the concept of the Anthropocene. Indeed, 'In naming the Anthropocene and attempting to understand it, those who refer to it are thus loudly proclaiming what we have done, what we are, what we are capable of, and what we could or should become' (Eckersley, 2017, p. 5). The Anthropocene mobilises political anthropology. It poses a question of historical anthropology (who have we been up until now, which has led to this impact?) but also a question of prospective anthropology: who might we, and who do we want, to become? Having experienced structural changes in our relationship with nature that have steered human development with the development of agriculture, the Industrial Revolution, and the spread of consumer societies, we have a renewed and unprecedented ability to define what we want to be.[30] Thus, it is necessary to identify an alternative anthropology to that of the inevitable Anthropocene narrative. Here, the idea of an anthropological shift, in order to consolidate the political, allowing humanity to exist sustainably on Earth, could be mobilised and constructed. In their search for prospective outcomes (if not solutions), Earth System researchers never consider the possibility of an anthropological shift other than transhumanism. An alternative political anthropology urgently needs to be constructed,[31] to deal with the threat we all now face.

At present, we have a real difficulty in considering the long term in our political action and 'It is as if our knowledge does not percolate through the public decision-making system' (Bourg, 2017). For this reason, anthropological reflection is

[30] This is a project that goes beyond the remit of this book. German anthropologist Christoph Wulf is currently working on a prospective anthropology (Wallenhorst & Wulf, 2022, 2023).

[31] The term 'the political', here, refers to the arena in which humans, with all their differences, co-exist and act in concert. The main object of politics, as it is conceived of here, is to lay the groundwork for the future, and work to ensure the world remains habitable and hospitable for generations to come. The political is a particularly equivocal concept whose most basic element is often the regulation of the ascendancy of humans over other humans. The function of politics in Aristotle, which differs greatly from Arendt's understanding of it, is to work to regulate humanity to reduce the domination of the weak by the strong, by instilling humanity in both. For the French political scientist Raymond Aron, more generally, all concerted action is political. This research draws upon 'politica civilis' (the sharing of existence between all humans) more than 'politica policy' (the strategic actions of social actors), or 'politica politics' (power and institutions) (Aron, 1962, p. 10). One of the objects of politics is to make something happen rather than nothing; to make something happen between people that allows them to support each other's shared being. This is an existential function of politics, which should not be concerned solely with individual rights and economic gain. In view of the threat the planet now faces, what is the function of politics? Is it to change humanity, to protect it from its own self-destruction? Or is it to preserve humanity as it is, even if it could be otherwise? For the British philosopher Michael Oakeshott, for example, the preservation of humans, as they are, is the purpose of politics (*Rationalism in Politics and Other Essays*, Oakeshott, 1991).

necessary. For Dutch political scientist Frank Biermann (2014) working on Earth system governance, a wholesale transformation of social behaviours is utterly indispensable. He reminds us that large-scale lifestyle changes are quite possible (as Gandhi, for example, brought about), and that there are times when we encounter social tipping points. It is therefore important not to rubbish the idea of anthropological shifts because it seems Utopian. Moreover, there are many authors who encourage us to think along these lines. The Franco-German economist Christian Arnsperger calls for an anthropological change (Arnsperger, 2009, 2010, 2011c); Hartmut Rosa expounds on the importance of a profound paradigmatic shift in our relationship to the world, hinging on the concept of resonance (Rosa, 2016; Rosa & Wallenhorst, 2017a, b, 2022). French philosopher Maurice Bellet, for his part, discusses the necessity of a rebirth (1993, 2013). The American philosopher Roy Scranton repeatedly says that 'we are the problem' (2015) and that we must collectively learn to die (through the end of our civilization) if we hope to live in the Anthropocene.[32]

Some of the foundations of this anthropological mutation have been worked on by the German philosopher Herbert Marcuse[33] who, along with Theodor W. Adorno and Max Horkheimer, was among the first Critical Theoreticians. In his book *Eros and Civilization*, published in English in the United States in 1955, greatly inspired by Karl Marx and Sigmund Freud, Marcuse criticises the dehumanising and irrational dimension of the capitalist principle of efficiency. As an opposing concept, he proposes the hope of liberation (based on the transformation of sexuality into eros – i.e. liberation of the drive for life, and elimination of alienating work). He calls for science and technology to be at the service of human beings.[34] Marcuse's ideas provide interesting foundations for thinking about the necessary lifestyle changes in the Anthropocene, against a background of radical anthropological changes. He believes that a non-repressive society is necessary, presupposing profound changes in technology and a different relationship with nature. In his view, a mere change in the relations of production would be insufficient to allow a different type of society to emerge. He continued his condemnation of the inegalitarian and totalitarian dimension of capitalism in his book *One-Dimensional Man – Studies in the Ideology of Advanced Industrial Society*, published in English in the United States in 1964.[35]

[32] In Scranton's view, entering the Anthropocene means imagining a radically new vision of human life. His proposal is simple: in a different world, we must change in order to go on living. This requires the death of our civilisation. In his book, he ponders what it means to be human in the Anthropocene.

[33] Herbert Marcuse (1898–1979) met Max Horkheimer and Theodor W. Adorno and joined the *Institut für Sozialforschung*, where Critical Theory was developed, in 1932. When the Nazis rose to power in 1933, he emigrated to Switzerland with his family; he subsequently spent a short time for France, and finally left for the United States, where he worked in New York for the *Institut für Sozialforschung* (which had emigrated from Frankfurt).

[34] Here, we find elements that were later worked on by Maurice Bellet – notably in *Le paradoxe infini* (2004).

[35] Marcuse was heavily involved with the widespread student protests of 1968, lending support to their revolt and their protest movement.

1.1.5 Hannah Arendt (1906–1975) and Critical Theory

So far, we have identified that in order to address the political anthropological issues raised by the dawn of the Anthropocene, it will be useful to draw upon the thinking of Critical Theroeticians such as Hannah Arendt. It is appropriate, at this point, to say a few words about how Arendt's thinking fits into a critical theory. Indeed, Arendt was not a member of the *Institut für Sozialforschung*, founded in Frankfurt in 1923, where 'Critical Theory' (as it was called in the early 1930s) was developed. There, from the mid-twentieth century onwards, the so-called Frankfurt School[36] came into being. The Frankfurt School's work can be understood as thinking about emancipation, and participation in efforts to bring it about.[37] There are several notable differences between Hannah Arendt and the early figures in Critical Theory, her contemporaries Theodor W. Adorno, Max Horkheimer and Herbert Marcuse. Arendt was never a Marxist,[38] and she is little influenced by psychoanalysis, unlike the Critical Theoreticians.[39] The gap between the work of Adorno and Horkheimer, on

---

[36] The *Institüt für Sozialforschung* was founded following a donation from a German-Argentine Jewish entrepreneur, Hermann Weil (at the request of his son Felix J. Weil, a doctor of political science with a passion for Marxism) and a contract with the Ministry of Education requiring the director to hold a chair at the University. At the institute, German Jews developed a way of thinking, inspired by Karl Marx, about the need to transform society by identifying the mechanisms of alienation and domination. In 1931 a subsidiary of the institute was set up in Geneva, and in 1933 it moved its headquarters to Geneva. At the same time, its journal, *Zeitschrift für Sozialforschung*, established in 1932, was located in France, then in the United States from 1941. Both the institute and its journal were then attached to Columbia University in New York until 1950 (the journal ceased publishing in 1941), when the institute returned to Frankfurt. While the subversive thinking of the critical theorists was welcomed at Columbia University, two decades earlier, several scholars at that university had resigned because of the excessive support of the university from the established political authorities. Indeed, in 1917, the State of New York imposed a loyalty oath on students and faculty members (which prohibited protesting against state laws). Several academics from Columbia University (including John Dewey) resigned and created The New School for Social Research in 1919 with the spirit of protest and subversion (where politics is at the centre). In keeping with this original vision, in 1933, the New School for Social Research created a department known as the University in Exile, which welcomed hundreds of expatriate scholars between 1933 and the 1950s. Most of these were Jewish. It was at this institution that Hannah Arendt, her German compatriot Hans Jonas, and the French anthropologist Claude Lévi-Strauss worked. Arendt taught the thinking of Walter Benjamin (died 1940), a close collaborator of the *Institut für Sozialforschung* and an important critical theorist.

[37] On this point, a critical theory of the Anthropocene necessarily apprehends emancipation with a certain distance, without considering emancipation in the context of our break away from nature and considering emancipation from the point of view of its collective character.

[38] Critical theorists distance themselves from two components of Marx's thinking: the idea that history should be organised on the basis of progress, and the equating of emancipation with socialisation of the means of production (this disidentification leaves space for emancipation to be interpreted in a variety of ways) (Iakovou, 2001).

[39] The *Institut für Sozialforschung* interacted with the *Frankfurter Psychoanalytisches Institut* (Frankfurt Psychoanalytical Institute), which was founded in 1929. Freudian thinking is understood as critical thinking about social matters.

the one hand, and Arendt, on the other, is primarily due to the fact they belong to two different intellectual schools of thought: the former to Marxism, and the latter to phenomenology (and the classics). One of the differences between Arendt and her contemporary Critical Theoreticians is that she believes freedom can be achieved, through concerted action.[40] Adorno, Horkheimer and Marcuse[41] focus more on the extent to which alienation and reification undermine the very possibility of our having a future. Hannah Arendt was strongly attached to the theories of the Ancients, upon which her thinking is based. Unlike the Critical Theoreticians, she cannot be considered a progressive, and is sometimes labelled a conservative because of her reference to the Greeks, and her emphasis on the stability of the world in which human cultural creations exist.[42] It is the traces which the past has left on the present world that mean we can live together, and that we can have a life which is human or dignified, Arendt sometimes suggests. It is also important to emphasise that Arendt's thinking is structured around those moments of rupture – revolutions – where collective action erupts, allowing the world to be made anew, for the benefit of living together. Her ideas cannot be boiled down to merely a form of conservatism; though difficult to classify, her thinking can be understood as republicanism, which also feeds into communitarianism because of her understanding of politics as that which emerges 'between us', '*inter-esse*' or '*inter homines esse*': the common being of men (Arendt, 1961/1983, p. 16).

One aspect of Hannah Arendt's non-collaboration with the *Institut für Sozialforschung* and non-membership of the tradition of Critical Theory, whose

---

[40] A few decades later, Rosa, in *Resonance*, does not exclude the experience of freedom. In fact, it even seems to be consubstantial with resonance.

[41] The assimilation to Critical Theory or to the Frankfurt School is particularly variable from one author to another. Thus, Max Horkheimer can be considered the real founder, and Theodor W. Adorno (about 10 years younger) the co-founder. Then other authors are more or less closely attached to this critical theory, conceptualised by Horkheimer in his article 'Critical Theory and Traditional Theory' published in 1937 in the *Zeitschrift für Sozialforschung* (first chapter of the review *Traditionnelle und Kritische Theorie* that he published in 1970). All of them have developed their own thinking in a very singular way. Herbert Marcuse was particularly involved and then distanced himself at the end of his career; the philosopher and literary critic Walter Benjamin (1892–1940) was a member of the institute, from which he received several small scholarships, but with profoundly singular academic aims; the German-born American sociologist and psychoanalyst Erich Fromm (1900–1980) was closely associated with the school's agenda for a period and then radically distanced himself from it in 1939; Ernst Bloch and his work on utopia intersected with the issues at stake for critical theory. Other important thinkers among the critical theorists of the *Institut für Sozialforschung* include, for example, the Hungarian philosopher and sociologist György ukács, the German lawyer and philosopher Franz Neumann (1900–1954), and the German thinker Günther Anders (1902–1992), who is regularly associated with Critical Theory. To these early authors we can add the names of those who followed in their footsteps, starting with Jürgen Habermas, though he distanced himself from the initial direction of Critical Theory (Assoun, 2016, pp. 19–26).

[42] It is important to point out that Arendt does not glorify tradition, but rather, is able to show the destructive components of tradition, in close proximity to Benjaminian thinking (e.g. pointing out that the American founding fathers were unable to translate their political experience into concepts because they were trapped in the tradition of political philosophy) (Iakovou, 2001: 276).

importance should not be downplayed, is that Adorno and Arendt seem to have disliked each other.[43] For example, in a letter to Jaspers, she described Adorno as '*one of the most repulsive human beings I know*' (Letter from Hannah Arendt to Gertrud and Karl Jaspers, April 18, 1966, letter #395, p. 634). Arendt criticises Adorno and Horkheimer' lack of responsiveness during the rise of Nazism, while Adorno seems to despise Arendt's reference to phenomenology (Genel, 2006). More importantly, he also despised her reference to Martin Heidegger.[44] The problem between Arendt and Adorno also seems to have to do with Walter Benjamin's legacy and the reception of his work. Benjamin (1892–1940) was a particular influence on both Arendt and Adorno. He was the cousin of the German philosopher Günther Anders, Arendt's first husband, and he represents the real link between Arendt and the Frankfurt School.[45] Hannah Arendt was very close to Walter Benjamin during their exile in France and their flight to the United States, where Benjamin entrusted Arendt with a 'manuscript more precious than his life' before taking his own life. The exchanges with Benjamin marked her thinking, and she included an essay entitled 'Walter Benjamin' in *Men in Dark Times*, published in English in the United States in 1955. Adorno was also a close friend of Walter Benjamin. He was particularly invested in his thought and worked hard to bring his friend's unfinished work to prominence.[46] Adorno wrote a series of articles and prefaces to Benjamin's works, collected in a book published in 1968, *Über Walter Benjamin*, with the addition of correspondence.

While Arendt is not directly associated with Critical Theory, there are several factors which mean her work is closely linked to the Frankfurt School's Critical Theory. One of the commonalities between Critical Theoreticians is the importance attached to intersubjectivity, or relationships, viewed as being one of the defining features of humanity itself.[47] With Arendtian thinking and Critical Theory (Adorno, Horkheimer and Marcuse), we are dealing with two critiques of modernity.[48] If the

[43] Arendt mentions 'the Horkheimer clique' with contempt. In a letter to her husband Heinrich Blücher (1 March, 1955), Arendt criticises the *Institut für Sozialforschung*'s inadequate support for Benjamin (Arendt, Blücher, *Correspondences – 1936–1968*, published in French in 1999).

[44] Arendt decries the fact that Adorno changed his Jewish name, Wiesengrund, during the rise of Nazism (as she writes in a letter to Gertrud and Karl Jaspers, April 18, 1966, letter no. 395, p. 634). In turn, Adorno and Horkheimer vilify Arendt's connection to Heidegger, whom they consider a Nazi (Muhlmann, 2001/2002).

[45] Benjamin and Arendt also had an important mutual friend, Gershom Scholem (1897–1982). Benjamin met Adorno in 1923.

[46] So much so that the reception of Benjamin's work has long been reduced to Adorno's interpretation of it, as shown by the contributors to the journal *Lignes* n°11, edited by Michel Surya, on the theme of 'Theodor W. Adorno and Walter Benjamin'.

[47] This is the case, for example, with Jürgen Habermas' theory of communicative action and Axel Honneth's struggle for recognition.

[48] We may legitimately wonder, as does the French psychoanalyst and philosopher Paul-Laurent Assoun, whether the Frankfurt School's power derives from the precarious situation of its thinkers at that time in history (2016, p. VI). This remark can easily be extended to much of Arendt's professional life.

work of Adorno, Horkheimer and Marcuse, on the one hand, and Arendt, on the other, pursue similar aims, it is because their lives were very similar in many ways. All four were German Jews who fled to Europe (Arendt and Benjamin to Paris; Horkheimer to Geneva; Adorno to Oxford and London; Marcuse to Switzerland) and then to the United States, where Arendt and Marcuse remained for the rest of their lives, while Adorno and Horkheimer returned to Frankfurt. All four were exiles whose work treated the rise of Nazism and anti-Semitism and the totalitarian forces of the Modern Era. Considering the danger that humanity poses to itself, their work muses on the role of politics, and about the very characteristics of thought itself.

What Arendt has in common with the great figures of the Frankfurt School is that they tried to understand humanity in the context of their time, and proposed a critical reading of it. These intellectuals sought to teach their time – that is, not only to contribute something *to* their time, but also to convey the knowledge *of* their time. Arendt, like the Critical Theoreticians, denounced the hegemony of economic logics and the resulting reification.[49] Arendt and the Frankfurt School alike critique philosophical tradition (though Arendt relies more heavily on the Ancients than do the Critical Theoreticians. More profoundly still, they believe the crisis of reason currently facing us is grounded somewhere in human history. Arendt's theoretical and political stance has similarities to that of the Critical Theoreticians.[50] This is what leads Greek political scientist Vicky Iakovou to say that there is a 'structural and intentional proximity between the Arendtian approach and that of the Critical Theoreticians' (2001: 267).[51] Modernity carries within it something monstrous, both Arendt and the proponents of Critical Theory tell us. This is what we see with the Anthropocene, which is a form of its product.[52] This confirms that it is absolutely essential to develop a critical theory for

[49] In a way, the critique is aimed, first and foremost, at domination.

[50] In the course of the history of the late nineteenth and early twentieth centuries, we see the emergence of a set of intellectual currents and theoretical schools that intersected in the United States and Europe in which Arendt and the Critical Theorists Adorno, Horkheimer and Marcuse were immersed. In the United States in the first part of the twentieth century, we have: the pragmatist school, the first and then the second Chicago school, the New School for Social Research and the Frankfurt School. There are many relationships between these different schools, centred around the identification of the different forms of alienation and reification in the modern capitalist system, and the proposal of alternatives to the economic hegemony that gives rise to various injustices. In addition to the links already mentioned between the New School for Social Research and the Frankfurt School, we can note that pragmatism deeply marked the work of Karl-Otto Apel (1922–2017) (the German philosopher who was a professor at the University of Frankfurt). It was the meeting with Karl-Otto Apel (in 1950) that introduced Jürgen Habermas (1929-) to American pragmatism. Moreover, the Chicago School was born as a direct extension of pragmatism and of the importance given to surveys as a basis for the development of subversive thinking about power.

[51] One of the fundamental differences, however, is Arendt's view of thought and action as being indissociable parts of the same gesture, where Critical Theorists view them as being separate (Iakovou, 2001, p. 268), and, as already mentioned, the different views of the concept of freedom.

[52] Political work about the Anthropocene inevitably touches on totalitarianism, because it considers the very real possibility of humanity's own self-destruction. The French politician Frédéric Lambert understands totalitarianism as a form of the unthinkable, which, in some respects, is a description that also applies to the Anthropocene: 'What "totalitarianism" gives us to think about

the Anthropocene, which must necessarily draw upon the anthropology, critical of modernity, developed by Hannah Arendt.

The fundamental commonality between Critical Theory and Arendtian thinking is the way in which they view human connection and how it can be opposed to both individualism and collectivism. This is a fundamental feature in the quest for a critical theory for the Anthropocene. Thus, in this book, we will draw on the tradition of critical thinking – especially the work of Hartmut Rosa and Andreas Weber, and what we see as one of the contemporary forms of Critical Theory, with the convivialist movement and its ongoing theorisation. It is plurality that defines human beings, as both Arendt and the Critical Theoreticians tell us.[53] Why does humanity descend into barbarism instead of pursuing humanisation? (Horkheimer & Adorno, 1974, p. 13). This is the question which the Critical Theoricians and Arendt both addressed, but what neither realised until late on was that non-humans must also be included in the humanising plurality.[54]

1.1.6 *Note on the Aim of Critical Theory as Seen by Hartmut Rosa*

In June 2016, we interviewed the German sociologist Hartmut Rosa (Rosa & Wallenhorst, 2017a, b, 2022) about this anthropological consolidation of politics and the shift from a Promethean to a post-Promethean logic, in light of the contemporary acceleration – the Anthropocene is known as the 'Great Acceleration', and results from the same phenomena. In that interview, Rosa gives a particularly interesting definition of his function as a social scientist and Critical Theoretician of the Frankfurt School. Rosa makes explicit the essential prospective function of the Social Sciences which inspires the stance adopted throughout this work. Below is a

is a reflection on the accomplishment of the unthinkable, which we can define with Kierkegaard as "the reef on which thought is shipwrecked". To think of totalitarianism through the prism of the unthinkable is to make sense of a tragic experience, that of man, blind to his own finitude and confronted with absolute evil, and to question the human capacity to annihilate his own humanity' (Lambert, 2014, p. 71). The question of totalitarianism has been studied, in particular, by the Rennes-based political scientist Bernard Bruneteau, who has notably highlighted the complexity of the uses and surrounding this concept (of particular note here is Bruneteau, 2010, or Baudouin & Bruneteau, 2014). Bruneteau distinguishes four paradigms in the uses of the term 'totalitarian' by the main theorists of totalitarianism: 'the one that focuses on describing static structures (Friedrich paradigm); the one that emphasises the movement defined by the logic of a historicist idea (Arendt paradigm); the one that privileges the imaginary constructed from the fantasy of the One (Lefort paradigm); and the one that suggests the interactions of the systems within the framework of an "epoch" (Nolte paradigm)' (Bruneteau, 2014, pp. 12–13).

[53] This is one of the strong points of Arendt's thinking, but also characteristic of the founders of Critical Theory, who show that it is by living with others that we become human.

[54] It should be remembered, however, that the question of nature is particularly present from the very beginnings of Critical Theory, with discussion of the problem of its rationalisation, its reification and the hegemony of its industrialisation.

long excerpt from this interview, because it describes the intellectual project outlined in this work:

'Some colleagues believe sociology is the study of society, as physics would be the study of planets and stars, for example. This is not exactly my conception of sociology. I believe that any society needs space for self-reflection, free from the weight of action. [Sociology] must facilitate introspection, which allows us to move forward in analysing our own selves. Societies are not only subordinate to the material world. They are also shaped by their faculty of self-analysis, to create an intellectual and spiritual horizon. To think of foregoing such self-analysis is problematic for at least three reasons, which could be represented by the images of mythical birds. Sociology or social philosophy, which I believe must work together, can be compared to a three-headed bird. (1) The first bird is what Hegel calls the storytelling owl of Minerva. We develop narratives about what modernity is, and what has developed (for example, we develop a narrative about acceleration). (2) The second bird is the Capitol goose. The geese that lived on the Capitoline Hill alerted the Romans to enemy attack. When enemies approached, the geese's cackling woke the Roman soldiers to repel the invaders. The social sciences can perceive and shine a light on developmental failures, before they develop into disasters. What I am trying to do, in collaboration with my colleagues in Jena, is to show that this dynamic stabilisation program is intrinsically contradictory and poses serious problems. This is exemplary of the warning function of the social sciences. I try to apply this with desynchronisation; high rates of technical and economic production create an ecological problem. In today's world, we face exactly this issue of desynchronisation: nature goes too slowly and we go too fast. This is enormously problematic because it also generates a democratic problem (...) This would be the Capitol goose. Today this role is not particularly difficult to fill, because many people, from an extremely diverse range of backgrounds, are warning of the mistakes of modern development. (3) The crucial question is whether we can also be a kind of phoenix, creating a vision of what a good life could be, and proposing a path to that goal which people want to walk. This is the really complex point that we all struggle with. What I think is essential in the Critical Theory of the Frankfurt School is the perception that there is something wrong with the way we exist in the world. The idea began with Karl Marx. The young Marx, with the concept of alienation, said that through the way in which we manage matters, we alienate ourselves, we cut ourselves off from work, from other humans, from nature and ultimately from ourselves. In the tradition of the Frankfurt School, this idea was taken up wholeheartedly – notably by Herbert Marcuse and Erich Fromm, in the development of the processes of alienation. Alienation is precisely what I would describe as a loss of resonance. Yet alienation is also a matter of reification, consisting of viewing nature merely as a "thing", and in particular, a deaf thing. Herbert Marcuse denounces Prometheism; Theodore Adorno and Max Horkheimer the fact that we impose our control upon things. This seems to me to refer to an interest in making the world tractable and deaf. From this observation, Adorno develops the idea of a mimetic relationship with the world – one of silent imitation rather than a relationship of control. Marcuse proposes an 'loving relationship with the world (or an erotic sense of reality), whereby we do not attempt to control or commodify the world. In Walter Benjamin, there is the idea of an aura, and an auratic relationship with the world (to tell the truth, it is not entirely clear what he means by this). Ultimately, I would say that the concepts of mimetic, erotic or auratic relations with the world are only vague proposals. Max Weber's work also speaks of a charismatic relationship, but that is not clear either. This was my goal in introducing the concept of resonance[55]: to propose a means of relating to the world. The idea is that a successful life is one that is lived in resonance with the world and with others. This has earned me criticism in Germany. Some say that the Critical Theory of the Frankfurt School develops an irreconcilable opposition to the existing conditions of our lives, strongly marked, as they are, by capitalism' (Rosa & Wallenhorst, 2017a, pp. 26–27).

In this long excerpt, we clearly perceive the aim of Critical Theory to propose a direction, organised around the possibility of a good life, starting from the observation that 'something is wrong in our relationship to the world'. To our minds, in the Anthropocene, what seems crucially important is not so much to imagine what a good life could be, but rather to consider the very possibility of a life together. As we identify in this excerpt, Rosa views his work as belonging to sociology. Throughout his book *Resonanz* (2016), Rosa repeatedly claims the epistemological and disciplinary positioning of his normative thinking as belonging to this discipline. This is a distinctively German approach to sociology. From our point of view, the development of a critical theory for the Anthropocene is not sociological, in the strictest sense of the word, but rather belongs to political theory.

1.1.7 Note on the Relationship of Critical Theory to Karl Marx's Thinking

Marxism is the main reference point for Critical Theory, but from the 1930s onwards, Critical Theory is identified as such (rather than simply as a form of Marxism), viewed as being a departure from Marxism. What Critical Theory claims is not affiliation with Marxism, but its own stance – a unique intellectual project.[56] Critical Theory thus rejects positivism and marks a break with the scientific dimension of Marxism. The critical Marxism of the Frankfurt School develops two aspects which mark that departure from Marx: 'the conceptual deconstruction of political economy that Marx makes possible, and the prospect of emancipation that presupposes a relationship with historical truth' (Assoun, 2016, p. 91). Among Critical

[55] This concept of resonance is expanded upon in the rest of this book.

[56] Marxism is the main point of reference in Critical Theory. This means that 'Criticism necessarily intersects with historical materialism in order to succeed in its passage to history (see *supra*, p. 24) and that Marxism is not a "system" that would short-circuit criticism, but only the tool to steer criticality. With this caveat, we can draw the connection between the Frankfurt School and Marxism – which explains why it has its natural place there, but with so specific a theoretical stance that it does not lend itself well to such a summary label' (Assoun, 2016, p. 79).

Theorists, emphasis is placed on the human dimension of the processes – i.e. 'on the social subject of work and practice, rather than on the technological and instrumental aspect' (Assoun, 2016, p. 92).

The Critical Theoreticians' most vehement critique is of the instrumental rationality that is responsible for domination in our societies, and whose historical destiny appears eminently problematic. Here, Marxism is no longer a resource that can be utilised to think about domination and the processes of reification, but one of the problems.[57] Marxism, which has a scientific ambition, has an ideological dimension that is widely drawn upon by the political forms of Stalinism. On this point, Critical Theory did not seek first to identify the gap between Marx's writings and the political use made of it, but worked towards re-evaluating his thinking. A fundamental conflict emerges: Marxism is both a weapon for criticising domination and a tool for consolidating the instrumental reason upon which domination is founded (Assoun, 2016, pp. 96–97). On this point, this critique intersects with Arendt's critique of the rationality of *homo laborans*.

It can be said that Critical Theory is both linked to and untethered from Marxism. For our part, in this work, all that remains of Marx's thinking is the critical and distanced reappropriation of it by Hartmut Rosa, and the project, here initiated, to transform the social world based on his analysis. Karl Marx is not an author analysed, as such, in this book.

1.2 Linking Land, Politics and Education to Prepare for the Future (Inventive Function)

1.2.1 Preparing for the Future

How can we prepare for tomorrow today? The French terms '*devenir*', '*avenir*' and '*futur*' are all translatable as *future* in English. The French philosopher Bernard Stiegler proposes to define '*devenir*' as that which is most probable (2015, p. 144). Here, the '*devenir*' is understood from a programmatic point of view, as the consequence of decisions made in the present. The '*avenir*', (which could be translated into English as '*what is to come*'), is the improbable, and is in line with the concept proposed by the German theologian Jürgen Moltmann as what happens as the result of our decisions marked by freedom. The future includes the political, with the possibility for humans to take collective action, and is thus unpredictable by definition. Using Arendt's terminology, we can say that the emergence of concerted action,

[57] Before he became interested in reading Hegel and then Marx, Horkheimer was more influenced by Kantian German idealism. 'For Horkheimer, on the other hand, the primary intuition is that of a collapse of knowledge, contemporary with (and indicative of) a formidable crisis of historical reason. Meanwhile, for Marxists of the previous type, there is only the reconstruction of theory and the world with the help of Marxism – the idealist and bourgeois past already being, in principle, a lapsed, even forgotten past. For Horkheimer, it is collapse as an ongoing process that is the most salient aspect' (Assoun, 2016, p. 87).

which is not the natural programmatic extension of the present, is the miracle of politics. The '*futur*' could be defined as that which encompasses '*devenir*' and '*avenir*'.

The '*devenir*' is a projection of the past, and refers in part to what is behind us. Today, political leaders have real difficulty in foreseeing and preparing for the '*futur*'. The problem lies in how to manage a dynamic, shifting present, which is the focus of political concerns. The '*futur*', with little investment from the public sphere, is increasingly becoming the focus of investment in the private sphere. In particular, the Silicon Valley giants – headed by Google, Amazon, Facebook and Apple – have set themselves the mission of designing for the '*futur*', but they focus only on the '*devenir*' aspect. The way in which these modern multinationals approach the matter of our '*devenir*' is by attempting to use techno-scientific and digital breakthroughs to increase life-expectancy indefinitely (even to the point of eliminating the hitherto-ever-present spectre of death). However, in doing so, they are creating an extension of the present – not a future.[58] This way of preparing for the '*futur*' is based on doing away with the '*avenir*', in a manner of speaking. Indeed, what is '*à venir*' (to come), starting with 'generations to come', is eclipsed by the present generations' determination to conquer death itself. Thus, politics fails to prepare for an uncontrollable '*futur*', and also to prepare for the '*avenir*', being too heavily embroiled in managing an unstable, fluid and ephemeral present. In addition, some of the main economic actors of today fail to prepare for the '*avenir*' by preparing a backward-looking '*devenir*', which is, in substance, an indefinite extension of the present. The '*avenir*' is that which is coming to us; that which is '*à venir*', and which does not depend directly on our programmatic intent. We must think about ways in which we can use the entry into the Anthropocene to open up new possibilities for the future, rather than closing it off or restricting it.

The entry into a new geological epoch of an anthropogenic nature invites a new way of thinking. Humanity is confronted with an uncertain 'future', with the very survival of the human adventure is under threat. The entry into the Anthropocene is indicative of a crisis of the political, understood in line with Arendt's thinking, as the tent sheltering the whole of humanity, in all its diversity, on the earth's surface. Arendt attaches a great deal of importance to the primacy of Earth over humanity: Earth is necessary above all: without the earth, there can be no habitable world. Given that the ultimate goal of politics is to shape the world to make it hospitable, the worsening conditions of habitability of the earth indicate a crisis in our taking responsibility for the world. Politics is in crisis, characterised by powerlessness, experiencing difficulties in carrying out collective actions and defining priorities. Where modernity is marked by a temporal mindset of projection into the 'future', underpinned by the ideology of progress, the postmodern Anthropocene world is marked by a future in serious jeopardy, dominated by the need to take action immediately, or as a matter of real urgency (Boutinet, 2004). While previously, human existence in the world could be thought of in terms of long time periods, that is no

[58] In a letter from Hannah Arendt to Martin Heidegger on 28 July 1970, she decries the fact that the present time is marked by 'pure and simple abrogation of the future' (Arendt & Heidegger, 2001, p. 197).

longer so; the existential threat we – and indeed the whole world – face is imminent. We need to restore the possibility of a future, so that the innumerable future generations have the chance at life.

Having entered the Anthropocene, humanity now faces an unprecedented challenge, with multiple aspects of uncertainty about the future: 'Living in the Anthropocene means inhabiting the non-linear and unpredictable world of the responses of the Earth system, or rather of history-Earth, to the disturbances we cause' (Fressoz, 2013, p. 223). Indeed, the entry into the Anthropocene raises a set of existential questions about the future. The historian of sensibilities Alain Corbin, anthropologist Jean-Jacques Courtine and historian Georges Vigarello even speak of 'Anthropocene fears' to signify the fear of natural disasters that have been emerging from the mid-twentieth century onwards (Corbin et al., 2017).

1.2.2 Forgetting the Earth in Politics

Has the Earth not been forgotten in politics? Is this not what we are powerfully reminded of by the dawn of the Anthropocene? Over the past ten years, a number of studies have been published which question the political from the standpoint of the Anthropocene (Bonneuil & Fressoz, 2013; Cochet, 2013; Karlsson, 2013; Dalby, 2016; Eckersley, 2017; Bourg et al., 2017; Sinaï & Mathilde, 2017; Beau & Larrère, 2018; Lesourt, 2018; Villalba, 2021; Wallenhorst, 2019, 2020a, b, 2021, 2022; Testot & Wallenhorst, 2023). These publications come in the wake of a long period in which, in political science, insufficient consideration has been given to the importance of the Earth, with political organisation having been thought of independently of its interactions with matter flows. It must be admitted that, in recent decades, the notional relationship between the Earth and politics has been organised around the notion of sustainable development. While this notion has the merit of having helped raise widespread awareness of the importance of taking the Earth into account in political machinations, it also has the great limitation of 'pledging allegiance to the frameworks that condition its integration' (Villalba, 2009, p. 363) – in particular, the economic framework. It has a chimeric component with a politicised representation, but does not sufficiently take political constraints into consideration (Villalba, 2009, p. 366). In view of the advent of the Anthropocene and the ecological and energy transitions that it necessitates, it is now possible to move beyond the crescentic paradigm of sustainable development, questioning 'the material conditions of existence of democracy, which has long perceived itself as external to nature' (Villalba, 2015, p. 56). This is a point that Corine Pelluchon has illustrated particularly clearly and attempted to address in *Les nourritures* (Pelluchon, 2015) by showing how the body politic relies upon bio-geo-chemical matter flows. The German philosopher and biologist Andreas Weber,[59] whose approach is similar to that of Critical Theory,

[59] Andreas Weber is not, strictly speaking, an academic. He works as a freelance journalist and writer and teaches at Leuphana University in Lüneburg and at the University of the Arts in Berlin.

emphasises the fact that humans ourselves are bio-physical beings, these elementary building blocks being common to all living things, and indeed to the world as a whole (Weber & Kurt, 2015; Weber, 2016a, b, 2017). In an extension of this idea, French philosophers Léo Coutellec and Jean-Philippe Pierron examine the existential foundations of the 'world of relationships we construct when we eat' (2017, p. 19) and the importance of relating to the environment in the emergence of collective action (Pierron, 2014b, p. 12). Another way of formulating this same idea is that 'Democracies only exist because of the interactions between the living world and human societies' (Villalba, 2015, p. 57). Thus, the Anthropocene jeopardises 'the very conditions of the continuation of the democratic agenda as it has been instituted for several centuries' (Villalba, 2015, p. 56). Indeed, the ideal of individual autonomy and emancipation from all natural constraints is consubstantial with the idea of democracy.[60]

On the other hand, one element of Arendt's anthropology appears particularly problematic in light of the tensions revealed by the Anthropocene and these recent contributions to the literature. The critical anthropology of modernity developed by Arendt in *The human condition* is based on the distinction between two opposing categories: life that needs to be maintained, referring to the economic sphere, and the world, which is the space in which politics emerges. Yet is life truly opposed to the world? Can the biological be oppositional to the political? Is human life, integrated into the fabric of the living world on which it depends (through breathing and eating, but also through the subtle interlinks between climate and biodiversity), not the very foundation of the body politic? Thus, it is important to think more about the connections than the separations between the anthropological categories proposed by Arendt.[61]

[60] On the question of the Earth in politics, the American sociologist John Bellamy Foster's reading of Karl Marx's thinking is particularly interesting. He shows that the critic of capitalist society, who has sometimes been considered a non-ecologist, is in fact particularly sensitive to the question of changes in our relationship with nature. In 2000, Foster wrote a book entitled *Marx's ecology: Materialism and nature*, in which he shows the centrality of ecological questions in Marx's thinking (in his analysis of agriculture or the circulation of nutrients in the soil, for example). This same idea is also explored by the French philosopher Henri Pena-Ruiz in *Karl Marx penseur de l'écologie* (2018), who goes so far as to show the presence of an accomplished ecological theory in Marx's writings. For example, Marx writes: 'It is not the unity of living and active men with the natural, inorganic conditions of their exchange of substance with nature, nor, therefore, their appropriation of nature, that requires explanation or is the result of a historical process, but the separation between these inorganic conditions of human existence and this active existence, a separation that has been posited as a total separation only in the relation of wage labour and capital' (Marx, 1857–1858, p. 489).

[61] We perceive, through her anthropological categories, that Arendt was marked by her experiences of the horror of the Holocaust and then the Cold War. Her contemporaries who really developed an environmental thought, such as Günther Anders, Hans Jonas and Ivan Illich, realised this in the late 1970s and early 1980s (Arendt died in 1975).

1.2.3 The Ecological Thinking of the Environmental Humanities: A Support in a Critical Theory for the Anthropocene

This work to develop a critical theory for the Anthropocene is part of the field of Environmental Humanities, which emerged in the 2010s, bringing together disciplines in the Humanities and Social Sciences. The aim in Environmental Humanities is to understand human beings and their environment as parts of a single, connected entity. Of particular note is the impact of the entry into the Anthropocene on thought, and in particular, the break with the dualism between nature and culture, so typical in modern western thinking (Descola, 2005).[62] Environmental Humanities, which have been consolidating as a recognised academic discipline for several years now, are the result of the way in which a number of disciplines (Sociology, Philosophy, Law, History, Political Science, Anthropology, etc.) interact with the environment.

The ecological branch of the Environmental Humanities is an academic discipline (within the Environmental Sciences) characterised by its concern with action and the formulation of tentative predictions. In recent years, departments and faculties of 'Environmental Sciences' have been set up at various universities. Environmental sciences, while they can be a resource for critical theory, deal with an object of study that is distinct from politics. This is not the case with the Environmental Humanities, which have a political object and are marked by an epistemology of action that is close to the stance of critical theory mentioned above. Signs of the emergence of the environmental humanities can be seen in a series of publications. Several recent publications help formalise of the ecological thinking of the Environmental Humanities: *La pensée écologique – Une anthologie* by Bourg and Fragnière (2014, *op cit.*)[63]; the *Dictionary of Ecological Thought*, edited by Dominique Bourg and the Swiss lawyer Alain Papaux (2015); the *Guide des Humanités environnementales*, published in 2016, edited by the French literary scholar Aurélie Choné, the French sociologist Isabelle Hajek and the French sociologist Philippe Hamman; *Science, conscience and environment*, edited by Swiss philosopher Gérald Hess and Dominique Bourg (PUF, 2016), developing, notably from the point of view of Environmental Humanities, an interesting critique of contemporary science; *Environmental Humanities: investigation and counter-investigation*, edited by the French historian Guillaume Blanc, the French anthropologist Élise Demeulenaere and the French philosopher Wolf Feuerhahn in 2017; or the Environmental Humanities blog (with institutional support from the University of Versailles St-Quentin-en-Yvelines).

---

[62] However, as Curnier (2017, p. 22) mentions, 'a number of disciplines apply their traditional analytical tools to the environment without refuting the nature–culture duality'.

[63] It is noteworthy that ecological thinking emerged in the nineteenth century, bringing its critique of anthropocentrism and its observations concerning the geological power that humanity had come to wield. Most of the precursors of ecological thought in the 19th and 20th centuries presented in this anthology have in common that they were brilliant, visionary and pioneering scientists, and that they struggled to open their contemporaries' eyes to the way in which current events close the door to the future rather than opening it.

In addition to the *Dictionary of Ecological Thought*, a number of environmental dictionaries have appeared in recent years, such as the *Handbook of sustainable development* (Atkinson et al., 2007), *Routledge international handbook of social and environmental change* (Lockie et al., 2014), following the *Handbook of environmental sociology* (Dunlap & Michelson, 2002).[64] At international level, the Environmental Humanities have a presence in various places. Stanford and Princeton Universities have initiated research programs in the field of environmental humanities: *Environmental Humanities Project* for the former and *Environmental humanities initiative* for the latter. There is also an environmental humanities laboratory at the Royal Institute of Technology in Stockholm, and the journal *Environmental Humanities* was established in 2012 at the University of Sydney. In recent years, there has been continuous development in the field of environmental humanities.

Ecological thinking is not consubstantial with Earth System Sciences, but is invariably based on them. The *Dictionary of Ecological Thought* lays the foundations of environmental humanities, at the intersection between the human and social sciences and the natural sciences, with a view to putting forward recommendations. The critical theory on environmental humanities fills the gap created by the failure to take account of the Earth in political decisions. Environmental Humanities cover particularly varied epistemological and methodological currents (environmental sociology, environmental history, environmental ethics and philosophy, green political theory, ecocriticism, ecological economics, etc.); the point of commonality between them is that the environment is understood 'as a material condition for the existence of human activities' (Szuba, 2015, p. 945). One of the anthropological contributions of ecological thinking is the awareness of how *humanitas* (human nature) originates and is received from *humus* (earth) and how damaging this *humus* directly affects man's capacity to *be* human.

Environmental Humanities, which are fundamentally interdisciplinary and emphasise the interdependence between humanity and nature, are set apart from two dominant paradigms in the human and social sciences: humanity's centrality and its independence. Ecological thinking is mostly critical of modern western technocentric society, and rejects the paradigm that man and nature are separate, independent entities (Bourg & Papaux, 2015, p. 757). Environmental Humanities therefore constitute a powerful and critical tool for the theoretical overhaul of the Humanities and Social Sciences as a whole. Indeed, an anthropocentric approach ultimately places humanity on the road to disaster. Where should the spotlight be placed: on the hospitable planet, or on its inhabitants whose survival depends on their environment? Ecological thinking puts humanity in its rightful place in relation to the historical precedence of the Earth and the cosmos, and advocates an anthropological shift rather than a technical evolution (or revolution) to allow

[64] There are also certain handbooks which deal directly with the Anthropocene, such as: the Encyclopedia of the Anthropocene (DellaSalla & Goldstein, 2017), the *Critical Dictionary of the Anthropocene* (Collective, 2020), the *Handbook of the Anthropocene* (Wallenhorst & Wulf, 2023).

humanity to break out of the ecological impasse in which we find ourselves.[65] Humans cannot get out of its predicament without changing, and beginning to take account of their own finitude: like every human being, humanity is mortal, and its adventure takes place within the bounds of a limited biosphere. Since modernity in the west has been marked by synchronicity between the development of humanism and technological progress against a backdrop of anthropocentrism, it is reasonable to hypothesise that the development of humanist thought in postmodernity will inevitably involve ecological thinking, breaking away from anthropocentrism. Ecological thinking refers to human beings in both a biological and a transcendental dimension, with the reflection centring on the biosphere and the flows of matter that make up human beings, and with the aesthetic function of metaphysical openness that nature represents. Humanity's continued survival is contingent upon linked to a wholesale shift in its relationship with planet Earth. Hitherto, a great deal of hope has been pinned on progress, but what can we hope for tomorrow? In today's world, is it possible to nurture intellectually reasonable hope?

The ecological thinking in Environmental Humanities feeds into a critical theory for the Anthropocene. How can humanity recover in order to undertake its adventure anew, in a different direction? This question is in line with that asked by French glaciologist Claude Lorius and French journalist Laurent Carpentier (2010): 'What do we want to do with this world of which we have become both the gravediggers and the guardians?' The challenge facing humanity is to devise and establish 'a democratic regime capable of applying a policy that is considerate of people, nature and their common future' (Guillaume, 2015, p. 35). Because this issue is at odds with the current dominant economic model, it requires an unprecedented 'political upheaval' that also takes account of 'the metaphysical challenges of our future' (Guillaume, 2015, p. 35). Under the guidance of the environmental humanities, politics can 'no longer ignore the ecological and material reality in which democracy is embedded' (Semal & Szuba, 2015, p. 990). This paradigm also allows us to think about how to give due consideration, in politics, to the finitude of the planet's environment (for example, through concrete measures to define a pace for the redistribution of resources that takes the Earth and its needs into consideration). One of the unique features of the environmental humanities is their focus on humanity's existence, enabled and facilitated by the host environment, with concern for the future and preparations for it. We can identify two environmental components of humanity: firstly, the environment allows the human adventure to become human. Secondly, because our changed environment today threatens human life, we have the opportunity to reconfigure politics around a collective dynamic, driven by hope for the future. The political novelty of ecological thinking is its long-term perspective. The Anthropocene is seen as a gateway to set a course towards preparing for of the future, and ensuring the world remains hospitable for generations to come.

[65] This is the most profound contribution of ecological thought, which we can find in different ways in the work of such diverse authors as French historian Jacques Ellul (1954/2008), Teilhard de Chardin (1955/2007), Austro-American philosopher Ivan Illich (1971a, b, 1973a, b), Hans Jonas (1990, 1998), Dominique Bourg and Christian Arnsperger.

Indeed, in some respects, public awareness of the tragic degradation of the biosphere and entry into the Anthropocene may be an opportunity for politics to regain consistency in preparing for the future, ensuring that future generations have a world to inhabit and a life to live (Bellet, 1993). Paradoxically, the urgency of environmental issues can serve as a spur[66] to force politicians to begin thinking in terms of the long timeframes that have not been taken into account in our postmodern world.

1.2.4 Can We Be Assimilated to Our Capacity for Instrumental and Calculating Reason?

In *Negative Dialectics*, published in 1966,[67] Theodor W. Adorno[68] focuses on the necessary emancipation of humanity in modern alienating societies. He denounces a form of totalitarianism, stemming from the Age of Enlightenment, which can be understood in the light of the Anthropocene. Adorno seems to be open to other types of rationality than thought, with the theory of negative dialectics holding that not everything can be appreciated through thought. We have here some solid foundations on which to base a critical theory for the Anthropocene. Before writing *Negative Dialectics*, he had written an important work with Max Horkheimer[69] (1895–1973), *Dialektik der Aufklärung* (published in Germany in 1947, after a first

[66] The ecological crisis can become a spur for politics, in the double sense of an indicator pointing to the political failure of the past several decades, and an arrow guiding the reorientation of political action.

[67] *Negative Dialectics* is the starting point for the intellectual work of both Jürgen Habermas (1929-) and Axel Honneth (1949-), two important figures in Critical Theory, who are identified as part of the second generation of the Frankfurt School (and neither Habermas nor Honneth are Jewish). On the basis of Adorno's work, Habermas proposed another important concept for understanding rationality. Habermas met Adorno, Horkheimer and Marcuse in the mid-1950s and was influenced by American pragmatism. Axel Honneth drew upon Adorno's thinking in creating the concept of the struggle for recognition (*Kampf um Anerkennung* was published in Germany in 1992), which distinguishes between affective recognition, legal recognition and cultural recognition.

[68] Theodor W. Adorno (1903–1969) was a musicologist whose work was marked by psychoanalysis. *Negative Dialectics*, the original German edition of which was published in 1966, was one of his most significant books, in which he criticised Heidegger.

[69] Horkheimer arrived at the University of Frankfurt in 1925, and soon after that, contributed to the founding of the *Institut für Sozialforschung*, becoming one of its directors. As mentioned previously, after fleeing the Nazis, he returned to Frankfurt in 1949 to resurrect the institute, again serving as director. His work was characterised by interdisciplinary research, as founder of the *Zeitschrift für Sozialforschung* (which publishes texts by philosophers, economists, psychoanalysts, sociologists and more).

publication in German under another title in the United States in 1944),[70] which made its mark on Critical Theory.

Horkheimer and Adorno, in *Dialektik der Aufklärung*, try to salvage the goals of the Age of Enlightenment, while showing that the barbarism of the twentieth century is connected to the Age of Enlightenment. The latter is understood as a process of civilisation in which progress and instrumental reason become sacrosanct, whereas all trace of myth and wonder are erased. *Aufklärung* was part of a process of human liberation or emancipation, but reason was instrumentalised in the process. From the purpose of knowledge, reason was employed with the aim of attaining domination over nature (becoming 'masters and possessors of nature', to borrow Descartes' phrase), identified as the result of the emancipation process of the Age of Enlightenment.[71] Horkheimer and Adorno attempt to understand the reasons for this descent of humanity into barbarism, made possible by the self-destruction of reason that is concomitant with the hegemony of instrumental reason. Historically, reason and barbarism are bedfellows.

In this book, the authors show how western reason has mutated into instrumental reason. Their analysis could be extended to the way in which that instrumental reason is continuing to mutate in the early twenty-first century, into a calculating reason marked by the predominance of algorithms in decision-making. Here, not only has wonder been entirely eliminated from the process of reason, but thought itself has been taken out of the realm of rationality. Importance is attached only to that which can be counted. It is in this context of shifting rationality that the Anthropocene has appeared.[72] We can therefore legitimately wonder what impact this continuing mutation of rationality will have. What kind of rationality should be promoted through education? Here, the thinking of Andreas Weber, whose intellectual approach is marked by interdisciplinarity and critical theory, is particularly interesting.[73] The approach he favours, in order for us to continue living in the Anthropocene (in his *Manifest für das Anthropozän*, published with Hildegarde Kurt in 2015) is distinctly romantic, drawing on the power of solidarity inherent in living things. In

---

[70] *Aufklärung* is translated as 'reason', but can more accurately be rendered as 'enlightenment'. In a very broad sense, *Aufklärung* refers to the progressive thinking of the Age of Enlightenment (which led to the Industrial Revolution).

[71] In our work, we have relied particularly on a set of contemporary authors, and less on the early Critical Theorists, who nevertheless worked particularly on this question of humans' relationship to nature.

[72] There are a variety of hypotheses for dating the entry into the Anthropocene. The earliest date back to the Stone Age with the creation of tools and the mastery of fire, which marked the beginning of the anthropisation of the environment and alteration of the chemical make-up of the atmosphere; however, no geological traces can be found in the sediment record to support this hypothesis). Another early dating hypothesis relates to the emergence of agriculture. Agriculture was made possible by the climatic stabilisation of the Holocene, which allowed for the organisation of agricultural surpluses, from which the great civilisations arose. However, the dating hypotheses which indicate an unambiguous change of geological epoch have in common that they point to *homo oeconomicus*' logic of maximisation of individual interests.

[73] Andreas Weber regularly refers to the theorists of the Frankfurt School.

another work, *Sein und Teilen*, published in 2017, he shows how it is possible to listen to nature, which is continuously 'talking' to humans. It may be that, after reason has ousted any element of wonder, keeping only resources from nature that can be consumed by humans, it is only possible for reason to continue on this path of reification, ousting thought to favour only that which can be computed – a task now handled by artefacts.

In *Dialektik der Aufklärung*, Adorno and Horkheimer show, on the one hand, how progress may be destructive, and on the other, how there is a self-destructive component to reason. These two characteristics are central to what the notion of the Anthropocene brings to contemporary thinking. This book is a rejection of positivism, and underlines the totalitarian aspect of the Age of Enlightenment, which seeks to suppress all forms of wonder and leave room only for reason, whose purpose is domination. *Dialektik der Aufklärung*, by delving into this totalitarian component of the Age of Enlightenment, and its erasure of all traces of wonder from our way of life, lays the foundations for a series of works, including Hartmut Rosa's *Resonance*, David Abram's *The Spell of the Sensuous* (*Comment la terre s'est tue*) and the works of Andreas Weber (perhaps especially *Sein und Teilen*), referring to the idea that the earth speaks – in a sense that cannot be reduced to the merely metaphorical.[74] Adorno and Horkheimer identify the return of mythology in the way in which humans begin to be seen as objects and commodities, rather than people. That is to say, *Aufklärung* as such is a myth in itself which, through rejecting myth, suppressed all those characteristics of humanity such as art, culture and thought, in order to impose the myth of reifying reason, by which humans are transformed into mere things.

1.2.5 What Paradigm Is Necessary for Education in the Anthropocene?

Numerous articles in Earth System Sciences regularly mention what should be done as a result of our entry into the Anthropocene, and offer a prospective or recommendations. However, very little mention is made of education: the education of the younger generations is not mentioned among the things that need to be done. The means of real, meaningful action are viewed as being political or economic. For example, Steffen et al. (2011a) present a set of recommendations relating to geoengineering or a global governance system, but fail to identify the anthropological shifts that are absolutely necessary in order for us to live in the Anthropocene, and the means of education to bring about those shifts. The fact that, in the view of authors like Crutzen and other Earth system researchers, the only means of action by which to deal with the Anthropocene is geoengineering, reflects a real pessimism

[74] The phenomenology of Maurice Merleau-Ponty is the common theoretical basis of these three works.

about human beings and their intentionality: 'A man [in the sense of an individual] is powerless to withstand Man [in the sense of humanity as a whole]. However strong his faith in the power provided by technology, he may doubt humans' ability to change' (Federau, 2017, p. 215). Given the depth of pessimism of the dominant naturalist narrative, it is essential to find means of political action and identify the style of education needed to bring it about.

The same is true of social scientists, who are aware of the scientific data on the entry into the Anthropocene. In literature on how to address the political issue of the Anthropocene, the first element that emerges is a reorganisation of the global governance system, and the setting up of new institutions to regulate human activity. In the literature on the Anthropocene, we regularly see analyses of the *Homo sapiens*' dominance in the biosphere. Yet it must be recognised that human beings are not a species 'like any other': they are characterised by unparalleled abilities to adapt and metamorphose. Indeed, humans are the product of their socio-cultural context, their particular experiences and their educational history. Whilst it is important, of course, to reflect on our institutions, it seems crucial to reflect on the kind of educational paradigm that is needed in the Anthropocene. For many social scientists, it is important to think in a radically new way about how we can live together on Earth, but they neglect to examine the way in which that peaceful coexistence can be achieved through education. Finally, even in Parts II and III of the IPCC Fifth Assessment Report[75] on how we can adapt to environmental change and how to mitigate climate change, education is barely mentioned (with the exception of a brief item at the end of Part II on human development).

In recent years, there have been a number of studies published which draw the connection between education and the Anthropocene. One such is Daniel Curnier's doctoral thesis, in Environmental Sciences, defended in 2017: *Quel rôle pour l'école dans la transition écologique?* [What role can schooling play in the ecological transition?]. The thesis describes a pre-set curriculum, based on the principle of sustainability, for the canton of Vaud in Switzerland. In the course of this book, we shall examine Curnier's thesis in detail, with a critical eye. Curnier's thesis is highly engaging. Education school has a key role to play in the ecological transition, and as such, it is important to provide children with knowledge of the Anthropocene at school, and break away from the traditional segmentation of knowledge into different disciplines, and instead offer a cohesive, coherent curriculum that inculcates citizenship. Curnier's thesis holds that education should be understood as a political means of developing a critical theory for the Anthropocene. German palaeontologist Reinhold Leinfelder published an article in 2013, entitled '*Assuming Responsibility for the Anthropocene: Challenges and Opportunities in Education*'. Leinfelder emphasises the importance of accepting responsibility – as a society and as individuals. In the article, education is viewed as 'one of the most powerful tools for transformation, to make the Anthropocene sustainable, equitable and worth living'

[75] The IPCC (Intergovernmental Panel on Climate Change) was established in 1988, after the first world climate conference in 1979. It brings together the countries of the United Nations, quantifies the risk of climate change and proposes guidelines for mitigating anthropogenic climate change.

(Leinfelder, 2013, p. 10). One of the merits of Leinfelder's article is its awareness of education's essential role in a new social contract, which facilitates a profound societal transformation. However, he does not propose a fundamental shift in educational or pedagogical thinking. Leinfelder sets out a model of education *for* the Anthropocene (rather than education *in* the Anthropocene) based on an understanding of the state of the planet, of our relationship with time, and learning from history. In this article, education is viewed primarily as a means of instruction, and Leinfelder's description is largely an overview of what the Anthropocene is and how it should be explained to children or adolescents. Leinfelder's educational model represents instruction for the Anthropocene, but does not propose how education can bring about anthropological shifts, to allow humanity to exist in increasingly limited space.[76] Ultimately, what the author proposes is to learn about the Anthropocene – that teachers convey the scientific knowledge that is currently available about this new geological era. Indeed, this is an important part of Anthropocene education, which will help define new curricular content. However, what kind of political education can we devise, with a view to educating education *in* the Anthropocene, which is not reduced solely to educating *for* the Anthropocene? This new geological epoch necessitates a new educational paradigm, beyond transmitting awareness of the environmental impact of human activities. We must think about the pedagogical implementation of political consolidation as one of the most valuable tools we have to help navigate the anthropological and civilisational *krisis* the Anthropocene represents. Here, education, in this most ambitious sense – understood in the sense of a transformation or metamorphosis (Moreau, 2012) – is presented as a means of choice in bringing about an anthropological shift.[77]

There is also another connection between the terms 'pedagogy' and 'Anthropocene' in the literature. The Anglo-Canadian geographer and environmentalist Simon Dalby mentions the 'political and pedagogical potential' (2016, p. 36) carried by the notion of the Anthropocene. We often see discussion of how the Anthropocene presents an opportunity for a sorely needed reassessment of our activities – it is a radical and irreversible change, which will inevitably affect everyone, and can therefore help raise awareness of environmental issues. However, only Dalby uses the term 'pedagogical' in relation to this learning opportunity. By its very nature, the Anthropocene is a dramatic event, and as such, carries the potential for learning.

[76] However, it should be noted that the author puts forward two interesting concepts. Firstly, participation – notably, citizens' participation in political decisions – is presented as one of the phenomena to be encouraged. If they are involved in the process, people are more likely to adhere to social changes. Secondly, his model of education for the Anthropocene includes an experience-based component, with each person performing an environmental 'good deed' every day.

[77] The Anthropocene is beginning to make its mark in the field of educational sciences, but even at this early stage, there are various relevant works that can be highlighted: Jagodzinski, 2018; Wulf, 2013, 2022a, b; Priyadharshini, 2021; Tannock, 2021; Hétier, 2021, 2022; Hétier & Wallenhorst, 2021, 2022; Garnier et al., 2021; Prouteau, 2021; Paulsen et al., 2022; Wallace et al., 2022.

Finally, the typical connections between the notions of education and the environment relate to environmental education, or education for sustainable development. Environmental education has always offered a new way of thinking about education, including a political component, with emphasis being placed on citizenship (Pineau, 1992; Sigaut, 2010–2011; Sauvé, 2009, 2015; Taleb, 2016). However, these approaches are not enough, because they do not necessitate major paradigmatic shifts which would completely reshape the way in which we think about education – and it is a complete reworking such as this which is absolutely necessary, in view of our entry into the Anthropocene. Daniel Curnier's proposal to educate *for* sustainable development, based on strong sustainability (2017), differs from simply educating *about* sustainable development, in that Curnier's approach is more radically engaged, aiming to bring about a real shift in environmental behaviour, and create a new type of society.

The educational approach that needs to be developed goes beyond simply providing knowledge about the Anthropocene. We need to foster and develop the capacity to act in concert, united by our shared existence on our one and only planet. In line with Hartmut Rosa's recent critical theory, education for resonance could be the necessary pedagogy of politics, integrating and going beyond this merely informational education about the Anthropocene. In fact, in German literature on education, we see a positive reception of Rosa's idea of resonance. Education in the Anthropocene aims to achieve an overhaul of educational concepts which, in the main, still largely hinge on individual development, laying the groundwork for contemporary neoliberalism. Education is viewed as a lever to effect the anthropological change which, according to certain environmentalists (such as Bourg, Arnsperger, Curnier and Federau) is so desperately needed, but whose educational foundations have yet to be lain. Education has always been a valuable political tool to bring about long-term change (for this reason, education is a central theme in Plato's work). In the context of the author's native France, and even in Europe as a whole, none of the political parties (whose positions range from the extreme left to the extreme right) disputes the fundamentally important role of education in preparing for the future. No-one is opposed to education. However, we must wonder what educational paradigm is needed.

1.3 Conviviality as a Paradigm for Political Education (Creative Function)

1.3.1 Multidisciplinary and Interdisciplinary Work

A critical theory for the Anthropocene must necessarily continue in the same vein of interdisciplinary research as Critical Theory, which may be said to be true 'transdisciplinarity in action' (Assoun, 2016, p. 1). Indeed, the work of the leading Critical Theoreticians has always straddled the boundaries between disciplines, so that

Critical Theory cannot easily be pigeonholed as philosophy, sociology or politics.[78] The combined perspectives of different disciplines facilitate the unique approach, identifying the root cause of the problem, and using that as a starting point from which to propose a new direction. Within Critical Theory, sociology (as the study of social facts) and social philosophy (as speculative and normative musings) are included in the same movement and the same critical output. Moreover, the entry into the Anthropocene represents a new departure, which means it is essential to explore a combination of disciplines to produce new perspectives and ideas.[79] Thus, a critical theory for the Anthropocene must be multidisciplinary and interdisciplinary (which entails taking certain risks).[80] The critical theory required presupposes the emergence of a new scientific perspective, looking at the relations between the Anthropocene, politics and education, on the basis of anthropological reflection. This political work will involve work in Earth System Sciences, in Environmental Humanities and in Educational Sciences. Thus, this research brings into play a whole raft of notions, including the Anthropocene, the Earth system, action, hospitality, responsibility, the German idea of *Bildung*, emancipation and citizenship. These generate notional creativity, with the idea of bringing about an anthropological shift allowing for the consolidation of politics, the notion of human adventure (which includes the profit mindset of *homo oeconomicus*, the responsibility mindset of *homo collectivus*, and the hospitality mindset of *homo religatus*), the pedagogy of resonance as a political pedagogical paradigm, and existential citizenship.

1.3.2 A Political Stance of Uprising and Consolidation

In this new geological epoch, humanity faces an unprecedented situation: 'The Anthropocene, and the procession of transcendental damage that it brings, thus places the City – that political construct founded by humans with the aim of sustaining their society – in a vulnerable situation. The City finds its very survival under threat' (Lesourt, 2018, p. 11). The political thread of this work is in keeping with the work of French philosopher Enzo Lesourt: 'to put the City back on the path to

---

[78] For example, Adorno was a musicologist, interested in psychoanalysis, whose philosophical work on the cultural industry provided very fertile soil for information and communication sciences. Max Horkheimer worked in philosophy and sociology, but was highly conversant with psychoanalysis. Herbert Marcuse's work straddles the boundary between Marxist sociology and psychoanalysis. Walter Benjamin's is as much historical as it is aesthetic, literary and philosophical.

[79] As Guillaume Blanc, Elise Demeulenaere, and Wolf Feurhahn note, 'True interdisciplinarity means working on oneself, with no guarantee that it will be fruitful, but it is worth trying' (Blanc et al., 2017, p. 275).

[80] Twenty-five authors of a recent collective work, *Des Sciences sociales à la Science sociale*, edited by Alain Caillé, Philippe Chanial, Stéphane Dufoix and Frédéric Vandenberghe (including contributions from French sociologists François Dubet, Danilo Martuccelli and Michel Wieviorka), call for the gradual construction of Social Science with anti-utilitarian foundations.

long-term survival – even if that means transforming it' (2018, p. 12). The theoretical work that this research represents is not concerned primarily with the City, but with humanity in its political component: it is a question of putting humanity back on the path to sustainability, starting with a political transformation. Therefore, as already mentioned, we think of an anthropological shift that allows for the consolidation of politics. This theoretical endeavour comes down to 'the art of ensuring the City's survival and long-term future' (Lesourt, 2018, p. 14). While that may appear utopian and unrealistic, we consider it to be '*the* fundamental aim of politics' (Lesourt, 2018, p. 14). In this approach, Hannah Arendt's questioning during a period of profound crisis about the future of the human condition after the discovery of the Nazi extermination camps, is a prime resource. As is beginning to become clear, the critical theory proposed in this book has its origins in the connections of the Anthropocene[81] with Arendtian political thinking, and with the political agenda of uprising and consolidation. However, these three elements alone are still not insufficient. Indeed, as has already been briefly mentioned, Arendt's anthropology, especially as set forth in *The Human Condition*, is based on an immutable opposition between the sphere of life (referring to economic activity) and the sphere of the world (referring to political action). Life and the world are two opposing categories. This division refers to the conflicting driving logics behind economics and politics, and the problematic hegemony of the former over the latter. On the other hand, the real limitation of this anthropology is that it does not consider the biological and bio-geo-chemical foundations of the body politic. Indeed, the Anthropocene highlights how various schools of thought are floundering. Pairs of concepts which, in the Modern Era, have traditionally been binary and irreconcilable, are now brought together. For example, where nature and culture have typically been on opposite ends of a spectrum, we now see the end of culture's war on nature. An even more extreme example is the dichotomy of life and death, where we are now seeing the fusion of living man with deathly machine (Wallenhorst et al., 2019).

Five other authors are important in developing a critical theory for the Anthropocene. The philosophical thinking of Dominique Bourg, and the anthropological work of Christian Arnsperger, have been crucial in undertaking this work. Indeed, Dominique Bourg, a great reader of the scientific works of the Anthropocene, is constantly suggesting that these works be taken into account as part of the foundation for contemporary philosophical thinking, organised around this fundamental political question: What can we do together? Christian Arnsperger's work on anthropological means of escaping from the capitalist hegemony in which we currently live, of which the Anthropocene is the strongest indicator, is important here. Secondly, the philosopher Maurice Bellet, whose entire body of work calls for and puts forward a proposal for an anthropological shift, is also an important resource.

[81] This striking theoretical creativity generated by the Anthropocene is particularly noticeable in three recent collective books: *Gouverner la décroissance – Politiques de l'Anthropocène III* (2017) edited by French political scientists Agnès Sinaï and Mathilde Szuba; *Penser l'Anthropocène* (2018), edited by French philosophers Rémi Beau and Catherine Larrère; and *Atlas de l'Anthropocène* by Gemenne et al. (2019).

His prospective and romantic idea of uprising, and of consolidation, can serve as a source of inspiration.[82] However, it should be noted that all of Bellet's work, published between the 1960s and 2019, is philosophical in nature without anchorage in the biological, or consideration of the scope of the environmental problem. Here, the work of Andreas Weber who proposes a poetics for the Anthropocene, will anchor the political project in the environmental. Moreover, Weber's work can anchor political anthropology in the world of the living without opposing the categories of life and politics, as Arendt does. Finally, the German sociologist and philosopher Hartmut Rosa brings the question of pedagogy and learning into the equation. Whilst it is patently obvious that we need to change, we need to identify how. This is what is made possible by the idea of resonance, which Rosa says is inspired by the Critical Theory of the Frankfurt School.

There is a fundamental connection between these different authors. Each of them contributes an indispensable stone to the construction of this critical theory. Bourg offers a philosophical and political interpretation of the work in Earth System Sciences. Bellet is the initiator of a romantic school of thought, which makes it clear that a profound anthropological shift can be achieved by thinking in terms of the 'between-us' mentality that Arnsperger described, and in the Social Sciences. Arendt lends weight, and genuine political meaning, to the idea of life shared 'between us'. Weber anchors the body politic and the emergence of concerted action in the solidarity of the living world. Finally, Rosa reflects on educational means of setting that anthropological shift in motion, and anchors the whole of the work in the tradition of critical thought. This theoretical ensemble is organised around the tenet of convivial coexistence, between humans, and with non-humans as well. As such, it stands in opposition to the hegemonic economic rationale which is so evident in the contemporary period, marked by neoliberal capitalism, of which the Anthropocene is the clearest reflection.

1.3.3 A Work of Convivialist Anthropology

The denunciation of the error of the Promethean race forward gives rise to the proposal of an alternative theoretical anthropology, organised around a post-Promethean 'between-us', rather than around a hypermodern individual.[83] This work questions what conditions humanity, such as space, time or the sharing of human existence with others. The Anthropocene represents a reconfiguration of the space inhabited by humans, animals and the entire fabric of the living world. This new geological

[82] Arnsperger describes Bellet's intellectual agenda as fomenting 'uprising in the very place where humans' humanity is decided' (Arnsperger, 2010, p. 35).

[83] This notional proposal is in line with the epistemological extension of the anthropology of finitude as a basis for political thought, developed at the University of Lausanne, notably by Dominique Bourg (2010, 2012, 2018) and Christian Arnsperger (2009, 2010, 2011a, b, c). It is also in keeping with the work of Jean-Philippe Pierron (2014a, p. 65; 2016) and Bruno Villalba (2016).

epoch is altering the living and mineral matter with which human societies interact and which is essential to human life. Additionally, though, the Anthropocene represents a temporal reconfiguration: the temporal processes of matter flows are being reorganised, and the bio-geochemical indicators of the Earth system necessary for human societies are evolving at an unusual rate. The Anthropocene, also called the Great Acceleration, is indicative of the indissociability of space and time. Finally, the third fact of the human condition, the sharing of existence with others, is at play. The Anthropocene, which is indicative of the domination of *homo oeconomicus*' logic of maximising individual interests, highlights the need to strengthen coexistence. This relates to other humans, of course, but the Anthropocene also highlights how problematic our inability to coexist with animals and the whole non-human world is. Climate and biodiversity are intrinsically linked: the destruction of life by humans affects the climate, which in turn alters life.[84] We coexist more than we exist, and this coexistence includes non-humans. The Anthropocene necessitates a profound and fundamental break with anthropocentrism. The Anthropocene is forcing humans to rethink our civilisations built on the mastery of nature, allowing for agricultural surpluses (so that some people can devote themselves to tasks other than merely eking out an existence). The Anthropocene, which is no longer characterised by the climatic stability of the Holocene, is reconfiguring the future, and we must learn to anticipate and project ourselves differently. Moreover, the Anthropocene is shifting imaginary projections and reorganising the great narratives of humanity.

In this research, we propose a 'post-Promethean *between-us*', resonating with the work of French biologists Pablo Servigne and Gauthier Chapelle, *L'entraide – L'autre loi de la jungle* (2017 traduced in English in 2021, *Mutual Aid: The Other Law of the Jungle*), who show that, in the kingdom of the living, solidarity takes precedence over competition. This is also the basis of the work of French anthropologist François Flahault in his latest book, *L'homme, une espèce déboussolée – Anthropologie générale à l'âge de l'écologie* (Flahault, 2018). This anthropological work, based on a biological analysis of the living world, serves as the foundation for the convivialist political reflection in which this research participates. Indeed, this work has some points in common with a body of work recently federated around the term 'convivialist'; the group includes Alain Caillé, Bernard Perret, Corine Pelluchon and Philippe Chanial, but also Edgard Morin, François Dubet and Dominique Bourg. Conviviality refers to living together with all political 'subjects' (in the philosophical sense of the term), animals included (Pelluchon, 2017a, b). The critical theory for the Anthropocene developed in this work contributes to the development of a political anthropology of education that could be described as convivialist anthropology.

[84] Scientists are talking about the possibility of a sixth mass extinction, with 75% of biodiversity disappearing (Barnosky et al., 2011).

1.3.4 Political Education in the Anthropocene

A political anthropology of concerted action includes a dynamic of educability – i.e. the idea that it is possible, through learning, to become in part something other than what we are. While there is a growing body of political research relating to the Anthropocene,[85] political reflections on education in the Anthropocene are rarer. As previously mentioned, we note the work done by Curnier (2017), as well as a few books and themed issues of scientific journals.

The link between education and the Anthropocene is different from environmental education, or education for sustainable development. It is a matter of thinking about certain fundamentals of a political education which (a) lays the foundations of interpersonal relationships that favour the emergence of concerted action, and (b) can lead to a different type of relationship with nature. This education for politics in the Anthropocene is not primarily underpinned by the drive to conserve the planet, which is generally one of the foundations of environmental education, but by the necessity of changing the way we live together in the world. We are capable of acting together among humans (of course), but perhaps and above all within, with and in solidarity with the living world. Much like the world itself, this living fabric of existence long predates humans, and will endure long after we are gone. The aim of the intended political education is ambitious: to enable the public sphere to regain ground from the private sphere. Education *in* the Anthropocene presupposes the development of new educational paradigms, based on a conception of politics as that concerted action which sometimes arises between humans. This education *in* the Anthropocene necessarily includes education *for* the Anthropocene, with the transmission of knowledge about the Anthropocene[86] and its implications (for example, in terms of understanding the Age of Enlightenment, or modernity and all its dualisms). The concept of the Anthropocene as theorised in Earth System Sciences carries with it a worldview marked by naturalism and the predominance of the notion of species in the accompanying narrative. The notion of species carries the dual characteristic of an undifferentiated *anthropos* without will or freedom, but with intelligence capable of achieving technical creations that enable it to work against and overcome the greatest geological forces on the planet.

The relationship of education to politics in this critical theory for the Anthropocene is twofold: it is instrumental (education is a political tool), but also teleological (the

---

[85] Of particular note is the interdisciplinary research programme 'Politiques de la Terre à l'épreuve de l'Anthropocène' "Earth Policies for the Anthropocene"), organised by the Université Sorbonne Paris Cité with a number of other institutional actors, and directed by Bruno Latour from Sciences Po Paris. There is also a series of books and thematic issues of journals (Sinaï, 2013; Bourg et al., 2017).

[86] The question of enhancing climate and environmental education is raised regularly, as noted in an article in *Le Monde*, by Audrey Garric and Marine Miller on 23 March 2019: '*Le dérèglement climatique est trop peu enseigné, de l'école à l'université*' (Climate change is too little taught, from school to university). The journalists point out that both teachers and researchers are increasingly critical of the inadequate place afforded to climate change in the various school curricula.

education worked on here is an education in politics). In line with what Arendt demonstrated, the purpose of education is to understand the world in order to take responsibility for it, and allow others to exist. Here, the world is understood as the prerequisite for the survival of the human adventure, which is now compromised by the entry into the Anthropocene. Yet this learning of responsibility for the world takes on a very particular form in today's world, since it is with regard to the flows of bio-geo-chemical matter that organise and allow life to exist that we need to exercise responsibility, going beyond the Arendtian dichotomy of life and the world.

This critical theory for the Anthropocene takes up the thread from Theodore W. Adorno's 1966 article '*Erziehung nach Auschwitz*' ('Education after Auschwitz'). We cannot (unfortunately) think about education *after* the Anthropocene here. We are left with the possibility of thinking about what educational paradigm is desirable *in* the Anthropocene, in order to survive in this new geological epoch, without the runaway change of the Earth system getting the better of the human adventure. Above all, though, it is important to think about how to guide the taking of responsibility for the world, the Earth and all living beings, when the educator is aware of the grave threat that the Anthropocene represents for the human adventure.

1.3.5 Criticism, Resistance and Utopia

Arendtian thinking may be said to be more consensual than that of the Critical Theoreticians. As has been mentioned, its characteristic is that it can be invested with a set of very different sensibilities. With the Critical Theoreticians, their neo-Marxism restricts the types of investment of their thought. It is important to point out that critical theory is not unique to German thinkers. The leading figures of Critical Theory were from Germany, because of the establishment of the *Institut für Sozialforschung* in Frankfurt, and it was Germans who first claimed this theoretical filiation. As mentioned, the strongest characteristic of Critical Theory is the drive to identify the social mechanisms of alienation and reification, with a view to proposing a way forward offering hope. The critique of modernity is not sufficient, on its own, to characterise Critical Theory. It is only the first step in this intellectual and political effort, whose goal is social transformation. The cerebral work required in developing a critical theory for the Anthropocene requires the identification of reasons for hope – as has regularly been the case, since Kant's work. In addition to being a challenge to reason, this is a challenge to the Anthropocene, the enormity of which can crush any attempt to inculcate hope. The identification of convivialism as an educational paradigm which will allow us to survive in the Anthropocene is the result of this need to preserve hope in spite of the Anthropocene (the educational style represents the narrative through which the new epoch is presented).

The great figures of the Frankfurt School never ceased to carry the Marxian hope for the transformation of society, despite the dark times in which they lived, marked by Nazism, the Holocaust, the Cold War, the reification of humans through industrial labour, and growing financial capitalism. It is this utopian hope, combined with the vigour of their critique, that interests us in the dark times of the Anthropocene, when it is important not to be carried away by the fatalism of a form of determinism grounded in the contemporary hegemony of individualistic economic drives.[87] Finally, the proponents of Critical Theory never ceased to actively resist the logics of reification, alienation and annihilation of humans, so typical of their time. It is this call for resistance that must be manifest in a critical theory for the Anthropocene. Thus, critique, resistance and utopia are the three functions by which the social utility of a critical theory for the Anthropocene must be measured.

Critical theory is not an ideological straitjacket, where reference can only ever be made to the founding theorists of the Frankfurt School. Indeed not; today, there are critical theorists working in different countries and various languages.[88] For example, the work of the French philosopher Fred Poché is an extension of Critical Theory because it updates that philosophical and political stance, mainly from the perspective of those who have been forgotten.[89] The same is true of the work of the convivialists, but their intellectual goal is the same as that of critical theory (Les convivialistes, 2013; Caillé et al., 2018; Wallenhorst et al., 2020).

1.3.6 A Work of Critical Theory with Proximity to the Work of the Rennes School of Political Science

The proposal of an alternative to Promethean anthropological approaches, in this book, is driven by normative rationale. The proposed approach differs here from the work of political sociology in the Anthropocene, which is particularly important

---

[87] It is up to us to identify reasons for hope (other than the fulfilment of the Promethean project of modernity). Marcuse, at the end of his book *One-Dimensional Man* (1963), refers to a statement by Walter Benjamin: 'It is only for the sake of those who are without hope that hope is given to us'.

[88] The four volumes (numbers 10/11, 12/13, 14/15 and 16/17) of the journal *Illusio* on the theme of 'Critical Theory of the Crisis' show this clearly.

[89] Fred Poché presents an interesting way of thinking about 'remaking democracy' (2014) based on 'ethics of the oppressed', highlighting the importance of standing with the 'voiceless' in the public arena (this is also the stance adopted by researchers in the north of France who have studied sobriety from the point of view of people in precarious situations – Villalba, 2016). Speaking out is one of the most important concrete ways of bringing democracy to life and breaking with 'loneliness'. Fragility – the 'difficulty of keeping a grip on one's existence' (Poché, 2013, p. 10) and on society, is particularly prevalent in today's societies. On the other hand, for Poché, fragilities can be organised into 'joyful resistance': through solidarity, fragility can be made into social foundations. Poché's project consists of developing action for dignity through the development of politics of fragility and ethics of the oppressed. For Poché, it is important to provide social actors with spaces for argumentation – the true foundation of politics and democracy. Thought allows for the creation of a space free of the dominant contemporary economic rationale, where individuals are invited to 'de-think' (p. 20) rather than 'think'. Poché's way of addressing the excesses of dominant neoliberalism is to highlight the dynamics inherent in the social reality, which he regards with a critical and positive eye. This analysis is hopeful and encourages the creativity of collective action, which can sometimes take the form of resistance to the logic of annihilation of human beings.

because it identifies the ongoing transformations, precipitated by the Anthropocene, within social and political practices, and more broadly, in social spaces and in the structuring of political acts (Ollitrault, 2001, 2008, 2016). In this work, the Anthropocene is not only viewed with detachment, without any involvement by the researcher other than analysis. The existential resources mobilised by work on the impacts of our entry into this new geological epoch encourage us to adopt stances, in line with the tradition of Critical Theory.[90] While this may seem unusual in the context of academic research, it should be noted that it is the very nature of critical theory to invest this intellectual register. In addition to the Frankfurt School, our work is rooted in the work of the 'Rennes School of Political Science', which developed in contrast to the assimilation of Political Science to the Bourdieusian Political Sociology that became institutionalised in the 1980s. The work of the Rennes-based political scientist Jean Baudouin contributed to the development of Political Theory, in line with the way in which it is understood by Arendt. The unique characteristic of Political Theory is the desire to 'rehabilitate actors (by endowing them with autonomy of volition and action), the ideas they carry, the regimes within which they act' (Portier, 2017, p. 14). This Rennes School of Political Science, which draws heavily upon Political Theory in Political Science, could be described, in line with Jean Baudouin's work, as granting primacy to the dynamics of politics, carried by actors driven by their own motives, over a form of sociological determinism. It consists of 'a singular way of articulating understanding and explanation, the actor and the system, interest and value' (Portier, 2017, p. 27). This took shape from 1995, when Jean Baudouin set up the *Centre d'études et de recherches autour de la démocratie*, where he was joined by Philippe Bénéton, Jacques Le Goff, Thuriane Séveno, and Philippe Portier (Portier, 2017, p. 23) then by Frédéric Lambert and Bernard Bruneteau. This team then forged relationships with other academics feeling the same need to think about the political and its own dynamics (whose development of thought is not axiologically neutral) outside the box of sociology.[91] The social is thought of as being shaped through experience, in the framework of comprehensive sociology (German sociologist Max Weber laid the foundations for such comprehensive sociology, and the sociologists Luc Boltanski and François Dubet then took up the baton and developed the approach further in France). The epistemological approach adopted in this work is in continuity with the understanding of politics of the Rennes school (including the works of Jean Baudouin, Philippe Bénéton, Philippe Portier, Bernard Bruneteau and Frédéric Lambert), though the subjects we address here are not the same as those earlier thinkers: we look at the questions of the Anthropocene and education.

---

[90] For example, Curnier, in the development of educational reflection in the Anthropocene (2017, p. 28), expresses a dual epistemological foundation: the complex thinking of French sociologist Edgar Morin, and the 'post-normal science' of the Argentine philosopher Silvio Funtowicz (working in Norway), the American philosopher Jerome J. Ravetz (working in Great Britain), and the Chilean environmentalists Cecilie Modvar and Gilberto C. Gallopín whose purpose is directly political: to influence political decisions.

[91] These include Philippe Raynaud, Marc Sadoun and Jean-Marie Donegani.

1.4 The Proposed Way Forward

1.4.1 Politics in the Anthropocene

In order to put the notion of the Anthropocene to work in developing a critical theory, we must first identify its multidisciplinary scientific content (spanning the fields of stratigraphy, chemistry, physics, biology, palaeontology, etc.). The first part of this book begins with a brief overview of the countless findings surrounding the notion of the Anthropocene in Earth System Sciences over the last 20 years.[92] The aim here is to uncover the underlying political anthropology, that will enable us to zero in on the type of education needed in the Anthropocene.

Chapter 2 consists of an introduction to the notion of the Anthropocene, whose origins are particularly interesting. Firstly, while this concept has a 20-year history, the idea of humanity as a geological force has been around for over 150 years. Secondly, this stratigraphic concept was put forward informally by a chemist, before being enthusiastically taken up by the Earth System research community, public opinion, and of course, the geological and stratigraphic community. This proposal to define a new geological epoch on the geological timescale is in the process of being officially validated by the international community of geologists.

Chapter 3, 'The Notion of Planetary Boundaries', explores an important component of the Anthropocene: the transgression of planetary boundaries as defined by the Swedish environmentalist Johan Rockström et al. This new geological epoch is characterised by the way in which humans have transgressed (and continue to transgress) 'natural' boundaries. By crossing several of the nine boundaries defined by these researchers, humanity has catapulted itself out of the safe zone, in which it had some room for manoeuvre, and into imminent danger. The conditions of habitability of the earth have been altered, and even if pushed no further, will take several thousand years to return to normal. In addition to the transgression of these limits, the Anthropocene is characterised by the exponential rate at which humans have extracted a range of resources from the biosphere since the 1950s, with the intensification of industrial production to meet the demands of a globalised consumer society. This is known as the 'Great Acceleration'.

The purpose of these first two chapters is to 'lay the groundwork' for the political anthropological reflection that follows, and then for the proposal of convivialist political education. The next three chapters examine the political implications. Indeed, there are contradictory interpretations of our entry into the Anthropocene. As such, Chap. 4 is devoted to the 'Political ambiguities surrounding the Anthropocene'. One interpretation confirms what many authors have been writing for decades: it is essential that we radically change our relationship with the earth we inhabit. Another

[92] We have published a more in-depth analysis of this body of work in a book introducing the notion of the Anthropocene and proposing a political interpretation of it, *L'Anthropocène décodé pour les humains* (Wallenhorst, 2019), as well as another book, *La vérité de l'Anthropocène* (Wallenhorst, 2020a), presenting the scientific content of the 'major articles' surrounding the Anthropocene.

interpretation, though, confirms the hope carried by progress and the seemingly unlimited mastery of matter through tools, offering potentially infinite power: 'If nature is unjust, change nature' (Cuboniks, 2016, p. 273); many authors appear to echo this sentiment, in reference both to nature and to human nature. Following criticism of the failure to differentiate between the Anthropocene and *anthropos*, in the early 2010s, we see a slight shift in scientific discourse emanating from certain researchers, such as the American geochemist Will Steffen and the Dutch geochemist Paul Crutzen, who won the Nobel Prize for chemistry in 1995, both of whom are heavily involved in the debate surrounding the Anthropocene. However, it must be noted that they have not really allowed their work to be influenced by the paradigms of the human and social sciences. Their contributions on differentiating between the levels of responsibility for the entry into the Anthropocene attributable to different regions of the world remain merely anecdotal. The conclusion of the techno-scientific narrative of the Anthropocene remains the same: we are hurtling ineluctably towards a deterioration of the conditions of habitability of planet Earth, and the only potential solution envisaged to apply the brakes is geoengineering. The summary of research on the Earth system, and the interpretations and narratives about the Anthropocene in this third chapter, helps to understand the environmental situation as it currently stands, and how it is linked to a political crisis (Rozès, 2005; Lambert and Lefranc, 2012, pp. 50–52; Batout & Constantin, 2014). Indeed, the political ambiguities of the Anthropocene that are highlighted are indicative of how politics is being misused, promoting the economic hegemony of the globalised world.

Chapter 5 is devoted to the different conceptions of political action in the Anthropocene, which are broadly organised into two opposing categories: Promethean and techno-scientific policies, on the one hand, and post-Promethean policies marked by the need for a major lifestyle change, on the other. Political outlooks marked by a form of post-Prometheism appear to be particularly relevant in a critical theory for the Anthropocene; what they have in common is that they attach great importance to politics, thought of as the organisation of concerted action. Looking at the literature on the Anthropocene, it becomes apparent that proposed solutions based on a techno-scientific approach are underpinned by an anthropology of inevitability, which plays down humanity's capacity for deliberate action, not viewing humans as capable political agents. This sentiment of inevitability, among Earth system scientists, seems to be due to a lack of faith in humans' ability to act together. To the question 'Why have we failed?' the American economist Dennis Meadows, one of the four authors of the famous 'Meadows Report' entitled *The Limits to Growth* (1972),[93] opines in a 2010 article that we act as if technical progress can solve problems. When we seriously examine these post-Promethean political orientations and reflections, we are led to question our anthropological conceptions and to conceive of humans as capable of acting together and capable of making choices about the direction in which humanity is to go.

Chapter 6, the last in Part One of this book, looks at the integration of the Anthropocene into civic debate. This is studied through the lens of the transfer of

[93] The famous Club of Rome report was updated by Meadows, Randers and Meadows in 2004.

knowledge of the anthropogenic systemic alteration of the Earth from scientists to lay citizens. That transfer can be seen in the profusion of political manifestos published in recent years (nine manifestos are studied). Indeed, we are seeing ordinary citizens return to the public debate surrounding environmental issues. This is partly due to the Anthropocene's intrinsic capacity to spur people into action (it should be noted that scientists have been involved in the drafting of many of these manifestos). We are beginning to see the first impacts of citizens' understanding of the alteration of the Earth system on political discourse. The manifestos may be viewed as being indicative of environmental awareness and awareness of the entry into the Anthropocene in civic debate. What is important here, is to orientate ourselves within the miasma of ideas thrown up by the Anthropocene, rather than to systematically analyse all these texts (with discourse analysis and a sociological analysis of the context of their writing).

1.4.2 A Consolidation of Politics Requiring an Anthropological Shift

Having invested a great deal of hope in progress, what can we hope for tomorrow? Can an intellectually reasonable hope be nourished in the contemporary period? On what would that hope be founded? This work proposes moving from thinking about humanity as a nature and as a condition to thinking about humanity as an adventure. This shift breaks with essentialised reflections on humanity, instead favouring a dynamic reflection marked by uncertainty as to its long-term survival, given the uncertainty by which the Anthropocene is characterised. Thus, humanity is seen as plastic – and therefore educable; we are faced with a *krisis* and must evolve to cope, due to the irreversible changes in the habitability of the biosphere.[94] This shift also breaks with the anthropological paradigm of humanity having elevated itself above the natural world (this is an idea which has been dominant since the Age of Enlightenment), in favour of an anthropology of humanity's immersion in nature (Papaux, 2015).[95]

[94] Daniel Curnier's definition of the biosphere is particularly clear: 'The concept of the Biosphere refers to the thin layer on the Earth's surface whose geochemical cycles interact with living processes. While the Anthropocene forces us to shift our temporal perspective, the Biosphere redefines the spatial dimension of our relationship with the world. This concept makes it possible to restore the position of the human species on the surface of a planet whose distance from the sun and geochemical composition offer the only known environment in the Universe that has allowed life to develop. (...) The concept of the Biosphere also recalls the biophysical laws on which human beings depend, and which limit the expansion of their activities' (Curnier, 2017, p. 221).

[95] Curnier points to this idea of humanity being 'uprooted' from nature as particularly widespread in anthropological representations in public opinion, and argues that it could lead 'to the collapse of civilisation, or even to the extinction of the human species' (2017, p. 99). His analysis of the responsibility of school is clear: 'From the point of view of the relationship between humans and the living world, school can be seen as symbolic of the student's uprooting from a social reality anchored in a biophysical substrate' (Curnier, 2017, p. 180).

The second part of this work studies four theoretical movements – one per chapter – that appear necessary for an anthropological shift, bringing about political consolidation with regard to the anthropological reasons for the advent of the Anthropocene. Chapter 7 continues the analysis of the nine manifestos presented in Chap. 6, starting from the two questions underlying their drafting. Firstly, if there is something fundamentally wrong with our relationship to the world, of which the Anthropocene is an indicator, then we need to change… but what needs to change profoundly in our relationships with one another? Secondly, it is important that we change (and this takes the form of a change in lifestyles), but it is also necessary to rethink the way in which we consider ourselves, and thus change our anthropological conceptions. The entry into this new geological epoch can thus constitute 'the basis for a new adventure of civilisation' (Perret, 2012, p. 161).

Chapter 8 is entitled 'Getting through the Contemporary Anthropological *Krisis*'. The contemporary period is regularly understood as marked by an environmental crisis. The characteristics of the Anthropocene described in Part I show that this period has nothing to do with the timescales of crisis, but rather, is to do with the transition to a new state, as discussed in Chap. 7. However, the notion of crisis, or rather, *krisis*, remains relevant for understanding the present and the anthropological and civilisational issues marked by this new structuring of the conditions of habitability of the biosphere in the next few thousand years. The passage through the *krisis* humanity now faces is apprehended in this chapter as the condition for preventing destruction.

The etymological root of the word crisis, *krisis*, which means decision, choice, judgement or discernment, has threefold meaning: (1) encountering difficulties with a certain degree of intensity; (2) questioning, distancing (criticism) and separation; and finally (3) judging (from *krinein*, meaning both separating and judging) with decisions to be made to bring about a change. These three meanings together constitute a temporal process; *krisis* has a beginning and an end: it is the moment of transition from one form to another. We prefer this notion to that of crisis, also because of the latter's overuse in the contemporary period (especially in reference to economic crisis). *Crisis* has partly lost its temporary aspect (now corresponding to a structural social characteristic) and does not refer to the imperative of effecting a paradigmatic change, which is what the notion of *krisis*, with its meaning of radical criticism, refers to. *Krisis* refers as much to experiential intensity, and to the impression of loss of reference points, as to this radical critical experience (of which it is also the etymological origin). It is on the basis of this anthropological and civilizational *krisis* that an intellectual gesture of uplift and consolidation is outlined, for which Maurice Bellet's thinking is an important source of inspiration.

Chapter 9, entitled 'From the (augmented) individual to a post-Promethean '"between-us"', is the third anthropological movement to be discussed. As highlighted in Chapter 3, the entry into the Anthropocene is characterised by the transgression of planetary limits. A Promethean dynamic marked by the quest for 'more and more', disregarding all limits, is at the root of the entry into the Anthropocene. Further transgression of biogeochemical limits will jeopardise the very possibility for humanity to survive. The integration of limits and commitment to refrain from transgressing them is a fundamental anthropological issue, and it appears necessary to enter a post-Promethean stage. Moreover, on reading the scientific work on the Anthropocene, we may legitimately wonder whether the changes wrought in the Earth system are anthropogenic in nature or whether they are, in fact, sociogenic. If the latter is the case, it means that there is something that needs to be thought of differently in the way in which human existence is shared within the Earth's expanse. It is thus a matter of allowing the emergence of other types of interactions and of re-imagining the 'between-us' (Lévinas, 1991) as a 'post-Promethean between-us'.

Thus, this chapter takes the exact opposite view to some research and theories currently circulating in the economic sector around human augmentation. Transhumanism is the goal of anthropological mutation based on the achievement of the Promethean agenda. While in transhumanist theories, the emphasis is placed on the shifting of individual thought without any relational context, consisting of a form of annihilation of politics, we emphasise the 'between-us' as a political space for human genesis.

Chapter 10 is entitled 'From the human condition to the human adventure' and studies another movement of anthropological consolidation of politics. This chapter proposes to model the adventure – a peril-fraught journey into the unknown – that humanity is facing. In order to do so, we will use Arendt's concept of the human condition, based on the three categories of labour, work and action, which refers to the different facets of our relationship to the world. It is on the basis of this conceptualisation of the human condition that Arendt criticises the alienating modern world, where the public arena loses importance, while the private arena becomes more important. For her, this is a worrying process, since it alters the freedom of humans who express themselves only in public. For Arendt, the category of labour refers to an activity close to that of animals, insofar as its sole purpose is to sustain life in its biological sense. For Arendt, life refers to the economic and social sphere and to movement that needs to be constantly maintained through consumption. In Arendt's thinking, labour is done by an *animal laborans* and refers exclusively to the private sphere. We can say that Arendt's three categories are hierarchical in relation to the degree of freedom and the way in which each category of activity helps to make humans human. The category of work, unlike labour, refers to a human activity marked by permanence and endurance over long timescales. Its purpose is not to be consumed but to be used and welcomed in the world that exists both before and after human life. In the production of the work, the human being moves away from bare necessity (though without actually leaving it behind entirely, since this activity is structured by its goal), and participates in an adventure that extends beyond his own lifetime, allowing a shared world to emerge. Work is produced by a *homo faber*; in a manner of speaking, it is at the junction between the private and public spheres: produced in the private sphere, its purpose is to perdure in the public sphere, helping to construct the world which humans share. Finally, the category of action refers to politics, and takes place only in the public space. For Arendt, action is free and marked by communication. It is powerful, and emerges among humans, with one indispensable condition: plurality. Action is rendered sterile by standardisation, and in this sense, stands in opposition to any form of totalitarianism (defined as the absence of politics). Action is never individual, but necessarily originates in a pluralist collective, and is therefore linked to hospitality. Action is performed by *zoa politika*.

Arendt's anthropological model of human relations with the world, and with life, based on these three categories of activity, is a critique of the economic hegemony that is so typical of modern societies. The omnipresence of labour over the last two centuries, with the domination of private space all around the globe, restricting public space, is one of the reasons for the planet's entry into the Anthropocene. This chapter offers an anthropological model of the human adventure in which Arendt's action is central, viewed as being the real anthropological issue of the Anthropocene. On the other hand, it is just as crucial to think of the body politic on the basis of its bio-geo-chemical substrate and the innumerable interactions between humans and the non-human. Life and the world cannot be understood as two opposing categories, since it is life that enables the world to be. The proposed anthropological model looks at the *homo* of action as being somewhere between the *homo collectivus* of plurality, within an agora, and the *homo religatus* within its environment, marked by complementarity to the hegemony of today's *homo oeconomicus*, creating anthromes[96] within the biosphere (corresponding in part to Arendt's concept of the *animal laborans*). By delving into the logics driving the activity of *homo collectivus* and *homo religatus*, it becomes possible to identify the anthropological shift of consolidation of the political, which is so necessary in order to allow action to take place.

---

[96] 39–50% of Earth's surface is currently used by humans, and at least 25% of Earth's photosynthetic output goes to serve human needs. This directly leads to a loss of biodiversity, including wildlife. The appropriation of the earth's surface by humans and the destruction of other animal habitats is the primary cause of biodiversity loss (Federau, 2017, pp. 34–35). The American geographer Erle C. Ellis et al., in 2011a, b, mapped the anthropogenic transformation of the biosphere between 1700 and 2000 (i.e. before and after the Industrial Revolution). They mapped what they call anthromes (i.e. anthropogenic biomes), by taking two characteristics into consideration: population density and land use. In their study, they show that in 1700, half of the biosphere was wilderness and 45% of the remaining biosphere was semi-natural. In contrast, 300 years later, only 25% of the biosphere is in a wild state and 20% in a semi-wild state. Only a quarter of the earth's surface not covered by ice can be considered wild – and indeed, every region of the globe is impacted by pollution and climate change. The Industrial Revolution is the event that the authors put forward as the explanation for the explosion of anthromes in the biosphere. Thus, between 1700 and 2000, the biosphere underwent a transition from a predominantly wild state to a predominantly anthropised state. Although human alteration of the biosphere has been significant since the development of agriculture, it is only since the Industrial Revolution, and then *a fortiori* during the intensification and globalisation of the Great Acceleration in the mid-twentieth century, that the biosphere has undergone a real transformation, sufficient to be quantified as a change of state, which differentiates it from the Holocene biosphere (Ellis, 2011b).

1.4.3 Conviviality as a Paradigm of Political Education

The fundamental anthropological problem to which this work of critical theory is attached is the Promethean *hybris* generated by the maximisation of individual interests. This is opposed by the conviviality of a post-Promethean 'between-us'.[97] The *Convivialist Manifesto* proposes four fundamental principles to define legitimate politics and acceptable ethics: the principles of common humanity, common sociality, individuation and mastered confrontation. In the update published as the *Second Convivialist Manifesto* (2020), the authors have added a principle of common naturalness[98] – the importance of which we identify in this work, particularly through Andreas Weber's research. The authors specify in the updated publication that the absolute imperative behind these five principles is to keep humans' *hybris* in check. This absolute imperative, and these five principles, provide the structure of the third part of this work, which is centred on identifying a style of education for politics in the Anthropocene.

To understand convivialism as a style of political education, in line with Christoph Theobald's (2007) notion of style, means the importance of the way of inhabiting the world. In our own interpretation of the concept of style, we consider the state of mind of convivial friendship with others, the world, non-humans and the earth itself, as the type of education needed in the Anthropocene. It is a matter of learning to inhabit the world in a certain way, in a convivialist style. Theobald builds, among other things, on Maurice Merleau-Ponty's understanding of style (which also inspired critical authors such as Hartmut Rosa and Andreas Weber. 'All style is the shaping of the elements of the world that orientate the world towards one of its essential parts' (Maurice Merleau-Ponty). By understanding convivialism as an educational style, we do not merely reduce it to knowledge to be transmitted, but rather view it as a different way of inhabiting the world. On the other hand, knowledge of the Anthropocene is necessary in order to learn – knowledge which must be worked on. This third part focuses on what it is necessary to learn, rather than primarily to transmit. Chapter 11 is entitled 'Learning convivial citizenship in the Anthropocene'. The goal here is to identify what it is necessary to learn, in view of our entry into the Anthropocene, to enable the human adventure to continue. The primary focus is not on the redesign of school curricula. Such a rethink is, of course, necessary, but it has been worked on in a relevant way by Daniel Curnier, in the context of the canton of Vaux in Switzerland (2017), highlighting the need to increase interdisciplinarity, in particular.

---

97 We edited a collective book on convivialism with French political scientist Anaïs Theviot and the infocomist Sandra Mellot: *Inter-connectés? Numérique et convivialisme* [Interconnected? Digital and convivialism] (2020).

98 The addition of this principle in a version worked on in 2019 is indicative of the diffusion, within the community of social scientists and philosophers, of scientific work related to human/animal differences (of degree rather than of nature), to the complex functioning of trees and their capacities for interaction, or to the planet's entry into the Anthropocene.

The critical discussion sparked by Curnier's thesis in Environmental Sciences highlights the importance of thinking about the type of citizenship that should be promoted in the Anthropocene (allowing us to go beyond the paradigm of education for sustainable development). We shall weigh this notion against the German notion of *Bildung* and its existential register, and thus examine the relevance of existential citizenship, before settling on the notional proposal of convivial citizenship (because convivialism is open to the whole non-human world as well as fellow humans) as one element of the necessary learning in the Anthropocene. This chapter identifies the way in which convivial citizenship involves *homo collectivus* working against the background of *homo religatus*, in opposition to the dominant *homo oeconomicus*. This chapter continues the effort to explain convivialism as a style of political education in the Anthropocene, and the way in which it allows us to think about another way of living together. 'Between-us' can be a space for post-Promethean learning, differentiated from the logics of optimisation and pre-emption.

Part III continues with Chap. 12, 'Resistant Education: Combatting Incipient *Hybris*'. Indeed, it is human *hybris* that emerges from the various scientific articles on the Anthropocene: *hybris* that needs to be objectified and identified in order to educate against it when it emerges. This chapter is about *hybris*, but convivialism as a whole, as an educational paradigm, is designed to oppose *hybris* by championing the sharing of existence. This chapter begins by questioning the aims of education in the Anthropocene, positioned within these two poles: learning to live and become oneself, and learning about the outside world. These two poles will then be revisited in light of the Anthropocene. The first shift will consist of moving from learning to live to learning to *make* live, while supporting ourselves within limits (Hétier, 2019, 2021, 2022), starting with the limits of the biosphere. The second will consist of moving from learning about the world – which brings perpetual acceleration – to learning about resonance with the world: the idea recently developed by Hartmut Rosa. The concept of resonance will then be of prime importance in Chaps. 13 and 14.

Thus far, along with the function of resistance, we have examined one of the three functions of education in the Anthropocene developed in this book. Moving forward, the concept of resonance provides a basis for thinking about the critical function of education in the Anthropocene, in Chap. 13 (entitled 'A Critical Education: We Are not Separate from the Earth, We Are the Earth'). The concept of resonance merits inclusion in a discussion about education in the Anthropocene, for several reasons. First of all, resonance, understood as the counterpart to acceleration, is a fertile source of inspiration when thinking about what it is necessary to learn in the geological epoch of 'the great acceleration' (Chap. 12). Secondly, in Rosa's thinking, nature is a sphere of resonance for the subject (in the philosophical sense), who is considered from the standpoint of an ineluctable link to nature. Finally, resonance is linked to an educational dynamic because it refers to relational learning. In the context of this chapter, Andreas Weber's thinking also highlights what we consider to be Rosa's minimisation of the Anthropocene. The radical thinking of these two German authors allows convivialism, as a paradigm of political education, to propose a break with the typical paradigm of modernity that postulates separation between humans and the non-human world.

Chapter 14, 'A Utopian Education: Earth and the World Speak' is the final chapter in this critical theory for the Anthropocene, which questions the separation between a silent non-human world and the world of humans who, alone, are capable of speech. In this light, one of the functions of education in the Anthropocene would be to enable the world and the Earth to sing again. In this chapter, we continue to appropriate the knowledge of the Anthropocene, on the basis of which to work towards this convivial friendship with others, the world, and the planet. First, we look at how Earth speaks about us, including the sediment records in the debate about how to date the entry into the Anthropocene. Secondly, we try to understand some of the reasons why, nowadays, it is becoming increasingly difficult to hear the earth speak. Finally, we question how education in the Anthropocene can help us learn to listen to the earth, the world, and those humans with whom we share existence.

Bibliographical References

Afeissa, H.-S. (2014). *La fin du monde et de l'humanité, Essai de généalogie du discours écologique*. PUF.

Arendt, H. (1983). *Condition de l'homme moderne*. Calmann-Lévy. (original edition 1958, French translation 1961), French translation.

Arendt, H., & Heidegger, M. (2001). *Lettres et autres documents 1925–1975*. Gallimard. (original edition 1998), French translation.

Arnsperger, C. (2008). L'angoisse dans la financiarisation: les causes existentielles de la crise. *Finance & Bien Commun, 31–32*, 94–102.

Arnsperger, C. (2009). *Éthique de l'existence post-capitaliste – Pour un militantisme existentiel*. Cerf.

Arnsperger, C. (2010). Changer d'existence économique: enjeux anthropologiques de la transition du capitalisme au post-capitalisme. *Revue d'éthique et de théologie morale, 258*, 23–50.

Arnsperger, C. (2011a). *L'homme économique et le sens de la vie*. Textuel.

Arnsperger, C. (2011b). Dépasser le capitalisme, mais par étapes. *Revue Projet, 324–325*, 73–81.

Arnsperger, C. (2011c, October 3). *Economie existentielle, critique sociale et plasticité anthropologique*. Texte accompagnant le séminaire "Anthropologie et Justice Sociale" au Centre Sèvre.

Arnsperger, C., & Bourg, D. (2017). *Ecologie intégrale, Pour une société permacirculaire*. PUF.

Aron, R. (1962). A propos de la théorie politique. *Revue française de science politique, 1*, 5–26.

Assoun, P.-L. (2016). *L'Ecole de Francfort*. PUF. (first edition 1997).

Atkinson, G., Dietz, S., & Neumayer, E. (Eds.). (2007). *Handbook of sustainable developpement*. Edward Elgar Publishing.

Aubert, N. (Ed.). (2005). *L'individu hypermoderne*. Eres.

Badré, M., Bourg, D., Brunetière, J.-R., Gadrey, J., Granjean, A., Perret, A., & Villalba, B. (2011). Les droits de l'avenir. *Esprit, 3*, 205–235.

Barnosky, A. D., Matzke, N., Tomiya, S., Wogan, G. O. U., Swartz, B., Quental, T. B., Marshall, C., McGuire, J. L., Lindsey, E. L., Maguire, K. C., Mersey, B., & Ferrer, E. A. (2011). Has the Earth's sixth mass extinction already arrived? *Nature, 471*, 51–57.

Batout, J., & Constantin, E. (2014). Croissance, crise et dépérissement de la politique. *Le Débat, 182*, 145–155.

Baudouin, J. (2002). *Les idées politiques contemporaines*. PUR.

Baudouin, J., & Bruneteau, B. (Eds.). (2014). *Le totalitarisme – Un concept et ses usages*. PUR.

Beau, R., & Larrère, C. (Eds.). (2018). *Penser l'Anthropocène*. Les Presses de Sciences Po.

Bellet, M. (1993). *La seconde Humanité. De l'impasse majeure de ce que nous appelons l'économie*. Desclée de Brouwer.

Bellet, M. (2004). *Le paradoxe infini*. Desclée de Brouwer.

Bellet, M. (2008). La contestation évangélique. *Etudes, 408*, 636–648.

Bellet, M. (2013). *L'avenir du communisme*. Bayard.

Biermann, F. (2014). The Anthropocene: A governance perspective. *The Anthropocene Review, 1*, 57–61.

Blanc, G., Demeulenaere, E., & Feuerhahn, W. (2017). Difficile interdisciplinarité. In G. Blanc, E. Demeulenaere, & W. Feuerhahn (Eds.), *Humanités environnementales* (pp. 271–275). Publications de la Sorbonne.

Bonneuil, C. (2014). L'Anthropocène et ses lectures politiques. *Les Possibles, 3*, 1–7.

Bonneuil, C., & Fressoz, J.-B. (2013). *L'événement Anthropocène*. Seuil.

Bourg, D. (2009, December). L'impératif écologique. *Esprit*, 59–71.

Bourg, D. (2010, July). L'éco-scepticisme et le refus des limites. *Etudes*, 29–40.

Bourg, D. (2012). Pour une anthropologie de la finitude. In J.-P. Pierron & M.-H. Parizeau (Eds.), *Repenser la nature: dialogue philosophique, Europe, Asie, Amériques*. Presses Universitaires de Laval.

Bourg, D. (2017). Gouverner le long terme. *lapenseeecologique.com*, Points de vue, *1*(1), PUF, 1–9.

Bourg, D. (2018). *Une nouvelle Terre*. Desclée de Brouwer.

Bourg, D., & Fragnière, A. (2014). *La pensée écologique – Une anthologie*. PUF.

Bourg, D., & Papaux, A. (2015). Pensée écologique. In D. Bourg & A. Papaux (Eds.), *Dictionnaire de la pensée écologique* (pp. 756–759). PUF.

Bourg, D., Augagneur, F., Blondiaux, L., Cohendet, M.-A., Fourniau, J.-M., François, B., & Prieur, M. (2017). *Inventer la démocratie du XXIème siècle*. Les liens qui libèrent.

Bourg, D., Prouteau, F., Hétier, R., & Wallenhorst, N. (interview). (2019). *Bad Anthropocene* et anthropologie politique postprométhéenne (*forthcoming*). *Raisons politiques*, Presses de Science Po.

Boutinet, J.-P. (1990). *Anthropologie du projet*. PUF.

Boutinet, J.-P. (1998). *L'immaturité de la vie adulte*. PUF.

Boutinet, J.-P. (2004). *Vers une société des agendas*. PUF.

Boutinet, J.-P. (2010). *Grammaires des conduites à projet*. PUF.

Brassier, R. (2016). Accélérer la raison. In L. de Sutter (Ed.), *Accélération !* (pp. 157–181). PUF. French translation.

Braud, P. (2002). *Sociologie politique*. LGDJ.

Bruneteau, B. (2010). *Le Totalitarisme. Origines d'un concept, genèse d'un débat. 1930–1942*. Cerf.

Bruneteau, B. (2014). *Les totalitarismes*. Armand Colin.

Caillé, A., Chanial, P., Dufoix, S., & Vandenberghe, F. (Eds.). (2018). *Des Sciences sociales à la Science sociale – Fondements antiutilitaristes*. Le Bord de l'Eau.

Cochet, Y. (2013). Trois modèles du monde. In A. Sinaï (Ed.), *Penser la décroissance* (pp. 49–72). Presses de Science Po.

Collective Work. (2020). *Dictionnaire critique de l'anthropocène*. CNRS Editions.

Corbin, A., Courtine, J.-J., & Vigarello, G. (Eds.). (2017). *Histoire des émotions, tome 3, De la fin du 19ème siècle à nos jours*. Seuil.

Coutellec, L., & Pierron, J.-P. (2017). Penser une éthique alimentaire. *Revue française d'éthique appliquée, 4*, 19–24.

Crutzen, P. J. (2002). Geology of mankind: "The Anthropocene". *Nature, 415*, 23.

Crutzen, P. J., & Schwägerl, C. (2011). Living in the Anthropocene: Toward a new global ethos. *Yale Environment 360*. http://e360.yale.edu/features/living_in_the_anthropocene_toward_a_new_global_ethos, 2011, consulted online on 25 October 2017.

Cuboniks, L. (2016). Accélérer le féminisme. In L. de Sutter (Ed.), *Accélération !* (pp. 253–273). PUF. French translation.

Curnier, D. (2017) *Quel rôle pour l'école dans la transition écologique? Esquisse d'une sociologie politique, environnementale et prospective du curriculum prescrit*. Doctoral thesis in

Environmental Sciences from the University of Lausanne, supervised by Dominique Bourg and Philippe Hertig.
Dalby, S. (2016). Framing the Anthropocene: The good, the bad and the ugly. *The Anthropocene Review, 3*(1), 33–51.
Dalrymple, G. B. (2001). The age of the earth in the twentieth century: a problem (mostly) solved. *Journal of the Geological Society of London (Special Publication), 190*, 205–221.
de Sutter, L. (Ed.). (2016a). *Accélération !* PUF.
de Sutter, L. (2016b). Introduction. In L. de Sutter (Ed.), *Accélération !* (pp. 7–25). PUF.
de Tribolet, M. (1886, January–April). Les animaux disparus depuis l'apparition de l'homme. *Revue savoisienne*, 45–114.
DellaSalla, D., & Goldstein, M. (Eds.). (2017). *Encyclopaedia of the Anthropocene*. Elsevier.
Descola, P. (2005). *Par-delà nature et culture*. Gallimard.
Dunlap, R. E., & Michelson, W. (Eds.). (2002). *Handbook of environmental sociology*. Greenwood Press.
Eckersley, R. (2017). La démocratie à l'ère de l'Anthropocène. *lapenseeecologique.com, 1*(1), 1–19. French translation.
Ellis, E. C. (2011a). The planet of no return: Human resilience on an artificial Earth. *The Breakthrough Journal, 2*, 37–44.
Ellis, E. C. (2011b). Anthropogenic transformation of the terrestrial biosphere. *Philosophical Transactions of The Royal Society, 369*, 1010–1035.
Ellis, E. C. (2015). Ecology in an anthropogenic biosphere. *Ecological Monographs, 85*(3), 287–331.
Ellis, E. C. (2017). Physical geography in the Anthropocene. *Progress in Physical Geography, 41*(5), 525–532.
Ellis, E. C. Prouteau, F., Hétier, R., & Wallenhorst, N. (interview). (2019). *Good Anthropocene* et anthropologie politique prométhéenne (*forthcoming*). *Raisons politiques*, Presses de Science Po.
Ellul, J. (2008). *La technique ou l'enjeu du siècle*. Economica. (original edition 1954).
Federau, A. (2017). *Pour une philosophie de l'Anthropocène*. PUF.
Fischer, E. (1915). Der Mensch al geologischer Faktor. *Zeitschrift der Deutschen Geologischen Gesellschaft, 67*(2), 106–148.
Flahault, F. (2005). Vers une nouvelle pensée sociale. *Revue du MAUSS, 26*, 377–382.
Flahault, F. (2008a). *Le crépuscule de Prométhée – Contribution à une histoire de la démesure humaine*. Mille et une nuit.
Flahault, F. (2008b). Comment l'homme peut-il être à la fois égoïste, bon et méchant? *Revue du MAUSS, 31*, 307–317.
Flahault, F. (2018). *L'homme, une espèce déboussolée – Anthropologie générale à l'âge de l'écologie*. Fayard.
Fressoz, J.-B. (remarks collected by T. Henneton and J. Confavreux). (2013). L'apocalypse et l'Anthropocène, *Vacarme, 65*, 202–233.
Garnier, A., Martinez, M.-L., Hétier, R., & Wallenhorst, N. (coord.). (2021). L'éducation en Anthropocène. *Recherches et éducation*.
Garric, G., & Miller, M. (2019, March 23). Le dérèglement climatique est trop peu enseigné, de l'école à l'université. *Le Monde*. https://www.lemonde.fr/planete/article/2019/03/23/le-climat-trop-peu-enseigne-de-l-ecole-a-l-universite_5440113_3244.html
Gemenne, F., Rankovic, A., Ansart, T., Martin, B., Mitrano, P., & Rio, A. (2019). *Atlas de l'Anthropocène*. Presses de Sciences Po.
Genel, K. (2006, January 11). Hannah Arendt et l'école de Francfort, deux critiques de la modernité. https://philolarge.hypotheses.org/files/2017/09/11-01-2006_genel.pdf
Guillaume, B. (2015). Anthropocène (point de vue 1). In D. Bourg & A. Papaux (Eds.), *Dictionnaire de la pensée écologique* (pp. 32–35). PUF.
Hess, G., & Bourg, D. (Eds.). (2016). *Science, conscience et environnement*. PUF.

Hétier, R. (2019). Apprendre à faire vivre en Anthropocène. In N. Wallenhorst & J.-P. Pierron (Eds.), *Eduquer en Anthropocène*. Le Bord de l'eau.
Hétier, R. (2021). *L'humanité contre l'Anthropocène. Lutter contre les effondrements*. PUF.
Hétier, R. (2022). *La liberté à corps perdu*. Le Pommier.
Hétier, R., & Wallenhorst, N. (coord.). (2021). L'éducation politique en Anthropocène. *Le Télémaque, 58*.
Hétier, R., & Wallenhorst, N. (2022). *Enseigner à l'époque de l'Anthropocène*. Le Bord de l'eau.
Horkheimer, M., & Adorno, T. W. (1974). *La Dialectique de la Raison – Fragments philosophiques*. Gallimard. (original edition 1947), French translation.
Hubelin, J.-J., Ben-Ncer, A., Bailey, S. E., Freidline, S. E., Neubauer, S., Skinner, M. M., Bergmann, I., Le Cabec, A., Benazzi, S., Harvati, K., & Gunz, P. (2017). New fossils from Jebel Irhoud, Morocco and the pan-African origin of Homo sapiens. *Nature, 546*, 289–292.
Huzar, E. (2008). *La fin du monde par la science*. Editions Ere. (original edition 1855).
Iakovou, V. (2001). La critique entre Hannah Arendt et l'école de Francfort. *Tumultes, 2*(17–18), 259–278.
Illich, I. (1971a). *Une société sans école*. Seuil.
Illich, I. (1971b). *Libérer l'avenir*. Seuil. French translation.
Illich, I. (1973a). *La convivialité*. Seuil. French translation.
Illich, I. (1973b). *Tools for conviviality*. Harper and Row.
Jagodzinski, J. (Ed.). (2018). *Interrogating the Anthropocene. Ecology, aesthetics, pedagogy, and the future in question*. Palgrave Macmillan.
Jonas, H. (1990). *Le principe responsabilité*. Cerf. French translation 1990 (or. 1979).
Jonas, H. (1998). *Pour une éthique du futur*. Payot & Rivages. French translation.
Karlsson, R. (2013). Ambivalence, irony, and democracy in the Anthropocene. *Futures, 46*, 1–21.
Lambert, F. (2001). *Introduction à l'histoire des idées politiques*. Armand Colin.
Lambert, F. (2014). Le 'totalitarisme': l'impensable dans la philosophie politique? In J. Baudouin & B. Bruneteau (Eds.), *Le totalitarisme* (pp. 69–81). PUR.
Lambert, F., & Lefranc, S. (2012). *50 fiches pour comprendre la Science politique*. Bréal. (4ème éd).
Leinfelder, R. (2013). Assuming responsibility for the Anthropocene: Challenges and opportunities in education. In H. Trischler (Ed.), *Anthropocene – Envisioning the future of the age of humans, RCC perspectives* (Vol. 2, pp. 9–28). Rachel Carson Center.
Les convivialistes. (2013). *Manifeste convivialiste*. Le bord de l'eau.
Lesourt, E. (2018). *Survivre à l'Anthropocène*. PUF.
Lévinas, E. (1991). *Entre nous – Essais sur le penser-à-l'autre*. Grasset.
Lockie, S., Sonnenfled, A., & Fischer, D. (Eds.). (2014). *Routledge international handbook of social and environnemental change*. Routledge.
Lorius, C., & Carpentier, L. (2010). *Voyage dans l'Anthropocène*. Actes Sud.
Magny, M. (2019). *Aux racines de l'Anthropocène – Une crise écologique reflet d'une crise de l'homme*. Le bord de l'eau.
Marcuse, H. (1963). *Eros et civilisation*. Editions de minuit. (original edition 1955), French translation.
Marsh, G. P. (1970). *The earth as modified by human action: A new edition of "Man and Nature"*. Arno Press. (original edition 1874).
McDougall, I., Brown, F. H., & Fleagle, J. G. (2005). Stratigraphic placement and age of modern humans from Kibish, Ethiopia. *Nature, 433*, 733–736.
McPherron, S., Alemseged, Z., Marean, C. W., Wynn, J. G., Reed, D., Geraads, D., Bobe, R., & Béarat, H. A. (2010). Evidence for stone-tool-assisted consumption of animal tissues before 3.39 million years ago at Dikika, Ethiopia. *Nature, 466*, 857–860.
Moreau, D. (2012). L'étrangeté de la formation de soi. *Le Télémaque, 41*, 115–132.
Muhlmann, G. (2001). Pensée et non-pensée selon H. Arendt et T. W. Adorno. *Tumultes, 17–18*, 279–319.
Notre affaire à tous. (2019). *Comment nous allons sauver le monde – Manifeste pour une justice climatique*. Massot éditions.

Oakeshott, M. (1991). *Rationalism in Politics and Others Essays*. Liberty Fund.
Ollitrault, S. (2001). Les écologistes français, des experts en action. *Revue française de science politique, 51*, 105–130.
Ollitrault, S. (2008). *Militer pour la planète, Sociologie des écologistes*. PUR.
Ollitrault, S. (2016). Engagement et trajectoires socio-spatiales à l'heure de la mondialisation: les bénévoles de Greenpeace et les agriculteurs biologiques. In D. C. Goirand, D. Garibay, & H. Combes (Eds.), *Les lieux de la colère – Occuper l'espace pour contester, de Madrid à Sanaa*. Karthala.
Palsson, G., Szerszynski, B., Sörlin, S., Marks, J., Avril, B., Crumley, C., Hackmann, H., Holm, P., Ingram, J., Kirman, A., Pardo Buendia, M., & Weehuizen, R. (2013). Reconceptualizing the 'Anthropos' in the Anthropocene: Integrating the social sciences and humanities in global environmental change research. *Environmental Science and Policy, 28*, 3–13.
Papaux, A. (2015). Homo faber. In D. Bourg & A. Papaux (Eds.), *Dictionnaire de la pensée écologique* (pp. 536–540). PUF.
Paulsen, M., Jagodzinski, J., & Hawke, S. M. (2022). *dir. Pedagogy in the Anthropocene. Re-wilding education for a New Earth*, Palgrave Macmillan.
Pelluchon, C. (2015). *Les nourritures*. Seuil.
Pelluchon, C. (2017a). L'éthique des vertus: une condition pour opérer la transition environnementale. *lapenseeecologique.com, PUF, 1*(1), 1–18.
Pelluchon, C. (2017b). *Manifeste animaliste – Politiser la cause animale*. Alma.
Pena-Ruiz, H. (2018). *Karl Marx, penseur de l'écologie*. Seuil.
Perret, B. (2010, January). Quand la valeur n'est plus monnayable. *Esprit*, 98–117.
Perret, B. (2012). L'urgence occultée de la crise climatique. *Etudes, 4162*, 151–161.
Pierron, J.-P. (2013). Ressource naturelle ou service écologique gratuit? Ce que l'eau nous donne. *Revue du MAUSS, 42*, 271–280.
Pierron, J.-P. (2014a, February). Insolite ou insolente idée de nature? *Etudes*, 55–65.
Pierron, J.-P. (2014b). *Mythopées*. Vrin.
Pierron, J.-P. (2016). L'écologie, lieu d'une renaissance spirituelle. *Revue Projet, 354*, 84–89.
Pierron, J.-P. (2018). *La poétique de l'eau – Pour une nouvelle écologie*. Editions François Bourin.
Pineau, G. (dir). (1992). *De l'air. Essai sur l'éco-formation*. Païdéia/Sciences et culture.
Poché, F. (2013). (entretien avec Piolot, F.), *A-t-on encore le droit d'être fragile ?* Chronique Sociale.
Poché, F. (2014). *Le temps des oubliés, refaire la démocratie*. Lyon.
Portier, P. (2017). Jean Baudouin et la science politique – Trajectoire d'une hétérodoxie. In B. Bruneteau, G. Châton, & P. Portier (Eds.), *L'aventure démocratique – Cheminement en compagnie de Jean Baudouin* (pp. 13–28). PUR.
Priyadharshini, E. (2021). *Pedagogies for the post-Anthropocene – Lessons from apocalypse, revolution and Utopia*. Springer.
Prouteau, F. (2021). *Odyssée pour une Terre habitable*. Le Pommier.
Reclus, E. (1881). *La Terre: Description des phénomènes de la vie du globe (1868–1869), in Les continents, t2*. Hachette.
Rosa, H. (2013). *Accélération. Une critique sociale du temps*. La découverte (ed. or. 2010), tr. fr.
Rosa, H. (2016). *Resonanz, Eine Soziologie der Weltbeziehung*. Suhrkamp.
Rosa, H., & Wallenhorst, N. (interview). (2017a). Apprendre à écouter le monde. *Chemins de formation, 21*, 19–31.
Rosa, H., & Wallenhorst, N. (interview). (2017b). Apprendre ensemble dans la société de l'accélération. *Bildungsforschung, 2017*(1), 1–7.
Rosa, H., & Wallenhorst, N. (interview). (2022). *Accélérons la résonance!*. Le Pommier.
Rozès, S. (2005). Aux origines de la crise politique. *Le Débat, 134*, 4–18.
Sauvé, L. (2009). Le rapport entre éthique et politique: un enjeu pour l'éducation relative à l'environnement. *Education relative à l'environnement, 8*, 147–162.
Sauvé, L. (2015). Education à l'environnement. In D. Bourg & A. Papaux (Eds.), *Dictionnaire de la pensée écologique* (pp. 376–379). PUF.

Scranton, R. (2015). *Learning to die in the Anthropocene*. City Light Publisher.

Semal, L., & Szuba, M. (2015). Théorie politique verte/Green political theory. In D. Bourg & A. Papaux (Eds.), *Dictionnaire de la pensée écologique* (pp. 988–990). PUF.

Semaw, S., Rogers, M. J., Quade, J., Renne, P. R., Butler, R. F., Dominguez-Rodrigo, M., Stout, D., Hart, W. S., Pickering, T., & Simpson, S. W. (2003). 2.6-Million-year-old stone tools and associated bones from OGS-6 and OGS-7, Gona, Afar, Ethiopia. *Journal of Human Evolution, 45*, 169–177.

Sigaut, O. (2010). L'éducation à l'environnement, entre politique et politiques publiques. *Education relative à l'environnement, 9*, 59–75.

Sinaï, A. (Ed.). (2013). *Penser la décroissance – Politiques de l'Anthropocène*. Presses de Sciences Po.

Sinaï, A., & Mathilde, S. (Eds.). (2017). *Gouverner la décroissance – Politiques de l'Anthropocène III*. Les Presses de Sciences Po.

Srnicek, N., & Williams, A. (2013). #ACCELERATE. Manifesto for an Accelerationist Politics. *Critical Legal Thinking*, 14 May 2013. http://criticallegalthinking.com/2013/05/14/accelerate-manifesto-for-an-accelerationist-politics

Steffen, W., Sanderson, R. A., Tyson, P. D., Jäger, J., Matson, P. A., Moore, B., III, Oldfield, F., Richardson, K., Schellnhuber, H.-J., Turner, B. L., & Wasson, R. J. (2004). *Global change and the earth system. A planet under pressure* (The IGBP Book Series). Springer.

Steffen, W., Crutzen, P. J., & McNeil, J. R. (2007). The Anthropocene: Are humans now overwhelming the great forces of nature? *Ambio, Royal Swedish Academy of Sciences, 36*(8), 614–621.

Steffen, W., Grinevald, J., Crutzen, P., & McNeill, J. (2011a). The Anthropocene: Conceptual and historical perspectives. *Philosophical Transactions of the Royal Society, 369*, 842–867.

Steffen, W., Persson, A., Deutsch, L., Zalasiewicz, J., Williams, M., Richardson, K., Crumley, C., Crutzen, P., Folke, C., Gordon, L., Molina, M., Ramanathan, V., Rockström, J., Scheffer, M., Schellnhuber, H. J., & Svedin, U. (2011b). The Anthropocene: From global change to planetary stewardship. *Ambio, 40*(7), 739–761.

Steffen, W., Broadgate, W., Deutsch, L., Gaffney, O., & Ludwig, C. (2015). The trajectory of the Anthropocene: The great acceleration. *The Anthropocene Review, 2*(1), 81–98.

Stiegler, B. (2015). Sortir de l'Anthropocène. *Multitudes, 60*, 137–146.

Szuba, M. (2015). Sociologie environnementale. In D. Bourg & A. Papaux (Eds.), *Dictionnaire de la pensée écologique* (pp. 942–946). PUF.

Taleb, M. (2016). Ecoformation. In A. Choné, I. Hajek, & P. Hamman (Eds.), *Guide des Humanités environnementales* (pp. 83–91). Presses universitaires du Septentrion.

Tannock, S. (2021). *Educating for radical social transformation in the climate crisis*. Palgrave MacMillan.

Teilhard de Chardin, P. (2007). *Le phénomène humain*. Seuil. (original edition 1955).

Testot, L., & Wallenhorst, N. (2023). *Vortex – Faire face à l'Anthropocène*. Payot.

Theobald, C. (2007). *Le christianisme comme style – Une manière de faire de la théologie en postmodernité*. Cerf.

Vernadsky, V. I. (1925). L'autotrophie de l'humanité. *Revue générale des sciences pures et appliquées, 36*(17–18), 495–502. French translation.

Villalba, B. (2009). Conclusion. Une appropriation politisée mais peu politique. In B. Villalba (Ed.), *Appropriations du développement durable – Emergences, diffusions, traductions* (pp. 363–369). Presses Universitaires du Septentrion.

Villalba, B. (2015). Au fondement matériel de la démocratie. *Revue Projet, 344*, 56–63.

Villalba, B. (2016). Sobriété: ce que les pauvres ont à nous dire. *Revue Projet, 350*, 39–49.

Villalba, B. (2021). *Les collapsologues et leurs ennemis*. Le Pommier.

Wallace, M. F. G., Bazzul, J., Higgins, M., & Tolbert, S. (Eds.). (2022). *Reimagining Science Education in the Anthropocene*. Palgrave Macmillan.

Wallenhorst, N. (2019). *L'Anthropocène décodé pour les humains*. Le Pommier.

Wallenhorst, N. (2020a). *La vérité de l'Anthropocène*. Le Pommier.

Wallenhorst, N. (2020b). La valeur de l'expression du désaccord dans l'espace de la classe. In A. Wagener & L. Nicolas (Eds.), *La valeur du désaccord*. Les presses de la Sorbonne. (*forthcoming*).

Wallenhorst, N. (2021). *Mutation. L'aventure humaine ne fait que commencer*. Le Pommier.

Wallenhorst, N. (2022). *Qui sauvera la planète?* Arles.

Wallenhorst, N., & Wulf, C. (Eds.). (2022). *Humains. Un dictionnaire d'anthropologie prospective*. Vrin.

Wallenhorst, N., & Wulf, C. (Eds.). (2023). *Handbook of the Anthropocene*. Springer.

Wallenhorst, N., Robin, J.-Y., & Boutinet, J.-P. (2019). L'émergence de l'Anthropocène, une révélation étonnante de la condition humaine? In N. Wallenhorst & J.-P. Pierron (Eds.), *Eduquer en Anthropocène*. Le Bord de l'eau.

Wallenhorst, N., Mellot, S., & Theviot, A. (2020). *Interconnectés? Numérique et convivialisme*. Le bord de l'eau. (*forthcoming*).

Waters, C. N., Zalasiewicz, J., Summerhayes, C., Barnosky, A. D., Poirier, C., Galuszka, A., Cearreta, A., Edgeworth, M., Ellis, E. C., Ellis, M., Jeandel, C., Leinfelder, R., McNeill, J. R., Richter, D. d B., Steffen, W., Syvitski, J., Vidas, D., Wagreich, M., Williams, M., Zhisheng, A., Grinevald, J., Odada, E., Oreskes, N., & Wolfe, A. P. (2016). The Anthropocene is functionally and stratigraphically distinct from the Holocene. *Science, 351*, 137–147.

Weber, A. (2016a). *Enlivenment. Eine Kultur des Lebens. Versuch einer Poetik für das Anthropozän*. Matthes und Seitz.

Weber, A. (2016b). *The biology of Wonder – Aliveness, Feeling, and the Metamorphoses of Science*. New Society Publishers.

Weber, A. (2017). *Sein und Teilen – Eine Praxis schöpferischer Existenz*. Transcript Verlag.

Weber, A., & Kurt, H. (2015). *Lebendigkeit sei! Für eine Politik des Lebens. Ein Manifest für das Anthropozän*. Klein Jasedow, Think Oya.

Williams, M., Zalasiewicz, J., Waters, C. N., Edgeworth, M., Bennett, C., Barnosky, A. D., Ellis, E. C., Ellis, M. A., Cearreta, A., Haff, P. K., Ivar do Sul, J. A., Leinfelder, R., McNeill, J. R., Odada, E., Oreskes, N., Revkin, A., Richter, D. d B., Steffen, W., Summerhayes, C., Syvitski, J. P., Vidas, D., Wagreich, M., Wing, S. L., Wolfe, A. P., & Zhisheng, A. (2016). The Anthropocene: A conspicuous stratigraphical signal of anthropogenic changes in production and consumption across the biosphere. *Earth's Future, 4*, 1–20.

Woeikof, A. I. (1901). De l'influence de l'homme sur la Terre. *Annales de Géographie, t1*(50), 193–215.

Wolff, E. W. (2014). Ice sheets and the Anthropocene. In C. N. Waters, J. Zalasiewicz, M. Williams, M. A. Elis, & A. Snelling (Eds.), *A stratigraphical basis for the Anthropocene* (pp. 255–263). Geological Society, Special Publications 395.

Wulf, C. (2013). *Anthropology. A continental perspective*. University of Chicago Press.

Wulf, C. (2022a). *Education as human knowledge in the Anthropocene. An Anthropological perspective*. Routledge.

Wulf, C. (2022b). *Human beings and their images. Imagination, mimesis and performativity*. Bloomsbury.

Zalasiewicz, J., Williams, M., Smith, A., Barry, T. L., Coe, A. L., Brown, P. R., Brenchley, P., Cantrill, D., Gale, A., Gibbard, P., Gregory, F. J., Hounslow, M. W., Kerr, A. C., Pearson, P., Knox, R., Powell, J., Waters, C., Marshall, J., Oates, M., Rawson, P., & Stone, P. (2008). Are we now living in the Anthropocene? *GSA Today, 18*(2), 4–8.

Zalasiewicz, J., Williams, M., Steffen, W., & Crutzen, P. (2010). The new world of the Anthropocene. *Environmental Science & Technology, 44*, 2228–2231.

Part I
The Tensions of Politics in the Anthropocene

Chapter 2
Introduction to the Anthropocene

Abstract The Anthropocene is gaining recognition as a new geological epoch following the Holocene – the last relatively warm interglacial period in 11,700 years – and is characterised by the impact of human activity on the Earth system as a whole. 'The impacts of human activity will likely be observable in the stratigraphic record for millions of years into the future, suggesting that a new epoch has begun' (Lewis and Maslin, Nature, 519:171–180, 2015, p. 171). This chapter is divided into four parts: (1) Towards a new geological era; (2) History of the concept of the Anthropocene; (3) The 'Anthropocene Working Group' for recognition on the Geologic Time Scale.

Keywords Anthropocene · Anthropocene working group · History of the Anthropocene · Geologic time scale

2.1 Towards a New Geological Epoch

The word 'Anthropocene' is a neologism made up of two Greek words: *Anthropos*, meaning human being, and *kainos*, meaning new. It refers to a new geological epoch marked by the impacts of human behaviour. The suffix 'cene' indicates that it has the status of an epoch.[1] The planet in the Anthropocene has a warmer climate, has less land and more ocean, and is dominated by humanity (or at least a portion of the planet is). In the eyes of one set of Earth system scientists, 'Human activity has clearly altered the earth's surface, oceans and atmosphere, and reordered life on Earth' (Lewis & Maslin, 2015, p. 172). Jan Zalasiewicz, a British geologist whose work is particularly widely read – not least because of his chairmanship of the official Anthropocene Working Group (2009–2020) – concludes an article published

[1] Ages, which refer to a shorter unit, usually have the suffix 'ian'; periods are longer units and contain several epochs; a set of periods constitutes an era, and a set of eras an aeon. The history of the Earth contains four aeons.

© The Author(s), under exclusive license to Springer Nature Switzerland AG 2023

N. Wallenhorst, *A Critical Theory for the Anthropocene*, Anthropocene – Humanities and Social Sciences, https://doi.org/10.1007/978-3-031-37738-9_2

with a collective of scientists in 2017 in the journal *Anthropocene* as follows: 'We conclude that human impact has now increased to the point where it has changed the course of Earth history by at least several millennia, in terms of predicted long-term climate effects (...), and in terms of significant and ongoing transformation of biota, incorporating a geologically unprecedented phase of accelerating species invasions and anthropogenic species extinctions' (Zalasiewicz et al., 2017, p. 57). The term 'Anthropocene' is currently much used within the international scientific community, and it is most likely the Anthropocene will be formally recognised as a new epoch in the history of Earth. The ecumene (Berque, 2009a, b), the inhabited part of Earth's surface, is set to shrink, and we do not know how much of Earth's surface will remain habitable for humans.

For the international scientific community, there is no doubt that we have permanently altered the habitability of the biosphere; on the other hand, the exact date of entry into this new epoch is still subject to debate, and several hypotheses are currently being worked on. The first of these refers to an early Anthropocene, which some believe dates back to the Stone Age (Doughty, 2013) or between 5000 and 8000 years ago with the development of agriculture and rice fields (Kaplan et al., 2011). Other hypotheses position the entry into the Anthropocene with the meeting of the old and new worlds (Lewis & Maslin, 2015) or with the Industrial Revolution in the nineteenth century (Crutzen, 2002). If we consider the Anthropocene to have begun with the Industrial Revolution, it is easy to identify that the speed with which we entered this new epoch is based on the extraction of resources in the imperial colonies by European powers (Sinaï, 2013, p. 31). Still other hypotheses identify the start of the Anthropocene with the great acceleration of consumption and of industrial production in the mid-twentieth century. For this reason, the new epoch is alternatively known as 'the Great Acceleration', as has already been discussed (Steffen et al., 2004, 2015; Waters et al., 2016). Other authors pinpoint the entry into this new geological epoch with the nuclear bomb detonation (Masco, 2010; Lewis & Maslin, 2015). Finally, for some scientists who are convinced of the anthropogenic global changes, it will be for future generations to decide on the entry into this new geological epoch, once the Anthropocene will have shown all its cards (Wolff, 2014). What is immediately apparent is that the Anthropocene is characterised by the combination of technology, fossil energy and acceleration.

The novelty of entering the Anthropocene means that humans will remain 'a major environmental force for millennia' (Crutzen, 2002, p. 23), comparable to other natural geological factors. The dawn of the Anthropocene causes 'the long timescale of the Earth and of life and the timescale of human history, separated by the modern industrial era, to collide' (Bonneuil, 2015, p. 38). The question of human power in the Anthropocene is enormously paradoxical: 'humans have never had so much power to shape nature through their techniques, but they have lost control of this transformation and may be powerless in witnessing the catastrophe they have brought about' (Chalier & Schmid, 2015, p. 6). The entry into the Anthropocene marks a decisive stage for humanity. In becoming a geological agent, for a set of Earth system researchers, humanity may 'modify or override certain processes of the planetary system, to the point of influencing the dynamics of the biosphere and affecting, in turn, the underpinnings of our societies, and even of our own anthropological condition' (Guillaume, 2015, p. 34). However, there is one set of findings that is universally accepted within the international scientific community: contemporary climate change is anthropogenic, and humanity has become a geological actor in the Earth system, and 'will remain a major geological force for many millennia, and perhaps millions of years' (Steffen et al., 2007, p. 618). For the scientific community, there is no doubt that we will see geological evidence of human activity in the future. Humankind has become a geological force in the truest sense of the word: humans have been transporting more rock than all 'natural' phenomena combined (Wilkinson, 2005). Since the beginning of the twentieth century, the surface area of anthropogenic erosion has been equivalent to all the arable land of the United States, and an area equal to the Amazon Rainforest has been desertified (Deléage, 2010, p. 22).

2.2 History of the Concept of the Anthropocene

2.2.1 *Publication of the Concept by Paul Crutzen in 2000*

The concept of the Anthropocene has several peculiarities. The first of these is that it refers to a geological epoch, which are usually defined by stratigraphers on the basis of their observations of soils. However, one of the first uses of this term was proposed to the scientific community by a Dutch geochemist working in Germany, Paul Crutzen, independently of stratigraphic observations. The term 'Anthropocene' was not originally proposed by geologists in reference to stratigraphic evidence, but refers to the alteration of Earth system processes as a result of human activity. Thus, in its origin, the concept of Anthropocene is a systemic concept rooted in the Earth System Sciences, before it took root in geology. This is unique to the Anthropocene, compared to other geological time units. As the debate on the dating of the Anthropocene progresses, it will never be separated from its systemic origins.

The second peculiarity is that it was suddenly proposed, in the course of an exchange at a symposium of the International Geosphere-Biosphere Programme (IGBP – an interdisciplinary federation for the study of the Earth system) in February 2000 in Mexico. We can go so far as to say that Crutzen's proposal of the term 'Anthropocene' in 2000 was an *ad lib* (Zalasiewicz et al. 2017, p. 56). In a 2003 interview in the journal *New Scientist*, Paul Crutzen recounts how it happened for him: 'I was at a conference where someone was talking about the Holocene, the geological epoch covering the period since the last ice age. I suddenly thought this was wrong. In the last 200 years humans have become a major geological force on the planet. For that reason, I said, "No, we're not in the Holocene any more: we're in the Anthropocene." I just made up the word on the spot. But it seems to have stayed there' (Keats, 2011, p. 19). After using the term at the IGBP symposium, Paul Crutzen contacted the American biologist Eugene Stroemer to write a paper with him in the IGBP journal, *Global Change Newsletter*, because of Stroemer's informal use of the term since the 1980s. The French philosopher working in Geneva, Jacques Grinevald (2007, p. 243), reports Stroemer's words: 'I started using the term "Anthropocene" in the 1980s, but I never formalised it until Paul [Crutzen] contacted me'. Although this one-page article is clear, one of the unique characteristics of the term 'Anthropocene' is that it was put forward before a scientific outline was given to it, and before a precise and exhaustive definition had been proposed. Crutzen then drew upon these elements in a short article in 2002 in the journal *Nature*, in which he summarised all the environmental changes humans have effected since the Industrial Revolution, attesting to the fact that we are entering the Anthropocene. In his 2002 article, Crutzen mentions a set of quantitative elements responsible for the entry into the Anthropocene: a tenfold increase in the world's human population between 1700 and 2000, and an equivalent increase in livestock; a depletion of fossil resources and the release of CO_2 into the atmosphere; and the intensification of the rate of species extinction by a factor of 1000.

The third notable feature is that Crutzen won the Nobel Prize for Chemistry in 1995, giving this new terminology particularly strong media exposure and scientific resonance. The fact that a word was coined without being precisely defined (even in 2011, Steffen, Grinevald, Crutzen and McNeil acknowledge the still-informal nature of the term 'Anthropocene') and without the usual academic boundaries being respected is a major factor in the success of this term, and its uptake by a very heterogeneous set of disciplines before stratigraphers in their turn adopt it. Thus, in line with Ellis (2017), we can say that the concept of the Anthropocene was created by Eugene Stroemer and then published by Paul Crutzen. Its scientific establishment, based in particular on stratigraphic observations, was the result of the work of a group of Earth system scientists, including Steffen and Zalasiewicz.

2.2.2 The Idea of Humanity as a Geological Force in the Nineteenth Century

The question of the impact of human activity on the biosphere as a whole considerably predates the proposal of the concept of the Anthropocene. This awareness of the global impact of human activities predates the 1972 Club of Rome report on the *Limits to Growth* or the NASA photograph showing the 'Earthrise' in 1969, highlighting humanity's technological prowess.[2] Awareness of the impact of human activity on the climate or on Earth as a whole has a long history, and the idea has

[2] The French geographer Michel Lussault, in *L'avènement du Monde* (2010, pp. 15–18), shows the importance of this photograph by American astronaut William Anders on 24 December 1968 during the Apollo 8 mission. In this photograph, the sunlit globe rises over the moon. This image, which reverses perspectives, has been identified as the most important environmental photograph ever taken. The power of this image is that it was taken by the hand of a man tearing himself away from Mother Earth.

been discussed by many scientists over the past two centuries. In the very concept of the Holocene, we find a trace of the idea of humanity's global impact. Originally, this concept incorporates the idea of human presence on Earth. Indeed, while this term, which etymologically means 'wholly new', was popularised by the French geologist Paul Gervais, it is a reworking and development of the term 'Recent Period' proposed by the Scottish geologist Charles Lyell (1833/1990), who noted the concomitance of the end of the Ice Age and the development of human civilisations (Lewis & Maslin, 2015, p. 172). After the dating of the entry into the geological Quaternary Period with the appearance of *Homo habilis*, the Holocene marked the second time that the human race played a part in a geological transition. The Anthropocene will be the third occurrence of humanity's involvement in a geological transition.

In 1778, the French naturalist, Georges-Louis Leclerc, Compte de Buffon, stated in *Les époques de la nature*, that 'today, the entire force of the earth bears the hallmarks of man's power'. This means that mankind is capable of modifying the climate by acting on their environment: they will be able to 'modify the influences of the climate they inhabit and fix its temperature, so to speak, at the point that suits them' (Buffon, 1778/1998, p. 237).[3] He even went so far as to write that 'the power of Man has assisted that of Nature' (1778/1998, p. 225). Then, between 1830 and 1833, Charles Lyell defined the contemporary age in *Principles of Geology* as 'the human age' (p. 52). 20 years later the English geologist and philosopher William Whewell wrote that 'the human epoch in the history of the Earth is different from all previous epochs' (1853/1990, p. 88). In 1854, the Welsh geologist and theologian Thomas Jenkyn was the first person to identify a geological stage of the Earth marked by human activity (1854), which he defined as 'the *human* epoch'. He writes: 'All recent rocks, called Post-Pleistocene in our last lesson, might have been called Anthropozoic – that is, rocks of human life' (1854, p. 313). Two years later, the English intellectual William Adams described his geological epoch on the basis of the detection of the traces of human activity in the sediments: 'The modern or human epoch is illustrated by the alluvial deposits, which are the effects of atmospheric and other more powerful local causes: these continue to operate, incorporating remains of man and inorganic matter' (1856, p. 247). Merely a decade later, the Irish Reverend Samuel Haughton published a *Handbook of Geology* in 1865, in which he defined the Anthropozoic as 'the age in which we live' (p. 138). Federau (2016, p. 64) notes in the 1880 edition of his textbook one of the consequences the reverend draws from this geological power: 'man must be at the "head of the system of life", because of his spiritual nature and his power to progress indefinitely [pp. 578-579].' The American geologist James Dwight Dana also published a *Manual of Geology* in 1863 in which he spoke of 'the age of spirit and the age of man' (p. 130). The Italian geologist and priest Antonio Stoppani also describes his contemporary period in the same way

[3] A few centuries earlier, in July 1494, Christopher Columbus already had the intuition of humanity's ability to influence the climate by managed deforestation (which he welcomed) (Fressoz, 2016).

(1873) using the expression 'Anthropozoic Era' (1873, vol. 2, p. 732). Then, the Russian geologist Aleksei Pavlov (1854–1929) in the last years of his life (in the late 1920s), used the expression 'Anthropocene' or 'Anthropozoic Era' to talk about his time, as noted by Shantser (1973, p. 140).

The second edition of *Man and Nature* in 1874 (10 years after the first edition), which the American diplomat and ecologist George Perkins Marsh entitled *The Earth as Modified by Human Action*, builds on Stoppani's work and his idea of an Anthropozoic era, and 'concludes that industrial societies are inherently incompatible with the balance of nature' (Federau, 2016, p. 61). Indeed, in Marsh's view, 'Earth is fast becoming an uninhabitable home for its noblest inhabitant' (1970, p. 44). Marsh's work had a significant impact. Lyell, writing in 1833, opined that humans were capable of transforming geography but as a minor geological agent. After reading Marsh's book 30 years later, he changed his mind and said so. Also, in *La pensée écologique – Une anthologie*, we read that in 1915, the German geologist Ernst Fischer viewed 'man as a geological factor' (Bourg and Fragnière, p. 137) with the capacity to modify the climate. The British geologist Robert Lionel Sherlock discussed the same idea in his 1922 book *Man as a Geological Agent*. All of these precursors were aware of humanity's impact on the Earth system centuries before it was theorised in the 2000s.

What these forefathers of the idea of the global impact of human activity on Earth have in common is that they approach this situation in a positive way. Their observations are based primarily not on stratigraphic evidence, but on the intuition that humans are fundamentally different from animals and have superior, global capabilities (Lewis & Maslin, 2015, p. 173). Federau, in his 'Genesis and Geology of the Anthropocene' (2016, pp. 53–86) explains that the view of humans as a geological force developed during the second half of the nineteenth century among a set of geologists – most of them churchmen. Their belief that humanity was purposefully created is one of the reasons for this idea of humans' power. At that time, it was not primarily the damage wrought upon the environment by humanity that was perceived, but the transformative power of the creative creature. For the American philosopher and geologist Joseph Le Conte, who defines the contemporary era as the Psychozoic, for example: 'Throughout geological history, Nature has always struggled upwards to achieve first life, then conscious life, then self-conscious immortal life. Man is the end, the achievement, the ideal of Nature's progress. But does progress stop here? In no way. Man himself takes charge of progress and continues it. When natural history ends, providential history begins. Redemptive history progresses again with a still higher purpose, and Nature, guided by the Divine Spirit, struggles again and attains divinity in Christ' (Le Conte, 1884, p. 282). For these nineteenth-century geologists, the Christian interpretation is that humanity has been granted dominion over nature, as reflected by Genesis 1:28: 'And God blessed them. And God said to them, "Be fruitful and multiply and fill the earth and subdue it, and have dominion over the fish of the sea and over the birds of the heavens and over every living thing that moves on the earth"'. These geologists link that Christian interpretation with scientific progress.

2.2.3 *Vladimir Vernadsky's Biosphere and Noosphere at the Beginning of the Twentieth Century*

The Ukrainian geochemist and naturalist Vladimir Vernadsky developed the concept of the biosphere, following in the footsteps of the Austrian geologist Eduard Suess, who coined the word and for whom it simply meant the space that hosts life on earth. Vernadsky then added complexity to the concept. For him, it is a cosmic phenomenon made possible by the sun's rays. The biosphere is organised as a system and is the main factor in the geological transformation of planet Earth. Man is part of the biosphere, acting upon it and, especially in industrial society, constituting a 'planetary geological force', as Vernadsky explains in his book *The Biosphere*, published in 1926. He proposes the idea of a relationship between the biosphere and human intelligence through the concept of the noosphere (from the Greek word for 'mind'), which emphasises the power of the human mind. For Vernadsky (1945), humans have the capacity to shape their own future by altering their environment. In his holistic approach, Vernadsky's thinking follows on from that of the German philosopher and educator Alexander von Humboldt (1769–1859). By developing the notion of the biosphere, studied from a biogeochemical perspective, Vernadsky made a founding contribution to the emergence of the Earth System Sciences. Vernadsky makes no religious or spiritual interpretation of humanity's geological power, but has great confidence in science enabling human beings to take good care of the planet. He believes deeply in progress, though is aware of humanity's capacity to destroy itself.

Steffen et al. (2011) recognise that the concept of the noosphere was the precursor to that of the Anthropocene. This idea of the noosphere resulted from Vernadsky's meeting with the Jesuit, palaeontologist and geology professor Pierre Teilhard de Chardin, and with the French philosopher Edouard Le Roy – also a Catholic with progressive thoughts. During Vernadsky's time in Paris between 1922 and 1925, these three thinkers socialised a great deal. They read Austrian geologist Eduard Suess's book *The Face of the Earth*, whose three volumes were published in 1885, 1888 and 1901. In a way, for Vernadsky, the problem is not in pinpointing a moment in time when a threshold is visibly crossed, signifying the planet's entry into a new geological epoch. In Vernadsky's thinking, what is important is identifying the moment when human beings became capable of reflecting upon their own impact on the biosphere. In his understanding of the notion of the biosphere, Vernadsky integrates human beings into the kingdom of the living. All that differentiates them from the rest of the kingdom of life is their self-reflexivity, with the notion of the noosphere. With the noosphere, Vernadsky indicates that we have entered a new age of collective responsibility. Human beings today, thanks to scientific knowledge, have entered the age of awareness of their own impact.

Teilhard de Chardin was rather enthusiastic, like his contemporaries, about the idea that humanity was acquiring power over its environment, and that the mind would play a determining role in the future. He went so far as to push the eschatological logic perceptible among the Christian geologists of the late nineteenth century to its conclusion. Le Roy believed humanity will control the planet's evolution, and his thinking is similar to that of Teilhard de Chardin. Indeed, for him, human beings, through their intelligence, are responsible for the adventure of life as a whole. As Steffen et al. (2011) point out, one of Le Roy's influences was the French philosopher Henri Bergson, whom he succeeded at the Collège de France in 1945 and who correctly perceived – in *L'évolution créatrice* (*Creative Evolution*) in 1907 – that humanity could be entering a new geological epoch: 'In thousands of years, when, seen from the distance, only the broad lines of the present age will still be visible, our wars and our revolutions will count for little, even supposing they are remembered at all; but the steam-engine, and the procession of inventions of every kind that accompanied it, will perhaps be spoken of as we speak of bronze or of the chipped stone of pre-historic times; it will serve to define an age' (1996, p. 140).

The development of the idea of mankind as a geological force was then interrupted by the two World Wars and resurfaced at a symposium at Princeton University in 1955, with a publication by the American geographer William M. Thomas (1956/1962), entitled: *Man's Role in Changing the Face of the Earth* (Steffen et al., 2011, p. 844; Robin et al., 2014). Finally, more recently, as identified by Steffen et al. (2011), a closely related term was used in 1992 by the American science journalist Andrew C. Revkin: '*Anthrocene*'. 'Perhaps earth scientists of the future will name this new post-Holocene period for its causative element—for us. We are entering an age that might someday be referred to as, say, the Anthrocene. After all, it is a geological age of our own making' (1992, p. 55).[4]

However, there is debate as to whether there have indeed been precursors to the concept of the Anthropocene. Australian philosopher and economist Clive Hamilton and Jacques Grinevald (2015) argue, instead, that the Anthropocene is a radically new concept that has had no precursors because of the current interdisciplinary and systemic understanding of the earth,[5] of which Stoppani, Vernadsky and Teillard de Chardin were unaware. (These earlier thinkers' ideas and analyses were about humans' impact on the face of the Earth, rather than humans' manipulation of the Earth system). Moreover, for Hamilton and Grinevald, the Anthropocene is set apart from the idea of helping the Earth along to a higher level of existence: 'Fundamentally, whereas Stoppani's Anthropozoic and Vernadsky's or Teilhard's Noosphere represent evolutionary extrapolations – i.e. based on the inevitable advance of progress – the Anthropocene is a lamentable break: not a regression, but a radical break away from any idea of advancing to a higher stage' (2015, p. 67).

---

[4] Other terms have been used to refer to the global importance of human activity (without referring to humanity's geological power). For example, Michael Samways, an entomologist from South Africa, coined the term 'Homogenocene' to mean the standardisation of the world. American biologist Michael Soulé refers to the contemporary period as the 'Catastrophozoic' era (Kolbert, 2015).

[5] In this 2015 article, Grinevald is partly at odds with what he published with Steffen, Crutzen and McNeil in 2011.

2.2.4 *From the Gaia Hypothesis to the Emergence of Earth System Sciences*

The British environmentalist James Lovelock takes up this idea of humanity as a geological force in his understanding of the Earth as Gaia (1979). He sees our planet as a self-regulating whole and even goes so far as to write that biota controls it. In his book for the general public, *Gaia: A new look at life on Earth*, the personification of our planet with the name of the Greek goddess of the Earth generated enormous success, and a set of esoteric references developed in spite of the author's reticence. Later on, the criticism he faced gradually led him to revise his positions and clarify that his statement 'the Earth is a living creature' was a metaphor (Federau, 2016, p. 80). While some symbolic elements of his theory may have produced a reaction, it remains a form of precursor to Earth System Sciences, including the idea of global biogeochemical cycles. Lovelock's ideas are complementary to Vernadsky's approach, which then gave the concept of the Anthropocene, which is profoundly systemic – that scientific sounding board and media sounding board, both in public opinion and in the social sciences.

The notion of the Anthropocene emerged within the Earth System Sciences. This is a multidisciplinary field that has emerged in recent decades, whose scientific focus is the Earth's entire complex system. The Earth system, according to the definition proposed by the German physicist Hans Joachim Schellnhuber (1999), can be defined as 'a single, planetary-level complex system, with a multitude of interacting biotic and abiotic components, evolved over 4.54 billion years and which has existed in well-defined, planetary-level states with transitions between them' (Steffen et al., 2016, p. 2). Following on from Chinese philosopher Lao Tzu, founder of Taoism's, idea that 'The world is a whole greater than the sum of its parts' in 600 BC, the German poet Johann Wolfgang von Goethe was one of the first modern thinkers to have this systemic intuition based on the idea that the parts are inseparable from the whole: 'those things which we call parts are inseparable from the Whole to such an extent, that they can only be conceived in and with the latter; and the parts can neither be the measure of the Whole, nor the Whole be the measure of the parts' (1784–1785, p. 8). During the 1990s, Earth System Sciences developed, with an understanding of the Earth as a systemic whole made up of biogeochemical flows. The interactions of biogeochemical cycles, integrating the circulation of matter and energy, provide the foundations for life on Earth and are the basis of our understanding of how the planet functions. The concept of the Earth system is an extension of Vernadsky's biosphere, and takes up the main logics he defined. For Schellnhuber (1999), the current beginnings of an understanding of the Earth system as a whole constitute a 'second Copernican revolution'. The fact that the term 'Anthropocene' was coined the following year and then developed into a concept fairly quickly made it possible to define this second Copernican revolution with a simple and accessible word.

One of the primary meanings of the Anthropocene is 'the anthropisation of the world'. It is possible, for example, to identify human impact on the environment through the creation of new materials such as plastic, or from radionuclides from nuclear bomb detonations, found in polar ice. It is possible to identify man-made artefacts that can be distinguished from nature that has not been altered by humans. However, as Federau (2016, p. 49), this is ancillary to the true scope of the Anthropocene, highlighting how humans have impacted the workings of the Earth system as a whole. What matters in the systemic understanding of the Anthropocene are the interactions between the biosphere (living things), atmosphere (air), cryosphere (ice), hydrosphere (water), and lithosphere (Earth's crust) (Cabanes, 2017, p. 21).

The notion of the Anthropocene has found success in the contemporary period due to its totalising power in the apprehension of the main characteristics of the current period. The Anthropocene is a concept that has captured the times well, as it is being used by thinkers who are increasingly sensitive to ecology, but also by a current of thought that is partly oppositional, with transhumanist theories and research. Its success is also due to the way in which it reconfigures the dualism of nature and culture or nature and society, central to the emergence of Western thought. The Anthropocene also raises a set of political, economic and social questions relating to the organisation of the sharing of human existence within the earth in the decades, centuries and millennia to come.

Just as there are journals entitled *The Holocene* or *Quaternary Science Reviews*, several scientific journals have been founded in recent years with the term 'Anthropocene' in their titles. The success of the concept of the Anthropocene has had a profound impact on the scientific debate and its publishing spaces. Finally, another international and interdisciplinary journal, combining Earth System Sciences with Humanities and Social Sciences, has been devoted to the notion of the Anthropocene since 2014: *The Anthropocene Review*. The journal *Earth's Future*, whose first issue was published in 2013, focuses on the 'Science of the Anthropocene', from a multidisciplinary perspective, with a particularly explicit aim: to try to apprehend the opportunities afforded by humans' domination over their environment. Two other scientific journals dedicated to the Anthropocene have been in publication since 2013: *Anthropocene*, and *Elementa – Science of the Anthropocene*.

2.3 The 'Anthropocene Working Group' for Recognition on the Geologic Time Scale

The Holocene has seen great climate and sea level stability, compared to hundreds of thousands of years ago (Petit et al., 1999). The Holocene is the last interglacial phase of the Quaternary and is classed as an epoch. It is the most stable period of the last 400,000 years, and thus the most stable period *Homo sapiens* has known. Up until this point, the transitions from ice ages to interglacial periods were due to shifts in Earth's orbit. There have been several hypotheses suggesting the Anthropocene should be classed either as an epoch or as a period. While it seems reasonable to declare the end of the Holocene epoch, it is still too early to declare the end of the Quaternary period.

Humans organise Earth's history into slices that are particularly recognisable in terms of climate, sea level and living species. Most boundaries are defined in two ways: by identifying a specific point on the Earth in a stratotype that marks a global geological time boundary. These are known as GSSPs: Global Boundary Stratotype Sections and Points. The term refers to rocks, sediments and glaciers that have developed during a given period. To find the location of this point, otherwise known as the 'golden spike', because it is marked by a golden spike driven into the rock, is the goal of the stratigraphic commissions. This golden spike refers to a singular physical manifestation following a global change. This is associated with another limit – a temporal boundary – defined by an absolute age, called a GSSA: Global Standard Stratigraphic Age.[6] The geological stages are always defined from the beginning – i.e. from the lower limit serving as a boundary. Together with the upper unit boundary, the GSSP defines a chronostratigraphic unit. The time interval defined contains stratigraphic signals that are interpreted from the GSSP of the lower boundary. The GSSP is intended to provide a set of signals for interpretation and consistency at global level, not just at local level. A famous example of a golden spike is the one located at El Kef in Tunisia, defining the lower limit of the Paleogene period 66 million years ago, ending the Cretaceous and marking the extinction of terrestrial dinosaurs. This GSSP boundary represents the iridium peak.[7]

According to the 2012 Geologic Time Scale, the official scale produced by the International Commission on Stratigraphy, we are currently 11,700 years into the Holocene (0.01165 million years BP – meaning 'before present', the present being taken as the year 1950). The GSSP is in the ice of northern Greenland, at a depth of 1492.25 meters and corresponds to a change in deuterium (^{2}H or D) levels,[8] which is indicative of global warming. This GSSP is associated with five other ancillary stratotypes that mark it as a global change (Walker et al., 2009).

As for the status of the Anthropocene on the geologic time scale, several hypotheses have been studied: the Anthropocene is the next epoch, following the Holocene; the Anthropocene is recognised as a period and succeeds the Quaternary period; or the Anthropocene is an age, which is an integral part of the Holocene. The hypothesis of recognition of the Anthropocene as an epoch being the most likely, in this book, we will speak of the Anthropocene as a new geological epoch in the process of official recognition.

Today, the Anthropocene does not appear on the geological time scale. However, the official process of studying the Anthropocene as a new geological epoch has begun. Formal work within the stratigraphic community began in 2008, with the Stratigraphy Commission of the Geological Society in London, which

[6] Not all geological boundaries on the geological time scale have a GSSP. For example, a GSSP has yet to be defined for the inner boundary of the Archean Aeon (about 4 billion years ago) and the Proterozoic Aeon (about 2.5 billion years ago).

[7] Iridium (^{77}Ir) is a member of the platinum group of metals. It is very rare on Earth, though present in large quantities in meteorites. For this reason, a meteor strike is the hypothesis considered; another hypothesis studied is that of massive volcanic eruptions. It is also possible that impact with a foreign body weakened the earth's crust and caused volcanic eruptions (Schulte et al., 2010; Renne et al., 2015).

[8] Deuterium is an isotope of hydrogen: ^{2}H (sometimes represented as 'D'). Isotopes are elements that have the same number of protons but a different number of neutrons – i.e. different types of the same element: different types of carbons, lead, etc.

agreed that there was a stratigraphic basis for the term 'Anthropocene'. This then led to the establishment of an Anthropocene Working Group (AWG) as part of the Subcommission on Quaternary Stratigraphy (SQS) of the International Commission on Stratigraphy (ICS), a member of the International Union of Geological Sciences (IUGS).[9] The Anthropocene Working Group started its work in 2009 and has integrated members that the Geological Society of London had summoned to take part in this task of formalisation. It is in charge of formalising the name of the Anthropocene. To officially recognise the transition from the Holocene to the Anthropocene, three steps are necessary. The first is a recommendation from the Anthropocene Working Group of the Subcommission on Quaternary Stratigraphy, which is currently underway. Secondly, the International Commission on Stratigraphy must vote, and ratify the transition by a large majority. Finally, the International Union of Geological Sciences must ratify the creation of a new geological stage before it can be included in the geological time scale.

Geologists and stratigraphers operate on long time periods – both in reference to the history of the earth, and to the working processes by which Earth epochs are formalised. For example, in 1840, the British geologist John Phillips proposed the mass extinction at the end of the Permian period as the beginning of the Triassic period (about 252 million years ago). It was not until 2001 that this proposal was finally endorsed with the definition of a GSSP. The last time the International Union of Geological Sciences ratified a new geological unit was in 2004 with the recognition of the Ediacaran (third Neoproterozoic period) from 635 to 541 million years ago.

In parallel with this work of stratigraphic formalisation, the term 'Anthropocene', formally proposed in 2000, has been conceptually refined within the Earth System Sciences, and within the Social Sciences. In a way, stratigraphers came after a group of researchers who found the term apt to cover and conceptualise the global impacts of human activity on the Earth system. There is no doubt that many lasting and irreversible changes in the Earth system processes are anthropogenic in nature, though they are still debated from a stratigraphic perspective. In some respects, the last few decades have shown that geology does not have a monopoly on writing the history of the earth, but that other sciences can have their say. While the Anthropocene is well on its way to formal recognition by the International Commission on Stratigraphy, we note that the scientific community has already adopted the concept to a large extent.

Towards Recognition of the Mid-Twentieth Century as the Official Date of Entry into the Anthropocene

The Anthropocene Working Group presented intermediate results and the first 'stratigraphic evidence' at the 35th International Geological Congress that met in August 2016 in Cape Town, South Africa. The conclusions of this first important

[9] The ICS is the largest scientific body of the IUGS, consisting of representatives from 50 countries, and composed of an executive and 17 subcommissions, each with about 20 voting members.

step in formalising the Anthropocene from a geological point of view were that its beginning would be dated in the mid-twentieth century, and that we were moving towards the finalisation of a GSSP point related to radionuclides of anthropogenic nature following nuclear detonations (Zalasiewicz et al., 2017). In a paper edited by Zalasiewicz in 2017, written by the members of the Anthropocene working group, the authors mention the following results at the 35th International Geological Congress, obtained following the presentation of their working group's intermediate findings:

- Of 35 voters, 34 voted in favour and one abstained on the question 'Is the Anthropocene real from a stratigraphic point of view?'
- To the question 'Should the Anthropocene be formalised?', 30 people voted for, 3 voted against and 2 abstained.
- As to the hierarchical level to be given to the Anthropocene: 2 people voted for it to be an 'era'; 1.5 for a 'period'; 20.5 for an 'epoch'; 1 for a 'sub-epoch'; 2 for an 'age'; 0 for a 'sub-age' (or 1 'if necessary'); 1 person for not assigning a level; 3 were unsure; and 4 abstained.
- To the question widely debated previously in the scientific literature, of dating the onset of the Anthropocene, the responses were as follows: there were 0 votes for −7000 years; 1.3 votes for −3000 years; 0 votes for 1610; 0 votes for +/− 1800; 28.3 votes for 1950; 1.3 votes for 1964; 4 votes for diachronic dating; 0 votes for 'uncertain'; and 0 abstentions.
- When asked whether the Anthropocene should be defined by a GSSP or a GSSA, 25.5 votes were for GSSP and 1.5 for GSSA (along with 8 'unsure' votes).
- Finally, on the last question about the best 'primary marker' for the Anthropocene, the votes were: aluminium: 0; plastic: 3; fuel and ash particles: 2; carbon dioxide concentration: 3; methane concentration: 0; carbon isotope change: 2; oxygen isotope change: 0; radiocarbon spike due to the bomb: 4; plutonium fallout: 10; nitrate concentration: 0; biostratigraphic extinctions and changes in composition: 0; other (lead, persistent organic pollutants, technofossils): 3; uncertain: 2; abstentions: 6.

The Anthropocene Working Group hopes to be able to complete the process of defining a GSSP by about 2020. The mid-twentieth century provides the most stratigraphically compelling evidence for the entry into the Anthropocene, in line with what some authors had already identified during years of intense scientific debate (Zalasiewicz et al., 2014a, b; Waters et al., 2014a, b) and which was subsequently confirmed (Syvitski et al., 2020).

This second chapter on the notion of the Anthropocene, its emergence, characteristics and formalisation, has highlighted the systemic component of the Anthropocene. The next chapter will explore this fundamental characteristic of the Anthropocene in greater depth, based on the notion of the planetary boundary proposed in 2009.

Bibliographical References

Adams, W. H. D. (1856). *The history, topography, and antiquities of the isle of wight*. Smith Elder.

Bergson, H. (1996). *L'évolution créatrice*. PUF. (original edition 1907).

Berque, A. (2009a). Les travaux et les jours. Histoire naturelle et histoire humaine. *L'Espace géographique, 38*, 73–82.

Berque, A. (2009b). *Ecoumène – Introduction à l'étude des milieux humains*. Belin.

Bonneuil, C. (2015). Anthropocène (point de vue 2). In D. Bourg & A. Papaux (Eds.), *Dictionnaire de la pensée écologique* (pp. 35–40). PUF.

Buffon, G. L. L. (1998). *Les époques de la nature*. Diderot éditions. (original edition 1778).

Cabanes, V. (2017). *Homo natura – En harmonie avec le vivant*. Buchet Chastel.

Chalier, J., & Schmid, L. (2015, December). Comment penser l'Anthropocène? *Esprit*, 5–7.

Crutzen, P. J. (2002). Geology of mankind: "The Anthropocene". *Nature, 415*, 23.

Dana, J. D. (1863). *Manual of geology: Treating of the principles of the science with special reference to American Geological history, of the use of colleges, academies, and schools of science*. Theodore Bliss & Co./Trübner & Co.

Deléage, J.-P. (2010). En quoi consiste l'écologie politique? *Ecologie & Politique, 40*, 21–30.

Doughty, C. E. (2013). Preindustrial human impacts on global and regional environment. *Annual Review of Environment and Resources, 38*, 503–527.

Ellis, E. C. (2017). Physical geography in the Anthropocene. *Progress in Physical Geography, 41*(5), 525–532.

Federau, A. (2016). *Philosophie de l'Anthropocène – Interprétation et épistémologie*. Doctoral thesis supervised jointly by the University of Lausanne and the University of Bourgogne, by Professors Dominique Bourg and Jean-Claude Gens.

Fressoz, J.-B. (2016). Colón también descubrió el cambio climático. *El Pais*. https://elpais.com/elpais/2016/06/09/ciencia/1465489189_275680.html, consulted online on 10 January 2018.

Grinevald, J. (2007). *La Biosphère de l'Anthropocène: climat et pétrole, la double menace*, Repères transdisciplinaires (1824–2007). Georg/Editions Médecine et Hygiène.

Guillaume, B. (2015). Anthropocène (point de vue 1). In D. Bourg & A. Papaux (Eds.), *Dictionnaire de la pensée écologique* (pp. 32–35). PUF.

Hamilton, C., & Grinevald, J. (2015). Was the Anthropocene anticipated? *The Anthropocene Review, 2*(1), 59–72.

Jenkyn, T. W. (1854). *Lessons in geology, XLIX*. Chapter V. On the classification of rocks section. *Popular Educator, 4*, 312–316.

Kaplan, J. O., Krumhardt, K. M., Ellis, E. C., Ruddiman, W. F., Lemmen, C., & Goldewijk, K. (2011). Holocene carbon emissions as a result of anthropogenic land-cover change. *The Holocene, 21*(5), 775–791.

Keats, J. (2011). Anthropocene. In J. Keats (Ed.), *Virtual words* (pp. 18–22). University Press of Oxford.

Kolbert, E. (2015). *La 6ème Extinction. Comment l'homme détruit la vie*. La librairie Vuibert.

Le Conte, J. (1884). *Religion and Science: A series of Sunday lectures*. D. Appleton.

Lewis, S. L., & Maslin, M. A. (2015). Defining the Anthropocene. *Nature, 519*, 171–180.

Lovelock, J. E. (1979). *Gaia: A new look at life on Earth*. Oxford University Press.

Lussault, M. (2010). *L'avènement du Monde – Essai sur l'habitation humaine de la Terre*. Seuil.

Lyell, C. (1990). *Principles of geology* (Vol. I, II, III). University of Chicago Press. (original edition 1830–1833).

Marsh, G. P. (1970). *The earth as modified by human action: A new edition of "Man and Nature"*. Arno Press. (original edition 1874).

Masco, J. (2010). Bad weather: On planetary crisis. *Social Studies of Science, 40*(1), 7–40.

Petit, J. R., Jouzel, J., Raynaud, D., Barkov, N. I., Barnola, J.-M., Basile, I., Bender, M., Chappellaz, J., Davisk, M., Delaygue, G., Delmotte, M., Kotlyakov, V. M., Legrand, M., Lipenkov, V. Y., Lorius, C., Pépin, L., Ritz, C., Saltzmank, E., & Stievenard, M. (1999). Climate and atmospheric history of the past 420 000 years from the Vostok ice core, Antartica. *Nature, 399*, 429–436.

Renne, P. R., Sprain, C. J., Richards, M. A., Self, S., Vanderkluysen, L., & Pande, K. (2015). State shift in Deccan volcanism at the Cretaceous-Paleogene boundary, possibly induced by impact. *Science, 350*(6256), 76–78.

Revkin, A. C. (1992). *Global warming: Understanding the forecast*. American Museum of Natural History, Environmental Defense Fund, Abbeville Press.

Robin, L., Avango, D., Keogh, L., Möllers, N., Scherer, B., & Trischler, H. (2014). Three galleries of the Anthropocene. *The Anthropocene Review, 1*(3), 207–224.

Schellnhuber, H. J. (1999). 'Earth system' analysis and the second Copernican revolution. *Nature, 402*, C19–C23.

Schulte, P., Alegret, L., Arenillas, I., Arz, J. A., Barton, P. J., Bown, P. R., Bralower, T. J., Christeson, G. L., Claeys, P., Cockell, C. S., Collins, G. S., Deutsch, A., Goldin, T. J., Goto, K., Grajales-Nishimura, J. M., Richard, A. F., Grieve, R. A. F., Gulick, S. P. S., Johnson, K. R., Kiessling, W., Koeberl, C., Kring, D. A., MacLeod, K. G., Matsui, T., Melosh, J., Montanari, A., Morgan, J. V., Neal, C. R., Nichols, D. J., Norris, R. D., Pierazzo, E., Ravizza, G., Rebolledo-Vieyra, M., Reimold, W. U., Robin, E., Salge, T., Speijer, R. P., Sweet, A. R., Urrutia-Fucugauchi, J., Vajda, V., Whalen, M. T., & Willumsen, P. S. (2010). The chicxulub asteroid impact and mass extinction at the cretaceous-paleogene boundary. *Science, 327*(5970), 1214–1218.

Shantser, E. V. (1973). « The anthropogenic system (period) ». In *The Great Soviet Encyclopedia* (Vol. 2, pp. 139–144). Macmillan.

Sinaï, A. (2013). Le destin des sociétés industrielles. In A. Sinaï (Ed.), *Penser la décroissance* (pp. 23–48). Presses de Science Po.

Steffen, W., Sanderson, R. A., Tyson, P. D., Jäger, J., Matson, P. A., Moore, B., III, Oldfield, F., Richardson, K., Schellnhuber, H.-J., Turner, B. L., & Wasson, R. J. (2004). *Global change and the earth system. A planet under pressure* (The IGBP Book Series). Springer.

Steffen, W., Crutzen, P. J., & McNeil, J. R. (2007). The Anthropocene: Are humans now overwhelming the great forces of nature? *Ambio, Royal Swedish Academy of Sciences, 36*(8), 614–621.

Steffen, W., Grinevald, J., Crutzen, P., & McNeill, J. (2011). The Anthropocene: Conceptual and historical perspectives. *Philosophical Transactions of the Royal Society, 369*, 842–867.

Steffen, W., Broadgate, W., Deutsch, L., Gaffney, O., & Ludwig, C. (2015). The trajectory of the Anthropocene: The great acceleration. *The Anthropocene Review, 2*(1), 81–98.

Steffen, W., Leinfelder, R., Zalasiewicz, J., Waters, C. N., Williams, M., Summerhayes, C., Barnosky, A. D., Cearreta, A., Crutzen, P., Edgeworth, M., Ellis, E. C., Fairchild, I. J., Galuszka, A., Grinevald, J., Haywood, A., Ivar do Sul, J., Jeandel, C., McNeill, J. R., Odada, E., Oreskes, N., Revkin, A., Richter, D. d B., Syvitski, J., Vidas, D., Wagreich, M., Wing, S. L., Wolfe, A. P., & Schellnhuber, H. J. (2016). Stratigraphic and Earth system approaches to defining the Anthropocene. *Earth's Future, 4*, 1–22.

Stoppani, A. (1873). *Corso di geologia, vol. 2, Geologia Stratigrafica*. Bernardoni Brigola Editori.

Syvitski, J., Waters, C. N., Day, J., Milliman, J. D., Summerhayes, C., Steffen, W., Zalasiewicz, J., Cearreta, A., Gałuszka, A., Hajdas, I., Head, M. J., Leinfelder, R., McNeill, J. R., Poirier, C., Rose, N. L., Shotyk, W., Wagreich, M., & Williams, M. (2020). Extraordinary human energy consumption and resultant geological impacts beginning around 1950 CE initiated the proposed Anthropocene Epoch. *Communication Earth and Environnement, 1*, 32. https://doi.org/10.1038/s43247-020-00029-y

Thomas, W. L. (Ed.). (1962). *Man's role in changing the face of the earth*. University of Chicago Press. (original edition 1956).

Vernadsky, V. I. (1945). Biosphere and noosphere. *American Scientist, 33*, 1–12.

Walker, M., Johnsen, S., Rasmussen, S. O., Popp, T., Steffensen, J.-P., Gibbard, P., Hoek, W., Lowe, J., Andrews, J., Björck, S., Cwynar, L. C., Hughen, K., Kershaw, P., Kromer, B., Litt, T., Lowe, D. J., Nakagawa, T., Newnham, R., & Schwander, J. (2009). Formal definition and dating of the GSSP (Global Stratotype Section and Point) for the base of the Holocene using the Greenland NGRIP ice core, and selected auxiliary records. *Journal of Quaternary Science, 24*, 3–17.

Waters, C. N., Zalasiewicz, J., Williams, M., Price, S. J., Ford, J. R., & Cooper, A. H. (2014a). Evidence for a stratigraphic basis for the Anthropocene. In R. Rocha, J. Pais, J. Kullberg, & S. Finney (Eds.), *STRATI 2013* (pp. 989–993). Springer Geology, Springer.

Waters, C. N., Zalasiewicz, J. A., Williams, M., Ellis, M. A., & Snelling, A. M. (2014b). A stratigraphical basis for the Anthropocene? *Geological Society (London, Special Publications), 395*, 1–21.

Waters, C. N., Zalasiewicz, J., Summerhayes, C., Barnosky, A. D., Poirier, C., Galuszka, A., Cearreta, A., Edgeworth, M., Ellis, E. C., Ellis, M., Jeandel, C., Leinfelder, R., McNeill, J. R., Richter, D. d B., Steffen, W., Syvitski, J., Vidas, D., Wagreich, M., Williams, M., Zhisheng, A., Grinevald, J., Odada, E., Oreskes, N., & Wolfe, A. P. (2016). The Anthropocene is functionally and stratigraphically distinct from the Holocene. *Science, 351*, 137–147.

Wilkinson, B. H. (2005). Humans as geologic agents: A deep-time perspective. *Geology, 33*, 161–164.

Wolff, E. W. (2014). Ice sheets and the Anthropocene. In C. N. Waters, J. Zalasiewicz, M. Williams, M. A. Elis, & A. Snelling (Eds.), *A stratigraphical basis for the Anthropocene* (pp. 255–263). Geological Society, Special Publications 395.

Zalasiewicz, J., Waters, C. N., Williams, M., Barnosky, A. D., Cearreta, A., Crutzen, P., Ellis, E., Ellis, M. A., Fairchild, I. J., Grinevald, J., Haff, P. K., Hajdas, I., Leinfelder, R., McNeill, J., Odada, E. O., Poirier, C., Richter, D., Steffen, W., Summerhayes, C., Syvitski, J. P. M., Vidas, D., Wagreich, M., Wing, S. L., Wolfe, A. P., Zhisheng, A., & Oreskes, N. (2014a). When did the Anthropocene begin? A mid-twentieth century boundary level is stratigraphically optimal. *Quaternary International, 30*, 1–8.

Zalasiewicz, J., Williams, M., & Waters, C. N. (2014b). Can an Anthropocene series be defined and recognized? *Geological Society (London, Special Publications), 395*, 39–53.

Zalasiewicz, J., Waters, C. N., Summerhayes, C. P., Wolfe, A. P., Barnosky, A. D., Cearreta, A., Crutzen, P., Ellis, E., Fairchild, I. J., Gałuszka, A., Haff, P., Hajdas, I., Head, M. J., Ivar do Sul, J. A., Jeandel, C., Leinfelder, R., McNeill, J. R., Neal, C., Odada, E., Oreskes, N., Steffen, W., Syvitski, J., Vidas, D., Wagreich, M., & Williams, M. (2017). The Working Group on the Anthropocene: Summary of evidence and interim recommendations. *Anthropocene, 19*, 55–60.

Chapter 3
The Notion of Planetary Boundaries

Abstract Certain notions have become symbolic of the Anthropocene and its systemic component, including the notions of planetary boundaries and the Great Acceleration, which have already been widely discussed in the literature. The over presented in this chapter is not intended to provide a new perspective on the Anthropocene, but to allow us to take the measure of the geoscientific data and systemic changes related to the Anthropocene, to illuminate the discussion in the remainder of this work. This is to elucidate the magnitude of the political anthropological problem we are facing, and the need to anthropologically consolidate the political.

Keywords Anthropocene · Planetary boundaries · Great acceleration · Threshold

The notion of a planetary boundary discussed in this chapter will be taken up again throughout this book, from the perspective of political anthropology, through the notional couple 'Promethean / post-Promethean', and the enjoinder to develop a 'post-Promethean between-us'. In the same way, the notion of Great Acceleration will be treated in connection with the threefold acceleration (technical, social changes and pace of life) described by Hartmut Rosa, to counter which he proposes learning resonance with the world, breaking away from the Promethean anthropological paradigm. The notion of the Great Acceleration will also contribute to a critical discussion with the political understanding of acceleration given by Alex Williams and Nick Srnicek in the *#Accelerate Manifesto*, mentioned in the introduction, published in 2013 and with a worldwide impact. In the third part, we shall revisit a set of scientific data on the Anthropocene that we feel are important to take into account in the context of Anthropocene education.

© The Author(s), under exclusive license to Springer Nature Switzerland AG 2023

N. Wallenhorst, *A Critical Theory for the Anthropocene*, Anthropocene – Humanities and Social Sciences, https://doi.org/10.1007/978-3-031-37738-9_3

3.1 A Safe Space for Humanity to Act

3.1.1 Threshold Effect and Tipping Point

The planet Earth functions as a system that has experienced *tipping points* in the course of its history, beyond which it has undergone structural changes resulting in a change in the conditions which support life. The notion of a threshold is particularly important in Earth System Science. The threshold is the point at which a tipping point occurs. This climatic concept was proposed by the American meteorologist Edward Lorenz in 1965, and has been adopted by Earth system researchers such as the American geologist Anthony Barnosky in relation to the biosphere. Ecosystems sometimes change from one state to another quite abruptly when they cross certain thresholds. The phenomenon has already been well documented. In an extension of this train of thought, Barnosky et al. (2012), in a famous article which notably inspired the film *Demain* [Tomorrow], directed by Cyril Dion and Mélanie Laurent in 2015, identify a set of indicators that attest that a change of state of the terrestrial biosphere as a whole is drawing dangerously close. For other authors such as Scheffer et al. (2009), it is not unlikely that we will soon be dealing with critical transitions caused by threshold crossings. Changes in the state of the biosphere have already taken place in the course of the Earth's history, and it would seem that humans are creating the conditions for what could be an abrupt and irreversible change (once a system has changed state, it does not return to the initial state). It is therefore necessary to anticipate global changes in the state of the biosphere, identify their local impacts, and make positive changes to our impact on biodiversity and ecosystems (Barnosky et al., 2012, p. 52).

Global state changes are difficult to anticipate because they are the result of a set of small-scale local changes.[1] To address this, Barnosky et al. (2011, 2012) and Barnosky (2008) offer a historical retrospective of these local changes that generated global changes.[2] Faced with the possibility of an abrupt change of state that the

[1] The Earth system is not the only complex system that can experience changes in state when critical thresholds are crossed. In medicine, for example, asthma attacks and epileptic seizures correspond to these tipping points; the same is true of global financial crises. The difficulty lies in anticipating these critical thresholds, despite the weak signals of change that the system may exhibit. A group of researchers have identified 'generic early warning signals' that can appear before these changes of state in various types of complex systems other than the climate (Scheffer et al., 2009). Other researchers have worked to aggregate a dataset of systemic changes already experienced by the Earth system over its history, in an attempt to identify early warning signals before a climate tipping point is reached (Lenton et al., 2012).

[2] Of particular note among the critical changes of state and transitions encountered on a global scale are the last glacial-interglacial transition, the five great extinctions of biodiversity, and the Cambrian explosion. (The Cambrian explosion took place 530–541 million years ago, marked by the diversification of bacterial, plant and animal species and the appearance of a group of multicellular animals). During the last glacial-interglacial transition, the Earth warmed up between 14,300 and 11,000 years BP, which led to major changes in the biosphere between 12,900 and 11,300 years BP, including the extinction of half of all species of large mammals, as well as the disappearance of large birds, reptiles and other types of mammals.

entry into the Anthropocene represents, there is widespread concern among the scientific community.

3.1.2 *Planetary Boundaries and Systemic Risk*

In order to avoid these tipping points, the Swedish environmentalist Johan Rockström et al. (2009a, b) propose to identify a planetary safe zone for human activities. This safe zone is delimited by nine planetary boundaries that must not be crossed. These are positioned in advance of nine biophysical tipping points beyond which 'irreversible and abrupt environmental change is almost certain to occur' (Rockström et al., 2009a, p. 32). Planetary boundaries 'define the safe space of action for humanity within the Earth system, combining the planet's biophysical subsystems or processes' (Rockström et al., 2009b, p. 472).

Planetary boundaries are set before the possible thresholds constituting non-linear transitions, which could be extremely sudden and extremely impactful in terms of the systemic relations between human beings and their environment. Planetary boundaries are safety measures that these scientists propose, and are defined by this group of researchers (based on the cumulative intelligence of a body of previous work showing human dominance over components of the Earth system – Vitousek et al., 1997). They are based on the quantitative data of a control variable positioned sufficiently far from a potential threshold. The authors emphasise that a planetary boundary is distinct from a tipping point or threshold crossing. Thus, the planetary boundary framework defines spaces. When a boundary is crossed, it means that we are leaving the arena in which we can safely take action. For this group of researchers, this necessitates an analysis of how contemporary human societies manage environmental risks. There is a set of choices on their part. For these authors, the planetary boundaries approach is a form of 'emerging science of warning signs' that gives 'warning of an approaching threshold or a decrease in the capacity of a system to persist under changing conditions' (Steffen et al., 2015b, p. 2). The planetary boundaries approach is directly political. Indeed, as its authors mention, 'for this science to be useful in a political context, it must allow sufficient time for society to respond to move away from an impending threshold before it is crossed' (Steffen et al., 2015b, p. 2). To date, though, only one planetary boundary out of the nine defined is beginning to attract interest from the public and from nation states: namely the climate-change boundary. It is particularly important to take account of dual inertia in the estimation of the timeframe between the crossing of a boundary and the passing of the tipping point: the inertia of the Earth system, and that of societies around the world.

Rockström et al. (2009a, p. 1) make it clear that 'the social impacts of transgressing boundaries will be a function of the socio-ecological resilience of the affected societies' and also that the proposed 'boundaries are rough, first estimates only, surrounded by large uncertainties and knowledge gaps'. The concept of planetary boundaries is intended to define a space for human action without risk to the species' survival: 'Planetary boundaries define, as it were, the boundaries of the "planetary playing field" for humanity if we want to be sure of avoiding major human-induced environmental change on a global scale' (Rockström et al., 2009a, p. 1). The challenges of the contemporary Anthropocene period thus necessitate a rethink of environmental sustainability; this is the ambition of the concept of planetary boundaries.[3]

In the 2009 paper, Rockström et al. identify two global boundaries for which they have not yet defined the prior boundary that should not be crossed in order to avoid the risk of a threshold or tipping point effect.[4] Three boundaries are identified as having already been crossed: climate change, the rate of biodiversity loss, and changes in biogeochemical cycles (both nitrogen and phosphorus).[5]

Steffen et al. in 2015 (2015b, p. 736) published an update and refinement of the planetary boundaries proposed by Rockström et al. in (2009a, b). In 2009, Steffen was the second author of the paper; in 2015, Rockström was the third author. The category of biosphere integrity now consists of two indicators: genetic diversity and functional diversity. While in 2009, the boundary in terms of phosphorus flux had not yet been reached, it had been largely exceeded by 2015 (notably due to a re-evaluation of the boundary, and to the way in which the control variable is calculated). In this new paper, they also propose a two-level security approach (global and regional) for several of the limits, allowing for regional heterogeneity: 'Several of the limits now have a two-level approach, reflecting the importance of cross-scale interactions and regional heterogeneity of the processes underlying the limits' (Steffen et al., 2015b). For one boundary where the indicator had not yet been quantified (atmospheric aerosol loading), the authors propose a regional quantification. Finally, they updated the quantified indicators of the planetary boundaries. Five global boundaries have a strong regional component: changes in biosphere integrity, biogeochemical flows, land-system change, freshwater use, and atmospheric aerosol loading. The authors show how the crossing of these regional boundaries has implications for the Earth system as a whole.

---

[3] The systemic component of the Earth's functioning is particularly important, and little is currently known. A range of elements contribute to the stability and resilience of the Holocene, such as the chemical circulation provided by the oceans, atmospheric chemistry, the physiology of ecosystems, the water cycle and biodiversity. To date, we do not know the extent to which human activity, such as deforestation, ecosystem acidification, or biodiversity loss, has contributed to the erosion of this resilience (Steffen et al., 2011b, p. 756). Rockström et al. (2009b, p. 472) state that 'Without pressure from humans, the Holocene is expected to continue for at least several thousands of years.'

[4] These are chemical pollution (a boundary that is titled 'Introduction of Novel Entities' in the 2015 update) and atmospheric aerosol loading.

[5] The first two of these breached boundaries are considered the 'core boundaries', because the greater the extent to which they are breached, the greater the probability of a global tipping point being reached. Climate and biosphere integrity have a special responsibility for the functioning of the Earth system. In the updated planetary boundary framework of Steffen (2015b), these two boundaries continue to be identified as core boundaries, and widely transgressed. Indeed, they are connected to all other planetary boundaries. Too great a transgression of either one of these two boundaries has the characteristic of a Holocene exit. These are, indeed, the main variables in Earth's history responsible for geological time changes (Steffen et al., 2015b, p. 7).

In January 2022, an international team of 14 researchers, led by Linn Persson, the chemist from the Stockholm Environment Institute, reported the crossing of a fifth planetary boundary that threatens to destabilise the Earth system: that of pollution, called 'introduction of novel entities', including plastics. In this article, entitled 'Outside the Safe Operating Space of the Planetary Boundary for Novel Entities', their observation is striking: 'Global [chemical] production increased 50-fold since 1950, and is projected to triple again by 2050. […] We recommend taking urgent action to reduce the harm associated with exceeding the boundary by reducing the production and releases of new entities, noting that even so, the persistence of many novel entities and/or their associated effects will continue to pose a threat' (Persson et al., 2022, p. 1). At the end of April 2022, another Swedish research team, from the Stockholm Resilience Centre, states in a paper entitled 'A planetary boundary for green water' that with so-called 'green water' (terrestrial precipitation, evaporation and soil moisture), a sixth planetary boundary is being crossed as a result of human activities (Wang-Erlandsson et al., 2022).

3.2 The Nine Planetary Boundaries

3.2.1 Climate Change

3.2.1.1 Anthropogenic Global Warming

Climate change in the twentieth century was almost imperceptible directly by human senses. This is less and less true in the twenty-first century, where we now noticeably perceive a range of signals, such as the increase in heatwaves, the increased frequency of weather disasters such as hurricanes, or the decrease in numbers of insects on car windscreens. The increase in the earth's temperature is the environmental issue most closely followed by public opinion and policymakers today and is the subject of ongoing analysis and prospective studies. The unit of measurement defined by Rockström et al. for this first global limit is the level of CO_2 in the atmosphere, and the defined boundary is 350 ppm.[6] However, the level of atmospheric CO_2 is not the only factor behind global warming. Indeed, it is primarily Earth's energy balance that determines its temperature; this incorporates solar radiation,[7] greenhouse gas forcing[8] and albedo (Steffen et al., 2016). Five gases are

[6] The amount of carbon dioxide in the atmosphere is measured in parts per million (ppm). A part per million is the fraction corresponding to one millionth. It is possible to measure CO_2 by mass or volume. Within Earth System Sciences, CO_2 is usually calculated on a mass basis; if the fraction is by volume, it will be specifically marked as such (ppmv).

[7] Solar radiation varies with solar flares, rather than being constant. The intensity of the sun's radiation has varied throughout the Earth's history and has been a factor in its evolution.

[8] Greenhouse gases alter the balance between solar radiation entering the atmosphere and infrared radiation leaving the atmosphere. The radiative forcing of greenhouse gases means that the atmosphere warms up because it retains more energy from the sun than it reflects back into space. This causes climate disruption.

particularly responsible for a greenhouse effect, causing an increase in the earth's surface temperature: CO_2 (carbon dioxide), CH_4 (methane), N_2O (nitrous oxide), halocarbons (gases that contain at least one halogen atom such as CFCs, HFCs or PFCs, which are responsible for the hole in the ozone layer[9]), and also H_2O (water vapour).

3.2.1.2 Prospective Scenarios Relayed by the IPCC

In the 1980s, a scientific consensus on the impact of human activity on the Earth system emerged, leading to the creation of the Intergovernmental Panel on Climate Change (IPCC), in 1988, by the World Meteorological Organisation and under the patronage of the United Nations Environment Programme (UNEP). According to the IPCC (2013), the global surface temperature increased by 0.9 °C between the beginning of the twentieth century and 2012. If the global temperature had not increased due to human actions, it is highly likely that the climate would have cooled in the current millennium (Dean et al., 2014, p. 277). The Earth's surface temperature will reach levels in the coming decades that the human species has never experienced.[10]

The IPCC associates its temperature increase ranges with four scenarios (correlated with population growth, economic growth and carbon consumption).[11] The Paris Accords at the end of COP21 set a target of between 1.5 °C and 2 °C maximum to control the increase in the earth's temperature. These political agreements are far removed from the scientific data currently in our possession. While we now know how low the probability of achieving this target is, we also know that a temperature increase of 2 °C by 2100 would lead to a temperature increase of about 3 °C by 2150, even if there were no further greenhouse gas emissions into the atmosphere between 2100 and 2150. Moreover, it is possible to estimate that the average temperature increase would be 5 °C in 2150 if the temperature increase in 2100 was 3 °C (Bourg, 2017, p. 4).[12] The current rapid climate change gives rise to

---

[9] Halogens are mainly the chemical elements fluorine (F), chlorine (Cl), bromine (Br) and iodine (I), belonging to group 17 in the Periodic Table. They have been used in air conditioners, refrigerators, aerosol cans, cleaners and foams.

[10] The American Earth system researcher John W. Williams et al. estimate, in a study published in 2007, that by 2100, between 12% and 39% of the Earth's surface will have climates never experienced by current living species.

[11] Irish-American sociologist and statistician Adrian E. Raftery et al. (2017, p. 637), on the other hand, believe these estimates are not sufficiently grounded in a statistical approach. According to these authors, the range of probability of temperature increase is between 2 °C and 4.9 °C, with a 50% probability of a 3.2 °C increase (and a 5% probability of keeping it 2 °C). According to them, it is carbon consumption, rather than population growth, which is the primary factor.

[12] Greenhouse gas emissions have an impact for millennia. According to the IPCC summary for policymakers: 'Cumulative emissions of CO_2 largely determine global mean surface warming by the late twenty-first century and beyond. […] Most aspects of climate change will persist for many centuries even if emissions of CO_2 are stopped. This represents a substantial multi-century climate

several consequent risks such as food, health, financial and geopolitical risks (Bourg, 2009). Already, the IPCC is blaming the increase in natural disasters in recent years on global warming. An increase of 5–6 °C would be catastrophic for the organisation of human life in society. Life would become a battle to survive, with every individual for themselves (Bourg & Bompard, 2012, p. 19).[13]

In 2018, the IPCC began its sixth assessment cycle. In August 2021, the IPCC released the first part of its 3949-page report – a synthesis report drawing upon 14,000 scientific publications (IPCC, 2021). Its main conclusions are as follows: (1) Human activities undoubtedly have an impact on the current climate disruption. (2) It is happening faster than we thought. (3) Runaway processes are underway. Human-induced climate change is intensifying greenhouse gas emissions from 'natural' environments (such as methane emissions from thawing Siberian soils or CO_2 emissions from dry forest fires). We are in danger of reaching tipping points whose consequences are unpredictable (but which we know will compromise, if not destroy, human life in society). (4) Radical human actions, to which we can hold fast, could keep the Earth system in a favourable state (with an increase in average temperatures below +1.5 °C).

In February 2022, the IPCC published the second part of its sixth report – 3675 pages. It offers a synthesis of 34,000 scientific articles on the impacts of climate change on human societies (IPCC, 2022a). Here, researchers study the vulnerabilities of our societies (as well as of all non-human beings) in terms of adapting to the current disruption. At the heart of the research behind this report, there is an apparent broad consensus in the international scientific community around one word: risk. Climate change poses global risks to ecosystems and societies alike: an increase in heat-related human mortality, tropical cyclones, megafires, loss of habitable land due to rising sea levels, droughts and floods, diseases, a growing scarcity of drinking water, and the weakening of human infrastructures.

---

change commitment created by past, current and future emissions of CO_2' (IPCC, 2013, p. 25). The IPCC experts continue, 'A large fraction of anthropogenic climate change resulting from CO_2 emissions is irreversible on a multi-century to millennial time scale, except in the case of a large net removal of CO_2 from the atmosphere over a sustained period. Surface temperatures will remain approximately constant at elevated levels for many centuries after a complete cessation of net anthropogenic CO_2 emissions. Due to the long time scales of heat transfer from the ocean surface to depth, ocean warming will continue for centuries. Depending on the scenario, about 15 to 40% of emitted CO_2 will remain in the atmosphere for longer than 1000 years' (IPCC, 2013, p. 26) (pages 27 and 28 in the English version). The consumption of all fossil fuels would generate an 1800 ppm increase in atmospheric CO_2, which would take 10,000 years to drop below 1000 ppm (Federau, 2017, p. 64).

[13] One of the problematic effects of the contemporary economic crisis is that its urgency masks the importance of the changes and the long-term issues; yet the problem of climate change is of an even more dangerous nature than that of debt (Perret, 2012, p. 159). To address the current global warming, France has set itself the goal of halving energy consumption by 2050 and quartering greenhouse gas emissions. To achieve this goal it would be necessary to decrease the amount of energy invested per production of one euro of GDP by 3% (currently we are experiencing a 1% annual increase) (Perret, 2014a, b, p. 36).

More than 18,000 scientific papers are synthesised in the 2913 pages of the third part of the sixth report on climate change mitigation options, released in April 2022 (IPCC, 2022b). It covers the range of options for limiting greenhouse-gas emissions and halting the current destructive runaway trend. The message is simple: (1) We know the solutions and they mainly revolve around a controlled energy regime. (2) We still have time to act. (3) We must do so immediately because the window for human action to have an effect on climate change is closing very quickly. Acting late will mean more serious effects. Acting now by investing is cheaper than doing nothing, which will involve 'repairs' that will soon become impossible.

3.2.2 Biodiversity Destruction and Extinction of Living Species

The 2015 paper by Steffen et al. proposes two indicators of developments in the biosphere and the boundaries that should not be exceeded. The first indicator, 'genetic diversity', refers to the 'information bank' that allows life to continue to evolve on Earth in the most resilient way possible. The control variable defined is the overall extinction rate of species. The second indicator, 'functional diversity', refers to the adaptive capacity of the biosphere in the long term. It measures the loss of biodiversity at the global and biome level – a biome being the set of ecosystems in a given geographical area. This indicator identifies the role of the biosphere in the functioning of the Earth system. The control variable defined is the *Biodiversity Intactness Index* (BII), with the pre-industrial rate set at 100%. It may be higher if human action favours the development of species abundance; on the contrary, it is lower if our actions reduce biodiversity (Steffen et al., 2015b, p. 3).

The biosphere has experienced two distinct periods: a microbial stage between 3.5 billion years ago and 650 million years ago, when life was present in the form of unicellular organisms (marked by an important moment 2.4 billion years ago when oxygenated photosynthesis began), followed by a metazoan stage since 650 million years ago, wthe a complexification of life forms in multicellular organisms, which then developed from 580 million years ago. 541 million years ago, complex animals began to take possession of the land (Williams et al., 2015, p. 2). British palaeobiologist Mark Williams et al. (2015) go so far as to suggest that the present biosphere shows distinct signs of the metazoan stage. They summarise them as follows: homogenisation of fauna and flora; the supremacy of a single species, *Homo sapiens*, which exploits between 25% and 38% of the Net Primary Production[14] of the biosphere. and uses other means of energy than photosynthetic energy, including fossil energy; the direct intervention of humans on other species; and an increase in

[14] Net Primary Production (NPP) is a measure of the net flow of carbon from the atmosphere to plants over a period of time. Various studies estimate that humans appropriate between 25 and 38% of NPP, leaving less and less room for other living species (Vitousek et al., 1997; Haberl et al., 2007; Williams et al., 2016). NPP appears to be relatively constant, independent of human activity (Lewis & Maslin, 2015, p. 172).

interactions between the biosphere and the technosphere.[15] For Williams et al. (2015), the Anthropocene biosphere has significant features that may have long-term structural impacts. Through toolmaking, habitat construction, and the engineering of increasingly complex social networks (Fox et al., 2017), humans are engineering important sociocultural niches that are the primary cause of the intensity of anthropogenic biosphere transformations. For Ellis, 'The "fluxes of nature" are rapidly becoming "cultures of nature"' (2015, p. 287). On the other hand, it is important to specify that these human niches, because they are socio-cultural, can be managed according to quite different rules that integrate the fundamental criterion of environmental sustainability.

The minimum scientific consensus regarding the destruction of biodiversity and the extinction of living species is that unless we act to prevent it, we can expect 'a further acceleration in the global rate of species extinction, which is already at least tens to hundreds of times higher than it has averaged over the past 10 million years' (IPBES, 2019, p. 12), in the words of what is sometimes called the 'IPCC of biodiversity', the Intergovernmental Science-Policy Platform on Biodiversity and Ecosystem Services (IPBES). Moreover, the collapse of biodiversity cannot be understood without considering climate change, and vice versa. This is the lesson that emerges from the joint work of IPCC and IPBES researchers in a report published in 2021, *Biodiversity and Climate Change*.

3.2.3 Biogeochemical Cycles

Four cycles fundamental to life have been disrupted over the last century: the carbon cycle, the nitrogen cycle, the phosphorus cycle and the sulphur cycle. The water cycle has also been very significantly altered directly by human activity. These cycles are interrelated, and are part of the complex function of the Earth system. The oxygen and carbon cycles are linked by respiration, and the nitrogen and phosphorus cycles are linked by the oceans. In their 2009 work on planetary boundaries, Rockström et al. propose boundaries for the phosphorus and nitrogen fluxes that should not be transgressed. In the 2015b update of that paperwork, Steffen et al. propose a more generic global limit of human influence on biogeochemical fluxes, identifying a range of markers expressing the boundary of biogeochemical fluxes which ought not to be transgressed in the coming years. The proposed limit of

[15] The impact capacity of the actions of *Homo sapiens* and their hominid ancestors on the Earth system has increased with the development of tools, and then the complexity of technological artefacts. Today, multiple authors speak of the impact of the technosphere on the biosphere (Haff, 2014; Williams et al., 2015). The technosphere can be defined as 'the global emergent system that includes humans, technological artefacts and associated social and technological networks' (Williams et al., 2015, p. 1).

anthropogenic reactive nitrogen modification[16] is 62 Tg of nitrogen per year[17]; currently, this is largely exceeded, with approximately 150 Tg of nitrogen per year. We tend to talk a great deal about disruption of the carbon cycle because of its impact on global warming, but the nitrogen cycle is far more affected by human actions than is the carbon cycle. The pre-industrial concentration of nitrous oxide (N_2O) in the atmosphere was 272 ppbv; this level increased to 288 ppbv[18] in 1950 (Steffen et al., 2007, p. 616) and 320 ppbv in 2010.[19]

Nitrogen is present in large quantities on planet Earth, since dinitrogen (N_2, an inert gas) accounts for 78% of the chemical make-up of the atmosphere. Nitrogen is important in agriculture because it is what allows plants to grow. The basic principle behind fertilisers is to enrich the soil with nitrogen. Nitrogen, however, is only transformed in particular situations, because of the amount of energy needed to break the bond between the two nitrogen atoms. Between 1909 and 1913, two Germans, the chemist Fritz Haber (who was awarded the Nobel Prize for Chemistry in 1918) and the engineer Carl Bosch, developed what is known as the 'Haber-Bosch process', which transforms atmospheric nitrogen into ammonia that was then widely used in fertilisers. This process profoundly altered the nitrogen cycle (Canfield et al., 2010).[20] The alteration of the nitrogen cycle has a series of disruptive effects, such as changes in ecosystems, including soil and water acidification. Too much nitrate[21] in the air and water is also toxic to humans, and causes illnesses including cancer. This anthropogenic modification of the nitrogen cycle also wors-

[16] Most of the nitrogen on Earth is in the form of dinitrogen in the atmosphere. Nitrogen is an inert gas, meaning that it reacts very little in the presence of other chemical elements. Nitrogen is reactive when it bonds with other elements – i.e. when the bond between the two nitrogen atoms is broken, which requires a large amount of energy.

[17] Tg is the symbol for teragram, a unit of mass measurement equivalent to 10^{12} grams, or one million metric tons.

[18] The unit ppb is based on the same principle as ppm, but it is the number of parts per billion (the symbol for billion is 'b'). The addition of the 'v' (ppbv) means that it is a volume unit, rather than a mass unit, as is the case in the measurement of CO_2 in the atmosphere.

[19] Methane (CH_4) in the atmosphere was 850 ppbv before the Industrial Revolution, rising to 1250 ppbv in 1950 (Steffen et al., 2007, p. 616) and 1835 ppbv in 2015.

[20] Nitrogen is extracted in different ways: through the Haber–Bosch process mainly used for fertiliser manufacture and gunpowder production, through the burning of fossil fuels, through the cultivation of legumes and through soil takeover (Federau, 2017, p. 78). Since 1965, humans have been responsible for more reactive nitrogen extraction than the rest of the natural world, especially by land and ocean plants (Gruber & Galloway, 2008). Humans have now become dependent on this mass transformation of nitrogen, carried out in huge chemical plants working to produce fertiliser. As Federau (2017, p. 81) notes, Czech-Canadian environmentalist Vaclav Smil estimates that *2/5* of humanity would not exist without the Haber-Bosch process and the ammonia industry. Similarly, American ecologist Joseph D. Cornell (2010) estimates that the Haber-Bosch process can feed 40% of the world's population. 'To feed the world's population, it is therefore impossible at present to do without synthetic fertilisers. A less meaty diet would alleviate the dependence on the Haber-Bosch process, but would not eliminate it' (Federau, 2017, p. 86). Given current food production practices, it is not inconceivable that this nitrogen transformation will double in the coming decades (Gruber & Galloway, 2008).

[21] Nitrates are an oxidised version of nitrogen (NO_3^-).

ens the greenhouse effect. Indeed, some of the nitrogen is transformed into nitrous oxide – a gas whose greenhouse effect is 298 times greater than that of CO_2. Nitrous oxide currently accounts for 8% of anthropogenic global warming (Federau, 2017, p. 88–91).[22]

3.2.4 Ocean Acidification

The control variable for the ocean acidification boundary proposed by Rockström et al. (2009a) and Steffen et al. (2015a) is the saturation state of aragonite (Ωarag), a form of calcium carbonate formed by marine organisms (which is directly correlated to the atmospheric CO_2 level). The limit defined states that it should not be less than 80% of the pre-industrial level. Currently, it is 84%, which is very close to the boundary. Thus, this boundary is likely to be crossed in the coming years or decades, given the inertia of the oceans in absorbing CO_2 from the atmosphere. If the limit of 350 ppm of CO_2 were respected (we are currently at 400 ppm), the rate of ocean acidification would remain within a safe limit for humanity.[23]

3.2.5 Introduction of Novel Entities

Pollution, or 'the introduction of novel entities', is another of the nine planetary boundaries that does not have a control variable. These are 'new forms of existing substances, and modified life forms that have the potential for unwanted geophysical and/or biological effects' (Steffen et al., 2015b, p. 7). Human activity does indeed introduce a range of chemicals and products that are not naturally

[22] However, in addition to nitrogen, another biogeochemical cycle has been significantly altered by human activity. During the twentieth century, phosphorus has also played an important role in the production of fertilisers, of which it has even become an indispensable element. In, 2014 we extracted 225 million tons of fossil phosphorus (Williams et al., 2016, p. 10), of which 90% were used in agriculture. Unlike nitrogen, which is produced from inert nitrogen in the atmosphere, phosphorus is a fossil resource mined in Asia, North America and Africa. With population growth and the rapid development of a range of countries, it is likely that phosphorus production will peak around the 2030s and that global reserves will be depleted very rapidly thereafter (Steffen et al., 2011a, p. 854). For these authors, it is important to think about the long-term management of phosphorus between different regions of the globe, otherwise food security may well be in jeopardy. The control variable for the modification of the phosphorus cycle is the transfer of phosphorus from groundwater to the oceans. At the global level, the limit defined by Steffen et al. (2015a) is 11 Tg of phosphorus per year; the current rate is 22 Tg of phosphorus per year.

[23] Ocean acidification is related to the rate of CO_2 emissions into the atmosphere, because CO_2 absorbed by the oceans causes a release of H^+ ions that lower the pH (thus acidifying the oceans): $CO_2 + H_2O = H^+ + HCO_3$. Over the last 200 years, the concentration of H^+ ions has increased by 30% as a result of the increase in atmospheric CO_2 (Steffen et al., 2015b, p. 4) and the pH is now 0.1 lower than in the pre-industrial period (Caldeira & Wickett, 2003).

encountered in the Earth system. An example of these 'novel entities' are CFCs (chlorofluorocarbons), which have had unforeseen negative effects on the stratospheric ozone layer. The number of these newly produced substances, including nanomaterials, is now estimated to be over 100,000. This number is constantly increasing with the development of the chemical industry and the globalisation of the economic market (Steffen et al., 2015b, p. 6).[24]

3.2.6 *Freshwater Use, Stratospheric Ozone Depletion, Atmospheric Aerosol Loading, and Land-System Change*

The other four planetary boundaries defined by Rockström et al. (2009a, b) and Steffen et al. (2015a, b) are freshwater use, stratospheric ozone depletion, atmospheric aerosol loading, and land-system change. The control variable for the boundary labelled 'freshwater use' is the use of so-called 'blue water' – i.e. water from renewable rivers, lakes, reservoirs, and groundwater reserves. The globally boundary limit is the use of 4000 km^3 of blue water per year; today, we use 2600 km^3 of blue water per year (Steffen et al., 2015b, p. 5). This is another aspect of the Earth system that is heavily impacted by humans. Too much blue water use disrupts the terrestrial and marine ecosystems that need it.

For several decades now, we have observed the thinning of the ozone layer in the stratosphere. Its thickness varies according to the time of year and the latitude (it is thicker in spring and at the poles). The main cause of stratospheric ozone depletion is the anthropogenic production of CFCs. The control variable for the stratospheric ozone depletion boundary is the stratospheric ozone (O_3) concentration. The boundary is set so that stratospheric ozone depletion is less than 5% of 290 DU.[25] This limit is transgressed only over Antarctica, which appears in spring over Australia (200 DU).

[24] Plastics offer an example of these new entities. Every year, we produce millions of tons of plastic, and the waste accumulates in the oceans, where it will remain for hundreds of years. As plastic waste decomposes, it turns into microscopic fragments that are then ingested by marine animals, and we currently do not know the health consequences of this contamination, and whether toxic substances then pass into the food chain (Thompson et al., 2004). The American environmentalist Jena R. Jambeck et al. (2015) conducted an estimation of the mass of plastic entering the oceans from the continents. In 2010, they identified the production of 275 million metric tons of plastics generated by 192 coastal countries. They estimate that between 4.8 and 12.7 million metric tons of plastic entered the oceans. Most plastic produced is for packaging purposes.

[25] DU stands for 'Dobson Unit' which is the unit for measuring the height of the ozone column. The so-called 'ozone layer' is located in the stratosphere (which is between 10 and 40 km above the earth's surface). This layer is vital for human life because it provides protection from the sun's ultraviolet radiation. On the other hand, low-level ozone is harmful to various forms of life (including humans) and is a pollutant.

The control variable of the planetary boundary for atmospheric aerosol loading[26] is the depth of optical aerosols, for which no quantified boundary has yet been defined.[27] We know that the concentration of aerosols in the atmosphere has doubled since 1750, causing disruption to cloud formation and the water cycle, thus impacting the radiation balance and contributing to global warming.

Finally, another boundary – land use – is defined because of the way its transgression alters biodiversity. Too much regional land use can generate systemic risk. The overall control variable for the land-use boundary is the area of forested land as a percentage of the original forest cover. The boundary is set at 75%, and we are currently at 62%.

3.3 The Notion of the 'Great Acceleration'

The notion of planetary boundaries is based on a systemic conception of the Earth. This approach offers a very clear diagram, easily read and understood by the general public, concerning the environmental risk to humanity. This visual is presented in Part III of this manuscript, given its importance in academic work on the Anthropocene. Another visual representing the Anthropocene in contemporary scientific literature is the 'Great Acceleration' graphs. The Great Acceleration graphs were first published in 2004 by Steffen et al. as part of the International Geosphere-Biosphere Programme (IGBP) synthesis project. They have subsequently been widely used and commented upon. These graphs show the trends of the Earth system and its interactions with the global socio-economic system between 1750 and 2000. The graphs have been updated with extensions to 2010 (Steffen et al., 2015a). They put forward 12 indicators for human action (on the themes of population, economic growth, urbanisation, globalisation, transportation, communication, and resource use) and 12 indicators for earth system functionality (on the themes of atmospheric composition, the climate system, stratospheric ozone, the water cycle, the nitrogen cycle, marine ecosystems, land use, tropical forests, and terrestrial biosphere degradation). Among these 12 indicators of Earth system functionality are those that identify the nine global boundaries. Steffen et al. describe this phenomenon as follows in 2004: 'The second half of the twentieth century is unique in the entire history of human existence on Earth. Many human activities reached take-off points sometime in the twentieth century and have accelerated sharply towards the end of the century. The last 50 years have without doubt seen the most rapid transformation of the human relationship with the natural world in the history of

[26] Aerosols are small particles suspended in the atmosphere. They are generated by human activity, as well as by 'natural' activities (volcanic eruptions, fires).

[27] An atmospheric layer can be measured by its degree of transparency; this is called the optical thickness. The optical thickness is high if the transparency is low. (Optical thickness can be measured for any type of material, and means the amount of light passing through it).

humankind' (Steffen et al., 2004, p. 131). The Great Acceleration graphs are now symbolic of the Anthropocene.[28]

The Anthropocene: A Scientific or Political Concept?
The Earth System Science data presented in this chapter highlight anthropological features (such as the manipulation of ecosystems to generate surpluses for humans, the maximisation of profit, and the difficulty of sticking to limits), whose implications are particularly problematic. Both the great acceleration of recent decades and the planetary boundaries (and the crossing of those boundaries) are crucial to understanding the Anthropocene. These will be particularly important in the remainder of this book. Having looked here at notion of planetary boundaries and their importance in understanding the systemic scope of the Anthropocene, the next chapter examines the political ambiguities surrounding the Anthropocene. One could be led to believe that the Anthropocene is a non-political and non-ideological scientific concept. Indeed, all of the arenas for publication of work on the Anthropocene are editorial spaces devoted to scientific work of the highest level (notably, the *Science Family of Journals*, and *Nature*). However, the next chapter shows the mistake made by Earth system researchers in depoliticising the *anthropos* who are responsible for the Anthropocene. It also examines the imbrication of the concept of the Anthropocene with politics (whether it is a question of the involvement of Anthropocene scientists in the City, or the use, for political ends, of the Anthropocene concept, which has, paradoxically, been depoliticised).

[28] Steffen et al. (2015a) identify 12 variables highlighting how the Earth system is impacted by and swept along by the great socio-economic acceleration. The first 12 indicators focus on socioeconomic development from 1750 to 2010. The first graph shows the evolution of the world population, the second the evolution of the real global GDP. The third graph identifies international direct investment (international capital movements measured in current US dollars). The fourth curve shows the evolution of the urban population; the fifth shows the evolution of world energy consumption and the sixth shows the global consumption of fertilisers (nitrogen, phosphate and potassium). The seventh graph shows the number of large dams (measuring more than 15 metres in height above the foundation). The eighth graph shows global water use (irrigation, domestic water, power generation and livestock consumption). The ninth diagram shows global paper production. The tenth shows the global number of new motor vehicles produced per year. The eleventh refers to the total number of wired telephone lines and mobile phone subscriptions. The last graph shows the number of international tourist arrivals per year. For more information on the sources used to produce these curves, see Steffen et al., 2015a. The following 12 indicators focus on the structure and functioning of the Earth system. Of the five gases most responsible for the greenhouse effect, the authors selected three as indicators: carbon dioxide, nitrous oxide and methane. These are the first three graphs of the Earth system indicators. The fourth graph shows the percentage of the total ozone column above the Halley Research Station (Antarctica). The fifth graph shows global surface temperature variances. The sixth graph shows ocean acidification measured using the concentration of hydrogen ions. The seventh graph shows the sum of marine species fished, except for mammals and plants (in millions of tons); the eighth graph shows the world production of shrimp in aquaculture. The next graph shows the anthropogenic disruption of the nitrogen cycle (assessed near the coast). The tenth graph shows the loss of tropical forests compared to 1700. The eleventh graph shows the increase in agricultural land. The twelfth graph shows the decrease (in percentage) in the abundance of terrestrial species (compared to the abundance of species in ecosystems not disturbed by humans).

Bibliographical References

Barnosky, A. D. (2008). Megafauna biomass tradeoff as a driver of Quaternary and future extinctions. *Proceedings of the National Academy of Sciences, USA, 105*(Suppl 1), 11543–11548.

Barnosky, A. D., Matzke, N., Tomiya, S., Wogan, G. O. U., Swartz, B., Quental, T. B., Marshall, C., McGuire, J. L., Lindsey, E. L., Maguire, K. C., Mersey, B., & Ferrer, E. A. (2011). Has the Earth's sixth mass extinction already arrived? *Nature, 471*, 51–57.

Barnosky, A. D., Haldy, E. A., Bascompte, J., Berlow, E. L., Brown, J. H., Fortelius, M., Getz, W. M., Harte, J., Hastings, A., Marquet, P. A., Martinez, N. D., Mooers, A., Roopnarine, P., Vermeij, G., Williams, J. W., Gillespie, R., Kitzes, J., Marshall, C., Matzke, N., Mindell, D. P., Revilla, E., & Smith, A. B. (2012). Approaching a state shift in Earth's biosphere. *Nature, 486*, 52–58.

Bourg, D. (2009, December). L'impératif écologique. *Esprit*, 59–71.

Bourg, D. (2017). Gouverner le long terme. *lapenseeecologique.com*, Points de vue, *1*(1), PUF, 1–9.

Bourg, D., & Bompard, J.-P. (2012, May). Comment verdir la démocratie? *Projet, 330*, 14–22.

Caldeira, K., & Wickett, M. E. (2003). Anthropogenic carbon and ocean pH. *Nature, 425*, 365.

Canfield, D. E., Glazer, A. N., & Falkowski, P. G. (2010). The evolution and future of Earth's nitrogen cycle. *Science, 330*, 192–196.

Cornell, J. (2010). Fertilizers. *Encyclopedia of the Earth*. https://editors.eol.org/eoearth/wiki/Fertilizers

Dean, J. R., Leng, M. J., & Mackay, A. W. (2014). Is there an isotopic signature of the Anthropocene? *The Anthropocene Review, 1*, 276–287.

Ellis, E. C. (2015). Ecology in an anthropogenic biosphere. *Ecological Monographs, 85*(3), 287–331.

Federau, A. (2017). *Pour une philosophie de l'Anthropocène*. PUF.

Fox, T., Pope, M., & Ellis, E. C. (2017). Engineering the Anthropocene: Scalable social networks and resilience building in human evolutionary timescales. *The Anthropocene Review, 4*(3), 199–215.

Gruber, N., & Galloway, J. N. (2008). An Earth-system perspective of the global nitrogen cycle. *Nature, 451*, 293–296.

Haberl, H., Erb, K. H., Krausmann, F., Gaube, V., Bondeau, A., Plutzar, C., Gingrich, S., Lucht, W., & Fischer-Kowalski, M. (2007). Quantifying and mapping the human appropriation of net primary production in Earth's terrestrial ecosystems. *Proceedings of the National Academy of Sciences, USA, 104*, 12942–12947.

Haff, P. K. (2014). Humans and technology in the Anthropocene: Six rules. *The Anthropocene Review, 2*, 126–130.

IPBES. (2019). *Summary for policymakers of the global assessment report on biodiversity and ecosystem services of the Intergovernmental Science-Policy Platform on Biodiversity and Ecosystem Services.*

IPCC. (2013). *Changements climatiques 2013 – Les éléments scientifiques, Résumé à l'intention des décideurs*. Contribution du Groupe de travail I au cinquième rapport d'évaluation du Groupe d'experts intergouvernemental sur l'évolution du climat.

IPCC. (2021). *Climate change 2021. The physical science basis. Summary for Policymakers*. Working Group I contribution to the Sixth Assessment Report of the Intergovernmental Panel on Climate Change.

IPCC. (2022a). *Climate change 2022. Impacts, adaptation and vulnerability*. Working Group II contribution to the Sixth Assessment Report of the Intergovernmental Panel on Climate Change.

IPCC. (2022b). *Climate change 2022. Mitigation of climate change*. Working Group III contribution to the Sixth Assessment Report of the Intergovernmental Panel on Climate Change.

Jambeck, J. R., Geyer, R., Wilcox, C., Siegler, T. R., Perryman, M., Andrady, A., Narayan, R., & Lavender Law, K. (2015). Plastic waste inputs from land into the ocean. *Science, 347*, 768–771.

Lenton, T. M., Livina, V. N., Dakos, V., van Nes, E. H., & Scheffer, M. (2012). Early warning of climate tipping points from critical slowing down: Comparing methods to improve robustness. *Philosophical Transactions of the Royal Society, London, 370*, 1185–1204.

Lewis, S. L., & Maslin, M. A. (2015). Defining the Anthropocene. *Nature, 519*, 171–180.

Perret, B. (2012). L'urgence occultée de la crise climatique. *Etudes, 4162*, 151–161.

Perret, B. (2014a). Transition écologique ou choc de la finitude? *Revue du MAUSS, 43*, 35–40.

Perret, B. (2014b). De la difficulté de penser un avenir sous contrainte écologique. *Transversalités, 130*, 69–81.

Persson, L., Carney Almroth, B. M., Collins, C. D., Cornell, S., de Wit, C. A., Diamond, M. L., Fantke, P., Hassellöv, M., MacLeod, M., Ryberg, M. W., Jørgensen, P. S., Villarrubia-Gómez, P., Wang, Z., & Zwicky Hauschild, M. (2022). Outside the safe operating space of the planetary boundary for novel entities. *Environmental Science and Technology, 56*(3), 1510–1521.

Raftery, A. E., Zimmer, A., Frierson, D. M. W., Startz, R., & Liu, P. (2017). Less than 2 °C warming by 2100 unlikely. *Nature Climate Change, 7*, 637–641.

Rockström, J., Steffen, W., Noone, K., Persson, Å., Chapin, F. S., III, Lambin, E., Lenton, T. M., Scheffer, M., Folke, C., Schellnhuber, H., Nykvist, B., De Wit, C. A., Hughes, T., van der Leeuw, S., Rodhe, H., Sörlin, S., Snyder, P. K., Costanza, R., Svedin, U., Falkenmark, M., Karlberg, L., Corell, R. W., Fabry, V. J., Hansen, J., Walker, B., Liverman, D., Richardson, K., Crutzen, P., & Foley, J. (2009a). Planetary boundaries: Exploring the safe operating space for humanity. *Ecology and Society, 14*(2), 1–33.

Rockström, J., Steffen, W., Noone, K., Persson, Å., Chapin, F. S., III, Lambin, E., Lenton, T. M., Scheffer, M., Folke, C., Schellnhuber, H., Nykvist, B., De Wit, C. A., Hughes, T., van der Leeuw, S., Rodhe, H., Sörlin, S., Snyder, P. K., Costanza, R., Svedin, U., Falkenmark, M., Karlberg, L., Corell, R. W., Fabry, V. J., Hansen, J., Walker, B., Liverman, D., Richardson, K., Crutzen, P., & Foley, J. (2009b). A safe operating space for humanity. *Nature, 461*, 472–475.

Scheffer, M., Bascompte, J., Brock, W. A., Brovkin, V., Carpenter, S. R., Dakos, V., Held, H., van Nes, E. H., Rietkerk, M., & Sugihara, G. (2009). Early-warming signals for critical transitions. *Nature, 461*, 53–59.

Steffen, W., Sanderson, R. A., Tyson, P. D., Jäger, J., Matson, P. A., Moore, B., III, Oldfield, F., Richardson, K., Schellnhuber, H.-J., Turner, B. L., & Wasson, R. J. (2004). *Global change and the earth system. A planet under pressure* (The IGBP Book Series). Springer.

Steffen, W., Crutzen, P. J., & McNeil, J. R. (2007). The Anthropocene: Are humans now overwhelming the great forces of nature? *Ambio, Royal Swedish Academy of Sciences, 36*(8), 614–621.

Steffen, W., Grinevald, J., Crutzen, P., & McNeill, J. (2011a). The Anthropocene: Conceptual and historical perspectives. *Philosophical Transactions of the Royal Society, 369*, 842–867.

Steffen, W., Persson, A., Deutsch, L., Zalasiewicz, J., Williams, M., Richardson, K., Crumley, C., Crutzen, P., Folke, C., Gordon, L., Molina, M., Ramanathan, V., Rockström, J., Scheffer, M., Schellnhuber, H. J., & Svedin, U. (2011b). The Anthropocene: From global change to planetary stewardship. *Ambio, 40*(7), 739–761.

Steffen, W., Broadgate, W., Deutsch, L., Gaffney, O., & Ludwig, C. (2015a). The trajectory of the Anthropocene: The great acceleration. *The Anthropocene Review, 2*(1), 81–98.

Steffen, W., Richardson, K., Rockström, J., Cornell, S. E., Fetzer, I., Bennett, E. M., Biggs, R., Carpenter, S. R., de Vries, W., de Witt, C. A., Folke, C., Gerten, D., Heinke, J., Mace, G. M., Persson, L. M., Ramanahan, V., Reyers, B., & Sörlin, S. (2015b). Planetary boundaries: Guiding human development on a changing planet. *Science, 347*, 736–746.

Steffen, W., Leinfelder, R., Zalasiewicz, J., Waters, C. N., Williams, M., Summerhayes, C., Barnosky, A. D., Cearreta, A., Crutzen, P., Edgeworth, M., Ellis, E. C., Fairchild, I. J., Galuszka, A., Grinevald, J., Haywood, A., Ivar do Sul, J., Jeandel, C., McNeill, J. R., Odada, E., Oreskes, N., Revkin, A., Richter, D. d B., Syvitski, J., Vidas, D., Wagreich, M., Wing, S. L., Wolfe, A. P., & Schellnhuber, H. J. (2016). Stratigraphic and Earth system approaches to defining the Anthropocene. *Earth's Future, 4*, 1–22.

Thompson, R. C., Olsen, Y., Mitchell, R. P., Davis, A., Rowland, S. J., John, A. W. G., McGonigle, D., & Russell, A. E. (2004). Lost at sea: Where is all the plastic? *Science, 304*, 838.

Vitousek, P. M., Mooney, H. A., Lubchenco, J., & Melillo, J. M. (1997). Human domination of Earth's ecosystems. *Science, 277*, 494–499.

Wang-Erlandsson, L., Tobian, A., van der Ent, R. J., Fetzer, I., te Wierik, S., Porkka, M., Staal, A., Jaramillo, F., Dahlmann, H., Singh, C., Greve, P., Gerten, D., Keys, P. W., Gleeson, T., Cornell, S. E., Steffen, W., Bai, X., & Rockström, J. (2022, April). A planetary boundary for green water. *Nature Reviews Earth and Environment, 26*, 1–13.

Williams, M., Zalasiewicz, J., Haff, P. K., Schwägerl, C., Barnosky, A., & Ellis, E. C. (2015). The Anthropocene biosphere. *The Anthropocene Review*, 1–24.

Williams, M., Zalasiewicz, J., Waters, C. N., Edgeworth, M., Bennett, C., Barnosky, A. D., Ellis, E. C., Ellis, M. A., Cearreta, A., Haff, P. K., Ivar do Sul, J. A., Leinfelder, R., McNeill, J. R., Odada, E., Oreskes, N., Revkin, A., Richter, D. d B., Steffen, W., Summerhayes, C., Syvitski, J. P., Vidas, D., Wagreich, M., Wing, S. L., Wolfe, A. P., & Zhisheng, A. (2016). The Anthropocene: A conspicuous stratigraphical signal of anthropogenic changes in production and consumption across the biosphere. *Earth's Future, 4*, 1–20.

Chapter 4
The Political Ambiguities Surrounding the Anthropocene

Abstract The Anthropocene is the result of anthropogenic alteration of the Earth system, and is particularly widely studied in the Earth System Sciences. However, many authors call for an understanding of the concept of the Anthropocene to be developed in Humanities and Social Sciences (Reid et al., 2010; Malm and Hornborg, 2014; Lövbrand et al., 2014; Federau, 2017), in the wake of the Amsterdam Declaration of 2001, which reinforced interdisciplinary cooperation on environmental issues in international programmes, integrating the Humanities and Social Sciences. Such is the goal of this next chapter, which looks at the concept of Anthropocene from a political perspective.

Keywords Anthropocene · Anthropocene narrative · Politics · Political theory

It is possible to misinterpret the concept of the Anthropocene as depoliticising the subjects it affects. Not all humans are in the same boat: 'In reality, rather than the age of humans, it would be better described as an "OligAnthropocene", the age of a few men (and even fewer women)' (Gemenne, 2015, p. 233). It is important not to forget that not all human beings have been the agents of Earth's transformations; many are merely victims of it.

At the Amsterdam Declaration on Global Change in July 2001, the four chairs of international programmes spoke of the worsening impact of human activities on planet Earth: the International Geosphere-Biosphere Programme (IGBP), the International Human Dimensions Programme on Global Environmental Change (IHDP), the World Climate Research Programme (WCRP), and the International Programme of Biodiversity Science, DIVERSITAS.

© The Author(s), under exclusive license to Springer Nature Switzerland AG 2023

N. Wallenhorst, *A Critical Theory for the Anthropocene*, Anthropocene – Humanities and Social Sciences, https://doi.org/10.1007/978-3-031-37738-9_4

4.1 Political Ignorance of the Anthropocene Narrative

4.1.1 An Undifferentiated Anthropos

Since the 2010s, a series of critiques of the Anthropocene concept have emerged from the Humanities and Social Sciences. To begin with, they question the *anthropos* on which the Anthropocene is blamed. Though Crutzen, in his 2002 article, specifies that the entry into the Anthropocene is the work of 25% of humanity and not all of it, in most articles by Earth system researchers, the *anthropos* responsible for the entry into this new geological epoch appears to be undifferentiated, which is most unlike the critical contributions of history and the social sciences. The narrative of the Anthropocene produced by the Earth System Sciences is built on a speculative and naturalistic foundation. As early as the Rio Earth Summit in 1992, the multi-governmental political declaration mentions 'common but differentiated responsibilities'. This nuance is not completely absent from the Anthropocene narrative, but it does not leave enough room for such differentiation.

The following quote from the main developers of the Anthropocene concept – Zalasiewicz et al. (2010) – reveals their analysis of the root cause of the entry into the Anthropocene: the linkage of human population growth to fossil fuel use since the Industrial Revolution: 'How have the actions of humans altered the course of the Earth's deep history? The answers boil down to the unprecedented rise in human numbers since the early nineteenth century—from under a billion then to over six billion now, set to be nine billion or more by midcentury. This population growth is intimately linked with massive expansion in the use of fossil fuels, which powered the Industrial Revolution, and allowed the mechanization [*sic*] of agriculture that enabled those additional billions to be fed.' While this argument may, at first glance, seem fairly obvious, it is worth questioning the reasons for the gradual rise to hegemony of fossil fuels at the heart of the Industrial Revolution. In a 2013 article, Swedish historian Andreas Malm shows how the choice of the steam engine over water power, during the nineteenth century in Great Britain, was due to the primacy of capital over labour. The steam engine had the characteristic of being able to be positioned close to towns and workers, without the dependency on the environment of water power plants, which must, obviously, be built on water courses. While, in the early decades, the use of fossil energy for steam engines produced much lower energy yields, capitalist industrialists preferred the steam engine because of the possibility of bringing the energy closer to the workers in order to make more judicious use of the available labour.

Was humanity as a whole – an undifferentiated *anthropos* – responsible for the development of the fossil economy? Andreas Malm and his fellow Swede Alf Hornborg, Professor of Human Ecology (2014), critique the Anthropocene discourse by highlighting 'intra-species inequalities' that must be examined in order to understand the contemporary situation, and pointing out the importance of social and societal forces acting on the Earth system.[1] In his short but influential 2002 article in *Science*, Crutzen points directly to the invention of the steam engine, patented by James Watt, as responsible for the entry into the Anthropocene. As Malm and Hornborg note, the steam engine is most consistently identified in the Earth System Science literature 'as the one artefact that unlocked the potentials of fossil energy and thereby catapulted the human species to full-spectrum dominance' (p. 63). Malm and Hornborg strongly criticise the dominant Anthropocene narrative, which holds that mastery of fire and then the invention of the steam engine are responsible for the entry into a new geological epoch. Indeed, according to Malm and Hornborg, 'The rationale for investing in steam technology at this time [during the late 18th and early 19th centuries] was geared to the opportunities provided by the constellation of a largely depopulated New World, African-American slavery, the exploitation of British labour in factories and mines, and the global demand for inexpensive cotton cloth.' They go on to say, 'Steam-engines were not adopted by some natural-born deputies of the human species: by the nature of the social order of things, they could only be installed by the owners of the means of production. A tiny minority even in Britain, this class of people comprised an infinitesimal fraction of the population of *Homo sapiens* in the early 19th century.' Finally, they conclude: 'Indeed, a clique of white British men literally pointed steam-power as a weapon – on sea and land, boats and rails – against the best part of humankind, from the Niger delta to the Yangzi [*sic*] Delta, the Levant to Latin America. Capitalists in a small corner of the Western world invested in steam, laying the foundation stone for the fossil economy: at no moment did the species vote for it either with feet or ballots, or march in mechanical unison, or exercise any sort of shared authority over its own destiny and that of the Earth system.' (Malm & Hornborg, 2014, pp. 63–64).

Malm and Hornborg point out that from 1850 to 2000, 'the advanced capitalist countries (or the "North") represented 18.8% of the world population, but were responsible for 72.7% of the CO_2 emissions. In the 2000s, the poorest 45% of the world's people were responsible for 7% of CO_2 emitted'; in the early twenty-first century, the richest 7% of people produced 50% of emissions. British environmentalist David Satterthwaite (who was a member of the IPCC) shows in a precise study (2009) that depending on where people live around the world and on their lifestyle choices, their responsibility for greenhouse gas emissions varies by a ratio of 1 to 1000. The emission rate per person in tons of carbon is 10.94 for Qatar; 4.71 for the United States; 1.57 for France; and 0.01 for Chad and Mali. For this reason, Malm and Hornborg argue that "As long as there are human societies on Earth – there *will* be lifeboats for the rich and privileged" (2014, p. 66). Thus, they ask the simple question, 'Are these basic facts reconcilable with the view of *humankind* as the new geological agent?' (p. 64).

[1] In his essay, *L'Anthropocène contre l'histoire – Le réchauffement climatique à l'ère du capital* (original version: *Fossil capital: The Rise of Steam Power and the Roots of Global Warming*), Andreas Malm identifies the divisions between humans (rich capitalists and lowly working peasants) as one of the root causes of the entry into the Anthropocene.

Historically, the UK has been the largest emitter of carbon. In 1850, it was responsible for 77% of CO_2 emissions (and is responsible for 5% today). If we consider only Western Europe and the United States – i.e. 10% of the world population – we can see that they are responsible for 41% of CO_2 emissions since the Industrial Revolution. In 1960, Great Britain and the United States were responsible for 65% of CO_2 emissions (Bonneuil & Fressoz, 2013, p. 134). This figure alone scuppers the possibility of attributing responsibility for the Anthropocene to *all* of humanity, but necessitates that we take a closer look. The initial *anthropos* of the early Anthropocene narratives goes back to the few Englishmen of the eighteenth century who developed the steam engine, who are then universalised in the Anthropocene narrative. Yet, "the Industrial Revolution is much more the mark of a deep divide within humanity itself" (Federau, 2017, p. 215). Furthermore, the increase in population between the Industrial Revolution and today is frequently noted as having contributed to the entry into the Anthropocene. As already mentioned: 'How have the actions of humans altered the course of the Earth's deep history? The answers boil down to the unprecedented rise in human numbers since the early nineteenth century—from under a billion then to over six billion now, set to be nine billion or more by midcentury' (Zalasiewicz et al., 2010, pp. 2228–2229). Yet between 1800 and 2010, CO_2 emissions 'increased by a factor of 654.8', while human population '"only" did so by a factor of 6.6 (Malm & Hornborg, 2014, p. 65). In their article (Malm & Hornborg, 2014), Malm and Hornborg highlight the sociogenic rather than anthropogenic causes of the advent of the Anthropocene. The Anthropocene must be thought of from the perspective of the inequalities of today: 'How can we think of humanity as a whole in an era characterised by global wealth gaps; an era in which 1% of the richest monopolise 43% of the world's wealth; an era in which one member of the world's richest 500 accumulates more income each year than over a million of his or her fellow human beings?' (Dubois, 2016, p. 230). After the earliest criticisms of the idea that an undifferentiated *anthropos* was responsible for the entry into the Anthropocene, Steffen et al. have gradually incorporated elements of differentiation into their analyses, but this differentiation is relatively superficial, and the discussion does not go as far as to propose a notion other than the Anthropocene (2011b). In addition, it is evident that, while differentiation within the *anthropos* is gradually making its way into the graphs, the accompanying narrative remains marked by indifferentiation. The famous article by Lewis and Maslin, 'Defining the Anthropocene' in *Nature* in 2015, for example, continues to understand view the *anthropos* with no form of differentiation.

Thus, the present and future changes in the Earth system cannot and must not be viewed simply as the product of human nature or the activity of an undifferentiated human species. There is no natural inevitability here, but social, political, economic and historical causes that can be identified. The Earth System Sciences alone are not capable of giving a full understanding of the profound causes of our entry into the Anthropocene. The *anthropos* as a whole is not necessarily at the centre of the functioning of the Earth system – or, at the very least, the question merits further analysis.

4.1.2 *The Naturalist Discourse*

Another feature of the narrative accompanying some of the objective evidence from Earth System Sciences is that it is an extension of a naturalistic and speciesist ideology, as has been highlighted by a few authors, including the Indian historian Dipesh Chakrabarty (2009), the French historians Christophe Bonneuil and Jean-Baptiste Fressoz (2013), Malm and Hornborg (2014), and Federau (2017). In this discourse, the entry into the Anthropocene is something inevitable: both directly related to humanity, yet without any real exercise of human freedom. The main actor in the story of the Anthropocene (the 'antihero', poetically speaking) is often the human species. The only character in the story has neither intentionality nor free will (the only names that may be mentioned are James Watt, for his patenting of the steam engine, and Fritz Haber, for his revolutionary chemical process). Federau (2017, p. 212) goes so far as to write that in its narrative, rather than scientific, dimension, 'the Anthropocene is thus more akin to a myth supporting a progressive ideology than to a geological epoch'. In his study of the naturalistic narrative, he highlights its limitations and the generalisation of the situation, by authors working in this field, to the whole of humanity. Indeed, 'what is immediately striking when one examines the composition of the Anthropocene Working Group is its ethnic and social make-up: white, western, middle-aged, academic males speaking for humanity' (Federau, 2017, p. 249). Federau notes an interesting paradox: the naturalistic recourse to the notion of species, which is almost universal within the Anthropocene narrative put forward by Earth System Science researchers, is at odds with distancing human beings from the natural biosphere, which appears to be heavily anthropised and whose processes are impacted by human activity.

The American biologist Paul Ehrlich described humanity's impact on the biosphere using the following formula: $I = P \times A \times T$. I represents the impact on the biosphere, P represents population, A represents affluence and T represents technology, . This formula can be found in the naturalised narrative surrounding the Anthropocene in articles by Earth System researchers. 'In one swift move, it explains human evolution and its impact on the biosphere, within a unified environmental history' (Federau, 2017, p. 202). The paradox inherent in the technoscientific and naturalistic narrative about the Anthropocene is that it is centred upon an undifferentiated *anthropos*, with no place for human beings. In his analysis, Federau shows how crude and erroneous this formula is. Indeed, he notes that 'the countries with the fastest-growing populations are those with the lowest growth in greenhouse gas emissions, and vice versa'; one-sixth of humanity even has a negligible carbon footprint (2017, p. 219). Federau sees the central figure in the naturalistic narrative of the Anthropocene as a form of Leviathan: 'The dynamic that drives Leviathan seems so powerful that there is no indication in the naturalistic narrative that it is possible to stop it, or even deflect it (...); the only course of action open to humans in the face of Leviathan is a palliative one' (Federau, 2017, p. 204). Technical progress is understand as an extension of a form of human nature.

4.1.3 *The Discourse on a Blinkered Human Species Finally Enlightened by Science*

Another characteristic of the Anthropocene narrative is that it is marked by lack of awareness of the seriousness of the environmental situation, on the part of humanity as a whole. Before it was finally enlightened by science, humanity could not be held accountable for its actions. This is a characteristic that is highlighted, in particular, by Bonneuil and Fressoz (2013). The dominant narrative of Anthropocene scientists is summarised as follows: 'We, the human species, have for two centuries altered the Earth system unawares, to the point of changing its geological trajectory. Then, towards the end of the 20th century, a handful of scientists finally made us aware of the danger, taking on the mission of guiding humanity who had gone down the wrong path' (Bonneuil, 2014, p. 2). For these authors, it is important to avoid the pitfall of drawing comparisons between modernity, when humans were unaware of environmental risks, with postmodernity, when we are very well aware of the effects of industrial modernity. They soundly criticise the idea that human beings 'did not know', and show how a group of intellectuals and then activists raised the alarm with their as early as the nineteenth century.

Humanity as a whole is not responsible for entering the Anthropocene; those who are responsible bear unequal shares of that responsibility. Thus, the French philosopher Frédéric Neyrat echoes Bonneuil and Fressoz. 'From Fourier, observing the 'material deterioration of the planet' to the scientists of the 20th century, for whom the Great Acceleration was perfectly visible, there has always been self-reflection. Sensitivity to the fragile *circumfusa* (surrounding things) in the 18th century, awareness of the relationship between deforestation and the possibility of climate change as early as the 1770s, awareness of the inevitable depletion of resources – *we knew*' (2014, p. 41). Bonneuil and Fressoz thus oppose Latour, who argues that humans were unaware of what was happening at the heart of the Industrial Revolution. One of the merits of the book *The Shock of the Anthropocene* is that it is necessary to think of modernity being politically divided rather than divided along temporal lines, between an initial period marked by naivety and a second marked by reflexive self-awareness (Neyrat, 2014, p. 41). Indeed, one faction of the actors of modernity has helped to drive the planet into the Anthropocene, while another faction has tried to put on the brakes.[2] The term 'Anthropocene',

[2] There are other accounts of the Anthropocene, highlighted by Bonneuil. These include the idea of the Anthropocene as collapse, marked by the politics of degrowth, or the eco-Marxist idea of the Anthropocene as unequal ecological exchange (Bonneuil, 2014, pp. 3–4). This reading denounces Britain's unequal ecological exchange with a range of other countries. Britain benefited from more than ten million hectares of exploited land in addition to its territory, which facilitated the entry into the Industrial Age. Christophe Bonneuil shows that from this eco-Marxist standpoint, the Industrial Revolution was not only the result of techno-scientific progress but also of a global geopolitical configuration (including wars, imperialism and slavery). Moreover, the eco-Marxist interpretation also views the history of capitalism in light of its unsustainable environmental component – an intrinsic characteristic of capitalism. Its advantages is that it allows us 'to inscribe the

hinging as it does upon the central *anthropos*, fails to reflect those perpetrators who are truly responsible for the entry into the Anthropocene, including the oil companies (Fressoz, 2013, p. 226).

Bourg and Fragnière, in their *La pensée écologique: Une anthologie*, point to a number of seminal texts of ecological thinking, written as early as the 1850s, that were particularly far-sighted. The dominant narrative about the Anthropocene ignores these authors. Indeed, Steffen et al. point out that 'We are the first generation with widespread knowledge of how our activities influence the Earth System, and thus the first generation with the power and responsibility to change our relationship with the planet' (Steffen et al., 2011b, p. 757). In another article, also from 2011, Steffen et al. mention that 'the emerging global environmental problems [have been] largely ignored' (Steffen et al., 2011a, p. 852) and they go on to mention that 'The implications of these emissions for the climate did not attract widespread attention until the 1990s, and the cautious scientific community did not declare, with any degree of confidence, that the climate was indeed warming and that human activities were the likely cause until 2001 [IPCC report]' (Steffen et al., 2011a, p. 852).

4.2 The Anthropocene as a Political and Engaged Concept

4.2.1 The Anthropocene as a Political Concept

In light of the above, it is necessary to be vigilant when the notion of the Anthropocene is presented in a depoliticised manner. American geologists Stanley C. Finney and Lucy E. Edwards (2016) suggest that stratigraphers should be very careful with this concept. Through direct human observation, it is possible to clearly perceive the anthropisation of the environment, and the alteration of certain elements of the Earth system. According to them, one should not read into stratigraphic records any more than they actually show. These two authors question the scientific community and ask whether it is really up to the International Commission on Stratigraphy to recognise the Anthropocene, which is more of a political statement. For Finney and Edwards, the current recognition of the Anthropocene does not follow the rules usually used for the recognition of other chronostratigraphic units. This difference is not unimportant in their view, because of the concept's popularity in the international scientific community, in the media and in public opinion. The Anthropocene as such is an awareness of human impact on the climate, the Earth's surface, the oceans and the atmosphere. For these authors, there is an anthropocentric and political component to the promulgation of the term 'Anthropocene'.

---

reality of material and energy flows and ecological processes in a critical history of capitalism' (Bonneuil, 2014, p. 6). It allows us to show the way in which the capitalist system appropriates hours of work as limited bio-geo-physical resources.

The science journalist William Laurence, in a 1946 article for the New York Times, proposed the expression 'the Atomic Age' to refer to the present time. For Finney and Edwards, the definition of the Anthropocene by Zalasiewicz – whose entry is dated at the first atomic bomb detonation in 1945 – carries identical content to the phrase 'Atomic Age'. From a strictly stratigraphic point of view, if the first nuclear detonation is recognised as the main marker of entry into the Anthropocene, it would be preferable to depoliticise and de-ideologise the term 'Anthropocene' in favour of 'Atomic Age'. Similarly, in the journal *Nature*, in 2011, the editorialists pose the political questions of what is at stake in the recognition of the Anthropocene. For stratigraphers, the question is whether it is wise for them 'to endorse a term that comes gift-wrapped as a weapon for those on both sides of the political battle over the fate of the planet'. Yet at the same time, scientific recognition 'would encourage a mindset that will be important not only to fully understand the transformation now occurring but to take action to control it'. For that reason, 'the first step is to recognize[sic], as the term 'Anthropocene' invites us to do, that we are in the driver's seat' (*Nature*, Editorial, 2011, p. 254).

For the political scientist Frank Biermann, there is no doubt that 'The Anthropocene is political' (2014, p. 57); The Anthropocene must also be understood as a global political phenomenon. First of all, the Anthropocene makes nation-states interdependent, regardless of their power, and leads to the creation of a network of international institutions and organisations for the purposes of cooperation. The Anthropocene also accentuates the interdependence between generations. The political scientist François Gemenne goes so far as to compare the Anthropocene to a war: 'a war waged against the most vulnerable populations on this planet. We humans have become the main agents of transformation of the Earth, which transformation is making the Earth less and less habitable for an increasing number of populations' (Gemenne, 2015, p. 237). The Anthropocene directly poses the political question of how to organise the continuation of human societies: 'The Anthropocene is unsustainable: it is a massive and high-speed process of destruction operating on a planetary scale, and its current direction must be reversed. The question and challenge of the Anthropocene is therefore the "Neganthropocene", that is, to find a pathway that will enable us to escape from this impasse of cosmic dimensions – which requires the elaboration of a new speculative cosmology in the wake of Whitehead' (Stiegler, 2015, p. 138). French philosopher Bernard Stiegler has a political and economic interpretation of the Anthropocene, which he views as an 'Entropocene' – 'a period in which entropy is produced on a massive scale, thanks precisely to the fact that what has been liquidated and automated is knowledge, so that in fact it is no longer knowledge at all, but rather a set of closed systems, that is, entropic systems' (Stiegler, 2015, pp. 137–138). Thus, the Anthropocene is not just a geological epoch; the concept is becoming a political frame of reference as to what needs to be achieved for the future of humanity, our relationship to technology and our prospects for civilisation (Dalby, 2016, p. 34).

In addition to being a political concept, the Anthropocene is also an ideological one, which contains a form of sacralisation of a segment of humanity that is capable of impacting the Earth system as a whole, with an interest in continuing to set humanity apart from nature, and to dominate the natural world.[3] One of the questions posed by the Anthropocene is how we can restore the political sphere by deciding together what we should do.

4.2.2 *The Political Question of the Date of Entry into the Anthropocene*

A group of Earth System scientists believe that the formal recognition of a new geological epoch will be a particularly significant event in the history of humanity's relationship with Earth. This is why the debate on dating the entry into the Anthropocene is so hotly contested. The dating of the entry into the Anthropocene is the subject of particularly explicit and head-on debate through scientific journals. Ellis, who is a member of the Anthropocene Working Group, signed an article in the journal *Nature* in 2016, with a group of colleagues entitled 'Involve social scientists in defining the Anthropocene', calling for the official involvement of social scientists in their working group in defining the Anthropocene. They denounce the fact that '[three] dozen academics are planning to rewrite Earth's history' (p. 192) without the involvement of those in the social sciences. They fear rushing the recognition of the Anthropocene on the basis of an imperfect understanding, and on the basis of a questionable stratigraphic dataset. In particular, they mention the fact that the use of fire and then agriculture were the first real human activities to impact the Earth system as a whole. 'How can a human-centred geological period be defined without characterizing the development of societies, urbanization, colonization, trading networks, ecosystem engineering, and energy transitions from biomass to fossil fuels?' (Ellis et al., 2016, p. 192). Humans have been creating a set of materials – such as ceramics, brick and later concrete – for thousands of years, which must not be ignored in a dating decision that sets the beginning of the Anthropocene in 1945. To these criticisms, the Chairman of the Anthropocene Working Group, Zalasiewicz, responds with two other colleagues, Waters and Head, in an article entitled 'Anthropocene: its stratigraphic basis' published in *Nature* (2017, p. 289), that they regret that the official Anthropocene Working Group has less power than Ellis and his colleagues. While noting the various criticisms levelled at the official recognition of the Anthropocene in the geological time scale (too little stratigraphic material compared to the definition of other geological time units and reliance on human history and its prospective scenarios), they again demonstrate the

[3] After the Enlightenment severed humanity from God and conferred ontological autonomy on humans, industrial modernity completed the process of empowerment by placing humanity a step apart from nature (Papaux, 2015). This disengagement appears to be an anthropological error. It refers to the refusal to acknowledge the limits of the biosphere, and our own finitude (Bourg, 2012; Bourg & Papaux, 2015). The decision is loaded with consequences and is not sustainable in the long term. The narrative generated by the Anthropocene gives the lie to a certain type of glorious discourse on the Industrial Revolution and the emancipation it facilitated.

stratigraphic evidence upon which the impending official recognition would be based. The issues underlying the vehemence of these debates are political. The speed with which the notion of the Anthropocene has been taken up and is already being used in social and political discourse causes some scientists to be very cautious, and others, on the contrary, to want to regularise the Anthropocene by bringing it onto the geological time scale.

The scientific debate is about stratigraphic evidence. However, the real stakes in understanding it lie elsewhere. First of all, there are strategic issues of reputation that give certain individuals and groups sway over others. Which groups of authors will be sufficiently perceptive and convincing to go down in history by having their hypotheses officially confirmed? It is also a question of political issues affecting the conduct of human societies and the direction of collective action. When will it be possible to speak of the Anthropocene as a proven fact, on the basis of which to propose a set of political arrangements and the organisation of collective actions that will allow humanity to survive?

The date chosen as the official entry into the Anthropocene is politically important, and will determine the possible representations of our actions and of our relationship with the environment; it will also expand or restrict the range of actions available to us. The question of the dating of the entry into the Anthropocene is directly political, because it conditions the narrative surrounding the Anthropocene. Thus, as Bonneuil mentions, dating the new epoch from the development of agriculture would mean a normalisation of the global environmental effects of human activity. If the Anthropocene began 10,000 to 5000 years ago (or earlier, if the Stone Age hypothesis is accepted), this points to responsibility common to all of humanity. The date of 1610 places emphasis on humanity's colonial past and the origins of global trade (Lewis & Maslin, 2015, p. 177). This date highlights how, today, we live in a postcolonial world.[4] A date concurrent with the Industrial Revolution points directly to the issue of carbon-dioxide emissions by industrialised countries and their liberal capitalism. In 1945, blame could be directed at the Cold War, as well as globalised consumerist productivism (with the excesses it brings with it). The choice of 1964 brings another type of narrative: that of development based on technoscience[5] having acquired the potential power to destroy the planet totally, but also the

[4] This is a particularly original dating proposed by Simon Lewis and Mark Maslin. It corresponds to the stratigraphically perceptible impact of the collision between the old and new worlds. This event marks the beginning of a global reorganisation of humanity on Earth, with common foodstuffs, and a reorganisation of animal and plant life. Most striking, though, is the fact that the arrival of Europeans in America in 1492 was accompanied by a significant decline in the world population, which can be seen in the sediment record. Indeed, America's population declined considerably, from between 54 and 61 million in 1492 to 6 million in 1650 – this collapse can be attributed to wars, slavery, diseases brought by the Europeans, and famines. This decrease in population led to a decrease in agricultural land and an increase in the area of forests, estimated at 50 million hectares. This resulted in a decrease in the level of CO_2 in the atmosphere by between 7 and 10 ppm, which shows in the Antarctic ice between 1570 and 1620.

[5] The term 'technoscience', used by the philosopher of science Gilbert Hottois for several decades, refers to a technical operationality, as opposed to the symbolic dimension of language. Hence,

possibility that human beings have to avoid that eventuality, as was achieved on 5 August 1963 with the partial nuclear test ban treaty signed in Moscow. Thus, we see that each choice of date has both historical and moral implications. The choice of a recent date of entry into the Anthropocene refers, in the imagination, to the possibility of acting, and does not completely negate the possibility of collective mobilisation. Moreover, with the hypothesis of a recent date, multidisciplinary collaborations appear simpler because the decision must be made by more than geologists acting alone (Eckersley, 2017, p. 5).

The two possibilities that are proposed by Lewis and Maslin – i.e. 1610, with the drop in atmospheric CO_2 corresponding to the meeting of the old and new worlds, or 1964 with the peak of $\Delta^{14}C$ corresponding to the nuclear bomb detonations – have the significant advantage of being correlated with political events, deliberately orchestrated by politicians, rather than economic or scientific events, like the Industrial Revolution or the beginning of agriculture, for which there was no exercise of political freedom.[6] 1610 refers to the political issue of relations with others and cosmopolitanism; 1964 refers to the issue of technological mastery. In the early years of the debate on the entry into the Anthropocene, the Industrial Revolution appeared to be the ideal candidate for dating the Anthropocene, but has since been supplanted by the Great Acceleration or the nuclear detonations.

The debate on the dating of the entry into the Anthropocene has important political ramifications. We can identify at least two political implications of this debate. Firstly, the date of entry into the Anthropocene points the finger at those most responsible for changing the global state of the planet. Secondly, this date also has implications for the possible and conceivable remedies to maintain Earth close to its current state, preventing too abrupt a departure from the Holocene conditions which have facilitated human life throughout the species' entire existence (this second political significance will be discussed in the next chapter).

4.2.3 The Political Irrelevance of the Anthropocene

The need to differentiate the responsibility for triggering the Anthropocene has generated a range of alternative terminologies. Christophe Bonneuil and Jean-Baptiste Fressoz (2013) propose several that allow users to delve into the conceptual complexity of the Anthropocene from a Social Science perspective: Thanatocene, Phagocene, Thermocene, Anglocene, Capitalocene… The Thanatocene is the period

technoscience refers to the lack of distinction between science and its applications, and thus to the departure from the axiological neutrality usually conferred on science. It seems important to regain the separation between science, technology and politics (Jacq & Guespin-Michel, 2015, p. 112). It is also important, in thinking about the contemporary period, to bear in mind that the considerable acceleration of production of all kinds of objects – especially technological ones – hinders our capacity to think, as identified by Gunther Anders (1956/2002).

[6] Strictly based on stratigraphic data.

marked by the technical capability of dealing death on a massive scale, as the twentieth century has largely shown us, but it also refers to the mass death of species. The Phagocene refers to the massive consumption of the planet's resources. The Thermocene means the wasting of energy. The Anglocene refers to the birth of the Industrial Revolution, in colonial England. If we consider the Anthropocene to have begun concurrently with the Industrial Revolution, does the cognitive event really date from the beginning of the twenty-first century, as suggested by Crutzen's proposed terminology? The terminology proposed by Bonneuil and Fressoz, based on a historical reinterpretation of modernity, shows the extent to which a group of actors were aware of what they were doing (for example, industrialists who produced products with planned obsolescence). However, Bonneuil and Fressoz show that the Anthropocene is also marked by a Phronocene, an age of prudence (*phronesis*), with a group of people sensitive to the environment alerting their contemporaries to the need to change their way of life. The Anthropocene is also a Polemocene (from the Greek *polemos*, war), marked by struggles against the destructive approaches of capitalism or technoscience. Concomitantly, it is also an Agnotocene, where a group of human beings continue to deny the environmental degradation. Malm and Hornborg (2014) also propose the term Capitalocene, which highlights the centrality of capital, at the heart of industrial modernity, in entering the Anthropocene. The American philosopher and biologist Donna Haraway adds the terms Plantationocene and Chthulucene[7] (2015). The fundamental questions underlying this diverse range of terms relate to the western lifestyles of our consumer societies, hinging on industrial production, and the omnipotence of *homo oeconomicus*.

As just indicated, the answer to the question usually posed by the Humanities and Social Sciences in relation to the Anthropocene – 'Who is the *anthropos* to which the Anthropocene refers?' – points to individuals: the whole of humanity, the colonists, the English thermo-industrial and productivist capitalists of the eighteenth century, the capitalists of the 'consumer society' on various continents seeking to propagate infinite trade marked by planned obsolescence, or the set of technoscientists and politicians. Among all the candidate dates for the entry into the Anthropocene, we see an identical anthropological causal characteristic (except for the Stone Age hypothesis, which has not been much addressed by stratigraphers). This uniting thread is the approach of dominating ecosystems in order to generate agricultural surpluses, which then allow for urban development and the rise of

[7] Haraway views the Anthropocene as a frontier rather than as a geological epoch; as she says, what comes after will not be like what came before (the Holocene). From this standpoint, 'our job is to make the Anthropocene as short/thin as possible and 'to cultivate with each other in every way imaginable epochs to come that can replenish refuge' (Haraway, 2015, p. 160). She believes that a new name is needed –one which reflects the convergence and collaboration of all possible forces to allow for the human adventure's continued survival. This is what she calls the Chthulucene, which refers at once to the past, the present and the future, and to the convergence of human and non-human forces. Haraway's Chthulucene is a form of Earth metaphor similar to the Gaia hypothesis, with emphasis being placed on the powers of the Earth and of all its inhabitants, human and non-human alike.

civilisations. This anthropological causal characteristic is to be distinguished from individuals, whose responsibility can be identified as determining the magnitude of the indicators of systemic disruption of Earth. It should be remembered that the liberal *homo oeconomicus* (as spoken of by Adam Smith and Karl Marx, and which is decried by the various terms mentioned above) is only the latest variant, in the west, of civilisations based on such domination of nature. The logic of transforming ecosystems to the advantage of human beings cannot be attributed solely to the western *homo oeconomicus*: it is to be found much earlier. It is therefore important, when we talk about the Anthropocene within the Earth System Sciences, not to act as if the responsibility for the Anthropocene rested solely on the shoulders of the latest avatar of this logic of transformation of ecosystems (the western *homo oeconomicus* – even though it is indeed this figure who is responsible for the change of scale and crossing of thresholds (causal magnitude) – and of individuals in particular. In other words, the mastery of nature and the radical and irreversible transformation of ecosystems throughout the Holocene is a characteristic that cannot be confined to the west (the first urban civilisations appeared in Mesopotamia, India, China, etc.). There is undeniably a civilisational component to be taken into account.[8] In some respects, the Anthropocene can be understood as the culmination of the human development of the Holocene, marked by the anthropological and metaphysical imaginary of the very first narratives – including, of course, that of Prometheus, to which we shall return later in this book.

The importance of technology in contemporary times has led the American geologist Peter Haff to speak of the technosphere, created by human beings and now having a form of autonomy. He even develops the idea that 'the functioning of modern humanity is the product of a system that is beyond our control and imposes its own demands on human behaviour. The technosphere is a system for which humans are essential yet subordinate' (Haff, 2014, p. 127). Haff defines the technosphere as all human infrastructure (energy extractors, industrial production, energy transmission and transportation), cities, communication and financial networks, governments and bureaucracies, etc. This technosphere represents a new stage in Earth's technological evolution. The various networked technological elements on Earth's surface are understood as a system, which is almost endowed with the status of a new actor whose interests are partly distinct from those of human beings. In the same vein, the American environmental historian Timothy James LeCain follows on from neo-materialist theories and highlights how 'humans and their cultures have been created by and with a powerful material environment' (2015, p. 1). In other words, he reverses the discourse highlighting the power of human beings in their technoscientific productions. He argues that 'The technologies of the thermo-industrial revolution are framed not so much as evidence of human power, but as evidence that the material world has a much greater power to shape human minds, cultures, and technologies than has heretofore been recognized by most scholars'

[8] Moreover, the notion of civilisation has the important advantage of being a non-ideologically-based descriptor – unlike the capitalist *homo oeconomicus*.

(LeCain, 2015, p. 1). Thus, human beings are indeed struggling to escape from the contemporary era marked by the hegemony of ubiquitous carbon-based fuels. This approach, which has the advantage of distancing itself from the anthropocentrism of the Anthropocene, has the limitation of considering technology and humanity as two opposing categories – which Bourg contested in *L'homme artifice* in 1996, in which he shows how artifices contribute to humanisation. LeCain's (2015) neo-materialist perspective views human beings as products of a material world. Thus, 'Our increasingly frenzied attempts to develop technological fixes to problems we failed to anticipate and even yet do not fully understand [suggest] that the modern technological world is making humans weaker rather than stronger, vulnerable rather than resilient' (2015, p. 23). Following on with that idea, LeCain proposes other terms that do not run carry this dangerous risk of anthropocentrism. The Carbocene refers to the importance of carbon in our lifestyles. The neo-materialist approach takes a different view of the planet's position in its relationship to human beings, and incorporates a dangerous and possibly hostile component to human life: 'Contrary to our naïve and often religiously rooted beliefs, the Earth may now be in the process of revealing itself to be deeply inhospitable to intelligent hominid life. (…) Rather than believing that humans are a threat to the Earth, what manner of ethics and behaviours might emerge from realizing that the Earth may well be a threat to us?' (LeCain, 2015, p. 24).

In their graphs illustrating the Great Acceleration, published in 2015, Steffen et al. incorporate social science critiques of the concept of the Anthropocene, which focus on the undifferentiated *anthropos* blamed for the entry into this new geological epoch, and emphasise how much more responsibility lies with certain groups of people or regions. Steffen et al. clearly mention that it is mainly human activity in OECD countries that has impacted the Earth system, these countries accounting for 18% of the world's population but 74% of GDP (2015, p. 91). These authors have since revised their discourse and partly altered their monolithic understanding of humanity, and now consider that the Great Acceleration was mainly driven by a small portion of the world's population living in OECD countries. We perceive, in Steffen et al.'s (2015) article, the differentiation between OECD countries, the BRICS countries (Brazil, Russia, India, China – this latter also including Macau, Hong Kong and Taiwan – and South Africa), and other countries around the world (Steffen and his team were unfortunately unable to produce graphs on primary energy use and international tourism). These graphical depictions directly highlight the role of the OECD and BRICS countries in the Great Acceleration. The items from other countries around the world that are particularly marked by the Great Acceleration are population growth and urban population growth.

Within this profusion of notions, many authors (Bonneuil & Fressoz, 2013; Malm, 2013; Malm & Hornborg, 2014; Haraway, 2015; Moore, 2016) would have preferred the term 'Capitalocene', certainly because of the differentiation of responsibilities between different segments of the *anthropos* for the entry into this new stage of the Earth system, but also to signify that the contemporary globalised capitalist system is not an inevitability, and is based on cultural conventions that can be modified (Arnsperger, 2005). As Hornborg mentions: 'Another world is possible.

The human species is not inevitably a cancer on the body of Gaia' (2014, p. 9). Of all the alternative concepts proposed, the Capitalocene is certainly the most unifying. While other terms could have been used to define this new geological epoch, the term 'Anthropocene' is well on its way to permanence, due to considerable attention from the scientific community and the extent of 'public and political resonance' (Gibbard & Lewin, 2016, p. 136) that the term is already experiencing.

4.2.4 The Anthropocene: An Engaged Concept

The concept of the Anthropocene is not just a unit of time, as mentioned by Zalasiewicz et al. (2010, p. 2229); its novelty also lies in the extent of international consideration given to this concept, which reflects the contemporary situation and can be easily communicated. Therein lies the ambiguity inherent in this scientific and philosophical concept, which is also used as a political tool to raise public awareness. The Anthropocene is a metaphor for contemporary environmental change, used for public and political communication purposes, and is also gaining recognition as an official geological epoch. The concept of the Anthropocene is not axiologically neutral: it incorporates its authors' political leanings, or potentially an ideological component, from which it is not easy to get away. This refers to the informal coinage of the term and to the way in which certain journalists quickly seized upon it to highlight a form of supremacy of humanity that has succeeded in wresting itself away from nature, as can be seen in the words of the science journalist Mark Lynas: 'Nature no longer runs the Earth. We do' (2011, p. 8). For Australian environmentalist Jeremy Baskin (2014), the Anthropocene is primarily a paradigm *presented as* a geological epoch, as opposed to the paradigm *of* a geological epoch. This is one of the clearest critiques of the Anthropocene concept, which brings together scientific data and normative prescriptions.

The emergence of the concept of the Anthropocene, and the scientificality it brings to the discourse of activists, journalists and social scientists, has caused a reaction from some members of the international community of geologists and Earth system scientists. For example, American geologists Whitney J. Autin and John M. Holbrook signed a paper with the provocative title 'Is the Anthropocene an issue of stratigraphy or pop culture?' (2012). Through this article, the two authors urge the stratigraphic community not to rush into declaring a new geological epoch before it has been stratigraphically proven and formally recognised. This question was immediately answered by the members of the International Commission on Stratigraphy's Anthropocene Working Group, which outlined the stratigraphic evidence thus far identified (Zalasiewicz et al., 2012).

At the same time, since the very earliest scientific work on the subject, the concept of the Anthropocene has been ambiguous, both in the articles published and in the ways in which scientific results are communicated. Indeed, in the following quote from Crutzen in 2002, we immediately perceive that the issue is not strictly scientific, but political: 'A daunting task lies ahead for scientists and engineers to guide society towards environmentally sustainable management during the era of the Anthropocene' (Crutzen, 2002, p. 23). It is worth noting that, even before he began working on the Anthropocene, Crutzen was known for ecological catastrophism. Crutzen always had the tendency to link his scientific research to sounding the alarm for international public opinion. In 1970, he highlighted the destructive power of CFCs and nitrogen oxides in the destruction of atmospheric ozone. This earned him the Nobel Prize in Chemistry in 1995, along with two other chemists who had worked on CFCs. In the 1980s, he warned of a possible nuclear winter. When he began working on the Anthropocene in the early 2000s, Crutzen joined forces with other scientists involved in the issue, such as Johan Rockström and Will Steffen. Each of them was well used to drumming up media coverage of their work and giving interviews in daily newspapers in order to raise public awareness of the issues, as widely as possible. All of the work published since the 2000s by the groups that they were able to mobilise, and then by the official Anthropocene Working Group, is aimed at a community much broader than that of geologists. The orientation of scientific work toward concrete action is what these Earth system researchers believe in, as can be seen particularly clearly in the following quotation from Will Steffen: 'Creative (and constructive) tensions around the [Anthropocene] concept can help the research community to move toward new conceptual syntheses and integrative action-oriented methodologies that are needed to understand the complexities of the Earth system, and are commensurable with the social and environmental challenges in front of us' (Brondizio et al., 2016, p. 318). Therefore, we can say that the Anthropocene 'has social and political appeal and is used as a media vehicle to propagate concern among a community' (Federau, 2017, p. 287). As we have seen, Rockström's concept of planetary boundaries particularly clearly shows this articulation between scientific work and engagement with the populace. Rockström thus speaks of the 'existential risk without historic precedent' (2015, p. 1) that humanity faces. As part of this engagement, working with industry appears to be essential, for many to work concretely on deploying solutions to manage the major issues facing planet Earth on a global scale, such as the stewardship of marine ecosystems: 'We describe [in this article] how we engaged with large seafood producers to coproduce a global science–business initiative for ocean stewardship. [...] Here, ocean stewardship is viewed as an adaptive and learning-based, collaborative process of responsibility and ethics, aimed to shepherd and safeguard the resilience and productivity of ocean ecosystems for human well-being' (Österblom et al., 2017, p. 1). This is what these authors, with Rockström as the fourth signatory, define as 'sustainability science'.

The Anthropocene: A Concept Marked by Involvement

Earth system researchers, like researchers in the Human and Social Sciences working on the concept of the Anthropocene, are part of a common paradigm that can be identified as that of Involved Sciences. Although, as we shall see in the following chapter, their recommendations diverge in terms of their anthropological foundations and the orientation given to the possibility or impossibility of political action, both involve a form of participatory courage which is at the root of the way in which responsibility is conceived in the Anthropocene. 'Involved science is a term for science which fully acknowledges its responsibility, that is aware of the need to pay attention to consequences, that opens up the possibility of questioning its purposes, a science that no longer claims axiological neutrality in order to assert its objectivity, a science that makes commitment (to a territory, to a particular aim, to a context) a central epistemic value, a science that organises the sharing of knowledge and of the powers linked to such knowledge' (Coutellec, 2015a, b, p. 19). The purpose of the Involved Sciences is to help 'build a common world in a context of vulnerability and uncertainty, rather than adding to the chaos for corporate or commercial purposes – a conclusive effect of contemporary productivism. There is no model for the Involved Sciences; above all, the discipline involves a propensity for reflection and action that can be embodied in many ways. The involved sciences put back, at the heart of the politics of knowledge and its transmission/appropriation, the concepts of values, commitment, responsibility and participation, while maintaining the thoroughness necessary in the science of a quest for truth in the confrontation with reality' (Coutellec, 2015a, b, p. 19). For the authors working on the Anthropocene, the entry into a new geological epoch jeopardises the possibility of humanity's long-term survival, and it is imperative that we prevent it. In this chapter, we discussed the political ambiguities surrounding the Anthropocene; the next chapter deals with the conceptions of political action in the Anthropocene, and the possible directions for political action. It is possible to identify two types of political action in the Anthropocene. The first, Promethean, gives predominance to techno-science, allowing for the continuation of the western ways of life that have developed since the Industrial Revolution; the second, which could be described as post-Promethean, postulates the need for a change in the way in which we live.

Bibliographical References

Anders, G. (2002). *L'obsolescence de l'homme, Sur l'âme à l'époque de la deuxième révolution industrielle*. Ivrea. (original edition 1956), French translation.

Arnsperger, C. (2005). La "moralisation du capitalisme": une approche du problème. L'expérimentalisme démocratique et la coévolution des normes. *Revue internationale des sciences sociales, 185*, 477–488.

Autin, W. J., & Holbrook, J. M. (2012). Is the Anthropocene an issue of stratigraphy or pop culture? *GSA Today, 22*(7), 60–61.

Baskin, J. (2014). The ideology of the Anthropocene? *MSSI Research Paper (Melbourne Sustainable Society Institute), 3*, 3–19.

Biermann, F. (2014). The Anthropocene: A governance perspective. *The Anthropocene Review, 1*, 57–61.

Bonneuil, C. (2014). L'Anthropocène et ses lectures politiques. *Les Possibles, 3*, 1–7.

Bonneuil, C., & Fressoz, J.-B. (2013). *L'événement Anthropocène*. Seuil.

Bourg, D. (2012). Pour une anthropologie de la finitude. In J.-P. Pierron & M.-H. Parizeau (Eds.), *Repenser la nature: dialogue philosophique, Europe, Asie, Amériques*. Presses Universitaires de Laval.

Bourg, D., & Papaux, A. (2015). Anthropologie de la finitude. In D. Bourg & A. Papaux (Eds.), *Dictionnaire de la pensée écologique* (pp. 44–46). PUF.

Brondizio, E. S., O'Brien, K., Bai, X., Biermann, F., Steffen, W., Berkhout, F., Cudennec, C., Lemos, M. C., Wolfe, A., Palma-Oliveira, J., & Arthur Chen, C.-T. (2016). Re-conceptualizing the Anthropocene: A call for collaboration. *Global Environmental Change, 39*, 318–327.

Chakrabarty, D. (2009). The climate of history: Four theses. *Critical Inquiry, 35*(Winter), 197–222.

Coutellec, L. (Ed.). (2015a). *Le temps des sciences impliquées, Ecologie et politique*, no. 51.

Coutellec, L. (2015b). Pour une philosophie politique des sciences impliquées – Valeurs, finalités, pratiques. *Ecologie et politique, 51*, 15–25.

Crutzen, P. J. (2002). Geology of mankind: "The Anthropocene". *Nature, 415*, 23.

Dalby, S. (2016). Framing the Anthropocene: The good, the bad and the ugly. *The Anthropocene Review, 3*(1), 33–51.

Dubois, J. (2016). Comprendre l'Anthropocène, entre données quantitatives et choix sociaux. *Espaces et sociétés, 164–165*, 227–233.

Eckersley, R. (2017). La démocratie à l'ère de l'Anthropocène. *lapenseeecologique.com, 1*(1), 1–19. French translation.

Ellis, E. C., Maslin, M., Boivin, N., & Bauer, A. (2016). Involve social scientists in defining the Anthropocene. *Nature, 540*, 192–193.

Federau, A. (2017). *Pour une philosophie de l'Anthropocène*. PUF.

Finney, S., & Edwards, L. E. (2016). The 'Anthropocene' epoch: Scientific decision or political statement? *GSA Today, 26*(3-4), 4–10.

Fressoz, J.-B. (remarks collected by T. Henneton and J. Confavreux). (2013). L'apocalypse et l'Anthropocène, *Vacarme, 65*, 202–233.

Gemenne, F. (2015). L'Anthropocène et ses victimes: une réflexion terminologique. In F. Gemenne (Ed.), *L'enjeu mondial* (pp. 233–240). Presses de Sciences Po, "Annuels".

Gibbard, P. L., & Lewin, J. (2016). Partitioning the quaternary. *Quaternary Science Reviews, 151*, 127–139.

Haff, P. K. (2014). Humans and technology in the Anthropocene: Six rules. *The Anthropocene Review, 2*, 126–130.

Haraway, D. (2015). Anthropocene, capitalocene, plantationocene, chthulucene: Making kin. *Environmental Humanities, 6*, 159–165.

Hornborg, H. (2014, September). Does the Anthropocene really imply the end of culture/nature and subject/object distinctions?, *Os mil nomes de Gaïa*, Rio de Janeiro, www.osmilnomesdegaia.eco.fr, pp. 1–16.

Jacq, A., & Guespin-Michel, J. (2015). Science et démocratie: une articulation difficile mais nécessaire. *Ecologie & politique, 51*, 107–120.

LeCain, T. J. (2015). Against the Anthropocene. A neo-materialist perspective. *International Journal for History, Culture and Modernity, 3*(1), 1–28.

Lewis, S. L., & Maslin, M. A. (2015). Defining the Anthropocene. *Nature, 519*, 171–180.

Lövbrand, E., Beck, S., Chilvers, J., Forsyth, T., Hedrén, J., Hulme, M., Lidskog, R., & Vasileiadou, E. (2014, September 6). Taking the human (sciences) seriously: Realizing the critical potential of the Anthropocene. In *ECPR General Conference*, Glasgow.

Lynas, M. (2011). *The god species: How the planet can survive the age of humans*. Harper Collins.

Malm, A. (2013). The origins of fossil capital: From water to steam in the British Cotton Industry. *Historical Materialism, 21*(1), 15–68.

Malm, A., & Hornborg, A. (2014). The geology of mankind? A critique of the Anthropocene narrative. *The Anthropocene Review, 1*, 62–69.

Moore, J. W. (Ed.). (2016). *Anthropocene or capitalocene? Nature, history and the crisis of capitalism*. PM Press.

Nature. (2011). Editorial. The human epoch. Official recognition for the Anthropocene would focus minds on the challenges to come. *Nature, 473*, 254.

Neyrat, F. (2014). Critique du géo-constructivisme. Anthropocène & géo-ingénierie. *Multitudes, 56*, 37–47.

Österblom, H., Jouffray, J.-B., Folke, C., & Rockström, J. (2017). Emergence of a global science-business initiative for ocean stewardship. *Proceedings of the National Academy of Sciences, 114*(34), 1–6.

Papaux, A. (2015). Homo faber. In D. Bourg & A. Papaux (Eds.), *Dictionnaire de la pensée écologique* (pp. 536–540). PUF.

Reid, W. V., Chen, D., Goldfarb, L., Hackmann, H., Lee, Y. T., Mokhele, K., Ostrom, E., Raivio, K., Rockström, J., Schellnhuber, H. J., & Whyte, A. (2010). Earth system science for global sustainability: Grand challenges. *Science, 330*, 916–917.

Rockström, J. (2015, April). Bounding the planetary future: Why we need a great transition. *Great Transition Initiative*, 1–14.

Satterthwaite, D. (2009). The implications of population growth and urbanization for climate change. *Environment & Urbanization, 21*(2), 545–567.

Steffen, W., Grinevald, J., Crutzen, P., & McNeill, J. (2011a). The Anthropocene: Conceptual and historical perspectives. *Philosophical Transactions of the Royal Society, 369*, 842–867.

Steffen, W., Persson, A., Deutsch, L., Zalasiewicz, J., Williams, M., Richardson, K., Crumley, C., Crutzen, P., Folke, C., Gordon, L., Molina, M., Ramanathan, V., Rockström, J., Scheffer, M., Schellnhuber, H. J., & Svedin, U. (2011b). The Anthropocene: From global change to planetary stewardship. *Ambio, 40*(7), 739–761.

Steffen, W., Broadgate, W., Deutsch, L., Gaffney, O., & Ludwig, C. (2015). The trajectory of the Anthropocene: The great acceleration. *The Anthropocene Review, 2*(1), 81–98.

Stiegler, B. (2015). Sortir de l'Anthropocène. *Multitudes, 60*, 137–146.

Zalasiewicz, J., Williams, M., Steffen, W., & Crutzen, P. (2010). The new world of the Anthropocene. *Environmental Science & Technology, 44*, 2228–2231.

Zalasiewicz, J., Caerreta, A., Crutzen, P., Ellis, E., Ellis, M., Grinevald, J., McNeill, J., Poirier, C., Price, S., Richter, D., Scholes, M., Steffen, W., Vidas, D., Waters, C., Williams, M., & Wolfe, A. P. (2012). Response to Autin and Holbrook on 'Is the Anthropocene an issue of stratigraphy or pop culture?'. *GSA Today, 22*(7), e21–e22.

Zalasiewicz, J., Waters, C., & Head, M. J. (2017). Anthropocene: Its stratigraphic basis. *Nature, 541*, 289–289.

Chapter 5
Conceptions of Political Action in the Anthropocene: Between Prometheism and Post-Prometheism

Abstract This chapter addresses conceptions of political action in the Anthropocene and the tension between a desire to accomplish the Promethean project of modernity and a post-Promethean approach.

Keywords Anthropocene · Prometheism · Post-Prometheism · Political ecologies · Political theory

5.1 Political Ecologies

The term 'ecology' was proposed by the German biologist Ernst Haeckel in 1866. It was a neologism created after mesology, the science of environments, constructed from the Greek terms *oikos* and *logos* meaning the science of the habitats of living species. Ecology is etymologically close to economics, both being habitat sciences; *oikos* meaning 'house'. One of the fundamental differences in the conception of the world or habitat between economics and ecology is epistemological: 'economics calculates while ecology relates; one is quantitative while the other is qualitative. Economics is the science of equivalence while ecology is the science of differences and complementarities (sexual, nutritional, etc.); economics reduces everything to the individual while ecology examines bodies within the framework of their mutual interdependence and their relationship to the global environment (to the ecosystem)' (Zin, 2010, p. 44). Moreover, ecology operates on long timescales, whereas economics operates on short timescales.

After the Second World War, political ecology emerged as a school of thought whose purpose was to preserve or prepare for the future, and to assume the consequences and responsibility for the collective acts currently being carried out by our societies: 'it is no longer the end of history that is targeted, but rather its continuation – its sustainability' (Zin, 2010, p. 46). This emerging political ecology calls into question the political and economic organisation of societies that aim for

© The Author(s), under exclusive license to Springer Nature Switzerland AG 2023

N. Wallenhorst, *A Critical Theory for the Anthropocene*, Anthropocene – Humanities and Social Sciences, https://doi.org/10.1007/978-3-031-37738-9_5

personal enrichment through increased production and consumption. Political ecology notes that western societies have reached an impasse, and that technology cannot give us a way out; its objective is to reconcile politics with the long-term view taken in ecology (Villalba, 2010, p. 95). The political ecology that developed at the end of the war was founded upon two main pillars: ecological thinking[1] and scientific research on anthropogenic changes to the biosphere (Bourg and Whiteside, 2017, pp. 2–3). It culminated in the Meadows report published by the Club of Rome, *The Limits to Growth* in 1972. The clarity of the thesis developed on the impossibility of unlimited growth in a finite world made the LTG report one of the most significant works of political ecology.

In their 2017 article, Écologies politiques: essai de typologie" ("Political ecologies: a typology essay," Bourg and Whiteside offer a typology of schools of thought in political ecology; they identify eight. The first is *Arcadian ecology*, mostly prevalent among French speakers, which questions the compatibility of material accumulation and the flourishing of humanity. This school of thought critiques industrial civilisation, presenting conviviality as an alternative (Illich, 1973a, b). This French political ecology, which developed in the 1970s, was not very critical of anthropocentrism and paid little attention to nature for its own sake. The second type is *authoritarian ecology*, which holds that democracy is not capable of dealing with ecological problems.[2] The third type of political ecology, *institutionalist ecology*, categorically opposes eco-authoritarianism, instead proposing ecological democracies that integrate new decision-making and institutional procedures. It is a true school of thought that demands the constitutional recognition of environmental rights, the consideration of future generations by representatives, the creation of new forms of ecological representation, and the strengthening of involvement of deliberation in decision-making processes. The fourth category is *catastrophist ecology*, whose primary object of study is catastrophe.[3] The approach of the French philosopher Jean-Pierre Dupuy, one of the representatives of catastrophism, is to think about the mechanisms that lead to catastrophe in order to avoid it. Next comes *grassroots ecology*, whose proponents aim to change society by adopting simpler lifestyles that are less marked by consumption, but by accepting that changes are

[1] Ecological thinking, on the other hand, predates political ecology, which it integrates into its core. It emerged in the nineteenth century. Bourg and Whiteside define it in terms of the following three characteristics. Firstly, technology will not necessarily solve our problems (which, on the contrary, are sometimes created by it). Secondly, increasing industrialisation is viewed critically because of the changes it generates in our relationship to nature. Thirdly, ecological thinking is critical of anthropocentrism.

[2] Bourg and Whiteside (2017, pp. 7–8) show that the American economist Robert Heilbroner, the American politician William Ophuls and Hans Jonas, who are usually thought of as belonging to this school of thought, and who are categorised as authoritarian, are not. They have never inspired authoritarian parties or formed a particular school.

[3] All currents of political ecology identify the probability of a catastrophe, but not all of them think it through to the end in order to identify the room we have for manoeuvre if we wish to stave off that catastrophe.

made 'from the bottom up', or with a gentle touch and subtle influence, without political change being institutionally decreed. The idea is, in a way, that the world will change as the result of individual changes, and it is a question of starting by changing oneself. This school of though has certain similarities with Arcadian ecology. Finally, *anarchist ecology* is particularly marked by the rejection of any form of domination, by human beings, either of each other or of nature. All these political ecologies aim to change lifestyles and call attention to biogeophysical environmental limits. What distinguishes them from one another is the means they propose to use to achieve this goal. It is possible to identify two other currents that differ from the first six in their pursuit of other goals: *Malthusian ecology*, which has adherents among a particular kind of degrowthist: it is the demographics of the poor that should be reduced because they represent a threat to the lifestyles of the richest; and finally, the *constructivist ecology* of the French philosopher Bruno Latour presented in his *Politics of Nature* (1999), whose foundations are quite different from those of other strains of political ecology, in that it pays little attention to the question of limits.[4]

In the contemporary debate between the proponents of the 'good Anthropocene' and their opponents, whom we might define as being attached to a 'bad Anthropocene' (though they do not define themselves in this way),[5] we perceive similarities with the two modalities of reception of theories on the limits to growth in the 1970s. Some were survivalists, who advocated ecological authoritarianism, while some were Prometheans, placing faith in the idea of technological progress (Eckersley, 2017, p. 6). Our reading of the Anthropocene can reveal the folly of unchecked human excess or of Promethean techno-scientific genius.

Political organisations in the Anthropocene are not self-evident, and it is important to be extremely vigilant about respecting democratic foundations. If democracy is to be rethought in view of the challenges facing us in the long term (Bourg et al., 2017), we must ensure it does not mutate into anti-democratic forms. Indeed, we can identify a set of authors in the Earth System Sciences, but also associative activists, for whom the governance of the Earth system necessitates a set of measures that do not necessarily have to go through the usual democratic channels. The measures must be taken in an authoritarian manner in order to guarantee humanity's continued survival; they are warranted by the entry into the Anthropocene, considered 'bad'. For subscribers to this mindset, western lifestyles must change, even if this means decisions must be taken in an authoritarian manner and modernity must

[4] In the face of the first six currents of political ecology, we have seen the emergence. in recent years, of modernist political thinking (which is closely related to constructivist ecology). In light of the entry into the Anthropocene, a resolutely Promethean direction is proposed. This is how we could describe the *#Accelerate Manifesto* of Srnicek and Williams (2013) and the *Ecomodernist Manifesto* (2015), mentioned previously and discussed further in the next chapter.

[5] As mentioned, we conducted an interview with Dominique Bourg on the Anthropocene, which is understood as 'bad'. As a counterpoint, we interviewed Erle C. Ellis about his concept of the 'Great Anthropocene'.

be forcibly overcome in order to move on to post-modernity. The *Breakthrough Institute*[6] puts forward an alternative vision of the Anthropocene, where lifestyles are not threatened.[7] We are dealing with a form of hypermodernity and a revamping of technical progress that could be directed to allow humanity to continue to live on planet Earth. From this perspective, it is the duty of Earth system scientists to exercise their particular power to experiment with technical devices on a global scale, to effect planetary change (Eckersley, 2017). In both scenarios, democracy in the Anthropocene is under serious threat.

Swedish politician Rasmus Karlsson highlights the political radicality needed to get through the turbulence of the Anthropocene in the organisation of human societies. The first of the two political options consists of searching for other planets that are hospitable to humanity and developing advanced techniques to get there and make those places habitable. This means going beyond planetary boundaries. The second option, on the other hand, is to enforce those boundaries in a very clear way. He uses two metaphors to refer to these two political sustainability strategies: Star Trek for the first and Ecotopia for the second.[8]

---

[6] Bruno Latour, whose political position has changed since 2015 and who is becoming more alarmist on environmental issues (Bourg & Whiteside, 2017), was a member of the Breakthrough Institute.

[7] There is another American institute whose ecological work is well known: the *Tellus Institute* in Boston. This institute was founded in 1976 by the American physicist Paul Raskin, who is still its president, and is developing three prospective scenarios: barbarisation concomitant with a great collapse, adaptation of the conventional economic market based on weak sustainability, or 'great transitions' based on strong sustainability, marked by a decrease in our energy consumption. The model developed for such great transitions is that of localised eco-communalism (Curnier, 2017, pp. 84–109). This is at odds with the technophilia of the Promethean modernism found in the eco-modernist manifesto and, as Curnier (2017, p. 100) notes, quite akin to the approach developed at the Institute of Geography and Sustainability at the University of Lausanne).

[8] This second category includes, for example, the *Convivialist Manifesto*, which is mentioned in the introduction and studied in Chap. 6. A similar current is Christian Arnsperger's post-capitalist ethics (we can consider that the figure of the individual put to work in the convivialist theses is that of the citizen, if not the existential citizen that Arnsperger describes). There are also resonances with the communionist approach of Maurice Bellet (notably author of *L'avenir du communisme*, 2013). These latter approaches share various points in common with Dominique Bourg's writings on sustainability and the need to further integrate anthropological limits. Continuing in this vein, the British economist Kate Raworth proposes a model for human sustainability which she calls 'doughnut economics' because of the visual representation she offers. She identifies 'a safe and just space for humanity' within a set of planetary limits that must not be crossed (in line with Rockström's nine planetary boundaries), and within human and social standards below which human existence (or at least a dignified human existence) is no longer possible, with access to water, work, energy, income, equal treatment between individuals, education, health, democratic participation and social resilience (Raworth, 2017).

5.2 Promethean and Techno-Scientific Politics

5.2.1 A 'Good Anthropocene'

One of the first authors to use the term 'good Anthropocene' was the American Earth system scientist and biologist Erle C. Ellis, in a 2011 article in the *Breakthrough Journal*. In their book, *Big World, Small Planet. Abundance Within Planetary Boundaries* (2015), Rockström and photographer Mattias Klum develop the idea of a 'good Anthropocene', within which humanity is both connected to the Earth while assuming its role as steward of the Earth system, and working towards economic growth and prosperity within limits that must not be crossed. This strange terminology does not imply that the environmental conditions of the Anthropocene are 'good' or favourable for humans or any other life form. It refers to a positive and enthusiastic conception of the *anthropos*, which has become a geological force through its technological mastery. With the 'good Anthropocene', we are dealing with a narrative that 'celebrates the death of nature as an external entity' (Bonneuil, 2014, p. 3), and fits in very well with the concept of green growth, developed by what Bonneuil identifies as technophile post-environmentalists. In this narrative, 'nature, but also the human species, [are conceived] as a socio-technico-economic construct opening the door to transhumanism' (Bonneuil, 2014, p. 3). The 'good Anthropocene' is the environment within which a *Homo deus* in the making evolves, as depicted by Israeli historian Yuval Noah Harari (2017). The Anthropocene, which comes in the wake of humanity's acquisition of power, is 'good' because that power will enable humanity to improve the world and the human condition. Proponents of the 'good Anthropocene' go so far as to consider terraforming – that is, transforming the atmosphere of Mars so that it becomes hospitable to human life.[9] We perceive a set of oppositions between modernism and anti-modernism in the political literature around the Anthropocene; on this point, eco-pragmatism has the characteristic of espousing modernist logics as well as depoliticising modernity and naturalising the process of technological modernisation. Technological evolution is thus understood to be inevitable. As long as progress is possible, it is invested in as something to be explored, without human beings having to ask the question of what is desirable. The eco-pragmatic discourse ignores the political responsibility of those who produce and market techniques and technologies that have destructive effects on the environment and a set of populations.

Erle C. Ellis, for example, has worked particularly productively on the Anthropocene. The anthropological and political consequences he foresees as the result of the entry into this new geological epoch remain underpinned by a Promethean paradigm, as can be read in the following quote from an opinion piece

[9] https://thebreakthrough.org/issues/energy/time-to-embrace-geoengineering, accessed 7 December 2022.

he wrote for the New York Times in 2013: 'The only limits to creating a planet that future generations will be proud of are our imaginations and our social systems. In moving toward a better Anthropocene, the environment will be what we make it'. Faced with this type of statement, Australian philosopher Clive Hamilton's analysis of the question of the 'good Anthropocene' is blunt: 'To the dismay of those who first proposed it, the Anthropocene is being reframed as an event to be celebrated rather than lamented and feared. Instead of final proof of the damage caused by techno-industrial hubris, the "ecomodernists" welcome the new epoch as a sign of man's ability to transform and control nature. They see it as evidence neither of global capitalism's essential fault nor of humankind's shortsightedness and rapacity; instead, it arrives as an opportunity for humans finally to come into their own' (Hamilton, 2015, p. 1).

5.2.2 Planetary Stewardship Through Geoengineering

Climate change is often the work of the gods, in literature or mythology. It is they who control the elements. Thus, in Homer's *Odyssey*, gods or goddesses alter the weather conditions for the benefit of Odysseus' exploits – or indeed to frustrate his efforts (Schneider, 2008, p. 1). However, in the twentieth century, human beings began considering the possibility of this type of control. For some 60 years now, large-scale environmental modifications have been imagined (some authors have envisaged increasing precipitation, injecting iron into the oceans, dispersing dust in the stratosphere, etc.) (Schneider, 2008, p. 1). The Russians Rusin and Flit, in 1960, in their book *Man versus climate*, mention, for example, that 'the Arctic ice is a great disadvantage, as are the permanently frozen soil (permafrost), dust storms, dry winds, water shortages in the deserts, etc. […] If we want to improve our planet and make it more suitable for life, we must alter its climate.'

The term 'geoengineering' was first published by the Italian physicist Cesare Marchetti in 1977 in the first issue of the journal *Climatic Change*, discussing options on how to control atmospheric CO_2. In the same year, the Russian climatologist Mikhail Budyko developed this theme, opining that it was humans' responsibility 'to develop a plan for climate modification that will maintain existing climatic conditions, despite the trend towards a temperature increase due to human economic activity' (Budyko, 1977, p. 244). Most of the contributions in Earth System Science on the Anthropocene consider the possibility of geoengineering on a global scale. Geoengineering can be defined as 'the deliberate manipulation of Earth system processes' (Steffen et al., 2011b, p. 752). It is a form of response to global warming, and the extent of anthropogenic interference within Earth system processes, which continues to be subject to debate.

After coining the term 'Anthropocene', Crutzen entered the field of environmental remediation. Indeed, he mentions that the Anthropocene 'will require appropriate human behaviour at all scales, and may well involve internationally accepted large-scale geo-engineering projects, for instance to "optimize" climate' (Crutzen, 2002, p. 23). Crutzen goes far in the way he redefines humanity and nature in the Anthropocene and the hierarchical relationship of one to the other. For example, he said in a lecture at Yale University, reported in an online article in 2011 co-authored with a science journalist specialising in the issue of the Anthropocene, 'It's no longer us against "Nature." Instead, it's we who decide what nature is and what it will be' (Crutzen & Schwägerl, 2011). For Ellis, who addresses the Anthropocene in a particularly Promethean way and advocates seeing it in a 'positive' way: 'we must not see the Anthropocene as a crisis, but as the beginning of a new geological epoch ripe with human-directed opportunity' (Ellis, 2011a, p. 43).

When the Great Acceleration is not defined as the date of entry into the Anthropocene, it is understood by some authors as its second stage (Robin, 2013; Steffen et al., 2011a; Robin et al., 2014). Steffen, Crutzen and McNeil raise the question of the third stage of the Anthropocene and call for humanity to take control of the Earth system as '*stewards*' in a 2007 article entitled *The Anthropocene: Are Humans Now Overwhelming the Great Forces of Nature?* According to these three authors, the next few decades will be decisive in the further evolution of the Anthropocene. The beginning of the twenty-first century shows a new facet of the Anthropocene, with responsibility for environmental impact no longer being confined to the OECD countries, as was the case during the Great Acceleration up until the end of the twentieth century, but now expanding to China, India, Brazil, South Africa and Indonesia (Steffen et al., 2011a). In recent years, we have seen the concept of planetary stewardship emerge in the literature, referring to our control of the Earth system and the assumption of control over key Earth system regulatory mechanisms (Steffen et al., 2011a; Crutzen & Schwägerl, 2011). In this context, stewardship refers to technical intervention on biogeochemical cycles, through geoengineering. For a number of authors, the use of climate engineering techniques should be a last resort. In addition to the question of feasibility, the problem of using geoengineering is well expressed ed. in Schneider's statement that it could be a 'cure worse than the disease' (2008, p. 14).

Two main types of geoengineering devices are being considered. The first concerns the control of greenhouse gases through intervention in the carbon cycle, known as CDR (*Carbon Dioxide Removal*). We do not currently know how to reduce the level of CO_2 in the atmosphere by storing carbon in an inert form.[10] The only thing that could be done would be to store carbon biologically in deep caves or on the ocean floor, but this has the limitation that it could return to the atmosphere quite quickly if changes were to occur in the management of the carbon cycle (Steffen et al., 2011b, p. 752).

[10] Oil consumption for vehicular travel is not the only issue. In fact, in 2004, 43% of the world's electricity production came from coal. Coal is a very cheap fuel, and we still have large reserves of it in the ground. Capturing and sequestering the carbon generated by coal consumption is a major challenge (Breeze, 2008).

The second type of geoengineering device involves SRM (*solar radiation management*) by controlling the heat that enters the atmosphere. Lovelock, the creator of the Gaia hypothesis, aged 88 in 2008, states: 'We appear to have exceeded the natural capacity to counter our perturbation and consequently the system is changing to a new and as yet unknown but probably adverse state. I suggest here that we regard the Earth as a physiological system and consider amelioration techniques, geoengineering, as comparable to nineteenth century medicine' (Lovelock, 2008, p. 3883). Lovelock is particularly critical and cautious of geoengineering techniques: 'Consider what might happen if we start by using a stratospheric aerosol to ameliorate global heating; even if it succeeds, it would not be long before we face the additional problem of ocean acidification. This would need another medicine, and so on. We could find ourselves enslaved in a Kafka-like world from which there is no escape' (2008, p. 3888). Lovelock prefers to propose a device whose effects are not irreversible, as we cannot be at all sure of its effect on the climate. This is a device for cooling the atmosphere from cooling the surface of the oceans by positioning large pipes 10 m in diameter and 100 metres deep in the oceans to mix surface water with water from the depths. A range of devices are being considered by scientists. The American astronomer Roger Angel, in 2006, proposed another type of solar radiation management, blocking 1.8% of the sun's radiation with the help of a huge parasol in orbit (made up of 800,000 transparent and extremely thin sheets of one square metre, each weighing about 1 g).

Crutzen, in an article published in 2006 in the same journal, *Climatic Change*, which published Marchetti in 1977 (the paper which first mentioned the concept of geoengineering[11]), puts forward the hypothesis of sending sulphur particles into the atmosphere to cool the temperature at the surface of the globe.[12] These particles have reflective properties and would reflect the sun's rays. The cooling effect would be greater if the particles were sent directly into the stratosphere, and would last between 1 and 2 years (Crutzen, 2006). In addition to directly reflecting the sun's rays, these sulphate aerosols also have an indirect effect by influencing the size of cloud droplets and extending their lifetime (Rasch et al., 2008a, b, p. 4010). The production of these sulphur particles could be envisaged using the same process as that which takes place during a volcanic eruption, with the oxidation of sulphur dioxide (SO_2). In fact, during volcanic eruptions, large quantities of sulphur are sometimes blasted into the atmosphere, generating a global cooling effect that stops when the sulphur disappears from the atmosphere. If sulphur particles are released into the stratosphere by humans, after a year or two, they would fall back into the troposphere (a layer of the atmosphere closer to Earth's surface than the stratosphere) and then have no real cooling effect. In a paper by American climatologist

[11] This article was written at the request of the journal, in light of Marchetti's still informal use of this term.

[12] This idea is not a new one. The first scientist to propose sending aerosols into the stratosphere was Russian geophysicist Mikhail Budyko in 1974. The proposal to send sulphur particles has been advanced both by Crutzen and by Hungarian-American physicist Edward Teller, one of the creators of the H-bomb (Guillaume, 2015, p. 468).

Philip J. Rasch, Paul Crutzen and American Danielle B. Coleman (2008a), which explores different scenarios for sending sulphur particles into the upper atmosphere, the authors show that the size of the particles would have an impact on the reduction of global warming. Large particles comparable to those released during volcanic eruptions are less effective than smaller particles. If the level of CO_2 in the atmosphere were to double, 1.5 Tg of sulphur per year of small particles might be needed – twice that if they are of comparable size to those from volcanic eruptions. This technical proposal is far from ideal. Sulphur particles are costly to produce, and they also have a significant impact on human health. The World Health Organisation (WHO) estimates that sulphur particles are currently responsible for the premature death of 500,000 people each year (Steffen et al., 2011a, p. 858)![13] In the 2008 study by Rasch, Crutzen and Coleman on atmospheric sulphation, the authors state that only the effects on the stabilisation of the Earth's temperature have been studied and that a range of impacts on the Earth system (such as ecosystems, the oceans or the cryosphere) still need to be worked on.

Although, as Steffen et al. (2011a) point out, a series of international summits have as yet produced no satisfactory outcome, certain innovative techno-scientific approaches are seen as hopeful (Steffen et al., 2011a, p. 856), particularly in relation to the NBIC convergence (nanotechnology, biology, information technology, cognition and artificial intelligence). The idea of a good Anthropocene is linked to theories and research on augmented humans, transhumanism or posthumanism. These techno-scientific advances in NBIC convergence are indeed impressive. We currently know how to chemically construct a genome and implant it into a bacterium to replace its DNA (Gibson et al., 2010). Artificial intelligence is making considerable progress and is able to 'learn on its own' on limited and circumscribed tasks with clear rules, such as a game (Silver et al., 2017). For Steffen et al. (2011a), one of the most consequential issues of the Anthropocene in the twenty-first century concerns the acceleration of the ability to produce life synthetically, in the wake of the work led by the American biologist John Craig Venter (Gibson et al., 2010). It is questionable, to say the least, that we are currently on the verge of creating new life forms, though what is certain is that we have destroyed much of the biodiversity.

While thinking about geoengineering very seriously and apprehending its results as 'very promising', Steffen et al. (2011a, p. 859) point out that a commitment to geoengineering would have such far-reaching consequences for humans that the alternative needs to be studied very carefully with a view to identifying the full range of undesirable effects. According to them, 'The need to achieve effective planetary stewardship is urgent. As we go further into the Anthropocene, we risk

[13] Whilst Crutzen advocates for the study of climate engineering devices, it is primarily because he despairs of the lack of political responsiveness to the climate problem. In his view, sulphation of the atmosphere ought to be considered only as a last resort (Guillaume, 2015, p. 469). In 2006, he mentioned his exasperation at the failure of societies to react to the danger of global warming. For Crutzen, the priority is to reduce CO_2 emissions: 'I repeat: the very best would be if emissions of the greenhouse gases could be reduced so much that the stratospheric sulfur release experiment would not need to took place. Currently, this looks like a pious wish' (Crutzen, 2006, p. 2017).

driving the Earth system onto a trajectory toward more hostile states from which we cannot easily return' (2011b, p. 739). Steffen et al. make a strongly worded plea to the international and scientific community to implement geoengineering: 'Effective planetary stewardship must be achieved quickly, as the momentum of the Anthropocene threatens to tip the complex Earth System out of the cyclic interglacial pattern during which *Homo sapiens* has evolved and developed. Without such stewardship, the Anthropocene threatens to become for humanity a one-way trip to an uncertain future in a new, but very different, state of the Earth system' (2011b, p. 757).

The planetary boundary approach (Rockström et al., 2009a, b; Steffen et al., 2015) does not propose social solutions to enable us to stay within the defined safety zone. Instead, it directly aims to inform the public, and policymakers, of the state of boundary transgression and the risks of crossing thresholds in the Earth system and the tipping points triggered. For Steffen et al., the direct impact of the increasing environmental risks is the need to turn our attention to stewardship of the Earth system. For these authors: 'There is a need for a truly global evidence base, with much greater integration among issues, in order to respond to these global challenges. New research initiatives [e.g. Future Earth, www.futureearth.org] provide evidence that science can respond to this need by applying Earth-system research to advance a new generation of integrated global analyses and to explore options for transformations toward sustainability. This is a clear sign that, as the risks of the Anthropocene to human well-being become clearer, research is maturing to a point where a systemic step-change is possible—and necessary—in exploring and defining a safe and just planetary operating space for the further development of human societies' (Steffen et al., 2015, p. 745). For the authors who developed the notion of planetary boundaries, the developmental paradigm is not altered. The identification of these limits and the risks involved does not result in a paradigmatic break in anthropological and economic conceptions.

Overall, the global stewardship and biogeochemical flow-control solutions proposed by these scientists have not yet garnered much response from decision-makers. For this reason, Federau says that the social effectiveness of engaged scientists is limited and that their scope for action is quite modest in the public arena (2017, p. 293). The 2013 IPCC report, in the summary for policymakers, ends with a very consistent warning about geoengineering methods:

> Methods that aim to deliberately alter the climate system to counter climate change, termed geoengineering, have been proposed. Limited evidence precludes a comprehensive quantitative assessment of both Solar Radiation Management (SRM) and Carbon Dioxide Removal (CDR) and their impact on the climate system. CDR methods have biogeochemical and technological limitations to their potential on a global scale. There is insufficient knowledge to quantify how much CO_2 emissions could be partially offset by CDR on a century timescale. Modelling indicates that SRM methods, if realizable, have the potential to substantially offset a global temperature rise, but they would also modify the global water cycle, and would not reduce ocean acidification. If SRM were terminated for any reason, there is high confidence that global surface temperatures would rise very rapidly to values consistent with the greenhouse gas forcing. CDR and SRM methods carry side effects and long-term consequences on a global scale (IPCC, 2013, p. 27).

However, there are proposals for 'environmentally friendly' geoengineering, such as seeding large, low-lying marine clouds with seawater particles.[14] This would provide condensation nuclei and activate new water droplets, which would increase the albedo of the clouds, and perhaps even their longevity. Latham et al. state that 'deliberate modification of clouds to produce sufficient cooling to balance the global warming resulting from the burning of fossil fuels is feasible' (2008, p. 3969). They suggest that 'this technique could thus be adequate to hold the Earth's temperature constant for many decades' (Latham et al., 2008, p. 3969). The great benefit of this work is that it offers a potential solution to maintain Earth's temperature which is both environmentally friendly and reversible.

One of the limitations of Earth System Science is its holistic component, with its proposed scientific remediation and stewardship of the Earth System. Earth System Science is underpinned by a form of 'modern society's firm belief in human rationality and control' (Lövbrand et al., 2009, p. 7). It is indeed a holistic scientific discipline that aims to understand the interactions of nature's processes with those of human societies in order to ensure human sustainability, with a focus on an Earth System toolkit (Steffen et al., 2004, p. 37). The use of mathematical modelling with powerful computational tools is central to the production of scientific results in Earth System Sciences. These researchers propose to take back control through Earth system management strategies. As noted by Swedish environmentalist Eva Lövbrand, Swedish political scientist Johannes Stripple, and Bo Wiman (2009), Earth system researchers regularly use mechanistic metaphors such as 'planetary machine' (Steffen et al., 2004, p. 9), or before that, 'the engine room of the Earth system' (Schellnhuber, 1999, p. 21), implying that this machine can be brought under human control and that by a good understanding, it is possible to optimise its operation. In a way, the Earth System Sciences lend scientific legitimacy to the concept of sustainable development, and to the more recent concept of green growth. Bourg and Whiteside (2017, p. 19) decry 'the naïve belief in the omnipotence of technology and the desire to make only marginal changes to the system', noting that these beliefs are currently 'stronger than they have ever been, with the association between transhumanist ideology and neoliberal world organisation'. Furthermore, Australian environmentalist Jeremy Baskin (2015) criticises the idealisation of instrumental reason underlying the notion of the Anthropocene as proposed by Earth System Sciences. He denounces the way in which this legitimises geoengineering. The Anthropocene narrative underlying the scientific papers of Earth System Sciences leads almost 'naturally' to a form of planetary stewardship that is not politically or democratically intentional. Particular caution is needed with technophile approaches that legitimise geoengineering as a result of the Anthropocene. One of the major questions posed by geoengineering is that of governance. Who can

[14] Clouds are important to the climate. They have the dual effect of raising and lowering the global temperature. On the one hand, they prevent the passage of some of the sun's (shortwave) rays, which they reflect back into space, but on the other hand, they keep some of the (longwave) radiation between the Earth's surface and the clouds, thus preventing it from escaping into space (Latham et al., 2008, p. 3969).

make the decision to impact the conditions of life in the biosphere as a whole in this way? Before envisaging a series of geoengineering projects, should we not rather study in depth the social, economic and political systems implemented by human beings and treat the Anthropocene as a much-needed wake-up call to force us to change our lifestyles?

5.3 Post-Promethean Policies and Changing Lifestyles

For French geographer Denis Chartier and the French sociologist Jean-Paul Deléage, it is necessary to reinvent politics in the Anthropocene (2010). The Anthropocene has 'major implications not only for how we theoretically grasp the complexity of the world, but also for the ways in which we must think about its political transformation' (Deléage, 2010, p. 23). The Anthropocene brings political questions to the forefront. There is a real gulf between the climate understood as the object of scientific study and the climate as perceived by elected officials (sometimes involving a denial of reality). When did politicians first learn about the consequences of human activities on the Earth system? Bourg (2017) points out that one of the first instances of scientific knowledge on anthropogenic global warming being passed on to politicians was in 1958, with a report to the US Parliament and then in 1965 with the submission of the President's Science Advisory Committee report to the President of the United States.

5.3.1 *The Long Term and Sustainability*

The timescale of Earth System Sciences covers millennia, millions and even hundreds of millions of years. This is what is meant by the notion of the long term, which refers to 'the evolution of the major physical and biological parameters of the Earth system, those to which the notion of the planetary boundary primarily refers' (Bourg, 2017, p. 5). Several factors that are currently being modified will have an impact on the next few thousand years: these include the temperature we will reach in the next century, the level of CO_2 in the atmosphere, the disruption of biogeochemical cycles, sea levels, the degree of ocean acidification, species extinction, land use, etc. The long term is structured by the biogeophysical elements that we know will remain present in the next few thousand years. The long term, as Bourg defines it, is thus different from projection, which identifies scenarios of societal organisation for the coming decades. Based on what we know about the long term, it is appropriate to define political priorities: 'The strictly political question of the long term pertains to the consequences of these parameters, and to the consequences of our knowledge of them for our current decisions'. Bourg goes on to propose a series of political decisions that can be taken now to prepare for what the long term has in store: 'For example, it is by acting now, by modifying current building standards and urban planning rules, by beginning a mass shift towards green cities, etc., that we can prepare for the heat waves that will continue to grow. Industrial and intensive agriculture is not very resilient to climate change, and therefore it is now that we need to organise the massive evolution of agriculture towards agroecology, in terms of training, financing, regulation, etc.' (Bourg, 2017, p. 6).

Steffen et al. (2016, p. 15) propose two possible scenarios for the future of the Anthropocene. The first hypothesis refers to the UN Sustainable Development Goals and the Paris climate agreements, mentioning the very rapid reduction of greenhouse-gas emissions and a different relationship of humans to the biosphere. In this state of the Anthropocene, part of the ice cap and some biodiversity is conserved. The second hypothesis includes an increase in anthropogenic pressures on the Earth system and the crossing of a series of tipping points. The climate is much warmer, polar ice extremely weak, sea level higher and the biosphere radically modified. This state could persists for millions of years. This second hypothesis is, plainly, to be avoided if at all possible.

The notion of sustainable development continues to contain within it humans' desire for growth – possibly without limit – while the notions of sustainability and ecological transition integrate more the concern for future generations in the way the conditions of a collective life integrating human finitude are founded. Several authors argue that sustainability should take precedence over development or so-called green growth. Bourg's critique of the notion of sustainable development is the trade-off between the necessary increase in the GDP of rich countries and the protection of the environment. The concept of sustainable development is based on the theoretical integration of a harmonious balance between economic, ecological and social factors. The notion of sustainable development is based on the idea of development (with its driving logic of infinitude) while that of sustainability is based on human finitude (Bourg, 2012); and this is even more true with regard to the notion of permacircularity recently developed by Arnsperger and Bourg (2017).

5.3.2 Political Responsibility for Preparing for the Future

Postmodernity is characterised by the singularity of its timescales where speed is more import than duration: 'With the right speed, one can consume the whole of eternity inside the continuous present of earthly life' (Baumann, 2013, p. 17). The contemporary period is marked by a particularly alienating sense of urgency. Aiming to become 'masters and possessors of time' through technological tools, humans have become more dependent on such tools than ever. On the political scene, primacy is given to the present over the future, which causes many difficulties. The contemporary postmodern period is marked by a reconfiguration of timescales and, in particular, the transition from a linear, relatively predictable, programmatic future to a more uncertain relationship with a future that can be prepared for without being programmed (Boutinet, 1990, 2004, 2010). Political figures are currently forced to react to events when they happen, rather than prepare in advance for a future that is becoming increasingly difficult to anticipate (Kemp, 2013). Nevertheless, the question of the future remains at the heart of politics in the contemporary period. Indeed, one of the differences between the world of the Ancients and that of the Moderns is the shift from a world of nature, marked by reflections of political philosophy and the question of defining the just order, to a world of history marked by historical knowledge and the question of the 'probable or desirable future' (Donegani and Sadoun, 2007, p. 25). This questioning of the future is one of the characteristics of modernity: the future is viewed as being the fruit of history and of what we construct in the present. The Enlightenment gradually did away with the religious idea of an imminent Second Coming and a day of reckoning; thus, people instead invested their energies in the ideology of progress of the Industrial Revolution. The entry into postmodernity seems to have removed all expectation and anticipation – though we can perceive the resurgence of a form of hypermodernity with transhumanist theories and research that look again at the linearity of progressive time.

The relationship of politics to time immediately raises the question of responsibility. To whom are we responsible, and is it possible to bear responsibility toward individuals who do not yet exist?[15] On this point, the German philosopher Karl-Otto Apel criticises his compatriot, the philosopher Hans Jonas, arguing that responsibility can only be assigned for what is controllable and sanctionable.[16] Jonas, though, believes the problems that will be faced by future generations must be taken into account, and the main political question to consider is what role the future plays in present decisions. Jonas also proposes a reformulation of the Kantian imperative by describing a moral imperative of responsibility towards future generations, hinging on ontological and metaphysical considerations: 'Act in such a way that the effects of your actions can be reconciled with the permanence of true human life on earth' and: 'Act in such a way that the effects of your action are not injurious to the future possibility of such a life' (Jonas, 1990/1979, p. 40). One of the criticisms Jonas makes of Kantian ethics is the integration of the subject of morality within a collective, working towards a future, rather than individual concerns in the present moment. Jonas, in his work, proposes an ethics of responsibility whereby we are under obligation us with regard to the life of future generations. For Jonas, humans have a responsibility towards future generations – that is to say, towards people who do not yet exist but whom we know will exist. Responsibility towards the future is a central theme in Jonas' work; it is addressed as an ontological imperative. Jonas' ethics of responsibility considers the long-term effects of our ways of being and acting. In a way, for Jonas, human beings must feel responsible for the future of humanity.

For her part, Arendt distinguishes between the perishable – the fruit of labour, destined to be consumed in order to maintain life – and the durable, referring to

[15] Israel and Hungary established an 'Ombudsman for Future Generations' in the 2000s (Bourg & Whiteside, 2017, p. 11).

[16] Responsibility implies obligations, including the obligation to the future. This implies an obligation to posterity without the usual reciprocity of obligation. For this reason, it is necessary to have an idea of the long-term effects of actions carried out in the present. The essence of Jonasian responsibility consists of the transformation of power 'over' into power 'for' (Jonas, 1990/1979, p. 189).

work destined to inhabit the world and reinforce its permanence.[17] The current reflection on sustainability indeed relates to the transition from the sphere of life to that of the world, from private space under economic hegemony to public space. Arendt's political conception of action refers to acting together in a common world. How can this be possible? Man is a breach in time through his thinking and his freedom of action. Through action, he signifies that he is not destined for death but for innovation and creation. Action and reflection are two sides of the same coin: that of the generation of a new beginning (Revault d'Allonnes, 2011, p. 203). Through natality, humanity has the ability to begin something decidedly new. Here, one can perceive a strong form of hope in Arendt – the hope that future generations will not repeat the horror of the Holocaust – or some other enterprise of mass destruction. For this reason, Arendt will be an author of choice for thinking, in Part II of this book, about the necessary consolidation of politics in the Anthropocene.

The political responsibility for preparing for the future is certainly one of the fundamental challenges posed by the Anthropocene. Indeed, 'The Anthropocene forces us to contemplate the possibility of an Earth without us' (Eckersley, 2017, p. 15). Human beings are reaching a stage in their history marked by a new responsibility. With the entry into the Anthropocene, we see how 'the very trajectory of our civilisation is eminently problematic' (Bourg, 2013b, p. 70). It is now up to us to think about the future in the knowledge of the finitude of our own power (Villalba, 2015, p. 59). The new responsibility incumbent on us is by no means individual – it affects humanity as a whole, which has become responsible for the Earth, and now has the power to transform nature irreversibly (Bourg, 2013b, p. 69). In this regard, 'Rather than an "ethics of responsibility" that is somewhat too individual, it would be better to speak of a "policy of collective responsibility" towards future generations and the consequences of our actions, with consideration of the negative impacts of our industry: pollution, nuisance, depletion of resources, disruption of ecological balances (global warming, deforestation, loss of biodiversity, GMOs, etc.)' (Zin, 2010, p. 47). For Roy Scranton, 'The choice is a clear one. We can continue acting as if tomorrow will be just like yesterday, less and less prepared for each new disaster as it comes, and more and more desperately invested in a life we can't sustain. Or we can learn to see each day as the death of what came before [...] If we want to learn to live in the Anthropocene, we must first learn how to die' (Scranton, 2013, p. 3).

At this point, it could even be said that the idea of future generations, which has underpinned ecological thinking since its inception, is now being reconsidered, since the entry into the Anthropocene has impacts on current generations. As Dupuy continuously points out, our contemporaries' responses are not equal to the environmental changes that are taking place: 'The main feature of the threats that I analyse

[17] In Arendt, there is a primacy of the world over life and in Jonas a primacy of life over the world. Jonas and Arendt were close throughout their intellectual lives. They met in Germany in Marburg in 1924 (Arendt was 21 and Jonas 18). They both began their philosophical work with a thesis on St Augustine: Arendt on the concept of love and Jonas on that of freedom. There is, in Arendt's thinking, a profound love of the world and of the diversity of life that inhabits it. In Jonas, we can perceive a great feeling of responsibility for the continuation of human life on Earth.

is precisely that they really exist and are serious, although they do not seem to arouse any fear in the populace. It is not that people are ill-informed. Knowledge about climate change, its causes and its likely consequences is available to anyone who cares about the subject. But who cares? The stranger has, if not a face, at least a likeness. But how can we imagine future disasters? The only thing we can know about them is that they will catch us off guard. How can we prepare ourselves to face the disturbing strangeness of the unrepresentable? As I have repeated *ad nauseam*, it is this lack of concern about the future in general, and the catastrophic future in particular, rather than uncertainty about it, that explains why we do not act when we still could' (Dupuy, 2012, p. 731). For Dupuy, the present must be organised and structured around the future: 'If we closed the door to the future, the whole meaning of the human adventure would be annihilated' (Dupuy, 2012, p. 737).

5.3.3 Towards Post-Capitalist Social Democracy?

British economist Tim Jackson has argued that we need to move beyond the age of irresponsibility, leaving behind the current growth-based economic model (2010). Following on from his thinking, for a range of other authors, the implications of environmental issues for the economy are clear: we need to produce and consume less with a view to achieving a steady state of production. Indeed, 'We are emerging from three centuries of abundance, with a background of stable natural conditions of existence, into an indeterminate era of shrinking ecumene and shrinking capacity for us to take action, marked by violence and instability of natural elements. […] We are entering […] an era of finitude, of stringent material constraints, probably with the emergence of symbolic constraints which will be no less so, even though our mental substratum remains marked by past abundance. We obviously cannot exclude the possibility that the trials that await us will involve the collapse of our very complex social systems, which are incapable of functioning for long without growth of flows of all kinds, blind to any reality other than monetary' (Bourg, 2013b, p. 68).[18] We must prepare ourselves in the Anthropocene for large-scale social issues. Ecology and social justice are inseparable because 'the poorest [will probably] have to pay the price of generalised human selfishness first' (Bourg, 2013a, p. 7). Today, the social contract of John Locke, a thinker seminal in liberalism (with the rule of law and the right to life, liberty and property), is inadequate to address the magnitude of ecological problems (Bourg, 2012).

World human population is currently seven billion, which will have grown to 9.5 billion in around 30 years' time. A number of raw materials will be exhausted if consumption does not decrease (several metals as well as fossil fuels). The distribution of wealth within our environment poses one of the biggest contemporary

[18] It is astonishing to note, for example, the extent to which Earth system scientists fail to identify the techniques that are currently being developed, such as permaculture, which run counter to production agriculture, though they do deliver higher yields.

political problems: the richest 20% of the world's people own 76.6% of the world's wealth (and among them, a mere 2% of the world's population owns 50% of the wealth), while the poorest 20% of people own 1.5%. 3 billion people – almost half of humanity – live on less than $2.50 a day (Biermann, 2014, p. 58). Also, regrettably, the global finance networks have little interest in the long term or in environmental preservation. As we have shown, a tipping of the ecosystems is by no means impossible, with a decrease in global food production that could ultimately decrease the number of human lives the planet can support.

The west has gradually come to prefer a binary anthropology of body and spirit, rather than the ternary body, soul and mind that was at the origin of the power of capitalism: since the mind is not taken into account, man has no other means than to live in fear of death, drawing comfort from materials and emotions (Arnsperger, 2010, p. 38). For Arnsperger, 'Capitalism was born of a promise – generalised opulence – and of the breaking of that promise: opulence constantly engenders its own "deferment", because the fear of old age, suffering, and death produces the idea of an imaginary lack that cannot be filled. Wealth itself becomes fictitious because we confuse the material with the spiritual – or rather, we mistake the material for the spiritual. We place the Infinite where there is only an infinite succession of finitudes' (Arnsperger, 2010, p. 39). On the basis of his existential analysis of the foundations of capitalism, Arnsperger proposes a possible exit route from capitalism and the principles that can lead to it. Two elements are fundamental for the development of post-capitalist societies[19]: 'to relate to our fear of lack (and therefore of death) in ways other than through material opulence, as in capitalist societies', and also 'to no longer base our solidarity solely on the fear of hardship, as in pre-capitalist societies' (Arnsperger, 2010, p. 39).

5.3.4 *Earth System Governance and Long-Term Governance*

Earth system governance differs from planetary stewardship in that the repertoire of potential actions to ensure the continuation of human life in the Anthropocene does not include geoengineering. Within Political Science, the notion of 'Earth System Governance' is emerging, both in an analytical and normative sense, in response to the notion of the Anthropocene. It was proposed in 2007 by Frank Biermann and is a resolutely political reaction to the entry into the Anthropocene, without any technophile flights of fancy. It is about working politically on five challenges related to the anthropogenic transformation of the Earth system: the 'persistent uncertainties' of the Earth system, the new intergenerational dependencies; the interdependencies of all the subsystems within the system; the spatial interdependencies with environmental and social consequences (degradation in one part of the globe can have

[19] For Tim Jackson, at $15,000 a year per person, the sense of well-being becomes dissociated from the accumulation of wealth.

global social consequences); and the possibility of catastrophic damage to humanity as a whole. The notion of 'Earth system governance' does not mean governance of the Earth, nor governance of the biogeochemical processes of the Earth system. The notion refers to human impact on the Earth system and the governance of human societies with a central focus on long-term effects on the Earth system (Biermann, 2014, p. 59). An 'Earth System Governance Project' was set up in 2009 and includes an analytical dimension through the study of current bodies (international organisations, activist groups, expert networks, national agencies, etc.) working in the field of Earth system governance. However, this project also includes a normative dimension, considering Earth system governance as a reformative political programme.

The integration of long-term issues into democratic life is a major challenge for the future. One of the major questions which the Anthropocene concept poses for politics is whether it can be a source of democratic renovation rather than democratic obfuscation (Eckersley, 2017, p. 4). A range of authors are looking for renewed ways to think about and implement democracy in the twenty-first century, given this close connection between human history and natural history: 'How should we Earthlings coexist and co-evolve with other Earthlings?' (Eckersley, 2017, p. 15). The concept of 'Long-Term Governance' (2017), proposed by Bourg as an extension of Earth-system governance, is interesting and an alternative to the concept of stewardship of the Earth system through geoengineering. Indeed, Bourg proposes to take note of the scientific data at our disposal concerning the shifts wrought in the Earth system and their implications for human life and our societies, in order to prepare ourselves to face what is coming. His approach is political and free from technophilia that favours geoengineering. Instead, it presupposes anthropological capacity for evolution. The concept of long-term governance restores politics instead of replacing it with the technoscience of Earth-system management. The entry into the Anthropocene, by virtue of the novelty to be taken into consideration in the organisation of political life, referring to the harmful long-term consequences of our lifestyles, may allow us to think differently about democracy and make it more dynamic. A number of authors advocate the creation of a third chamber, in addition to the usual two on which parliamentary systems are typically built. Such a third chamber would be characterised by deliberative method, and be tasked with protecting the interests of future generations.[20] After sketching out, in *We Have Never Been Modern* (1999), an idea of a democratic reorganisation around an upper and a lower chamber, Latour developed this idea further in 1999, in *Politics of Nature*. The American politician John Dryzek and the Australian politician Simon Niemeyer propose the establishment of a 'Chamber of Discourse' that would allow citizens to be better represented and would focus on environmental issues (2008). In the same vein, the French historian Pierre Rosanvallon muses on the idea of an

[20] This difficulty in integrating the concerns of the long term into the functioning of democracy is nothing new, as Rosanvallon notes. Condorcet, as early as 1789, noted the dangers of an 'immediate democracy' (2010, p. 151). In 1910, Alfred Fouillée pointed out that 'the interests of the present may run counter to the interests of the future' and that it was therefore necessary to set up a senate made up 'of more men yet to be born than of men already born' (Rosanvallon, 2010, p. 155).

academy of the future (2010). The British philosopher Rupert Read also proposes a chamber devoted to the challenges of the long term and the protection of future citizens of the United Kingdom (2008). Bourg and Whiteside develop and delve into the idea of a 'Chamber of the Future' on several occasions (2011, 2017; and Bourg et al., 2017).

Let there be no misunderstanding, a third chamber would not solve the issues raised by the crisis or transformation of representation, which is one of the primary political characteristics of our time and constitutes one of the most important objects of Political Science.[21] Indeed, the main challenge posed by representation today is that of a political unit integrating plurality (Lambert & Lefranc, 2012, pp. 46–53). On the other hand, one of the advantages of this idea of a third chamber is the way it proposes to organise part of the political debate around beings other than ourselves: humans who have not yet been born, but also the rest of the living world. We see this decentring as one of the conditions for pursuing a humanising adventure. In particular, it circumvents a small portion of the adverse effects of the elective system, where the candidate (or the elected official) seeks to satisfy his or her electors, even at the cost of sacrificing preparation for the future.

Recently, this third chamber is presented by Bourg et al. in *Inventer la démocratie du XXIe siècle* [Inventing the Democracy of the twenty-first century] (2017, p. 43) as a 'Citizens' Assembly of the Future', offering 'a space for reflection on all citizen initiatives, for the valorisation of citizens' contribution to the ecological transition, and for democratic experimentation with paths towards a desirable future'. This third chamber helps strengthen the organisation of citizen representation and the voice of citizens in the public space. It is not a legislative assembly with the capacity to make laws. Its power is different: it is to reinject debate into public life and to ensure the respect of constitutional principles relating to the long term. However, it could also officially make overtures to the government, the National Assembly or the Senate and has the power to approach to the Constitutional Council if the law about to be passed violates the constitutional principles of the environment. The third chamber would be responsible for organising debates on long-term issues. Its primary function would be to involve citizens in the legislative work integrating long-term data and issues. It is to be thought of as a form of institutionalised voice of citizens and 'an interface between citizen intelligence and the decision-makers' (Bourg et al., 2017, p. 36). The opinion of the citizens' assembly of the future is advisory, but if the debate is intense with a high level of citizen participation, this would obviously have an impact on the way the political topics in question are approached.

Representation in this third chamber would not be based primarily upon open elections, as happens for the other chambers. It would, first and foremost, be a question of stimulating the collective intelligence and mobilisation of citizens to gradually adopt a different view of the world, to comprehend the possibility of

[21] Similarly, moving from a fifth to a sixth Republic will not heal the rift between citizens and elective politics.

implementing new ways of life and to take the issues of the future into consideration to a greater extent in present decisions. The proposed Citizens' Assembly of the Future is made up of three colleges: the first college of 50 citizens appointed by drawing lots from the general population; the second college comprising 50 environmental specialists drawn by lot from a list compiled by environmental non-governmental organisations and approved by Parliament; and a third college of representatives of organised civil society (in line with the way in which the Economic, Social and Environmental Council currently operates). The political idea is that decisions should be shaped not be political alliances but by the quality of democratic debate based on scientific data.

Democratic institutions have not changed much in the last century or two, and are still organised around representations whereby economic and political life are treated as being separate from nature. The proposal for a third chamber rearranges the relationship between political life and nature, giving a place to scientific knowledge, not with decision-making directly delegated to scientists, but the authors of *Inventer la démocratie du XXIe siècle* propose that such knowledge be at the heart of debates so that it can gradually be appropriated by citizens. An important element in the organisation of the third chamber is the cognitive and scientific component. The Chamber of the Future is backed by a High Council for the Long Term, which produces studies on the long term and the factors that need to be taken into consideration now. This is a central point of this institutional arrangement. The High Council for the Long Term works to disseminate scientific data that illuminate the issues at stake in the debates; it is a science that the authors define as 'enlightening' as opposed to 'acting'. The authors are circumspect with regard to 'active science' that could have a totalising component, as observed in one part of this work. The High Council for the Long Term is designed in the same way as the IPCC, which produces a synthesis of scientific knowledge but is not directly responsible for producing new scientific data. Its mission is a very specific one, and it does not have to make prescriptive public policies, which is the direct responsibility of the Citizens' Assembly of the Future. The scientists on this High Council for the Long Term would be seconded from their research institution for a 3-year term, renewable once, and come from a variety of scientific disciplines, including Philosophy and the Humanities and Social Sciences (Bourg et al., 2017, p. 62).

The proposal for a third chamber developed by Bourg et al. (2017) is designed first and foremost so that those who cannot vote, because they do not yet exist, can be represented in our democratic system. Indeed, at present, in modern democracies, the future of the young generations or of future generations is not politically defended. This proposal for a third chamber is an example of an overhaul of democracy brought about by the awareness of the magnitude of environmental issues. In addition to taking long-term issues into consideration, such a third chamber, by organising citizen participation, would offer a form of democracy that does not currently exist.

The question of the long term obviously includes environmental issues but also those relating to biological engineering. It is a question of ensuring that decisions concerning humanity's genetic future are made in the public domain rather than the private domain: 'It is no more acceptable to authorise a few individuals to emit greenhouse gases on a massive scale than it is to authorise individuals, through anthropotechnics, to produce a new species within the human race' (Bourg et al., 2017, p. 77). This Citizens' Assembly of the Future would restore politics to its rightful place and allow the public space to regain ground on the immense private space extended to the planetary scale, made possible by the economic hegemony of today. Moreover, the recognition of the crime of ecocide when planetary boundaries are breached is one of the important legislative elements for this collective of authors to work on more in preparing for the future. We perceive, with this small collective work, that the transition from the notion of sustainable development to that of the Anthropocene generates paradigmatic ruptures in the conception of possible political organisations.

5.3.4.1 Towards a Progressive Integration of the Anthropocene Into the Citizen Debate

A study of the Earth System Sciences literature on the Anthropocene highlights the extent to which the *anthropos* at the heart of the Anthropocene is marked by the power of the capitalist logic put in place with the spread of the Industrial Revolution, and reinforced by the technoscience of the twentieth century. Given the hegemony of *homo oeconomicus*, there is a need for the emergence of a responsible, forward-looking citizen, as we have just identified through the discussion on long-term governance. The conceptions of political action in the Anthropocene are enormously diverse, and we are seeing the Anthropocene gradually enter the civic debate, which is perceptible in the various manifestos written in recent years.

Bibliographical References

Angel, R. (2006). Feasibility of cooling the Earth with a cloud of small spacecraft near the inner Lagrange point (L1). *Proceedings of the National Academy of Sciences, USA, 103*(46), 17184–17189.

Arnsperger, C. (2010). Changer d'existence économique: enjeux anthropologiques de la transition du capitalisme au post-capitalisme. *Revue d'éthique et de théologie morale, 258*, 23–50.

Arnsperger, C., & Bourg, D. (2017). *Ecologie intégrale, Pour une société permacirculaire*. PUF.

Baskin, J. (2015). Paradigm dressed as Epoch: The ideology of the Anthropocene. *Environmental Values, 24*, 9–29.

Baumann, Z. (2013). *La vie liquide*. Fayard. French translation.

Bellet, M. (2013). *L'avenir du communisme*. Bayard.

Biermann, F. (2014). The Anthropocene: A governance perspective. *The Anthropocene Review, 1*, 57–61.

Bonneuil, C. (2014). L'Anthropocène et ses lectures politiques. *Les Possibles, 3*, 1–7.

Bourg, D. (2012). Transition écologique, plutôt que développement durable. *Vraiment durable, 1*, 77–96.
Bourg, D. (2013a, July). Position. La guerre écologique. *Esprit*, 5–7.
Bourg, D. (2013b). Peut-on encore parler de crise écologique? *Revue d'éthique et de théologie morale, 276*(HS), 61–71.
Bourg, D. (2017). Gouverner le long terme. *lapenseeecologique.com*, Points de vue, *1*(1), PUF, 1–9.
Bourg, D., & Whiteside, K. (2011). Ecologie, démocratie et représentation. *Le Débat, 164*, 145–153.
Bourg, D., & Whiteside, K. (2017). Ecologies politiques: essai de typologie. *lapenseeecologique.com, PUF, 1*(1), 1–26.
Bourg, D., Augagneur, F., Blondiaux, L., Cohendet, M.-A., Fourniau, J.-M., François, B., & Prieur, M. (2017). *Inventer la démocratie du XXIème siècle*. Les liens qui libèrent.
Boutinet, J.-P. (1990). *Anthropologie du projet*. PUF.
Boutinet, J.-P. (2004). *Vers une société des agendas*. PUF.
Boutinet, J.-P. (2010). *Grammaires des conduites à projet*. PUF.
Breeze, P. (2008). Coping with carbon: A near-term strategy to limit carbon dioxide emissions from power stations. *Philosophical Transactions of the Royal Society, London, 366*, 3891–3900.
Budyko, M. I. (1977). *Climatic changes*. American Geophysical Union. (original edition 1974).
Chartier, D., & Deléage, J.-P. (2010). Mise à jour des écologies politiques pour une politique de l'Anthropocène. *Ecologie & Politique, 40*, 15–20.
Crutzen, P. J. (2002). Geology of mankind: "The Anthropocene". *Nature, 415*, 23.
Crutzen, P. J. (2006). Albedo enhancement by stratospheric sulfur injections: A contribution to resolve a policy dilemma? *Climate Change, 77*, 211–219.
Crutzen, P. J., & Schwägerl, C. (2011). Living in the Anthropocene: Toward a new global ethos. *Yale Environment 360*. http://e360.yale.edu/features/living_in_the_anthropocene_toward_a_new_global_ethos, 2011, consulted online on 25 October 2017.
Curnier, D. (2017) *Quel rôle pour l'école dans la transition écologique? Esquisse d'une sociologie politique, environnementale et prospective du curriculum prescrit*. Doctoral thesis in Environmental Sciences from the University of Lausanne, supervised by Dominique Bourg and Philippe Hertig.
Deléage, J.-P. (2010). En quoi consiste l'écologie politique? *Ecologie & Politique, 40*, 21–30.
Donegani, J.-M., & Sadoun, M. (2007). *Qu'est-ce que la politique?* Gallimard.
Dryzek, J. S., & Niemeyer, S. (2008). Discursive representation. *American Political Science Review, 102*(4), 481–493.
Dupuy, J.-P. (2012). On peut ruser avec le destin catastrophique. *Critique, 783–784*, 729–737.
Eckersley, R. (2017). La démocratie à l'ère de l'Anthropocène. *lapenseeecologique.com, 1*(1), 1–19. French translation.
Ellis, E. C. (2011a). The planet of no return: Human resilience on an artificial Earth. *The Breakthrough Journal, 2*, 37–44.
Ellis, E. C. (2015). Ecology in an anthropogenic biosphere. *Ecological Monographs, 85*(3), 287–331.
Federau, A. (2017). *Pour une philosophie de l'Anthropocène*. PUF.
Gibson, D. G., Glass, J. I., Lartigue, C., Noskov, V. N., Chuang, R.-Y., Algire, M. A., Benders, G. A., Montague, M. G., Ma, L., Moodie, M. M., Merryman, C., Vashee, S., Krishnakumar, R., Assad-Garcia, N., Andrews-Pfannkoch, C., Denisova, E. A., Young, L., Qi, Z.-Q., Segall-Shapiro, T. H., Calvey, C. H., Parmar, P. P., Hutchison, C. A., III, Smith, H. O., & Venter, J. C. (2010). Creation of a bacterial cell controlled by a chemically synthesized genome. *Science, 329*, 52–56.

Guillaume, B. (2015). Géoingénierie. In D. Bourg & A. Papaux (Eds.), *Dictionnaire de la pensée écologique* (pp. 468–470). PUF.
Hamilton, C. (2015, June 22). *The theodicy of the "Good Anthropocene" a talk to the breakthrough institute dialogue*, Sausalito. https://read.dukeupress.edu/environmental-humanities/article/7/1/233/8202/The-Theodicy-of-the-Good-Anthropocene
Harari, Y. N. (2017). *Homo deus, une brève histoire de l'avenir*. Albin Michel. (original edition 2015), French translation.
Illich, I. (1973a). *La convivialité*. Seuil. French translation.
Illich, I. (1973b). *Tools for conviviality*. Harper and Row.
IPCC. (2013). *Changements climatiques 2013 – Les éléments scientifiques, Résumé à l'intention des décideurs*. Contribution du Groupe de travail I au cinquième rapport d'évaluation du Groupe d'experts intergouvernemental sur l'évolution du climat.
Jackson, T. (2010). *Prospérité sans croissance: la transition vers une économie durable*. De Boeck-Etopia. French translation.
Jonas, H. (1990). *Le principe responsabilité*. Cerf. French translation 1990 (or. 1979).
Kemp, P. (2013). *Le Prince*. Seuil.
Lambert, F., & Lefranc, S. (2012). *50 fiches pour comprendre la Science politique*. Bréal. (4ème éd).
Latham, J., Rasch, P., Chen, C.-C., Kettles, L., Gadian, A., Gettelman, A., Morrison, H., Bower, K., & Choularton, T. (2008). Global temperature stabilization via controlled albedo enhancement of low-level maritime clouds. *Philosophical Transaction of the Royal Society, 366*, 3969–3987.
Latour, B. (1999). *Politiques de la nature*. La Découverte.
Lövbrand, E., Stripple, J., & Wiman, B. (2009). Earth system governmentality: Reflections on science in the Anthropocene. *Global Environmental Change, 19*(1), 7–13.
Lovelock, J. (2008). A geophysiologist's thoughts on geoengineering. *Philosophical Transaction of the Royal Society, London, 366*, 3883–3890.
Marchetti, C. (1977). On geoengineering and the CO_2 problem. *Climatic Change, 1*, 59–68.
Rasch, P. J., Crutzen, P. J., & Coleman, D. B. (2008a). Exploring the geoengineering of climate using stratospheric sulfate aerosols: The role of particle size. *Geophysical Research Letters, 35*, 1–6.
Rasch, P. J., Tilmes, S., Turco, R. P., Robock, A., Oman, L., Chen, C.-C., Stenchikov, G. L., & Garcia, R. R. (2008b). An overview of geoengineering of climate using stratospheric sulphate aerosols. *Philosophical Transactions of the Royal Society, London, 366*, 4007–4037.
Raworth, K. (2017). *Doughnut economics: Seven ways to think like a 21st century economist*. Chelsea Green Publishing.
Revault d'Allonnes, M. (2011). Hannah Arendt penseur de la crise. *Etudes, 415*, 197–206.
Robin, L. (2013). Histories for changing times: Entering the Anthropocene? *Australian Historical Studies, 44*(3), 329–340.
Robin, L., Avango, D., Keogh, L., Möllers, N., Scherer, B., & Trischler, H. (2014). Three galleries of the Anthropocene. *The Anthropocene Review, 1*(3), 207–224.
Rockström, J., Steffen, W., Noone, K., Persson, Å., Chapin, F. S., III, Lambin, E., Lenton, T. M., Scheffer, M., Folke, C., Schellnhuber, H., Nykvist, B., De Wit, C. A., Hughes, T., van der Leeuw, S., Rodhe, H., Sörlin, S., Snyder, P. K., Costanza, R., Svedin, U., Falkenmark, M., Karlberg, L., Corell, R. W., Fabry, V. J., Hansen, J., Walker, B., Liverman, D., Richardson, K., Crutzen, P., & Foley, J. (2009a). Planetary boundaries: Exploring the safe operating space for humanity. *Ecology and Society, 14*(2), 1–33.
Rockström, J., Steffen, W., Noone, K., Persson, Å., Chapin, F. S., III, Lambin, E., Lenton, T. M., Scheffer, M., Folke, C., Schellnhuber, H., Nykvist, B., De Wit, C. A., Hughes, T., van der Leeuw, S., Rodhe, H., Sörlin, S., Snyder, P. K., Costanza, R., Svedin, U., Falkenmark, M., Karlberg, L., Corell, R. W., Fabry, V. J., Hansen, J., Walker, B., Liverman, D., Richardson, K., Crutzen, P., & Foley, J. (2009b). A safe operating space for humanity. *Nature, 461*, 472–475.

Rosanvallon, P. (2010). Le souci du long terme. In D. Bourg & A. Papaux (Eds.), *Vers une société sobre et désirable* (pp. 151–162). PUF.
Schellnhuber, H. J. (1999). 'Earth system' analysis and the second Copernican revolution. *Nature, 402*, C19–C23.
Schneider, S. H. (2008). Geoengineering: Could we or should we make it work? *Philosophical Transaction of the Royal Society, 366*, 1–20.
Scranton, R. (2013, November 10). Learning how to die in the Anthropocene. *The New York Times*.
Silver, D., Schrittwieser, J., Simonyan, K., Antonoglou, I., Huang, A., Guez, A., Hubert, T., Baker, L., Lai, M., Bolton, B., Chen, Y., Lillicrap, T., Hui, F., Sifre, L., van den Driessche, G., Graepel, T., & Hassabis, D. (2017). Mastering the game of go without human knowledge. *Nature, 550*, 354–359.
Srnicek, N., & Williams, A. (2013). #ACCELERATE. Manifesto for an Accelerationist Politics. *Critical Legal Thinking*, 14 May 2013. http://criticallegalthinking.com/2013/05/14/accelerate-manifesto-for-an-accelerationist-politics
Srnicek, N., & Williams, A. (2015). *Inventing the futur – Postcapitalism and a World without work*. Verso.
Steffen, W., Sanderson, R. A., Tyson, P. D., Jäger, J., Matson, P. A., Moore, B., III, Oldfield, F., Richardson, K., Schellnhuber, H.-J., Turner, B. L., & Wasson, R. J. (2004). *Global change and the earth system. A planet under pressure* (The IGBP Book Series). Springer.
Steffen, W., Grinevald, J., Crutzen, P., & McNeill, J. (2011a). The Anthropocene: Conceptual and historical perspectives. *Philosophical Transactions of the Royal Society, 369*, 842–867.
Steffen, W., Persson, A., Deutsch, L., Zalasiewicz, J., Williams, M., Richardson, K., Crumley, C., Crutzen, P., Folke, C., Gordon, L., Molina, M., Ramanathan, V., Rockström, J., Scheffer, M., Schellnhuber, H. J., & Svedin, U. (2011b). The Anthropocene: From global change to planetary stewardship. *Ambio, 40*(7), 739–761.
Steffen, W., Richardson, K., Rockström, J., Cornell, S. E., Fetzer, I., Bennett, E. M., Biggs, R., Carpenter, S. R., de Vries, W., de Witt, C. A., Folke, C., Gerten, D., Heinke, J., Mace, G. M., Persson, L. M., Ramanahan, V., Reyers, B., & Sörlin, S. (2015). Planetary boundaries: Guiding human development on a changing planet. *Science, 347*, 736–746.
Steffen, W., Leinfelder, R., Zalasiewicz, J., Waters, C. N., Williams, M., Summerhayes, C., Barnosky, A. D., Cearreta, A., Crutzen, P., Edgeworth, M., Ellis, E. C., Fairchild, I. J., Galuszka, A., Grinevald, J., Haywood, A., Ivar do Sul, J., Jeandel, C., McNeill, J. R., Odada, E., Oreskes, N., Revkin, A., Richter, D. d B., Syvitski, J., Vidas, D., Wagreich, M., Wing, S. L., Wolfe, A. P., & Schellnhuber, H. J. (2016). Stratigraphic and Earth system approaches to defining the Anthropocene. *Earth's Future, 4*, 1–22.
Villalba, B. (2010). L'écologie politique face au délai et à la contraction démocratique. *Ecologie & Politique, 40*, 95–113.
Villalba, B. (2015). Au fondement matériel de la démocratie. *Revue Projet, 344*, 56–63.
Zin, J. (2010). Qu'est-ce que l'écologie politique? *Ecologie & Politique, 40*, 41–49.

Chapter 6
Integration of the Anthropocene into the Citizen Debate

Abstract In the introduction to this work, we mentioned the publication of a series of manifestos over the past few years, most written by academics. They are written by groups of two or three authors, or signed by collectives that are structured as think tanks. They express demands, denunciation, opposition and the dream of seeing the emergence of alternatives to the contemporary hegemonic economic models hinging on capitalism. Another world is possible, these manifestos tell us… or, at any rate, another world would be desirable. Nine manifestos are examined in this chapter: the *Manifeste des économistes atterrés* (Manifesto of the Appalled Economists, 2010) and the *Nouveau manifeste des économistes atterrés* (New Manifesto of the Appalled Economists – 2015); the *#Accelerate Manifesto: For an Accelerationist Politics* (2013); the *Ecomodernist manifesto* (2015); the *Manifest für das Anthropozän* (Manifesto for the Anthropocene, 2015); The *Commons Manifesto* (2018); the *Animal Manifesto* (2017); the *Manifesto for Climate Justice* (2019); the *Integral Ecology Manifesto* (2019); and the *Convivialist Manifesto* (2013).

Keywords Anthropocene · Manifesto · Citizen debate · Political theory · Narratives

The titles of many texts emerging nowadays include the word 'manifesto' or 'advocacy'. Other texts could have been chosen.[1] The main selection criterion we have

[1] The environmental question is the starting point for the *New Manifesto of the Appalled Economists*, the *Accelerationist Manifesto*, the *Ecomodernist manifesto* (which explicitly addresses the Anthropocene), the *Anthropocene Manifesto*, the *Climate Justice Manifesto* (2019), the *Integral Ecology* manifesto (2019), and the *Convivialist Manifesto* (2013). The *Animalist Manifesto* opens with one of the main questions of the Anthropocene: our relationship with the rest of the living world. The *Commons Manifesto* does not directly address an environmental issue but is an important text in the economy of this chapter because it makes use of technology in a non-Promethean way, addresses the question of the commons raised in the *Manifesto for the Anthropocene* and the *Convivialist Manifesto*, and puts to work an anthropological shift that could easily be described as convivialist.

© The Author(s), under exclusive license to Springer Nature Switzerland AG 2023

N. Wallenhorst, *A Critical Theory for the Anthropocene*, Anthropocene – Humanities and Social Sciences, https://doi.org/10.1007/978-3-031-37738-9_6

used is the way in which they explicitly mention the concept of the Anthropocene or its content: the anthropogenic alteration of the Earth system. In addition, these nine manifestos have attracted our attention either because of the extent of their adoption, the number of authors involved in writing them or who subsequently signed them, the innovative and stimulating nature of the proposals or the fact that they are contradictory to one another. Through a comparative reading of these various manifestos, we shall be able to gain a clearer understanding of convivialism, which that will be mobilised in Part III of this book. These manifestos have been published since 2010, have strong theses, and illustrate how the Anthropocene is at the heart of a war of ideas concerning the anthropological shift necessary to ensure the human adventure's survival. 'How are we going to get through the dark times that lie ahead at the beginning of the twenty-first century?' is the question addressed in each of these texts, and is taken up again in Chap. 7 of this book.

The variety and number of these manifestos are indicative of the dissatisfaction of many social actors and academics in our world, firstly, and a desire for change, on the other. Each one puts forward a specific point of view; each one tries to pull the thread which could potentially bring about a global change. The ideas they expound are numerous: a transformation of the world of finance and economic rules in general (*Manifesto of the Appalled Economists*), the acceleration of technological innovations (*Accelerationist Manifesto*), the possibility of creating a good Anthropocene through economic liberalisation, allowing the emergence of new technologies (*Ecomodernist Manifesto*), the establishment of a politics and poetics of life (*Anthropocene Manifesto*), Peer-to-Peer technology (*Commons Manifesto*), the recognition of animals as political subjects (*Animalist Manifesto*), the fight against productivism (*Manifesto for Climate Justice*), the possibility of renewing politics (*Manifesto of Integral Ecology*), and the control of *hybris* based on the conviviality of '*entre nous*' (between-us) (*Convivialist Manifesto*).[2]

Geological time and civilisation are directly correlated. Indeed, it is the entry into the period of climatic stability that is the Holocene which facilitated the control of ecosystems and the development of agriculture, whose management of agricultural surpluses was the key to the emergence of the great civilisations. It is possible that the entry into the Anthropocene will sound the death knell for our civilisation and that we will now need to think of other foundations on which to base our societies. This is what these different manifestos are about. What they have in common is that they are critical: they denounce injustices and decry the lack of rationality in the functioning of our societies. They are also concrete calls to resist. 'We must fight, say no, oppose', say the various signatories. The final characteristic is their utopianism. Another world is possible… and these manifestos, each in their own way, call for the advent of a new world not dependent on capital (except for the *Ecomodernist*

[2]We could also add gender equality (*FEMEN Manifesto*), the life drive and capitulation to desire (*Hedonist Manifesto*), etc.

Manifesto).[3] Resistance, critique and utopia: these are the foundations of the Frankfurt School's Critical Theory, which combines analysis, normative proposals and projections. These different manifestos, which sometimes contain errors or illusions that are more distracting than mobilising, feed into the progressive development of a critical theory for the Anthropocene.

This chapter has several objectives. Firstly, it offers an in-depth examination of the polarisation between Promethean and post-Promethean conceptions of politics in the Anthropocene. Here, a critical reading of the various manifestos published in recent years will be an interesting means of identifying the modes of mobilisation of the Anthropocene (three manifestos have already been briefly presented in this work: the *Convivialist Manifesto*, the *Ecomodernist Manifesto* and the *Accelerationist Manifesto*). Secondly, this chapter identifies the rise of the concept of the Anthropocene in politics and the importance of the paradigmatic ruptures that have been opened up. Thirdly, this chapter allows us to gain a deeper understanding of the idea of convivialism, in a comparison with other texts proposing a similar intellectual act. Fourthly, this last chapter of Part I allows for a transition to Part II, on the idea of an anthropological shift. Indeed, several of the manifestos agree on the need for changes in lifestyles that would profoundly alter the human adventure.

6.1 A Productivist and Growth-Oriented Alternative to Neoliberalism: The *Manifesto of the Appalled Economists* (2010) and the *New Manifesto of the Appalled Economists* (2015)

6.1.1 The Lack of Ecological Thinking in the 2010 Manifesto

In 2010, a few years after the start of the subprime mortgage crisis and against the backdrop of continuing liberal European policies, four French economists, Philippe Askenazy, Thomas Coutrot, André Orléan and Henri Sterdyniak published the *Manifesto of the Appalled Economists*. This quickly became a great success, attracting signatures from a group of other economists and intellectuals. In 2015, the collective '*Les économistes atterrés*' (The Appalled Economists), with more than 10,000 signatories, published the *New Manifesto of Appalled Economists*.

[3] Is it not possible to perceive, through these different texts, the sign of a return of the status of citizens as opposed to mere consumers, after decades characterised by domination of consumption in the private space over the public space? Indeed, citizens have apparently been lulled to sleep over the last few decades by excessive focus on their personal happiness – a form of apotheosis which can be seen in positive psychology, which is completely depoliticised; the only thing that matters is to be happy in order to be effective in the international competition of our economic market.

What 'appals' these economists is the observation that after the global economic and financial crisis of 2007–2008, European policies have done nothing to curtail the power of finance in the conduct of world affairs, instead continuing to follow a liberal path. These authors therefore believe that a paradigm shift in economic policies is necessary in Europe, and that all of economic thinking must be recast in light of the economic and financial crisis of the early twenty-first century (2010, p. 8). The manifesto denounces ten falsehoods, including: the belief that financial markets are efficient, that they are favourable to economic growth, that they are good judges of the solvency of States, that it is necessary to reduce public spending so that public debt is not passed on to our grandchildren, that the European Union defends the European social model, and that European States are economically united.

These economists are attentive to solidarity and social justice. On the other hand, they do not propose to break with the productivist and growth-oriented paradigm. The authors' proposal lies in strengthening State regulations in economic activities due to the failure of deregulation: 'The neoliberal doctrine, which is based on the now indefensible assumption of the efficiency of financial markets, must be abandoned' (2010, p. 59). The authors' position is clear: markets are inefficient and do not allow for sensible allocation of capital, as the crisis of the early twenty-first century attests. It is a question of countering the liberal madness and its excessive financialisation of the market economy – without, however, thinking beyond growth.

This manifesto consists mainly of denouncing the economic errors behind contemporary public policies. These are important, as can be identified in the following example: 'The rising public debt in Europe or the USA is not the result of Keynesian expansionist policies or of expensive social policies but rather of a policy favouring the privileged classes: the "fiscal expenditure" (tax and contribution cuts) increases the disposable income of those who need it least, who can then increase their investments even further, particularly in Treasury bonds, which are remunerated in interest by the tax levied on all taxpayers. Ultimately, a reverse redistribution mechanism is put in place, with wealth flowing from the working classes to the wealthy classes, *via* the public debt whose counterpart is always private rent' (2010, pp. 34–35). The authors denounce the supposed rationality of individuals apprehended as economic agents: 'In the still dominant neoliberal vision, which describes a world comprising individualistic and far-sighted economic agents in competition with each other, the general liberalisation of markets should mathematically lead to maximum growth and the economic optimum. This vision is certainly a convenient fiction for writing mathematical models, but not a reliable tool for guiding the decisions of the people's elected representatives' (2010, p. 9). Whilst the denunciations of improper economic function based on 'liberal doctrine' are clear, the authors do not identify the anthropological roots of the problem.

The *Manifesto of the Appalled Economists* remains strictly a work of economics, though the issues it deals with are political. The authors are aware of the limitations of their 2010 text. They state, at the end of their introduction: 'We are aware that the current crisis is much more than an economic crisis. It is also a social crisis, against a backdrop of ecological and geopolitical crises, which undoubtedly attests to a historical rupture. Our text is a long way from addressing all these issues' (2010,

p. 10). In the first manifesto, the term '*atterrés*' is *not* a play on words in which our relationship to the Earth would show an alternative path to the illusion of decoupling from the biosphere of economic liberalisation. The ecological component is mentioned only very infrequently in the text.[4] Ecological awareness is extremely low.

6.1.2 The Ambivalent Ecological Thinking of the 2015 Manifesto

The 2015 text is not signed solely by the four authors of the 2010 manifesto, but by 'The Appalled Economists Animation Collective'. The stance in relation to the state of planet Earth is very different. From the front cover and the beginning of the introduction, ecological issues are taken as the entry point to this *new manifesto*. Overall, the tone is more aggressive than in the 2010 version, and we find explicit accusations that were not there before, such as 'the government has thrown itself headlong into a futile supply-side policy relying on exemptions from social security contributions, while companies are primarily faced with the sluggishness of their order books and the demands of capital profitability. This is the result of a collusion, and even a confusion, between the high-level State apparatus and the financial bourgeoisie' (2015, p. 11). The political dimension is reinforced, as the authors mention in the introduction, with the statement of some of the convictions on which this manifesto is based: the fact that democracy takes precedence over the market and that democracy is inseparable from equality, the importance of citizen initiatives for the shared management of the commons or the need to respect the ecological boundaries of our societies.

The ecological component is a thread running through the *New Manifesto*, with the promotion of a far-reaching programme of ecological and energy transition which recurs in each of the chapters. On the other hand, the Earth is viewed solely as a frontier (which must be respected) to the development of contemporary societies: 'If ecology is our new frontier, it is important to draw its contours: to reduce the consumption of certain materials, to ban certain substances or certain energy-consuming or polluting production processes, or on the contrary to impose new processes' (2015, p. 22). On the other hand, the dominant paradigm of the book is still founded on productivism, with a chapter devoted to the reinvention of industrial policy. For example, we read: 'In the future, growth will be less due to demographic constraints and slowing productivity gains. Fortunately, France, which has retained a certain demographic dynamism, will continue to experience a slight increase in its active population. Productivity gains, for their part, are difficult to predict. In many service sectors, they are small or difficult to measure. In other sectors, they are

[4] On page 10, at the end of the introduction, the authors state that the financial crisis is also rooted in an ecological crisis; on page 20, they mention 'preferential rates for socially and environmentally priority activities'; on page 31, they mention the need to invest in ecological conversion.

currently being slowed down by sluggish growth, itself caused by austerity policies. Yet growth promotes productivity gains' (2015, p. 63). A perspective that only considers the gains of one's own country or continent at the expense of others is highly problematic in the context of the systemic globalisation of the Anthropocene.

The *New Manifesto* is bold and clearly displays its political aim of informing citizens so that they can take back control of their future. On the other hand, the thought process it displays does not address the anthropological roots of neoliberalism and attempt to provide a response that is commensurate with the depth of those roots.

6.2 Speed in Politics from the *Manifesto for an Accelerationist Politics* (2013)

6.2.1 Srnicek and Williams' Accelerationist Thesis

The *Manifesto for an Accelerationist Politics*, written by Nick Srnicek and Alex Williams, two then PhD students at the *London School of Economics* and the *University of East London*, published in 2013 on the website *Critical Legal Thinking* (*#ACCELERATE. Manifesto for an Accelerationist Politics*). From the very outset, the manifesto confronts the ecological threat: 'the most critical of the threats which face humanity' (paragraph 01.2). In the space of a few months, these 20 pages produced worldwide ripples.[5] For the French lawyer and philosopher working in Belgium, Laurent de Sutter, who edited a reception book on accelerationism (*Accélération!*, 2016), the proposals of the *Manifesto for an Accelerationist Politics* by Srnicek and Williams ushered in 'a new age of political theory' and their follow-up book, *Inventing the Future: Postcapitalism and a World without Work*, published in 2015, is for him, as for Aaron Bastoni (2015), 'the most important book of the year' (de Sutter, 2016, p. 25) because of the new ways of thinking about capitalism in order to overcome it.[6]

[5] The *Accelerationist Manifesto* provoked a worldwide polemic, polarised between enthusiasm for the new Marxist theories suitable for the postmodern period and the vehement rejection of a text outside the usual canons of the contemporary left, which Srnicek and Williams do not spare. One of the reasons for the success of this manifesto is that it illustrates the extent to which politics relies on judicious management of speed, finding a balance between acceleration and deceleration. One of the benefits of the *Accelerationist Manifesto* in our reflection is that it makes use of the idea of acceleration from a political perspective, very different from Rosa's sociological analysis, and allows us to integrate the concept of acceleration into the debate. In the face of the accelerationist capitalist problem that is largely responsible for the entry into the Anthropocene, Srnicek and Williams propose the solution of acceleration!

[6] The collective book *Accélération!* (2016) consists of the French translation of the article by Nick Srnicek and Alex Williams, and several international texts responding to their *Manifesto*. It opens up this intense debate to the French-speaking world, adding two previously unpublished texts: the introduction by the editor of Sutter's book (2016) and '*Accélérer l'écologie*' (Accelerating

The future is what motivated the authors of the *Accelerationist Manifesto* to speak out. Current policies are incapable of transforming our societies, as they must be transformed in light of the threat of annihilation of the future. What is getting in the way of the future, according to these two authors, is 'the paralysis of the political imaginary' (2013, paragraph 02.3). The *Accelerationist Manifesto* works towards the reinvigoration of this imaginary: a postcapitalist and modern future is possible. What the authors of the *Accelerationist Manifesto* find problematic is that we remain 'within a strictly defined set of capitalist parameters that themselves never waver' (2013, paragraph 02.2). The point is to accelerate: not to go faster and increase our growth within the same framework, but to drag ourselves out of the restrictive framework that is capitalism. Technological change is a tool of choice for bold post-capitalist politics to get us out of this framework.

In the face of the ever-increasing pace of capitalism, these two authors – and all accelerationists around the world – propose acceleration as a solution, in order to overturn the world system. It is a question of bringing about a workless post-capitalist society that allows everyone to live.[7] This *Manifesto* develops neo-Marxist thinking, updating the ideology of progress and considering the environmental problem as the greatest problem of the present time: 'The choice facing us is severe: either a globalised post-capitalism or a slow fragmentation towards primitivism, perpetual crisis and planetary ecological collapse' (2013, paragraph 03.23).[8] The direction, on the other hand, is decidedly productivist: 'Accelerationists want to unleash latent productive forces' (2013, paragraph 03.5). Such productivism, which aims to build on 'the material platform of neoliberalism'[9] (2013, paragraph 03.5),

---

Ecology) by Swiss philosopher Yves Citton. Eight other texts are presented: 'Accélérer la politique' (Accelerating Politics) by Italian philosopher Antonio Negri; 'Accelerating Capitalism' by the British philosopher Nick Land; 'Accelerating Revolution' by the British politician Mark Fisher; 'Accelerating Humanity' by the Iranian philosopher Reza Negarestani; 'Accelerating Reason' by the British philosopher Ray Brassier working at the American University of Beirut in Lebanon; 'Accelerating Automation' by Italian politician Tiziana Terranova; 'Accelerating University' by Austrian philosopher Armen Avanessian; and finally, 'Accelerating Feminism' by the international feminist collective Laboria Cuboniks. The grouping together in one book of texts written for other fields and in other cultural and national contexts is an interesting project, because of the sometimes contradictory conceptions of the meaning of acceleration, of the emerging accelerationist political theory and of the ways in which it is implemented – though readers may be confused by the juxtaposition of these texts, which are sometimes not very well connected to one another.

[7] One of the assumptions the authors make is that 'All of us want to work less' (2013, paragraph 3.2). Is this true? It is debatable. Perhaps it could be said that most people would like to see the constraints of their work reduced. On the other hand, their analysis a little later in their text has the ring of truth: 'What has instead occurred is the progressive elimination of the work-life distinction, with work coming to permeate every aspect of the emerging social factory' (2013, paragraph 3.2).

[8] At no point do the authors make the case that accelerationism is ecological.

[9] The authors believe it is necessary to organise some form of fundraising on a massive scale in order to accelerate the entry into a postcapitalist social era. Indeed, in their view, a 'postcapitalist technosocial platform' (2013, paragraph 3.19) is needed. These funds must come from 'governments, institutions, think tanks, unions or individual benefactors' (2013, paragraph 3.20).

appears to be a vision that is hardly compatible with the human adventure's sustainable survival in the Anthropocene. The *Accelerationist Manifesto* is based on a relationship to nature understood as a resource for technical development and as a space to be conquered – a far cry from the type of critical theory developed in the present book. The breadth of the chasm between these two perspectives is the reason why, here, we offer an in-depth analysis of this manifesto.

The *Accelerationist Manifesto* is vivid and challenging. The authors desire a 'future that is more modern', but make it clear that they hope for an 'alternative modernity' (2013, Paragraph 03.24). This alternative component refers to social spaces that are no longer characterised by their capitalist foundations. The political stance is very assertive. The authors set the scene: 'At the beginning of the second decade of the Twenty-First Century, global civilization faces a new breed of cataclysm. These coming apocalypses ridicule the norms and organisational structures of the politics which were forged in the birth of the nation-state, the rise of capitalism, and a Twentieth Century of unprecedented wars' (2013, paragraph 01.1). The positioning favours hypermodernity characterised by confidence in the linear timescales of technological progress. Technological evolution is a tool of choice for bold postcapitalist politics. The orientation is resolutely Promethean and technicist: 'Given the enslavement of technoscience to capitalist objectives (especially since the late 1970s) we surely do not yet know what a modern technosocial body can do. Who amongst us fully recognizes what untapped potentials await in the technology which has already been developed?' (2013, paragraph 03.6). Then the authors go on to make a real leap of faith: 'Our wager is that the true transformative potentials of much of our technological and scientific research remain unexploited, filled with presently redundant features (or pre-adaptations) that, following a shift beyond the short-sighted capitalist socius, can become decisive' (2013, paragraph 03.6).[10] In the accelerationist view, technological change is a means to a political end and does not supplant political action: 'We want to accelerate the process of technological evolution. But what we are arguing for is not techno-utopianism. Never believe that technology will be sufficient to save us. Necessary, yes, but never sufficient without socio-political action' (2013, paragraph 03.7).

One of the interesting points about this manifesto lies in the strangeness of its theses, which fall outside of both the academic codes and mainstream political thinking. The main thesis is 'that the left, obsessed with degrowth and resistance to capitalism, had forgotten the possibility of overcoming it – that is, of overcoming the game whose rules had been set by it, and which was killing our world' (de

[10] What is surprising, to say the least, is that a few lines later the authors criticise the leap of faith that a post-capitalist system could come about as a result of a revolution: 'The faith placed in the idea that, after a revolution, the people will spontaneously constitute a novel socioeconomic system that isn't simply a return to capitalism is naïve at best, and ignorant at worst' (2013, paragraph 3.8). A little further on, they also criticise the faith that was at the root of the 2008 financial and economic crisis: 'The 2008 financial crisis reveals the risks of blindly accepting mathematical models on faith' (2013, paragraph 3.9). They do not seem to recognise that their own vision of the future is based on a faith-based combination of technical and political vision.

Sutter, 2016, p. 9). The manifesto presents a critique of the left by the left. In contrast to the usual call of the left to resist capitalism and its accelerationist logic, the two young intellectuals instead advocate acceleration to overthrow capitalism. Accelerationism, whose political ideology has its paradigmatic foundations in the work of Karl Marx (1867) and the British philosopher Nick Land (2011), must thus allow the emergence of an 'alternative modernity' that cannot be the result of neoliberalism. The authors of the manifesto seem to choose a form of radicality as opposed to what they consider political bluntness. Thus the new social movements that emerged after 2008 are understood to 'expend considerable energy on internal direct-democratic process and affective self-valorisation over strategic efficacy, and frequently propound a variant of neo-primitivist localism' (paragraph 01.5). Strength is the common thread running through the *Accelerationist Manifesto*, and anything that is not directly effective in the pursuit of power is rejected: 'We do not believe that direct action is sufficient to achieve any of this. The habitual tactics of marching, holding signs, and establishing temporary autonomous zones risk becoming comforting substitutes for effective success' (2013, paragraph 03.12).[11] The authors of the *Accelerationist Manifesto* are not democrats. They are in favour of a strong State: 'the overwhelming privileging of democracy-as-process needs to be left behind' (2013, paragraph 03.13). They continue with this same mindset of power wielded by the few: 'The fetishisation of openness, horizontality, and inclusion of much of today's "radical" left set the stage for ineffectiveness' (2013, paragraph 03.13), and then a few lines later: 'We need to posit a collectively controlled legitimate vertical authority in addition to distributed horizontal forms of sociality, to avoid becoming the slaves of either a tyrannical totalitarian centralism or a capricious emergent order beyond our control' (2013, paragraph 03.14). The authors attempt to rehabilitate authority: 'But this is not to align ourselves with the tired residue of postmodernity, decrying mastery as proto-fascistic or authority as innately illegitimate' (2013, paragraph 03.21).

While the Anthropocene is characterised by the Great Acceleration, the *Accelerationist Manifesto* proposes, in contrast to deceleration, to continue accelerating. The innovative core of Srnicek and Williams' proposal is to move beyond capitalism, which is intrinsically linked to acceleration through its attachment to growth, by going faster than it.[12] Instead, it is about changing the parameters of acceleration and breaking away from 'the increasing speed of a local horizon, a simple brain-dead onrush' (paragraph 02.2) in favour of 'an acceleration which is

[11] This is a far cry from Hannah Arendt's use of process to identify the question of the birth of the actor at the heart of concerted action.

[12] The last few decades have been marked by a reversal of the temporal markers of politics. Traditionally, progressivism has been more accelerationist and conservatism more reticent towards acceleration. However, for the past few decades, conservatism has been associated with movements of socio-economic liberalisation favourable to technological acceleration. Progressivism, on the other hand, is less driven by the ideology of technical progress and has regularly addressed environmental issues, sometimes championing decelerationist degrowth. Accelerationism here restores the original temporal markers of politics.

also navigational, an experimental process of discovery within a universal space of possibility' (Srnicek & Williams, 2016, p. 32). Their understanding of acceleration entails the increase of horizontal discoveries of what is possible. Yves Citton offers two interesting images of acceleration in the accelerationist logic: to accelerate 'means first of all to "change gears": to pedal slower in order to go further' (Citton, 2016, p. 2016). Accelerating ongoing transformations is what will allow the human adventure to 'reach a certain speed of liberation that allows us to get out of the capitalist rut, like rockets (when everything goes well) manage to escape the grip of Earth's gravity' (Citton, 2016, pp. 206–207).

The idealism of founding a postcapitalist leisure society breaks with the political and organisational pragmatism of these two young intellectuals. Thinking is by no means enough – it is important to give organisational form to postcapitalist ideas so that they can bring about the vision they present. Thus, as Laurent de Sutter points out, these two authors attribute the triumph of contemporary neoliberalism much more to the organised diffusion of its ideology within networks than to the change of framework in the post-war period. Hence, the post-capitalist intellectual infrastructure needs to be thought about and better organised than it is now. The strong praxeological component of this neo-Marxism makes it 'pragmatic communism' (de Sutter, 2016, p. 21), one of the innovations of which is the emphasis on the institutionalisation of ideas. One of the strengths of the *Manifesto* is the way in which institutional organisations for the federation of alternative ideas are thought of in order to generate a phenomenon of overturning capitalist thinking by making it a minority.

6.2.2 Post-Capitalism as a Collective Glimmer of Hope

This vigorous left-wing thinking compensates for the reactionary tendency that has been prevalent on the contemporary left in various parts of the world for several decades. In fact, for over 30 years, when the left has not been directly involved in economic and financial liberalisation, it has been content to react to neoliberalism, to try to slow down the unbridled capitalist race, to cling to threatened social gains having lost all hope of real social improvement, or to search for a lost paradise where more authentic humanity exists. These left-wing reactionaries, while they have the important function of situating contemporary political debate in relation to history, have not produced any social, cultural, political or educational innovations. The main contribution of the *Accelerationist Manifesto* and its reception texts – like the contributions of *Accélération!* – is positive, non-reactionary, leftist political thinking, assuming that we can only make it to the future by moving forward. In the view of these authors, the left has forgotten that capitalism is its enemy. It is a matter of ending the left's 'melancholic attachment' to failure and 'its sense of powerlessness in relation to the overwhelming forces of capitalism' (de Sutter, 2016, p. 23). This book, while it may seem to be an extension of Arnsperger's work on the search for post-capitalist means of existence (Arnsperger, 2009, 2010a, b, 2011a, b), or that of Bourg (Bourg, 2010a, b, 2012), is based on completely different anthropological presuppositions and makes different political recommendations.

When, on 18 December 1865, Abraham Lincoln succeeded in passing the abolition of slavery in the United States, he knew he was making a historic political statement for the future. His political objective was clear and simple. On the other hand, the contemporary postmodern period is marked by complexity that makes it difficult to think and prioritise actions. Also, though – and this is even more serious and is the consequence – the present time is characterised by a difficulty to mobilise collectively. The contemporary period is permeated by a form of collective inertia. The quest to emancipate the human adventure from its many forms of alienation is no longer invested in, so to speak, and we have accepted the disenchantment of the contemporary period. One of the merits of Srnicek and Williams is their commitment to a struggle for 're-enchantment' through the definition of a collective goal which can serve to motivate everybody. One of the main contributions of this manifesto lies in the way its authors reactivate the virtue of hope.

6.2.3 A Resolutely Promethean Politics

The individual in the *Accelerationist Manifesto* refers to the fulfilment of the modern promise of the Promethean individual, which is always partly linked to *homo oeconomicus*. It seems that there is a real contraction here. Can the Promethean component not be intrinsically linked to the economic hegemony also characterised by maximum control (of one's own interests)?

These two authors decry the fact that we have abandoned the mobilising ideology of progress – particularly, scientific and technical progress. Srnicek and Williams pursue the Promethean logic to its natural conclusion; they 'declare that only a Promethean politics of maximal mastery over society and its environment is capable of either dealing with global problems or achieving victory over capital' (2013, paragraph 03.21). The driving idea behind the *Accelerationist Manifesto* is that capitalism does not seek to facilitate scientific progress, but instead to hinder it in order to prevent the advent of a society of leisure, replacing labour – which could be achieved by machines. This neo-Marxist stance is opposite to that of the degrowth theories, favouring an acceleration of the use of technological means. Accelerationist Prometheism is persuasive. Thus, it is appropriate to '[recover] the dreams which transfixed many from the middle of the Nineteenth Century until the dawn of the neoliberal era, of the quest of Homo Sapiens towards expansion beyond the limitations of the earth and our immediate bodily forms' (2013, paragraph 03.22). The postcapitalism to which the authors of the *Accelerationist Manifesto* aspire is based on a logic of conquest as strong as colonial capitalism: '.

After all, it is only a post-capitalist society, made possible by an accelerationist politics, which will ever be capable of delivering on the promissory note of the mid-Twentieth Century's space programmes, to shift beyond a world of minimal technical upgrades towards all-encompassing change. Towards a time of collective self-mastery, and the properly alien future that entails and enables. Towards a completion of the Enlightenment project of self-criticism and self-mastery, rather than its elimination'. (2013, paragraph 03.22). This particular point is a mistake: the techno-industrial development that came with the Enlightenment is precisely what has brought us squarely into the Anthropocene. On the contrary, we believe it is necessary to envisage a radical departure from these types of logic.[13]

6.2.4 *Categorical (Possibly Violent) and Insufficiently Radical Thinking*

Srnicek and Williams's goal is 'to generate a new left global hegemony' (paragraph 01.6), which 'entails a recovery of lost possible futures, and indeed the recovery of the future as such' (paragraph 01.6). While this search for hopeful and compelling collective goals is absolutely essential, it is important to be very vigilant when the ideas at work are categorical. The manifesto – and indeed some of the reception articles in response to it – is driven by particularly categorical convictions. The authors advocate 'an accelerationist politics at ease with a modernity of abstraction, complexity, globality and technology' (2013, paragraph 03.1). The 'unabashed' component can be accompanied by sterile violence. The strength of the political convictions of these texts in response to contemporary uncertainty and dissatisfaction should in no way lead us to the development of political thinking which risks negating others – be it capitalist, soft, lukewarm or liberal thinking. We must be particularly vigilant with the political implementation of Marxist thoughts based on categorical simplism and voluntarist accelerationism. The approach is voluntarist and collectivist; it calls for the establishment of a strong authority to make decisions on behalf of the whole. Force is the means of implementation. Though this text is interesting because it proposes the development of a way of thinking that is 'outside the box', 'off the beaten track' and outside the usual canons of the left, we must ask ourselves: fundamentally, does it propose anything other than communist force? Our two British authors are well aware of the limitations of voluntarist approaches since, while they mention the need to reinforce vertical authority, they articulate it with 'distributed horizontal forms of sociality', specifying that 'the command of the Plan must be married to the improvised order of The Network' (paragraph 03.14). Thus, they call for an ecology of autonomous organisations that are interconnected – but structured around a strong power.

[13] It should be noted that, for Rosa, the *Accelerationist Manifesto* makes a good point about the ridiculousness of the idea of slowing down, pointing out that slow movements are always rare (such as 'slow food') and that slowness is not desirable: 'A slow arrival of an emergency doctor or the fire brigade is not a good thing. Likewise, a slow internet connection is absolutely terrible' (Rosa & Wallenhorst, 2017, p. 21). Acceleration only becomes a problem for Rosa when it mutates into alienation: 'from the moment we can no longer appropriate things, when we can no longer experience resonance with the world' (Rosa & Wallenhorst, 2017, p. 21).

Accelerationist thinking is categorical. On the other hand, despite a strong affirmation that characterises the *Accelerationist Manifesto*, there is nothing radical about it, in the etymological sense of the word, *radix* (root): accelerationist thinking does not alter some of the currently problematic anthropological roots, such as the Promethean *hybris* of *homo oeconomicus*. On the contrary: the *Accelerationist Manifesto* consolidates the *hybris* of transgression that demultiplies the enjoyment of power and might. The anthropological roots of the contemporary period remain unchanged.

6.3 The Ecological Prometheism of the *Ecomodernist Manifesto* (2015)

As discussed in Chap. 5, the *Ecomodernist Manifesto* was published in 2015 by nineteen intellectuals,[14] twelve of them affiliated with the thinktank, the *Breakthrough Institute*. The authors claim to 'affirm one long-standing environmental ideal, that humanity must shrink its impacts on the environment to make more room for nature' while explicitly rejecting another ideal: 'that human societies must harmonize with nature to avoid economic and ecological collapse' (2015, introduction). While there is a need to reduce environmental impact, we do not need our societies to change lifestyles for greater harmonisation with nature to do so. The authors are aware of the anthropogenic alteration of the Earth; for example, they state in the first part that 'human flourishing has taken a serious toll on natural, nonhuman environments and wildlife'. In speaking of a 'good Anthropocene', the *Ecomodernist Manifesto* (2015) understands the Anthropocene as the anthropisation of the Earth, without recognising the far more consequential aspect and far greater threat to human sustainability that is the disruption of biogeochemical cycles, with a systemic impact on Earth. The question of threshold effects and tipping points is not mentioned by these authors. Ecomodernists believe that we have the capacity to cross a set of previously accepted boundaries, implicitly assuming that any boundary crossing is desirable. In some ways, the *Breakthrough Institute* goes beyond politics and the quest for international climate agreements. The logic is resolutely Promethean. This is a particularly surprising point insofar as Erle C. Ellis – a member of the official Anthropocene working group and author of famous articles on biomes and anthromes – is a signatory of the *Ecomodernist Manifesto*.[15]

[14] Most of them are American but some come from India, Canada or Australia.

[15] This commitment appears difficult to understand within the European intellectual context and perhaps makes more sense in an American context where the aim is to maintain dynamism that fosters hope for the future and encourages entrepreneurship. This is the impression we got with François Prouteau and Renaud Hétier during the interview with Erle C. Ellis.

6.3.1 Growth-Oriented Ecology

The main environmental problem of the organisation of human societies identified by the authors of the *Ecomodernist Manifesto* is that human populations over-rely on ecosystems.[16] To remedy this, only one option is considered: to develop our technical tools in order to optimise our ecological footprint: 'To embrace these new technologies is to find paths to a good Anthropocene' (2015, part 2). The authors reject the claim that there are limits to human growth: 'there is still remarkably little evidence that human population and economic expansion will outstrip the capacity to grow food or procure critical material resources in the foreseeable future' (2015, Part 1). Thus, they remain firmly committed to the economic paradigm of growth: 'To the degree to which there are fixed physical limits to human consumption, they are so theoretical as to be functionally irrelevant' (2015, Part 1). Economic growth continues to be the goal of human societies, 'to power a growing human economy' (2015, part 4). Here, it is noteworthy that the authors make a mistake in the sixth and final part of the *Ecomodernist Manifesto* when they mention a link between economic productivity and the ability to meet human needs: 'More-productive economies are wealthier economies, capable of better meeting human needs while committing more of their economic surplus to non-economic amenities, including better human health, greater human freedom and opportunity, arts, culture, and the conservation of nature' (2015, part 6.). This correlation of technical progress with human and social progress enabled by the economic growth of industrial modernity was demonstrable once, but the mechanism is now broken, since the entry into post-modernity. The enrichment of the richest does not allow the human needs of the most vulnerable to be met.

The regularly mentioned risk of crossing a threshold in the functioning of the biosphere as a whole[17] that would jeopardise global food security is dismissed out of hand: 'With proper management, humans are at no risk of lacking sufficient agricultural land for food' (2015, part 1). On the other hand, the authors do not explain what proper management would look like, and do not demonstrate why it is that we do not have to worry about this risk. The authors mention that 'Current and future generations could survive and prosper materially on a planet with much less biodiversity and wild nature'[18] (2015, part 5). This claim has yet to be verified.

[16] The authors of this manifesto write, for example, 'Ecosystems around the world are threatened today because people over-rely on them: people who depend on firewood and charcoal for fuel cut down and degrade forests; people who eat bush meat for food hunt mammal species to local extirpation. Whether it's a local indigenous community or a foreign corporation that benefits, it is the continued dependence of humans on natural environments that is the problem for the conservation of nature' (2015, part two). Though the dynamic described is in part the one we have seen from the development of agriculture and sedentarisation, it should be noted, on the other hand, that there are indigenous populations who are careful to give back to nature what it gives them, as David Abram emphasises in his book *The Spell of the Sensuous* (1997).

[17] Particularly since the 2012 paper by Barnosky et al.

[18] On this point, the authors make it clear that they do not place blind trust in market capitalism to achieve this decoupling of human societies from nature that they call for. This transition must be led 'by human societies' (2015, part six).

Indeed, not only do Earth System Sciences teach us that climate and biodiversity are intimately linked to each other, but also that the biosphere functions systemically and that a significant alteration of one of its components could lead to sudden and irreversible structural changes.

In outlining their thinking, the authors rely on the identification of current trends to avoid an environmental issue (in the name of the blind faith that this issue is dwindling into nothingness). Human population growth and food needs is an issue that is brushed aside because 'Given current trends, it is very possible that the size of the human population will peak this century and then start to decline' (2015, Part 2).[19] The same is true of meat consumption, which 'has peaked in many wealthy nations and has shifted away from beef toward other protein sources that are less land intensive' (2015, part 2). First of all, this argument is limited, because beef consumption is increasing in Asia – the most populous continent in the world, with 60% of the world's population. Above all, the authors fail to see that this decrease in meat consumption in some wealthy countries is attributable to a lifestyle change due to awareness of the environmental impact of consumption practices, based on the principle of harmonisation with nature – a principle which the authors say they reject.

6.3.2 *Faith in Technology as the Dominant Rationality*

A reading of the *Ecomodernist Manifesto* gives the impression that the authors are on the horns of a dilemma from which they cannot escape. Indeed, they propose to rely on technology to get out of the rut in which the development of technological mastery has put us: 'The modernization processes that have increasingly liberated humanity from nature are, of course, double-edged, since they have also degraded the natural environment' (2015, part 3). This is a contradiction to which the authors do not provide a satisfactory answer. Thus, the eco-modernists are aware of the consequences of environmental degradation on human health. They mention in particular that air and water pollution lead to premature deaths. On the other hand, the watchword seems to be 'don't worry', human genius will get us out of this, the power which humanity has accrued will allow us to improve the world and the human condition: 'we write with the conviction that knowledge and technology, applied with wisdom, might allow for a good, or even great, Anthropocene' (2015, introduction). It all appears to come down to the question of the technological challenge, against the backdrop of the power of human ingenuity: 'Meaningful climate mitigation is fundamentally a technological challenge. By this we mean that even dramatic limits to per capita global consumption would be insufficient to achieve significant climate mitigation' (2015, part 4). The issue of changing lifestyles is repeatedly dismissed. The authors are resolutely technophile. As an example,

[19] Moreover, since the human population is mostly urban, the environmental impact is less than when the population is evenly distributed around the globe.

nuclear power is presented as a real fulcrum for energy production despite the threat it poses. The same is true of industrial agriculture, which is lauded for its efficiency compared to other forms of agriculture: 'Suburbanization, low-yield farming, and many forms of renewable energy production, in contrast, generally require more land and resources and leave less room for nature' (2015, part 3).

In the *Ecomodernist Manifesto*, we quite clearly perceive a refusal to acknowledge the aesthetic and transcendental component of nature. The authors denounce the 'explicit aesthetic or spiritual reasons' sometimes put forward for inefficient and insufficiently rational environmental preservation (2015, part 3), and give primacy to technical and calculating rationality.[20] On the other hand, their entire argument is based on faith in the power of technology and human genius, of which they say nothing. This is, for example, perceptible in the following quote: 'Looking forward, modern energy may allow the capture of carbon from the atmosphere to reduce the accumulated carbon that drives global warming' (2015, part 4). Another example is: 'Significant climate mitigation, therefore, will require that humans rapidly accelerate existing processes of decarbonization' (2015, part four).[21] This leap of faith is summarised at the end of the manifesto in the following formulation: 'we embrace an optimistic view toward human capacities and the future' (2015, part 7). Is this leap of faith not excessive, having seen how technical development is linked to a destructive *hybris* that has contributed to the systemic alteration of the Earth and the entry into the Anthropocene?

6.4 The Uncontrollable Vitality of the *Manifest für das Anthropozän* (2015)

Andreas Weber and Hildegard Kurt published a *Manifesto for the Anthropocene*[22] which is quite clearly in the tradition of the Frankfurt School, not only because of the question that underlies this work about what a good life can be and how it might be possible to live with reference to this aim, but also because of the references regularly used (notably Max Horkheimer and Theodor W. Adorno). Of the various manifestos studied in this chapter, this is the only one that explicitly draws on the major articles about the Anthropocene mentioned in Chap. 3 of this book.

The manifesto begins with the idea of the end of the separation between humanity and nature. It starts with the premise that humans have had the illusion of total domination over the living matter of which we are composed. In our way of life today, we have transformed everything into culture and developed a position of

[20] Meanwhile, they make it clear that nature – which they claim to have a deep love for and emotional connection to – is essential to 'psychological and spiritual well-being' (2015, part 5).

[21] The manifesto then mentions that the means by which such a goal could be achieved are as yet unclear.

[22] The full title is *Lebendigkeit sei! Für eine Politik des Lebens. Ein Manifest für das Anthropozän.*

superiority over nature: 'humans have conquered life in a hostile takeover'[23] (Weber and Kurt, 2015, p. 9). Instead, the authors understand human beings as one more part of a system, completely immersed in nature and inseparable from it. Weber and Kurt denounce a worldview based on efficiency-driven mechanisms. Instead, their worldview is 'a process of creative and interpenetrating relationships for experiencing and expressing'[24] (p. 11). This view is not Utopian; indeed, it is rooted in the current revolution in biological thinking 'comparable in scope to the revolutions in physics triggered by the theory of relativity and quantum mechanics'[25] (p. 11). Indeed: 'Man and nature are one, for creative imagination and the expression of feelings are natural forces'[26] (p. 11).

Above all, it is vitality that is celebrated in this manifesto: 'What living nature brings is not romantic salvation, but the fact that nothing in it is subject to the control of an individual with a set goal; on the contrary, living nature is subject to a creative process that obeys only the urge of life'[27] (p. 12). In this quote, we perceive the political component of Weber and Kurt's conception of life. Life contains an uncontrollable and subversive power and creativity that no amount of control will be able to contain and direct. Moreover, it is the vitality of others (human and non-human) and of our environment that makes our own existence possible. In this manifesto, we are dealing with a text that celebrates humanity not in its exceptionality from the rest of the living world, but in its solidarity with the whole living fabric. Basically, life is beyond our control, these two authors tell us… and that is as it should be. Indeed, it is our immersion in the solidarity of the living fabric that will enable us to weather the storm of the Anthropocene. This power of the vitality and solidarity of living beings can be a source of hope to us, and it is important that we do not position ourselves outside this fabric. In a way, this is the main message that Weber and Kurt take from the entry into the Anthropocene. The *Manifesto for the Anthropocene* outlines a conception of life as a practice of creativity that is opposed to technicist conceptions, in which life is a technique of existence to be understood in order to orient it. It is possible to identify two possible orientations of the impact of the entry into the Anthropocene: indissoluble solidarity with the living world, from the viewpoint of human non-exceptionality, or, on the contrary, the power of human techniques that can rend humanity free from the constraints of nature. The authors of the *Manifesto for the Anthropocene* apprehend every living thing within the same movement, without the opposition between nature and culture.

---

[23] 'Vielmehr hat das Humane in einer feindlichen Übernahme das Leben erobert' (Weber and Kurt, 2015, p. 9).

[24] 'Ein Prozess schöpferischer Beziehungen und Durchdringungen auf dem Weg zu Erfahrung und Ausdruck' (Weber and Kurt, 2015, p. 11).

[25] 'deren Tragweite vergleichbar mit den in der Physik durch Relativitätstheorie und Quantenmechanik ausgelösten Revolutionen ist' (Weber and Kurt, 2015, p. 11).

[26] '‘Mensch und Natur sind eins, weil schöpferische Imagination und fühlender Ausdruck Naturkräfte sind' (Weber and Kurt, 2015, p. 11).

[27] 'Das Ware einer lebendigen Natur ist nicht das romantisch Heilsame, sondern dass nichts in ihr der Kontrolle eines Zwecke setzenden Subjekts unterliegt, vielmehr alles schöpferischer Prozess ist, der einzig dem Drang folgt, dass Leben sei' (Weber and Kurt, 2015, p. 12).

This manifesto demonstrates the way in which it is possible to think beyond this philosophical heritage. The manifesto highlights the need to rethink human beings in their articulation with the adventure of life – thus starting from their biological foundations. This is absolutely necessary: 'Without a thorough revision of representations of human beings – and thus representations of life – industrial modernity will reinforce anthropocentrism in devastating ways in the Anthropocene period'[28] (Weber & Kurt, 2015, p. 14).

It is in relation to the Enlightenment that the *Manifesto for the Anthropocene* takes a stance. What we need to do, say Weber and Kurt, is move from the '*Enlightenment*' characteristic of the past, in which the Industrial Revolution was rooted, to '*Enlivenment*' for the Anthropocene. It is a question of supplementing the Enlightenment by accompanying contemporary societies towards reintegration into the living world. The intellectual proposal is that of a transition from enlightenment to enlivenment. The term *Enlivenment* is proposed by Andreas Weber, who wrote a book on the subject in 2016: *Enlivenment. A culture of life. Towards a Poetics for the Anthropocene* (2016).[29] In each of his books written in German, the term *Enlivenment* is left in English (undoubtedly to emphasise the proposed parallel with the term *Enlightenment* and to spotlight the importance of the paradigm shift for thinking and for the organisation of human societies). It refers to a form of intrinsic dynamic of the living world that should be allowed to express itself and spread, and take the human adventure with it.

6.5 *Peer-to-Peer* Digital Technology as a Medium for Developing a New Relational Style (*The Commons Manifesto*, 2018)

The Commons Manifesto was written in 2018[30] by three researchers: Belgian computer scientist Michel Bauwens (founder of the P2P Foundation), Greek economist Vasilis Kostakis[31] and Greek PhD student Alex Pazaitis. The particularly concrete

[28] 'Ohne eine gründliche Revision des Menschenbildes – und damit des Lebensbildes – der Industriemoderne wird der Anthropocentrismus im Zeitalter des Anthropozän weitere verheerende Steigerungen erfahren' (Weber and Kurt, 2015, p. 14).

[29] *Enlivenment. Eine Kultur des Lebens. Versuch einer Poetik für das Anthropozän.*

[30] The full title is *The Commons Manifesto – Peer to Peer*. We received a PDF version from the 'Club convivialiste' mailing list. The website www.p2plab.gr states that the book is forthcoming (Westminster University Press, London). In the following presentation of this manifesto, and the pagination of the references, we rely on the document received in PDF (it is likely that the printed version assumes a different pagination of the references). This manifesto includes parts already published in the *Manifesto for a true participatory economy – Towards a society of the commons*, published in French in 2017 by Editions Charles Leopold Mayer. We can assume that Alex Pazaitis participated in the writing of Chap. 4: P2P and the Structure of World History, which does not appear in the 2017 version.

[31] Vasilis Kostakis is a professor of *P2P Governance* in Estonia and also teaches at Harvard University; he founded the P2P Lab conducting research on 'Peer-to-Peer' connection.

starting point of this manifesto is interesting: *Peer to Peer* (P2P) technology. P2P appears to be a dynamic for devising and creating new societal foundations. Contrary to the other manifestos studied, it is not primarily a problem – a threat – that has prompted the drafting of this manifesto, but on the contrary, a fundamental movement in the process of taking place, which has the potential to help bring about a societal transformation. P2P is viewed as the bearer of the promise of building a common world. First and foremost, P2P is a technological infrastructure: it is a network where anyone and everyone can be connected with everyone else without going through a central server. Thus, interests are shared between the members of the network. Because it allows the creation of horizontal relationships between individuals where interests are not dictated by the owners of capital, it can facilitate the entry into a post-capitalist world centred on the commons. Four characteristics of P2P have the potential to profoundly change human societies:

1. P2P is a type of *social relations* in human networks, where participants have maximum freedom to connect.
2. P2P is also a *technological infrastructure* that makes the generalization and scaling up of such relations possible.
3. P2P thus enables a new *mode of production and property*.
4. P2P creates the potential for a *transition* to an economy that can be generative towards people and nature.

(Bauwens et al., 2018, p. 6).

6.5.1 An Understanding of Technology as a Relational Style

The tone of the *Commons Manifesto* is resolutely political, based on a proposal for economic organisation. On the other hand, it is only very weakly environmental. One can assume that a development of the commons will have positive environmental impacts, but this is not the authors' entry point or concern.[32] However, this manifesto deserves to be studied in depth as part of a critical theory for the Anthropocene, because the reorganisation of social spaces that it proposes is particularly appropriate given the entry into the Anthropocene (moreover, the relationship with digital technology is particularly interesting).

The authors refuse to accept any technological determinism. It is all in the use: the same tool can allow the emergence of opposite socio-political spaces. P2P technology can be used in different ways, as is the case with Facebook (and, in the same vein, Uber or Bitcoin) or Wikipedia (but also with Enspiral,[33] Farm Hack,[34]

[32] Only in rare instances do they mention the issue of environmental goals, such as on page 20 with an example on permaculture.

[33] Enspiral is a network of professionals focused on social projects.

[34] Farm Hack is a community of farmers who build and modify their own machines.

Sensorica,[35] WikiHouse[36] or open-source software). The former embraces the forces of capitalism and uses the power of networking and cooperation to increase its capital – from the sale of digital traces of human intimacy – while the latter is a stone in the construction of an alternative world.

The technophilia of the authors of the *Commons Manifesto* is not blind faith in technology[37]: at no point does the book suggest that technology will save us. The authors simply mention that through these weakly regulated networks where there is little hierarchical constraint, cooperation is possible and can help to alter the capitalist hegemony. In this manifesto, 'P2P is thus primarily a mode of relationship that allows human beings to be connected and organized in networks, to collaborate, produce and share' (Bauwens et al., 2018, p. 7). What seems particularly important here is the possibility to cooperate without asking for permission. Here, digital technology is at the heart of a social and political reorganisation.

The authors of the *Commons Manifesto* see the development of the Internet as a historic emancipatory opportunity: '[like] the invention of the printing press before it, the Internet has created a historical opportunity for reconfiguring production, exchange, and the organization of society at large. The core emancipatory feature of the Internet lies in its capacity to massively scale up many-to-many communication, and therefore, in its capacity to lower the cost of self-organization and create and distribute value in radically new ways' (Bauwens et al., 2018, p. 33). This manifesto allows us to look at processes other than those in place with Facebook, Uber, AirBnB, Amazon, etc. where we see instead that the internet is perpetuating inequalities and the accumulation of non-redistributed wealth. When Bauwens, Kostakis and Pazaitis refer to P2P, it is to signify a type of social relationship including particular modalities of distribution of goods, but also to evoke a type of socio-technological infrastructure allowing a particular mode of production. This combination of social relations and infrastructure, but also of production and distribution of goods, should allow the advent of a post-capitalist model, which is the true purpose of P2P (2018, p. 12). As such, the development of 'peer-to-peer relationships', based on cooperation and collaboration, autonomy, creativity and the horizontality of the network, is an aim of the *Commons Manifesto*. This 'relational style', which is intended to have a profound impact on the future of the human adventure, is the real object of this manifesto.

[35] Sensorica is a collaborative network producing sensors and sensing systems. Sensorica is one of the structures studied by the authors of the *Commons Manifesto*. Sensorica has set up a system for valuing the work of each of the members of the collaborative network, with revenues redistributed proportionally to each member's contribution. In addition, an external partner owns the network's infrastructure and shared resources and is linked to the collaborative network by a 'non-dominium' agreement – i.e. non-control over the shared resources. Thus, the value produced can be distributed according to rules deemed fair by the members, without interference from the owner of the capital.

[36] WikiHouse is an open-source house building kit that allows you to build a house without any special building skills (using 3D printers or digitally controlled machines).

[37] At no point does the *Commons Manifesto* claim that technology will save humanity, as do, for example, the *Ecomodernist Manifesto* (2015) and, to a lesser extent, the *Accelerationist Manifesto* (2013).

6.5.2 *The Commons: A Type of Production That Differs from Capitalism*

The authors of the *Commons Manifesto* refer to the definition of the commons given by the American David Bollier, who works with Michel Bauwens. A commons is 'a shared resource, co-governed by its user community according to the rules and norms of that community' (Bauwens et al., 2018, p. 7). The commons refer either to so-called 'rivalrous goods and resources' that cannot be used simultaneously by two individuals or so-called 'non-rival goods' when they are inexhaustible. What are described as 'gifts of nature', such as the water and land, are thus part of these commons (Bauwens et al., 2018, p. 8).

The authors criticise what they call 'the extractive model of cognitive capitalism' (p. 36) which commodifies a set of information, in the form of data, collected on networks. This is a new form (if not a new stage) of capitalism where what matters more than material production to maximise profits is the control of information and networks. This type of capitalism uses P2P digital platforms: individuals interact with each other in a relatively free or autonomous way, but everything is under surveillance and everything happens within a platform that has an owner, whose goal is to create value (for themselves) from these exchanges. It all looks like a form of social cooperation (so individuals interacting on Facebook are not paid for it) when ultimately it is directly exploited. This is what enables the creation of wealth. This is a form of capitalism described as 'netarchical', where control has moved from the production tool to the entire network. It has therefore expanded considerably by taming the commons.[38]

With commons, we are dealing with the realisation of a production, but this production differs from capitalist production on the three levels that characterise it. To private property and the control of the means of production, commons production opposes collective property and the management of resources; to the control of work through the establishment of a hierarchy, it opposes horizontal coordination; to value creation through market surplus value, it opposes the production of social value. Thus, the authors propose to reconsider the question of value and of value production: 'Not all exchange of value is capitalist exchange-value' (p. 17). They distinguish between exchange value and use value, with a social utility. In the functioning of the commons, peers create value and resources that can be directly reinvested and utilised by peers to create new values and resources. This is the same mechanism as capitalism, but open to peers (Bauwens et al., 2018, p. 10). One of the important merits of the *Commons Manifesto* is that it does not only think about the question of distributing wealth and organising satisfactory social justice, but the creation of value of a different nature than capitalist value – without rejecting wealth creation entirely.

[38] With Uber or Airbnb, for example, we do not see the creation of commons; instead, we are dealing with a social space where everyone has to fight for their own survival, without any real sharing.

In the *Commons Manifesto*, it is capitalist investment in P2P that enables societal transformation. Indeed: 'Capital flows towards P2P projects, and even though it distorts P2P to use it to prolong the dominance of the old economic models, it simultaneously creates new ways of thinking in society that undermine that dominance' (Bauwens et al., 2018, p. 11). This being the case, once the capital has been invested, it is a matter of the *commoners* (which here means the peer participants in the commons) being skilful in gradually extracting the commons from the power of the capitalist economic model. The aim then is to arrive at a position 'where the balance of power is reversed: the commons and its social forces become the dominant modality in society, which allows them to force the state and market modalities to adapt to its requirements' (Bauwens et al., 2018, p. 11). It is a matter of creating a society centred on the commons, marked by 'a new post-capitalist value regime' (Bauwens et al., 2018, p. 12) and guiding the shift from microeconomic communities to the reorganisation of macroeconomic foundations.

However, there is a threat to this aim: the capitalist co-option of the commons, to serve the development of capital. The causality between the generalisation of a technical tool and this societal outcome is not obvious. All possibilities are open, and the P2P socio-technological infrastructure can also reinforce the power of capital.

6.5.3 *Commons-Based Peer Production* (CBPP)

P2P is part of the *commons-based peer production* (CBPP). CBPP is a new means of value creation and distribution, because there is no rivalry in relation to these commons (examples include Wikipedia, Linux, Mozilla Firefox and WordPress). These peer-produced commons have new ecosystems made up of three interlinked institutions: the productive community, coalitions of entrepreneurs focused on the commons[39] and the for-profit association.[40]

The *Commons Manifesto* is resolutely pro-entrepreneurship.[41] On the other hand, it is based on a break between the quest for profit and entrepreneurship. The entrepreneurial function is understood as the property of a collective rather than a form of innate individual characteristic.[42] One of the aims of entrepreneurship is '*the empowerment of commoners*' (Bauwens et al., 2018, p. 19). This means the ability for the commoners (the contributors to the commons or members of it) to take

[39] It is not only volunteers who work in the peer-generated commons. It is, of course, possible to have a paying role. Wikipedia, for example, funds developers to keep the encyclopaedia alive, but when people are paid, it is in contrast to individuals who produce the commons content on a voluntary basis.

[40] For-profit associations are separated from the commons and the productive community. They have no control over the production process.

[41] This manifesto is written by people who have set up structures of some sort – laboratories or international foundations.

[42] Entrepreneurship is therefore not seen as exclusive to capitalists.

power over their own existence, like political power in the public space. '*The empowerment of commoners*' is perhaps the most politically significant expression. The authors propose a semantic evolution by singling out not the entrepreneurs, but the entredonneurs.

6.5.4 Towards a New Type of Civilisation?

Bauwens, Kostakis and Pazaitis use a meaningful metaphor to illustrate P2P: 'institutional seedlings prefiguring a new social model' (2018, p. 12). In this imagery of a slow-growing plant, we see that the authors do not anticipate a revolution that occurs by force, but a gradual transition based on the coming together of a large number of individuals. It is a form of fundamental movement without concerted organisation and deliberate support. This transition would take place in the manner of an 'anarchist revolution' where the accumulation of alternative devices leads to the crossing of a threshold. The means of socio-political change is more that of the diffusion of autonomy than the force of a revolutionary takeover. In the P2P process, autonomy is a fundamental value. It seems to be different from the concept of freedom associated with liberalism, and refers to the way in which groups or organisations, when left to their own devices without hierarchical constraints, organise themselves around a good management of and responsibility for the commons. The commons must become a structuring point of reference in all social forms, including that of the State (p. 12).

P2P relationships are not a novelty made possible by a technological advance. On the contrary: they have existed and were the norm before *homo sapiens* became sedentary. The authors identify four modes of exchange between individuals. The first is that of gift/reciprocal gift within the same community; the second is that which takes place within a State marked by rule and protection; the third is that of the exchange of goods within the market; and the fourth is that of the association specific to the commons, which goes beyond the power of the State and the individualised interests of the market. Are we on the threshold of a new type of civilisation, based on associationist rather than capitalist exchange? The characteristic of associationist societies is that they are organised around communities that have the dual characteristic of being integrated into a larger, more complex whole and of preserving individual freedoms. We are not talking here about closed communitarianism or a system comparable to nomadism. With the extension of the commons to associationism, the authors of the *Commons Manifesto* give their subject a directly political scope: 'one of the central goals of the P2P theory is to investigate the transition from social forms based on the domination of the market forces (capitalism) to social forms based on P2P network dynamics' (Bauwens et al., 2018, p. 44). Thus, the aim is to allow associationist logics to become dominant within our globalised world. The aim is not to allow the establishment of a few pockets of post-capitalism, but to allow other logics of exchange to become dominant in order to allow us to enter a new type of civilisation. It is a question of succeeding, where the socialist and communist revolutions have failed, in positioning the commons at the centre of societal organisation. The steps which will lead to this advent are as follows: first, the possibility for all citizens to contribute to the commons from the network, made possible by the Internet (this is mainly cognitive work); then a set of 'entrepreneurial coalitions' will surround these commons, allowing the production of means of subsistence; and finally, these entities organised around the commons will progressively constrain markets to the demands of reciprocity.

In the associationist vision, the State does not disappear completely, but becomes a 'partner state' (p. 46) of largely autonomous organisations (it supports and promotes them). Thus, it is imperative that the State does not simply redistribute resources based on a form of acceptance of the rules of the game of capital and its domination. What is expected of the State is to create the conditions for the autonomy of citizens and not their submission to capitalism.[43] It would thus be a question of distributing the resources necessary for autonomy beforehand and not of redistributing a small portion of the resources created, *a posteriori* (Bauwens et al., 2018, p. 51). In this transition from a welfare state to a partner state, the authors consider it necessary to de-bureaucratise.[44]

One of the aims of the Commons Manifesto is to progressively transform the State in order to 'commonify' it (p. 47) (and, consequently, to deprivatise it). The vision of associationism is both reformist (it works in its current form) and revolutionary (it is a question of transforming the extraction system).[45] The authors thus propose a twofold strategy: control of capitalism, on the one hand, and its progressive erosion, on the other.

6.6 The Recognition of Animals as Political Subjects: The *Animalist Manifesto* (2017)

Unlike the other manifestos studied, the *Animalist Manifesto* ('*Manifeste animaliste*') is written by a single author: Corine Pelluchon. Although Corine Pelluchon is a political philosopher with an acute awareness of environmental issues – which she has notably worked on in *Les nourritures* (2015) – this manifesto addresses not the Anthropocene as a whole, but one of its main components: the destruction

[43] Within this associationist transition of the commons, cities also have a special place. Thus, the authors analyse the situation of the city of Barcelona, with its radical municipalism, and citizen platforms bringing about bottom-up changes. The city of Bologna is also studied as an interesting case in the development of institutional processes for the development of the commons.

[44] In this sense, their arguments are in line with David Graeber's anarchist critiques. On the other hand, in their approach, they do not reject the State, which has a necessary part to play to avoid the risk of sacralisation of individual freedom inherent in a purely anarchist approach.

[45] In their theorisation of reform, the authors of the *Commons Manifesto* draw upon the work of André Gorz, whereby a revolutionary reform must be acceptable to the system in place while creating the conditions for a profound transformation. A concrete example of revolutionary reform could be the allocation of a universal basic income, which would break with the commodification of work and free up time for the production of common goods.

of biodiversity. Pelluchon focuses on one of the problems of this destruction, which is the use of animals considered as mere things, and which humans kill in industrial slaughterhouses with little scruple. For Corine Pelluchon, the animal cause is the great problem of our contemporary society.[46]

6.6.1 *Politicising the Animal Cause*

Corine Pelluchon's *Animalist Manifesto*, published in 2017, is subtitled 'Politicising the Animal Cause'. Pelluchon is known for being an ardent animal advocate. In a hard-hitting philosophical essay from 2015, *Les nourritures*, within which she shows the biological component of the body politic, she had already made a stand for us to resolutely enter into a different kind of relationship with animals. The manifesto begins with the accusation that 'humanity is losing its soul' (Pelluchon, 2017a, b, p. 9). The statement is incisive: 'The cages where billions of rabbits, chickens, ducks, sows, foxes and minks, mice and monkeys, dogs and cats are locked up to produce meat, to be butchered or to be used as experimental subjects, dolphinariums and circuses where cetaceans, elephants and wild animals, broken by deprivation of freedom, are forced to make a spectacle of themselves in exchange for a little food or for fear of the whip, are the picture of our common shame. No description can capture its infinite sadness' (p. 9). Any limit (legal, moral or emotional) in our use of living things can be crossed for the sake of gain and profit. Our relationship with animals, then, becomes an indicator of the power and folly of capitalism.

Pelluchon defines animalism in her manifesto: 'Animalists are anti-speciesists, and their convictions lead them to veganism. Aware that their struggle is part of the history of the fight against all forms of discrimination, against slavery, racism, sexism, against the exploitation of humans by other humans and of nations by other nations, they draw no distinction between the defence of animals and the defence of human rights' (Pelluchon, 2017a, b, p. 63). Pelluchon's animalism has a philosophical and political depth because the animal cause, in allowing us to reconcile with animals, also allows us to reconcile with ourselves. By defending the cause of animals, we are able to combat violence against all discriminated social categories. The attachment to animals does not imply political or social disengagement – a withdrawal from the struggles of the human adventure. On the contrary: 'To fight against animal abuse is to fight against all violence inflicted on humans or non-humans, whether legal or illegal' (Pelluchon, 2017a, b, p. 63). Corine Pelluchon's thesis is that the animal cause carries a set of issues and struggles of the contemporary period. The commitment of this fight for others (animals) is part of the history of other collective fights of recent decades, including the fight against slavery or

[46] We perceive a set of indicators currently around the awareness that 'something is wrong' in our relationship to animals, with the development of veganism or the presence of a French 'animalist list' in the 2019 European elections. We can assume that these elements are part of the preparation for the progressive entry of the Anthropocene onto the political scene.

against fascism.[47] By politicising the animal cause, we take a stand against all forms of exploitation, be it that of women, children or slaves: 'Our relationships with animals are thus a reflection of our relationship with one another, and mistreatment of animals is very often a harbinger of violence towards humans, especially towards the weakest, such as children, women, people with disabilities, prisoners and, in the past, slaves' (Pelluchon, 2017a, b, p. 15). Based on the identification of the issues of the animal cause, the *Animalist Manifesto* aims to bring about a social, political and spiritual reconstruction.

The *Animalist Manifesto* becomes a resolutely political offensive when Corine Pelluchon asks 'who benefits from the crime' (p. 32): 'Who benefits from the massive and daily consumption of animal products that have a harmful impact on people's health, are factors that increase the risks of diabetes, obesity, cardiovascular disease and cancer? Who benefits from deforestation in South America and the use of land to grow soya for American and European livestock, when we know that the demand for animal products is partly responsible for the extreme poverty, famine and malnutrition suffered by around three billion human beings? […] Who benefits from factory farms, when we know that farm workers have to castrate piglets alive, that they put in earplugs so as not to hear their cries, turn into killers to eliminate those who do not "conform" and that slurry causes groundwater pollution and the profusion of algae dangerous to health?' (Pelluchon, 2017a, b, pp. 32–33). The fight for animals is a political fight; it is a fight against the hegemony of economic logics in the functioning of our world, where living creatures can be exploited like any other raw material. Expanding the horizon of one's existence to include others and not living just for oneself is fundamental in a period marked by an evaluation of life success based on what each individual manages to amass (Pelluchon, 2017a, b, p. 36).

The *Animalist Manifesto* is a form of awakening or reawakening of our sensitivity. How can we be so insensitive to the suffering of other sentient beings? What remains of our humanity, when we look at the way we are able to behave with sentient living beings? Not to face up to the way in which we treat animals 'is to accept being contaminated by evil'(!) Corine Pelluchon tells us (2017a, b, p. 12). The root of the problem is not animosity against animals, but a form of capacity for intrapsychic compartmentalisation. The aim of the *Animalist Manifesto* is to attack the root of our evil, perceptible with such clarity and magnitude in our relations with animals. The author identifies this root in the following way: 'the way we think about our condition and accept our finitude and vulnerability'[48] (Pelluchon, 2017a, b,

[47] The abolition of slavery is mentioned several times in the manifesto, underscoring the hope that 1 day, animals will have rights and will no longer be treated as mere commodities.

[48] Awareness of our own fragility is a necessary condition for allowing ourselves to be touched and affected by animal suffering: 'In order to feel in one's heart and flesh the cries of panic and anguish of animals, their mutilated and bruised bodies, the immensity of their frustrations, one must be able to present oneself, naked and exposed, before animals who are naked and exposed beings, deliv-

p. 15). In an extension of this idea, the proposal is situated at the intersection of a political theory and an anthropology redefining our relationship to animals and to living beings in general.[49] What drives this fight for the animal cause is the hope for a world that is more just, less violent, and more human.

6.6.2 Animals are Political Subjects

The *Animalist Manifesto* is resolutely anti-speciesist. It assumes the same consideration of interests for humans and non-humans. Animals are understood as living beings capable of feeling – i.e. they are sentient beings, and thus also have rights. The speciesism which is at the heart of our societies, postulating an irreducible difference between humans and animals that allows for the possibility to treat animals as marketable objects, is viewed as deeply unjust. The anti-speciesism of the *Animalist Manifesto* allows us to measure the extent of the anthropocentrism of contemporary societies. The manifesto implicitly relies on the results of the research from the past several decades showing that the difference between humans and animals is a difference of degree and not of nature.

It is not only a question of identifying the political stakes of our relationship with animals, but of considering animals as political subjects. The thesis is bold and unambiguous. Ultimately, what Corine Pelluchon is telling us is that we have a world in common with animals. This is the political basis of the politicisation of the animal cause.[50] We do not inhabit the earth alone; we cohabit with all living things in the same space. This space must be thought of from the perspective of this coexistence. Pelluchon uses the term 'zoopolitics' to refer to the hybrid community that we form with animals. Thus, animals are political subjects, even if they are not fellow citizens because they are not aware of this belonging. From this starting point, it is a matter of thinking about the rules of life that allow humans and animals to coexist in an equitable way, and this must be translated into law. The commons must not be thought of only for humans, but also for animals who need them to live.

---

ered almost defencelessly to humans armed with machines and tools. This initial exposure can be the breach through which human beings learn of this suffering, while they live in a system that is extremely efficient at hiding it' (Pelluchon, 2017a, b, p. 19).

[49] Corine Pelluchon identifies, very precisely, when awareness of animal mistreatment is reached. This is the first necessary step: 'the step that causes someone to decide to change their life is marked by a painful experience – a revelation that strikes with amazement. There can be no awakening without this hurt' (Pelluchon, 2017a, b, pp. 16–17). This realisation is a true revelation and acts as a conversion for the person: 'As soon as this truth penetrates the consciousness, the air becomes unbreathable. A silence arises all around one; it carries loneliness, shame, and the certainty that never again will it be possible to live as before' (Pelluchon, 2017a, b, p. 17).

[50] As mentioned in the introduction, this is an aspect which that Hannah Arendt did not work on in her time, centred on the common world that humans share. This is easy to understand in view of her time, marked by the discovery of the extent of the Holocaust, the disaster of the Second World War, the Cold War and the looming threat of a nuclear conflict. The inhumanity was so perceptible in the relationships between humans that it was difficult to focus the narrative and the struggle on relationships with non-humans.

Though Corine Pelluchon is particularly radical – if not 'hair-raising' in comparison to the usual type of political thinking – she does not confuse animals with humans, and distinguishes them regularly throughout her manifesto. The main difference, in her view, is that 'Only humans can be responsible for all living beings and feel that they belong to a political community with them' (Pelluchon, 2017a, b, p. 52). Animals must therefore 'enter politics' despite their asymmetry with humans: it is humans who can grant them rights – they cannot have rights without the mediation of humans (Pelluchon, 2017a, b, p. 57).

To bring animals into politics, Pelluchon envisages a representative device comparable to what was envisaged by several authors concerning a Chamber of the Future, or a representation of generations to come, as discussed in Chap. 5. Indeed, we are dealing with the same issue: representation of individuals who are unable to defend their own rights. After the State has been entrusted with the duty to improve the animal condition, animal representatives would sit in both chambers and would have a right of veto when laws are unfavourable to the animal condition. These representatives would be drawn by lot from a list of volunteers who have proven themselves in the defence of the animal cause.

One of the main barriers to the creation of a more just society, respectful of the life and rights of animals, in which we can truly coexist, is people working in the various sectors of animal exploitation (who would be obliged to change careers). The concrete measures envisaged are: firstly, an end to captivity; secondly, a ban on animal fighting for sport; thirdly, the abolition of hunting with hounds; fourthly, a ban on fur and foie gras. The transformation of our food practices and the reduction or elimination of animal consumption is a major challenge. On the other hand, there is growing awareness of the need to consume less meat for the environmental reasons of limiting global warming, and this goal is viewed as being achievable.[51]

6.7 The *Manifesto for Climate Justice*'s Call for Political and Legal Action (2019)

The short text *How we will save the world – Manifesto for Climate Justice* was published in March 2019 and signed by '*Notre affaire à tous*' – Everybody's Business – the citizen movement to force States or large corporations to reduce their greenhouse-gas emissions.

[51] The last lines of the *Animalist Manifesto* voice a particularly strong appeal: 'Animalists of all countries, parties and faiths, unite. Unite your forces so that the condition of animals improves here and now, and that 1 day their exploitation ceases. Unite also with those who are not animalists. Fight against animal abuse, spread the love of the living, human and non-human, and of justice. The animal cause is universal; it belongs to everyone. By doing justice to animals, we save our souls and secure our future. We have a world to gain' (Pelluchon, 2017a, b, p. 98).

6.7.1 A Call to Resist

This manifesto opens with particularly strong words. The authors begin with the fact that climate change and environmental changes are now evident to human senses. As scientists have been telling us for several decades, all spheres of our lives will now undergo radical upheaval. The sustainability of the human adventure is in jeopardy; living together will become increasingly complex. As mentioned in the introduction, the rise in drama is increasingly powerful: It is becoming impossible not to think about it. Not to know. The Earth is hunting us. It cannot take it any more. Our presence is too painful. In a century of industrial acceleration, we have captured so much energy, and used so many resources, that we have altered the balance of the planet to our sole advantage. So it is heating up, rumbling, sliding into the unknown, towards another version of itself, another ecosystem that is already shaking the foundations of our world' (Notre affaire à tous, 2019, p. 3). It is the Anthropocene – though that name is not explicitly mentioned – that is evoked through this personification of Earth.[52]

The *Manifesto for Climate Justice* is constructed in two parts. The first part focuses on the scope of contemporary environmental problems, where it is possible to identify the systemic alteration of the Earth by human activity. The second part highlights the generalisation of awareness and the possibility of winning environmental battles, as an extension of battles already won: 'A new generation is here, not defined by its age, but by its way of rethinking everything. A web is being woven around the planet: it is the awakening, it is life, the determined step of youth demonstrating and shouting their concern for the future' (Notre affaire à tous, 2019, p. 4). In order to fuel this energy with a struggle that can be won, the authors of the *Manifesto for Climate Justice* evoke several battles that have been won so far: Sri Lanka, the first country to ban the use of glyphosate; Costa Rica, a country on the way to becoming 100% renewable; Indonesia, which no longer subsidises kerosene; the hole in the ozone layer that is closing; the return of humpback whales off the coast of New York, etc. Through all these examples, they call for a non-aggression pact to protect nature. Ultimately, their project consists of drafting a *Universal Declaration of Planetary Rights*, a counterpart to the *Universal Declaration of Human Rights*.

The authors of the *Manifesto for Climate Justice* soundly denounce how governments have failed to take the 1972 Club of Rome report on the limits to growth seriously. They point to the way in which the logic of economic efficiency has taken over and ignored research, science and common sense in preparing for the future. The *Manifesto for Climate Justice* is a wide-ranging call to engage in a struggle: 'Let's indict all the accomplices of global warming. Let's not let them drape themselves in the fine cause of the environment to redesign their product packaging, to

[52] Christophe Bonneuil, co-author of *L'événement Anthropocène* (2013) with Jean-Baptiste Fressoz and director of the 'Anthropocene' collection at Éditions du Seuil, is one of the scientific advisors for this manifesto.

make a show at the official podiums, or to make ends meet with a simple green tax. Let's not give them the time they're asking for. We don't have it any more' (Notre affaire à tous, 2019, p. 4).

6.7.2 *A Fight Against Productivism and Financial* hybris

Our current relationship with our planet is viewed, quite simply, as suicide (Notre affaire à tous, 2019, p. 11). The tone is accusatory, but the recrimination is not aimed at citizens who have shifted into consumers in recent decades. Only the big companies and politicians of the last 50 years take the blame: 'Fear? Yes, we are afraid. But not afraid of the deficit. Not afraid of foreigners. Not afraid of youth. Not afraid of the poor. Not fear of revolt. All these fears with which governments play upon, for lack of a social project to propose, all these fears that allow them to set people against one another to remain at the centre of the game, at the end of our anxieties, of our fingers as we put our ballot slips into the ballot box. We have to end it. We are afraid of their inertia. Of their blackmail. In retrospect, of their inability to react when there was still time. Fear, yes. Enough of this! Enough of their litany of growth. Our jobs, they say. Never do they talk about our lives' (Notre affaire à tous, 2019, p. 14). The authors of the *Climate Justice Manifesto* call for fear to change sides – that is, for the world's leaders to start being afraid: afraid of climate change as well as of angry citizens demanding immediate and radical action. In this sense, their position embodies Hans Jonas' heuristics of fear.

The tone of the *Manifesto for Climate Justice* is resolutely against industrialisation at all costs and the fact that humans are seen as mere robots on an assembly line, interchangeable. It is about connecting struggles to give strength to current citizen movements: climate justice is social justice. Indeed, it is the poor who are the first to be exposed to climate change and who are the most vulnerable: 'It is the fragile roofs that the storms tear off. It is the poor farmers who are pushed into refugee camps by soil impoverishment because of the lack of crops. It is the low-income earners who go to toxic jobs or junk-food shelves' (Notre affaire à tous, 2019, p. 15).[53]

The authors of the *Manifesto for Climate Justice* denounce the way in which the disease of human *hybris* has contaminated the whole world, which in turn has contaminated our Earth. The banks are in the crosshairs of these authors – they are considered irresponsible, insofar as they finance the big oil and mining groups: 'In France, in 2017, the six largest banks increased their financing to fossil fuels, and decreased that devoted to renewable energy. In Britain, HSBC estimates that its assets would lose 40–60% of their stock-market value if the 2°C target were met' (Notre affaire à tous, 2019, p. 18). After being lulled to sleep by the promises of the

[53] Here, aside from the accusatory and particularly vehement tone, we can see the awareness of the scientific and geopolitical data of the Anthropocene in this manifesto.

economy (which has turned tyrannical), citizens are waking up and fighting back. One of their weapons is the law.

Never does the *Manifesto for Climate Justice* call for armed struggle or violence. Instead, it is a call to disobedience, to engage in a struggle that can sometimes take the form of breaking the law: 'We will return to steal chairs from the branches of banks that practice tax evasion and feed planetary destruction. We will lay our bodies in the path of those who defile the planet. We are everywhere, under various names, in various languages, we are international disruptors, we enter into rebellion and we swarm. We will become their nightmare, since they do not give us time to dream. We will base our happiness on the fight' (Notre affaire à tous, 2019, p. 29).

6.7.3 A Challenge to Teachers

One particularly interesting element at the end of the *Climate Justice Manifesto* is worth noting. Its authors issue a call to teachers. It is a question of making a real break with education for sustainable development, no longer leading students to believe that the situation is under control and that it is being taken into consideration by world leaders and entrepreneurs. The situation is serious, and it is important that teachers take a stand. In fact, it is difficult, if not impossible, to be moderate when one identifies the magnitude of the contemporary environmental problem. When the knowledge of the Anthropocene has taken hold within a teacher who works with it, it brings about seriousness and radicalness in the positions taken. The authors of this manifesto seem to be saying that it is necessary for these positions to be expressed in the classroom – even if this means that the classroom must temporarily become a militant space.

6.8 The Thunderous Entry of the Anthropocene Into Politics with *Integral Ecology – The Manifesto* (2019)

Delphine Batho, who signed this text, is a French politician. She is a Member of Parliament for Deux-Sèvres and was Minister of Ecology, Sustainable Development and Energy during the early days of French President François Hollande's 5-year term, from June 2012 to July 2013. This short text has an interesting stance, compared to other manifestos: it is written by a person in a position of power with elective concerns at stake, rather than by citizens questioning the ways in which power is exercised. It was published a few months before the European elections in May 2019. The manifesto is prefaced by Dominique Bourg, who joined Delphine Batho in her candidacy for the European elections. We can easily imagine that Dominique Bourg provided strong guidance in the drafting of this text, insofar as it contains all the themes on which he typically works. What was the initial political purpose of this manifesto? Indeed, it could have been believed that this text was a campaign tool for Delphine Batho for the European elections. In the end, it was Dominique Bourg who led the '*Urgence écologie*' list, which included the former minister. This is one of the first times that the Anthropocene has entered into French political discourse. This is the main point of interest of this text for the first part of this work on politics in the Anthropocene. The Anthropocene is seized upon here for its power to challenge and highlight the need to change our ways of life.

6.8.1 Policy Overhaul

The tone is combative. The style, here again, is that of emphasis and of the rise in power in catastrophic dramaturgy: 'The present manifesto is a hymn to life on Earth, to the love of Nature and to respect for the conditions of human existence. The challenge of preserving a habitable planet for humanity now supersedes all others. Ecology has become a matter of life and death' (Batho, 2019, p. 9). This is one of the first texts written by an elected official in a position of political responsibility who dares to clearly state the seriousness of the environmental situation. Delphine Batho writes: 'Forget everything that you have loved, imagined, dreamed of for yourself and for your children's future, in the more or less assured comfort of a modern life. Everything is destabilised. Everything can disappear. Our most basic needs – to breathe, to drink, to feed ourselves – are threatened by climate chaos and the mass destruction of life on Earth' (Batho, 2019, pp. 9–10). In this manifesto, the author mentions that the fight she is going to present integrates all others and is the most fundamental of all: 'If we fail to conserve what makes harmonious life on Earth possible, what good are the other issues?' (Batho, 2019, p. 10).

The political perspective is that of a break with socialism and liberalism, which it sees as two sides of the same coin. These two ideologies have indeed contributed to massive destruction of nature through the generalisation of productivism. Delphine Batho realises a failure of the traditional parties who are unable to recognise the way in which they have participated in the contemporary disaster. We are now at 'the last station before the crash' (Batho, 2019, p. 12) and we have before us a desperately simple choice to make between barbarism and ecology. This manifesto thus proposes an 'integral democratic ecology'. The tone is radical and democratic.

The *Integral Ecology* Manifesto begins with a chapter on the Anthropocene, where we can recognise Dominique Bourg's style behind many expressions. This chapter, like the rest of the book, is well informed and refers to a bibliography at the end of the book, which is quite rare for a book written by a political figure. We can see the importance here, given the radical nature of what is said, of referring to scientific studies and articles. The concept of the Anthropocene is used throughout the book as a proven fact. The entry into this new geological epoch generates a set of radical paradigmatic ruptures, including the transition from the idea of the welfare state to that of the resilience state. The first pages of the chapter on the Anthropocene are particularly anxiety-provoking: global warming, extinction of biodiversity, disruption of human hormonal functioning, depletion of natural resources... The Anthropocene is thus, for the first time in a political discourse, treated as an unquestionable objective fact (beyond the question of the dating of its entry), and the starting point from which any policy must proceed. The Anthropocene is mainly characterised by the transgression of planetary boundaries and by what that brings in its wake: war. We are not talking here about future wars, but about wars that have already begun in relation to access to food and water. Based on studies (not cited here), Delphine Batho estimates that there will be 213.9 million climate refugees between 2008 and 2016, and one billion by 2050.

The dichotomy between the right and the left is considered obsolete because of the denial of planetary limits and the same pursuit of growth marked by productivism. Delphine Batho is one of those few French elected officials who denounce the race for growth and show how destructive it is: 'growth is the indicator of our habitat, the Earth' (Batho, 2019, p. 31); or again: 'massacring Nature, which increases GDP!' (Batho, 2019, p. 31). Batho denounces greenwashing, or the fallacy of green growth.

In the Anthropocene, politics is organised around a cleavage: an opposition between the Earthlings, who protect the conditions necessary for human existence, and the Destroyers (all the rest!). The Earthlings are aware of their deep immersion in the living world, unlike the Destroyers. We find here, under another name, the distinction made by Bruno Latour between humans and earthlings. Delphine Batho does not hesitate to identify Donald Trump, Jair Bolsonaro, Vladimir Putin and Xi Jinping among the Destroyers (p. 34): those people currently in power who make people believe that they are working for their good, whilst ravaging the planet. The denunciation is particularly virulent for a non-extremist political figure.

6.8.2 *The Triumph of Alternative Lifestyles*

In the face of the contemporary disaster, Delphine Batho relies on the concept of integral ecology, developed by Dominique Bourg and Christian Arnsperger (2017). This concept means that the democratic issue and the ecological issue are one and the same fight. It is a matter of democratically taking back power and radically transforming our ways of life. At the heart of this concept is the figure of citizens who participate, wherever they are, through various commitments, in in-depth transformations (AMAPs, recycling centres, zero-waste objectives, complementary currencies, collaborative economy, zero long-term unemployment, etc.). In a way, Delphine Batho stands for the triumph of alternative lifestyles. She affirms the objective of permacircularity,[54] characterised by a neutral ecological footprint.

[54]Permacircularity is at the heart of integral ecology by Dominique Bourg and Christian Arnsperger (2017).

6.9 Conviviality as the Political Foundation of the *Convivialist Manifesto* (2013)

The Convivialist Manifesto opens with a list of the threats of the present time, the first four of which are ecological: 'Global warming, disasters of all kinds, and the huge migrations global warming will cause'; 'The sometimes irreversible weakening of natural ecosystems'; 'Air pollution that makes the air in many large cities, such as Beijing, New Delhi or Mexico City, increasingly unbreathable'; 'The persistent risk of a nuclear disaster' on an even greater scale than Chernobyl or Fukushima; 'The scarcity of energy (oil, gas), mineral (rare earths, in particular) and agricultural resources that had enabled growth; also, wars for access to these resources' (Les Convivialistes, 2013, p. 8). The Anthropocene as such is not specifically named, but it is indeed the threat identified by the convivialists.

6.9.1 The Federation of Alternative Thoughts

The *Convivialist Manifesto* emerged in the wake of the MAUSS (Mouvement Anti-Utilitariste dans les Sciences Sociales), in reference to Marcel Mauss, and as a form of French extension of the Critical Theory of the Frankfurt School. It was published in France in 2013 and then translated into a dozen languages. Beyond the date of publication of the *Manifesto*, it is in the success of its reception that lies the true starting point of what is at once a political stance and a movement of ideas whose purpose lies as much in the renewal of social and political thought as in the transformation of contemporary globalised societies. The *Convivialist Manifesto* has now been signed by more than 3700 intellectuals and people in various fields of activity in contemporary societies.

From the preamble of the *Convivialist Manifesto*, convivialism presents itself as the coming together of a set of alternative ideas, and as what can be identified as the lowest common denominator between these different ideas – within which conviviality is central. In a way, convivialism represents a political form of the MAUSS academic movement. With the term 'convivialism', this current, of which Alain Caillé is the initiator and one of the main animators, is part of the intellectual heritage of the Austrian-American thinker of political ecology, Ivan Illich. In his 1973 book, *Tools for Conviviality*, he criticised the ideology of growth. It is in this direct extension of the denunciation of growth and, more globally, of the hegemony of economic logics that convivialism was institutionalised (firstly among sociologists and economists). The definition given of convivialism in the manifesto is as follows: 'Convivialism is the name given to everything that, in doctrines and wisdom, existing or past, secular or religious, contributes to the search for principles that allow human beings to compete without massacring each other in order to cooperate better, and to advance us as human beings in a full awareness of the finiteness of natural resources and in a shared concern for the care of the world' (Les convivialistes,

2013, p. 25). French management researcher Pascal Glémain identifies the convivialist movement as a 'social movement that wants a "different" civilisation – a civilisation of conviviality fighting against the inhumanity of the world' (2017, p. 27). The Italian philosopher Francesco Fistetti defines convivialism as 'a political philosophy of living together' (2016, p. 247).

6.9.2 Togetherness and the Sharing of Freedoms

Etymologically, "conviviality" is made up of *con* meaning with, and *vivere*, victuals. Conviviality refers to the sharing of life, and food – it takes place around a meal! This sharing of food is a conception of existence in which coexistence is given primacy. In the main, existence is not lived individually, but shared with others. *Vivere* in fact also means living or inhabiting. Thus, it is the sharing of food (and therefore of resources) that is the basis for sharing existence with others. Convivialism proposes to understand the political on the basis of the means of sharing existence in the world. Another semantic characteristic of convivialism is the tone of the term 'conviviality', which conveys the idea of empathy, of a good time, of warmth, of support, of a moment shared with loved ones. There is, obviously, in convivialism the sharing of sympathy which can be here a spur for the regulation of our freedoms so that their exercise does not lead to the massacre (physical or psychological) of others. *Hybris*, identified as one of the fundamental problems, is opposed by an art of living together at the foundation of a new political imaginary.[55]

'What to do with our freedoms' is one of the fundamental political questions. How can our freedoms be deployed? How can we regulate our freedoms so that their exercise does not result in the massacre of others? In convivialism, freedoms are understood as being shared and as being able to meet or join. Thus, trust is placed not in individuals, but in the relationships between individuals.[56] The centre of gravity is not the individual subject, but intersubjectivity. In that sense, it is a political thought marked by the contemporary philosophy of intersubjectivity and the deconstruction of the subject. Rather than thinking about the State, convivialism makes use of togetherness between individuals and between constituted groups. We can

[55] The authors of the *Convivialist Manifesto*, aware of the threats of the present time, seek to nourish a form of hope for the future. 'What every individual is entitled to hope for is to see him or herself recognised with equal dignity with all other human beings, to reach sufficient material conditions to realise his or her vision of a good life, whilst respecting other people's views' and thus to seek to enjoy recognition by others by effectively participating, if he or she so wishes, in political life and in the making of decisions that affect his or her future and that of his or her community' (Les convivialistes, 2013, p. 29).

[56] Among the proposals the manifesto makes, we can note the aim of building a society of *care* (Les convivialistes, 2013, p. 23), which represents the mutual dependence and need we have for each other.

live together, convivialism tells us.[57] The necessary opposition between us (so that our freedoms can be respected and flourish) will not inevitably lead to a massacre.

Convivialism is presented as a proposal to go beyond Marxism, rather than as a neo-Marxist doctrine. Similarly, convivialism is not anarchism in the strictest sense, in spite of the faith placed in the 'between-us' which is the basis of the political. Convivialists do not seem to reject the centralisation of State power. It is a question of allowing a reinstitutionalisation that makes it possible to govern whilst keeping *hybris* under control. Convivialism is presented by Alain Caillé as a contemporary political ideology that proposes, by subsuming all of them, to go beyond the four political ideologies of modernity: liberalism, socialism, anarchism and communism (Caillé, 2011). Convivialism attempts to go beyond these four ideologies – especially on the ecological and moral question – by rescuing the democratic ideal. Currently, democracy is subordinated to an economy that is running out of steam and is bound to collapse, because it is unsustainable. It is a question of freeing democracy from the grip of the economy and of re-politicising it on the basis of 'between-us'. The political understanding of convivialism is very much in line with Hannah Arendt's conception.[58]

6.9.3 *Dealing with* hybris *– The Mother of All Threats*

The *Convivialist Manifesto* proposes an alternative way to deal with the mother of all threats – that combination of violence and madness: 'Humanity has been able to achieve stunning technical and scientific progress, but has remained as powerless as ever to solve its essential problem: how to deal with rivalry and violence among human beings' (Les convivialistes, 2013, p. 12); or indeed, 'How to hinder the accumulation of power, now unlimited and potentially self-destructive, over humans and nature' To deal with this destructive capacity, convivialists choose cooperation. Humans are seen as capable of giving the best of themselves and of opposing each other without killing each other.

Hybris is the anthropological root of many of the threats we currently face. Cooperation is presented as the counterpoint to *hybris*: 'A healthy society is one that knows how to cater for everyone's desire for recognition, and acknowledge the role of rivalry, of aspiration to continuously surpass oneself, and of openness to the risks that it carries, preventing it from turning into excess, into *hybris*, instead favouring cooperative openness towards others' (Les convivialistes, 2013, p. 14). Conflict

[57] In achieving this aim, courage is needed (Les convivialistes, 2013, p. 30). This echoes the courage of participation, defined by the German theologian who emigrated to the United States Paul Tillich, in his book *The Courage To Be*, published in 1952 in the USA.

[58] This is the reason why we have found a form of echo in the *Convivialist Manifesto*, its reception texts and the works of the various authors who refer to it. Indeed, Arendt understands politics as the art of association – which can be said to be at the heart of the *Convivialist Manifesto*, and which we also find in the *Commons Manifesto*.

must thus be understood as a life force rather than a lethal threat. In a new version of the *Convivialist Manifesto* currently being drafted, the goal of controlling *hybris* is reinforced and is identified as an 'imperative': 'The first condition for rivalry to serve the common good is that it be devoid of desire for omnipotence, excessive, *hubris* (and a fortiori *pleonexia*, the desire to possess ever more). On this condition, it becomes rivalry to cooperate better'.

6.9.4 *Four Principles at the Root of Politics for the Contemporary Period*

In addition to denouncing the *hybris* of the hegemonic *homo oeconomicus* in our contemporary societies,[59] the *Convivialist Manifesto* identifies four principles which are essential for our capacity to live together: common humanity, common sociality, common individuation and controlled opposition. Hence, the political question is directly moral: 'Good politics is therefore that which allows human beings to differentiate themselves by accepting and mastering conflict' (Les convivialistes, 2013, p. 27). What is forbidden to the individual is 'to tip over into excess and into the infantile desire for omnipotence (the *hybris* of the Greeks) – i.e. to violate the principle of common humanity and to endanger common sociality by claiming to belong to some superior species, or by monopolising such a quantity of goods or quantum of power as to jeopardise the social existence of those on a lower stratum' (Les convivialistes, 2013, p. 29).

A second version of the *Convivialist Manifesto* is currently being written. The coordinator of this rewriting, Alain Caillé, who was also the author of the first version of the manifesto before it was amended by a group of intellectuals, has sent us the second chapter, entitled 'Convivialism'. In this new version, a new principle appears: the principle of common naturality. The benefit of common naturality, which did not appear in the first version of the manifesto, is that it opens the door to conviviality with nonhumans: 'Humans do not live outside a nature, of which they should become "masters and possessors". Like all living beings, they are part of it and are interdependent with it. They have a responsibility to take care of it. If they do not respect it, it is their ethical and physical survival that is at risk'. This fundamental principle of convivialism reinforces the ecological component of the *Convivialist Manifesto*.

[59] The authors of the *Convivialist Manifesto* are in favour of the introduction of a basic income or living wage: 'More specifically, legitimate states guarantee all their poorest citizens a minimum of resources, a basic income, whatever its form, which keeps them safe from the abjection of extreme poverty, and progressively prevent the richest, by the introduction of a maximum income, from tipping over into the abjection of extreme wealth by exceeding a level that would render inoperative the principles of common humanity and common sociality. This level can be relatively high, but not beyond what is implied by a sense of common decency' (Les convivialistes, 2013, p. 31).

6.9.4.1 What Kind of Anthropological Shift Is Needed?

By studying these different texts, it is possible to identify the emerging contours of the citizen debate based on the human alteration of the Earth (and also on the very concept of the Anthropocene, in the case of certain manifestos). These texts analyse a fundamental problem in our relationship to the world, and several of them implicitly postulate both the need to change the way we are, and to think about the human adventure in a new way. The next chapter offers a comparative analysis of the manifestos studied, both from the point of view of anthropological shift and from that point of view of an evolution of anthropological conceptions. These two dimensions constitute the common thread of Part II of this book, entitled 'A consolidation of politics requiring an anthropological shift'.

Bibliographical References

Arnsperger, C. (2009). *Éthique de l'existence post-capitaliste – Pour un militantisme existentiel*. Cerf.

Arnsperger, C. (2010a). Changer d'existence économique: enjeux anthropologiques de la transition du capitalisme au post-capitalisme. *Revue d'éthique et de théologie morale, 258*, 23–50.

Arnsperger, C. (2010b). Monnaie, dette et croissance sans prospérité: portée et limites du 'tournant' jacksonnien. *Etopia, 8*, 109–116.

Arnsperger, C. (2011a). *L'homme économique et le sens de la vie*. Textuel.

Arnsperger, C. (2011b). Dépasser le capitalisme, mais par étapes. *Revue Projet, 324–325*, 73–81.

Arnsperger, C., & Bourg, D. (2017). *Ecologie intégrale, Pour une société permacirculaire*. PUF.

Bastoni, A. (2015, October). Inventing the future. *Novara Media*.

Batho, D. (2019). *Ecologie intégrale – Le manifeste*. Les éditions du Rocher.

Bauwens, M., Kostakis, V., & Pazaitis, A. (2018) *The Commons Manifesto – Peer to Peer, A paraître*. http://www.p2plab.gr/en/archives/117, version 2018.

Bourg, D. (2010a, July). L'éco-scepticisme et le refus des limites. *Etudes*, 29–40.

Bourg, D. (2010b). Un système qui ne peut répondre au défi environnemental. Interview with Pierre Le Hir, *Le Monde*, 31 October 2010.

Bourg, D. (2012). Transition écologique, plutôt que développement durable. *Vraiment durable, 1*, 77–96.

Caillé, A. (2011). Au-delà du libéralisme, le convivialisme. *Revue Projet, 324–325*, 94–97.

Citton, Y. (2016). Accélérer l'écologie. In L. de Sutter (Ed.), *Accélération !* (pp. 205–223). PUF.

de Sutter, L. (2016). Introduction. In L. de Sutter (Ed.), *Accélération !* (pp. 7–25). PUF.

Fistetti, F. (2016). Le convivialisme, 'contre-mouvement' du 21ème siècle. *Revue du MAUSS, 48*, 247–258.

Glémain, P. (2017). Penser le convivialisme en économie sociale contemporaine. *RECMA, 4*(346), 27–41.

Land, N. (2011). *Fanged Noumena: Collected writings 1987–2007*. Urbanomic.

Les convivialistes. (2013). *Manifeste convivialiste*. Le bord de l'eau.

Notre affaire à tous. (2019). *Comment nous allons sauver le monde – Manifeste pour une justice climatique*. Massot éditions.

Pelluchon, C. (2015). *Les nourritures*. Seuil.
Pelluchon, C. (2017a). L'éthique des vertus: une condition pour opérer la transition environnementale. *lapenseeecologique.com, PUF, 1*(1), 1–18.
Pelluchon, C. (2017b). *Manifeste animaliste – Politiser la cause animale*. Alma.
Rosa, H., & Wallenhorst, N. (interview). (2017). Apprendre à écouter le monde. *Chemins de formation, 21*, 19–31.
Srnicek, N., & Williams, A. (2016). #Accelerate. Manifeste pour une politique accélérationniste. In L. de Sutter (Ed.), *Accélération !* (pp. 27–47). PUF. French translation.
Weber, A., & Kurt, H. (2015). *Lebendigkeit sei! Für eine Politik des Lebens. Ein Manifest für das Anthropozän*. Klein Jasedow, Think Oya.

Part II
A Consolidation of Policy Requiring an Anthropological Shift

Chapter 7
The Idea of an Anthropological Shift

Abstract After having read each of these manifestos separately, we will examine the differences between their different anthropological aims. This will be the starting point of our reflection on the idea of an anthropological shift, which will then be developed in the course of the different chapters that make up Part II.

Keywords Anthropocene · Earthling anthropology · Anthropological crisis · Anthropological shift

All of the authors of the manifestos discussed above, except for those of the *Ecomodernist Manifesto*, identify a problem in our relationship to the world. This problem is fundamental and threatens our very existence. It is possible that this awareness is becoming more and more acute. Note that the manifestos written in recent years mark a break in style. The interpellation of the citizens is more and more lively. So it is with the *Animalist Manifesto* (2017), the *Manifesto for Climate Justice* (2019) and the *Manifesto for Integral Ecology* (2019). The *Commons Manifesto* (2018), meanwhile, is particularly radical but less virulent in tone.

Each of the manifestos studied proposes an overarching vision of the world. When a struggle is identified, it is understood as something that can enable us to enter a new world, profoundly transformed. The struggle for climate justice in the *Manifesto for Climate Justice* is presented as containing within it all the other fundamental struggles of the twenty-first century: 'We are the future. We are the last generation, the last chance for the world. This struggle does not cancel out any of the others – it contains them all. To fight against the enslavement of the Earth is to fight against those who dominate it. Dominate us' (Notre affaire à tous, 2019, p. 4). This approach apprehending a struggle that integrates all the struggles for justice, solidarity and dignity is in keeping with that of Corine Pelluchon's *Animalist Manifesto*, holding that the fight for the animal cause is also a fight for everything.

© The Author(s), under exclusive license to Springer Nature Switzerland AG 2023

N. Wallenhorst, *A Critical Theory for the Anthropocene*, Anthropocene – Humanities and Social Sciences, https://doi.org/10.1007/978-3-031-37738-9_7

As can be seen from these nine manifestos, the main problem of the present time – the systemic alteration of the Earth by humans[1] – gives rise to proposals for a world view that is both very different and has many points in common with today's. Our reading of the extent of Earth's alteration can reveal the folly of unchecked human excess or Promethean techno-scientific genius. Thus, the *Ecomodernist Manifesto* celebrates human greatness and the 'extraordinary capacities of the human species' (2015, introductory paragraph), while the *Convivialist Manifesto* denounces the *hybris* that is at the heart of the human adventure today.

Taking account both of the importance of technology in the contemporary period and of the need for change, three fields of prospective reflection are currently being explored, both within these manifestos and beyond. The first is geoengineering, which consists of using technology to drive a change in the biosphere on a global scale. As mentioned in Part I, this path seems to us to be a dead end; though it is widely debated, it is a given for a number of authors that technology will never be able to repair environmental damage on a global scale. The second is transhumanism, where technology is used to enhance human beings, reinforcing all their physical and cognitive capacities. This potential avenue will be studied in Chap. 9. However, is transhumanism, driven by the desire to push back the limits and in particular that of death, not in fact posthumanism? Does it not represent the end of humanity as we know it, which is characterised by its finite lifespan? Thirdly, might another shift not be possible, consisting of guiding *Homo faber* to become what they have the potential to be – *Homo sapiens*, – accepting their own finitude and working with responsibility to prepare for the future and ensure generations to come inherit a hospitable environment? Ought humanity not learn to use its power differently, starting with an integration of finitude – both anthropological and environmental?

The second part of this research puts to work the idea of an anthropological shift (Arnsperger, 2010, 2016; Bellet, 1993, 2013) – a consolidation of politics in a post-Promethean perspective. It is thus necessary to steer profound changes in humanity to strengthen its political dimension. This idea may appear quite theoretical, but we see it emerging in most of the manifestos studied in Chap. 6. The second part of this research begins with a comparative analysis of these different manifestos. The first section of Chap. 7 is devoted to the way in which we identify this idea of an anthropological shift within these manifestos that materialise the entry of the Anthropocene into the civic debate. The second section questions the anthropological conceptions behind these texts. The third and concluding section highlights the importance of understanding an earth anthropology in the Anthropocene.

[1] As mentioned, this is not the gateway to the *Commons Manifesto*, nor to the first *Manifesto of the Appalled Economists*, but to the second, the *New Manifesto of the Appalled Economists*.

7.1 The Integration of the Anthropocene Into the Citizen Debate and the Question of an Anthropological Shift

7.1.1 Manifestos That Do Not Target Anthropological Change: The Manifesto of the Appalled Economists, the New Manifesto of the Appalled Economists and the Ecomodernist Manifesto

Of the nine manifestos studied, two of them do not aim at anthropological change. The *Manifesto of the Appalled Economists* (as well as the *New Manifesto of the Appalled Economists*) aims to challenge political power in order to institute more socially efficient economic policies and to curb the growing power of financial speculation logics. While the Earth is not mentioned in the first manifesto, it has a very strong presence in the second. On the other hand, a better relationship with the Earth is not identified as the path by which an anthropological shift can be brought about. The Earth does not show us the path to follow; rather, it is the boundary that limits and frames our consumption. It is not understood in terms of its function of humanisation, of participation in relationships of solidarity between humans, nor in terms of its transcendental or spiritual function.[2]

In the *Ecomodernist Manifesto*, too, there is no ambition for anthropological change, because the authors do not identify a fundamental problem in our relationship to the world. It is a text which encourages economic development such as we currently have – but which proposes some readjustments. The authors' main message seems to be: 'have confidence; human genius is capable of ensuring a *Great Anthropocene*'. It does not challenge the citizens in their power to act. At best, it is addressed to one entrepreneur or another, or to a scientist, to help optimise energy research or intensive agriculture.

7.1.2 The Promethean Acceleration of the Accelerationist Manifesto

In their *Accelerationist Manifesto*, Srnicek and Williams propose to go beyond the opposition between 'authentic life' and 'inauthentic life': human life is in no way authentic – it is linked to technology, which it has gradually integrated throughout history. This has allowed us to extend the duration of our time on Earth and to reduce the daily time spent on work. Now, though, capitalism is 'a system that holds back progress' and 'our technological development is being suppressed by

[2] On the other hand, we note a sentence at the end of the first chapter that weighs our analysis slightly, since the authors mention that preserving climate balance and biodiversity 'will be a way of building social well-being' (The Manifesto of Appalled Economists, 2015, p. 24). However, this is the only formulation of this type.

capitalism, as much as it has been unleashed' (paragraph 03.22). The accelerationist political project is both Promethean and humanistic, but as the Italian philosopher Antonio Negri points out, it is a humanism that is open to posthumanism, and finds collective mobilisation in scientific utopia.[3]

We discussed this choice of Prometheism and transhumanism in the *Accelerationist Manifesto* (of which he was aware) with Hartmut Rosa in our 2016 interview. Rosa declares himself firmly opposed to accelerationism, which includes, among other things, the fallacy of the distanced acceptance of technology and, as we can identify in the following long excerpt from that interview, 'all its possibilities for the realisation of a better life and a better society, without any conception of what a good life is.' Rosa continues:

> We need a compass to identify what a better life and a successful community look like (…) We end up with a blind utopia of what is feasible with technology (…) It is an exclusively Promethean way of approaching the world. The *Accelerationist Manifesto* and the ideology it develops amplify the problem instead of solving it. Herbert Marcuse writes, in this regard, that the world becomes a place of aggression. The world presents itself to us and we want to own it, to get our hands on it and to coerce it. Control of the world is the cultural component of this dynamic stabilisation. But the other side of the coin is that the world becomes mute: what I have in my possession, what I control and what I have at my disposal no longer speaks to me. This is why I believe that the Prometheistic claim of accelerationism is a mistake. Instead, we need to have a different relationship to the world. This is what Herbert Marcuse calls an erotic relationship to the world, not based on a desire for commodification. Instead, it is a matter of making the world capable of speaking again. Accelerationists are not interested in a successful relationship with others, or in a relationship with the world based on resonance and not on commodification. Thus, they continue to aggravate the problem (Rosa & Wallenhorst, 2017, pp. 22–23).

The Promethean post-capitalist prospect proposed by accelerationism is insufficient to effect an anthropological shift that is capable of dealing with the challenges posed by the Anthropocene.[4]

---

[3] In the wake of the manifesto, the chapter 'Accelerating Reason' by the English philosopher Ray Brassier, working in Lebanon (2016), in the book *Accélération!* edited by Laurent de Sutter, for example, very explicitly chooses Prometheism in a distancing from his Heideggerian and Arendtian (1958/1983) critiques, as well as from the more recent ones by Ivan Illich (1971a, b, 1973a, b) and Jean-Pierre Dupuy (2007, 2009).

[4] However, the accelerationists have very different positions from one another, and some of the contributions in the collective work *Accélération!* are based on anthropological conceptions that differ radically from each other, especially in terms of humans' relationship to nature and the environment. The French philosopher and literature professor Yves Citton defines humanity in terms of its relationship with the Earth. He proposes that humanity come back down to Earth, and stay there, firmly rooted! He encourages us to become earthlings – or at least, to rediscover that identity anew – and invites us to distance ourselves from the ideal of progress. On the contrary, Ray Brassier enthusiastically takes up this ideal and emphasises the importance of techno-scientific Prometheism; for him, the ambition to improve and transform the human condition is part of his definition of humanity (Brassier, 2016). The 'xenofeminist' collective Laboria Cuboniks, made up of six women, goes further still, suggesting: 'If nature is unjust, change nature!' (Cuboniks, 2016, p. 272) (this is the most radical text in this collective work, and develops a novel, challenging, provocative and equivocal treatise on feminism and gender); in their 12 eyes, humanity is defined by Prometheism and its capacity to assume control over nature.

Srnicek and Williams' political thought unfolds against the backdrop of an anthropology based on humanity's separation from nature, rather than humanity's immersion in nature. In the reception pieces responding to this manifesto collected in *Accélération!* (de Sutter, 2016a, b) (except for Yves Citton's)[5] is the idea that humans may not be dependent on nature, and several of the authors are tempted by an anthropology based on bodily transformation. The limits of nature or of the human body are not viewed as concrete and insurmountable, structuring the data when thinking about the advent of a post-capitalist society, but instead as obstacles to be overcome. Is this not an anthropological as well as a political deception? It would seem that the *Accelerationist Manifesto* contains an anthropological fallacy that opens its political recommendations up to ecological critique. Though Srnicek and Williams are aware of the seriousness of the environmental situation, which is viewed as the worst of contemporary threats, they restrict their recommendations to questions of political economy, failing to recognise that environmental issues are an important anthropological indicator. They do not perceive that the entry into the Anthropocene is indicative of an anthropological problem in the relationship to limitations. Whilst we are in agreement with most of the authors referring to accelerationism, as to the urgent need for a profound anthropological shift, we question the type of shift envisaged. The Manifesto's proposal is to push back the limits rather than to integrate them further. Is not profound deep and radical acceptance of our environmental and anthropological limits – not least the fact that we are flesh-and-blood mortals – of fundamental importance? Can transhumanism and Prometheism have any other consequence than worsening the ecological problems which threaten to actually end the human adventure? Are they not simply dead ends? As Negri points out, the *Manifesto* builds 'ever more effective defensive walls to protect us against death, and more generally against all the hazards of life'

[5] In the first part of the *Accelerationist Manifesto*, the authors mention the importance of taking account of the contemporary environmental situation; Yves Citton's contribution in the collective book *Accélération!* (de Sutter, 2016a, b) is devoted entirely to this issue. For this author, the situation is clear: 'the damage is already done, even if we struggle to see it, measure it and give it the attention it deserves. The catastrophe is not to come: it is already here, already done and still in the process of being done.' (Citton, 2016, pp. 215–216). Citton points out that 'there is no Planet B and [that] the fate of our living environment is necessarily our common concern. Thus, what we need to accelerate – and this is already unquestionable – are the transformations that will reverse the current course of our so-called 'economic development' which has, for two centuries, involved trashing the very conditions upon which our future well-being – and even our survival – depend' (2016, p. 207). For Citton, the *Manifesto* denounces the 'dire insufficiency' (2016, p. 209) of all the token gestures made at local level to benefit the environment, and the absolutely vital need to ramp up our actions: 'Yes, ultimately, it will indeed be the widespread adoption of such small gestures in our daily habits that will set our lifestyles on paths that are both sustainable and emancipating. And yes, this is undoubtedly where each of us can start at the individual level, since changing our modes of consumption, communication, production, sharing, and attention is more immediately achievable than changing the "world order". But no, a movement cannot be reduced to its first and last steps. It is just as much the in-between that is decisive. And it is this in-between – necessarily collective, inescapably global, and with intense media coverage – that we must imperatively accelerate' (Citton, 2016, pp. 209–210).

(Negri, 2016, pp. 58–59). But do we need to be protected against death? Is our own mortality not one of the most effective motivators which spurs us into action?

It would seem that anthropological finitude, including the idea of limits, is the blind spot of most accelerationist authors,[6] who appear not to realise that one of the problematic roots of capitalist society – though so much denounced – is precisely in the erasure of limits in favour of the unlimited pleasure of consumer accumulation. Is a Promethean and progressive critique powerful enough to overthrow capitalism?

Following on from Srnicek and Williams, the authors of *Accélération!* outline the contours of a post-capitalist society and the need for political acceleration. They stress the inadequacy of the autonomous alternative initiatives of groups of individuals that have arisen here and there. The processes of societal change envisaged are poorly developed and are thought of against a background of a revolutionary and categorical shift in our behaviour. Yet, how can we envisage lasting political change which is not deeply rooted in individuals' consciousness, learned and established over time? Everything goes fast in the *Accelerationist Manifesto*. Orientations for the direction of travel are introduced by decree and implemented, rather than learned over time, through teachable situations.[7]

7.1.3 Anthropological and Political Implications of the Primacy of Life in the Anthropocene Manifesto

In his work, Andreas Weber proposes a critical theory that thinks of the conditions for an anthropological shift by transitioning from *enlightenment* to *enlivenment*. What is particularly interesting about Kurt and Weber's philosophical/political stance (and perhaps Andreas Weber's work especially) is that the Anthropocene is not cause for despair. On the contrary, the vitality of the living world offers a strong possibility to revitalise the way in which we live, and restore solidarity – in other words, to politicise our existences. In contrast to the enthusiasm of the *Ecomodernist Manifesto*, which celebrates the power of technology and humans' incredible capacity to dominate their environment, it is the living world and the Earth itself that point the way in the Anthropocene. We are not the masters of the Earth; rather, the Earth – and the fabric of life – is our master. In this sense, the *Manifesto for the Anthropocene* proposes a break with all forms of anthropocentrism.

After laying the scientific foundations of their conception of life, the authors of the *Manifesto for the Anthropocene* outline what humans can become. They describe

[6] Again, apart from Citton, who distances himself from the Promethean attitude.

[7] The only allusions to education are made by Negri in his article 'Accelerating Politics', where he mentions that the political aim described in the manifesto needs to be re-appropriated through education (2016, p. 55), and by Laurent de Sutter in his presentation of the accelerationist theses, where he mentions that 'If we really wanted to get rid of capitalism, then we had to start by learning to go faster than it' (de Sutter, 2016a, b, p. 10), and where we understand that learning is necessary.

the human adventure having a potential that is the opposite of its technicist reduction. The tone is somewhat romantic, if not spiritual, and we understand from reading it that an anthropological shift is not merely *possible* – it is absolutely *necessary*. Such a shift can be made with reference to the soul, the heart, or even Erich Fromm's 'spiritual nature' or Gary Snyder's irreducible wild.[8] Education in the Anthropocene has a prominent place in Weber and Kurt's manifesto, though they do not explicitly refer to it in this way. For them, it is about facilitating the development of human skills: 'An ecologically and socially equitable future will only be possible in the Anthropocene through emphatic self-understanding'[9] (Weber & Kurt, 2015, p. 15). This 'emphatic self-understanding' refers to believing in one's own abilities and putting them to work. These abilities lie in the existential register of empathy, a sense of justice, a sense of proportion, self-awareness, imagination, relational strength, and the capacity to bring peace.[10]

When we experience the world as a living, unknowable entity, it becomes impossible to understand others and materials as mere resources. In this sense, this scientific revolution of the living with the power of solidarity and this organic totality functioning in a systemic way allows us to develop fundamental political conceptions, which are necessary if we are to think, live and act together in the Anthropocene. This understanding of the living is oppositional to capitalism. Thus, the authors state that 'the most important task in the Anthropocene is to rethink vitality and produce it anew'[11] (Weber & Kurt, 2015, p. 13).

The manifesto ends by proposing a politics of life, which refers to this proposal of '*Enlivenment*'.[12] This is a politics of civilisation that can come about through the establishment of a culture of vitality: 'A politics of life is guided by the idea of a civilisation whose principles, institutions and economic practices obey the principle of enlivenment. This ethic is not achievable in the short term. It requires a commitment comparable to that for the implementation of human rights'[13]

[8] In the chapter '*Serons-nous un jour, enfin, indigènes? Permaculture et éducation des profondeurs*' (Will we 1 day be indigenous? Permaculture and Education from the Depths), Christian Arnsperger draws on the thinking of the North American poet Gary Snyder to describe the type of anthropological shift needed in the Anthropocene.

[9] 'Eine ökologisch und sozial gerechte Zukunft wird es im Anthropozän nur von einem emphatischen Selbstverständnis ausgehend geben können' (Weber & Kurt, 2015, p. 15).

[10] They identify a series of authors from the twentieth century upon whose work it is possible to draw in order to refine this 'emphatic self-understanding'. They evoke Amartya Sen's and Martha Nussbaum's approach to capabilities; Manfred Max-Neef's approach to human needs; and *Albert Camus: Pensée du midi*. They also evoke the thinking of Edgar Morin, Michel Serres or the Commons movement.

[11] 'Deshalb besteht die wichtigste Aufgabe im Anthropozän darin, Lebendigkeit neu zu denken und neu zu erzeugen' (Weber & Kurt, 2015, p. 13).

[12] Note that many of the ideas sketched out in the *Manifesto for the Anthropocene* (2015) would later be developed further in *Enlivenment* (2016) and then in *Sein und Teilen* (2017).

[13] 'Eine Politik des Lebens orienert sich an der Idee einer Zivilisation, deren Prinzipien, Institutionen und Wirtschaftspraktiken dem Leitsatz folgen, das Lebendigkeit sein. Dieses Ethos ist nicht kurzfristig erreichbar. Es erfordert einen Einsatz vergleichbar dem für die Durchsetzung der Menschenrechte' (Weber & Kurt, 2015, p. 18).

(Weber & Kurt, 2015, p. 18). The purpose of a politics of life is particularly ambitious, but its mission is to enable us to move through the Anthropocene[14]: 'The purpose of a politics of life is that all beings – all, without exception! – have the right to be alive, which means the right to be fully themselves and at the same time fully in relations with others. Such a task can only be accomplished by engaging the community in deep solidarity over many generations'[15] (Weber & Kurt, 2015, p. 18).

In the conclusion of the *Manifesto for the Anthropocene*, it is possible to perceive the authors' hope: 'Only in such solidarity with all living things can the Anthropocene become an age of humans worthy of the name'[16] (Weber & Kurt, 2015, p. 20). Here we identify a reversal of human power. Their authors do not deny that humans have the capacity to accomplish great things, but hold that they must focus on weaving a web of solidarity with all living things. In other words, after having greatly contributed to the destruction of life and a considerable alteration of biodiversity, humans have the possibility of recreating this fabric, not primarily through the power of technology, but through a profound change of attitude. The anthropological shift we are aiming for consists of breaking away from the idea that the *anthropos* is a rational, individual being, and realising that we are, profoundly, feeling, sensitive beings, connected to others.

7.1.4 Digital Technology as a Means of Non-transhumanist Anthropological Evolution as Proposed by the Commons Manifesto

Just as the so-called 'Big Four' (GAFA) want to invent the future and make human beings into something new, the authors of the *Commons Manifesto* are convinced that digital commons have this power. Digital technology is at the heart of a social and political reorganisation; it even has the possibility to bring about an anthropological alteration, far from that outlined by transhumanist theories and research. Whilst there are similarities between the ways in which various authors propose to

[14] Among the goals the authors propose for a politics of life is an educational one: 'An education that does not normalise the pursuit of abstract knowledge, amazing technologies, and the study of a dead world, but that reduces evaluation and aims to instil wisdom and the art of living' (Weber & Kurt, 2015, p. 20). ('Eine Bildung, die nicht abstraktes Wissen, verzweckende Technik und eine Tote Welt zum abfragbaren Standard macht, sondern Bewertung reduziert und die Weisheit einer Lebens- und Bindungskunst zu ihrem Ziel macht' (2015, p. 20)).

[15] 'Es ist das Ziel eine Politik des Lebens, dass allen Wesen – ausnahmslos allen! – das Recht darauf zustehen soll, lebendig zu sein, und das heißt, ganz sie selbst und zugleich ganz in Verbindung sein zu dürfen.. Eine solche Aufgabe kann nur über ein für viele Generationen weitergetragenes Engagement im tiefer Daseinssolidarität gelöst werden' (Weber & Kurt, 2015, p. 18).

[16] 'Erst in diese Solidarität mit allem, das lebendig ist, wird das Anthropozän zu einer Menschenzeit, die den Namen verdient' (Weber & Kurt, 2015, p. 20).

achieve such an evolution, the visions of humanity in reference to which the proposed shift is to take place[17] are diametrically opposed. With the *Commons Manifesto*, we are dealing with the vision of an anthropological shift that places the focus not on the individual, but on the relationship between individuals. This relational space, which is the fulcrum of a post-capitalist society, is what needs to change gradually. Individuals have not only needs, the authors of the *Commons Manifesto* suggest, but also desires: they may want to participate in the creation of a collective adventure, widely useful and open to others.[18]

The way in which digital technology and the power of human technology are used and understood is unique here. It is not technology as such that will allow an anthropological shift and the advent of another type of society. On the other hand, it can support in the widespread adoption of a different mode of relating to others.[19] In this manifesto, there is no sense of wonder attached to technological tools (unlike, for example, the way in which they are viewed in the *Ecomodernist Manifesto*), but they are seen as being a necessary support.

The aim proposed by the *Commons Manifesto* is interesting, but it should be noted that this manifesto says nothing about *hybris*, the quest for power, the imperious desire to have more (whatever one possesses) and the drive to dominate. The unthinking of *hybris*, like the unthinking of Prometheism, are the main limits of this text. The further one reads in the *Commons Manifesto*, the clearer the aim of socio-political transformation and a transition from capitalism to a post-capitalist era becomes. On the other hand, the authors omit an in-depth anthropological analysis of the capitalism which they hope to leave behind. The post-capitalist aim must be accompanied by a new vision of what human beings are, which must be gradually embodied and learned. Mere experimentation with a technological tool – even if it is particularly interesting for the development of a relational style that is necessary in the Anthropocene – is insufficient.

7.1.5 *A Profound Transformation Through Shared Feeling, as Proposed in the* Animalist Manifesto

For Corine Pelluchon, the fight for the animal cause is a fight against Promethean obsession. Contrary to a technicist perspective and maximisation of our power, the animal cause requires that we enter into a relationship with the 'other' that is non-humans, and that we have an acute awareness of our vulnerability and our

[17] The term 'anthropological shift' is not mentioned as such by the authors of this manifesto.

[18] The dynamic of participation in a collective adventure has much in common with the Arendtian conception of action.

[19] The cardinal value appears to be autonomy rather than freedom. Moreover, the conception of freedom is a form of creative and relational freedom that is not the same as the individual freedom to maximise profit.

dependence on the living fabric of which we are part. We must profoundly transform the mode of relationship that we have established with animals, which has considerably deteriorated since the dawn of the industrial period and the development of a meat industry.

The author denounces this Promethean root of *hybris* and the maximisation of individual interests which allows us to trade with everything that 'passes under our hand', including animals, with whom we share a sentient existence. Corine Pelluchon defines sentience as follows: it 'is not reduced to an individual's capacity to suffer; it includes the desire to live and flourish, the fear of death and resistance to the conditions imposed on it, as well as the expression of pleasure and one's willingness to cooperate or form bonds' (Pelluchon, 2017a, b, p. 50). Defending the animal cause will involve abandoning the anthropological characteristics of domination and Prometheism (the limitations of which are now becoming painfully clear, with the entry into the Anthropocene): 'In contrast to the dehumanisation engendered by the need for mastery, the fear of others, and the will to dominate, [animalists] propose an ethic of vulnerability combining the acceptance of finitude, responsibility towards nature and towards all living beings, and the love of life' (Pelluchon, 2017a, b, p. 63). Pelluchon calls for a rehumanisation or a new humanisation based on a profound transformation of our relationship to animals. Animalism is thus a political movement based on an anthropological shift[20] which involves the overhaul of democracy and the abandonment of capitalism.

7.1.6 *The Invention of a 'Different We' in the* Manifesto for Climate Justice

The *Manifesto for Climate Justice* is a call to action and an encouragement for activists to continue their commitment because of the many battles already won in recent years. Strictly speaking, its authors do not aim for an anthropological transformation, but rather to win very specific political battles. On the other hand, the authors write: 'Let us invent a different We' (Notre affaire à tous, 2019, p. 20). They then continue: 'Let us create a We of resistance and vigilance'. Through this 'new We', we identify a form of anthropological alteration or anthropological consolidation of the political – although the manifesto does not explicitly define the extent or depth of this change. On the other hand, it seems to us indispensable here to add a 'Utopian We' to this 'We of resistance' and this 'critical We', so as not to descend into a form of violence. It is a question of continuing to dream and allowing dreams to nourish commitment and to balance the gravity of growing critical knowledge.

[20] This term is not used by Corine Pelluchon; however, she makes explicit the need for a new vision of humanity and a new philosophy of the subject.

7.1.7 *Ecofeminism as a Pillar for the Anthropological Shift Proposed in the* Integral Ecology *Manifesto*

A chapter of the *Integral Ecology* manifesto is devoted to political ecofeminism, which identifies an analogy between human domination over nature and the domination of men over women. Ecofeminism is the fulcrum for thinking about an 'anthropological revolution' (Batho, 2019, p. 85). The tone is radical and reflects the need for an anthropological shift. The author hopes for a form of anthropological mobility by calling on every woman and man to become 'Earthling' in order to enter 'A new age of humanity' (Batho, 2019, p. 89). This text manifests a real and profound awareness of the need for an anthropological shift. Integral ecology 'will draw its strength from an inner ecology that represents a new stage in human evolution' (Batho, 2019, p. 90). This manifesto is in line with Andreas Weber's radicalism: 'The mental revolution we need to accomplish is to embrace the whole fabric of living things as part of ourselves' (Batho, 2019, p. 90). It is to leave behind our utilitarian relationship with nature and enter into a relationship of immersion in and deep dependence on nature.

In the last two manifestos published (in 2019), the *Integral Ecology Manifesto*, and the *Manifesto for Climate Justice*), the authors build on recent research on the interactions between trees (which protect each other against aggressors). The living world is apprehended as a contiguous unit of which we are part (Batho, 2019, p. 91). The tone is resolutely one of coexistence, it is a matter of entering a new age of humanity where relationships between humans and between humans and nature are marked by symbiotic relationships of cooperation rather than relationships of rivalry: 'In this sense, it is an age of mutual aid and turns the page on individualistic values representing society as a war of all against all' (Batho, 2019, p. 95). The manifesto ends with these sentences: 'The new narrative of integral ecology is one of reconciliation with nature, with our deep nature as living beings, and with our fellow human beings. We view it as being a continuation of the French Revolution, as a new stage of human progress' (Batho, 2019, p. 95).

7.1.8 *From the Satisfaction of Needs to the Pursuit of Desires* (Convivialist Manifesto*)*

Convivialism denounces a single economic reading (with this logic of maximising individual interests) applied to everything in existence, making *homo oeconomicus* the only frame of reference in thinking about needs, remedies, and the aspirations of humanity. Given that humans have individuality and a concern for their interests, it is wrong to see them only as *homo oeconomicus*: they are capable of giving, loving, exchanging and sharing.[21] Continuing in this vein, the authors denounce the

[21] In this sense, convivialism is the political materialisation of the MAUSS movement, which has been working since the early 1980s to put economics back in its rightful place, notably by working on donation.

contemporary madness of the race for growth: 'With an average growth rate of 3.5% per year, for example, the world's GDP would be multiplied by 31 in a century. Can we imagine 31 times more oil, uranium, or CO_2 consumed in 2100 than today?' (Les Convivialistes, 2013, p. 18); or, 'The ecological state of the planet makes it necessary to seek all possible forms of prosperity without growth' (Les convivialistes, 2013, p. 33). 'Are material scarcity and the difficulty of satisfying material needs really the cause of conflict among humans?' seems to be the question posed by the *Convivialist Manifesto*. Here, the convivialists propose to move from an anthropology based on the satisfaction of needs to an anthropology based on the pursuit of desires.[22] For the advent of the new civilization that convivialists call for, self-limitation seems to be the key. However, this is a treat that is learned gradually.[23] Convivialism cannot do without educational thinking to accompany the anthropological and civilisational shift it calls for: 'This is what is at stake in a new conception of political emancipation and human emancipation: to create concrete forms of "good life", sober and convivial, capable of tempering self-interest with interest in others, and where all goods, including economic goods, are first and foremost generators of social bonds' (Fistetti, 2016, p. 256).

The authors of the *Convivialist Manifesto* emphasise that the widespread financialisation of the world was preceded by a set of ideas built around the gradual perception of individuals as separate from each other and each seeking to maximise their individual interests. The progressive rise to hegemony of *homo oeconomicus* in the sphere of ideas during the twentieth century has allowed this domination of market logic. This is the reason why it is possible not to reject out of hand the anthropological shift that convivialism draws and calls for. Putting ideas – even Utopian ones – into practice is not a waste of time.

7.2 Comparative Reading of the Anthropological Conceptions of These Nine Manifestos

7.2.1 *The Advent of Post-Promethean Social Spaces*

Whilst the political aims of the *Accelerationist Manifesto* and the *Ecomodernist Manifesto* are opposites (planned postcapitalist economy *versus* liberalised capitalist economy), both their approaches are hypermodern, proposing to accomplish the Promethean project of modernity. On the other hand, accelerationists apprehend productivism differently from ecomodernists, since it will allow the dawn of a

[22] This intellectual approach is very similar to that of Maurice Bellet, who proposed that desire should be altered.

[23] The authors do not explicitly use the term 'anthropological shift', but this is the drive of the manifesto: 'It is a new humanism, radicalised and extended, that needs to be invented, and this involves the development of new types of humanity' (Les convivialistes, 2013, p. 23).

post-capitalist era[24] (whereas for ecomodernists, the continuation of productivism does not involve a rupture). The *Manifesto of the Appalled Economists* proposes a rethinking of finance based on a logic of social justice, but its objective is not to overcome capitalism.[25] The main characteristic that differentiates the other manifestos from these first three is their post-capitalist aim, which, even more fundamentally, can be described as post-Promethean. The last six manifestos (*Manifesto for the Anthropocene*, *Commons Manifesto*, *Animalist Manifesto*, *Manifesto for Climate Justice*, *Manifesto for Integral Ecology*, and *Convivialist Manifesto*) appear post-modern, characterised by a break with modernity and its linear temporalities of progress. We are entering a new world, these manifestos tell us, and it is important to learn to live differently. The overhaul of democracy and the escape from capitalism appear in the vast majority of manifestos as necessary steps in preparation for the future and the advent of a better and fairer world.

The *Accelerationist Manifesto* is an interesting indicator of the post-Promethean component underlying the other manifestos. We can see how accelerationists are voluntaristic and committed to control. For example, the authors write: 'The new social movements which emerged since the end of the Cold War, experiencing a resurgence in the years after 2008, have been similarly unable to devise a new political ideological vision. Instead they expend considerable energy on internal direct-democratic process and affective self-valorisation over strategic efficacy, and frequently propound a variant of neo-primitivist localism, as if to oppose the abstract violence of globalised capital with the flimsy and ephemeral "authenticity" of communal immediacy' (2013, paragraph 2.5). This reading of the 'Arab Spring' overlooks the 'miracle' of the emergence of concerted action and points to insufficient use of force for efficacy. In contrast, we perceive the extent to which the political approach of the *Anthropocene Manifesto* and of the *Convivialist Manifesto* is romanticised, favouring an existential interpretation of life within a community. Should efficiency take precedence in achieving a political goal? Continuing to use the *Accelerationist Manifesto* as an indicator of the post-Prometheism of the other manifestos, we note that the Accelerationists regularly mention the goal of collective self-mastery: 'a Promethean politics of maximal mastery over society' (paragraph 3.21); and 'a time of collective self-mastery' (paragraph 3.22). The term 'self-control' is also used by the convivialists,[26] but here, in contrast, it means the identification and observation of incipient *hybris*. It is not so much a question of 'becoming masters and possessors of ever more powerful techniques, as masters and possessors of ourselves' (Caillé, 2011, p. 97). The meaning of mastery here is diametrically

[24] Their argument on this point is not convincing.

[25] The appalled economists on the other hand, have a point in common with the convivialists in their fight against 'the profiteering and speculative excesses of the financial economy which are the main cause of capitalist excess' (Les convivialistes, 2013, p. 33).

[26] Alain Caillé, in an article published in 2011 in the journal *Projet* entitled '*Au delá du libéralisme, le convivalisme*' (Beyond Liberalism, Convivialism) talks about the need to develop such collective self-control.

opposed to that of the ecomodernists, where the objective is the attainment of power through technical mastery.

The *Commons Manifesto* has a consequent interest in the reading of the other five post-Promethean manifestos because it shows that technology can be utilised to achieve this end. The *Commons Manifesto* views digital technology as that which facilitates the transition to a post-capitalist social and political system, as in the *Accelerationist Manifesto*, but without flirting with transhumanism, nor with possibly violent political voluntarism. The advent of post-Promethean social spaces thus in no way signifies anti-modernism or the use of revolutionary violence. The *Commons Manifesto* and the *Convivialist Manifesto* are very close in their shared pillars. For example, the convivialists point out that a post-growth society will not come about without a commitment to the defence and development of the commons – be it digital, educational or environmental.[27] Conversely, the authors of the *Commons Manifesto* highlight the fact that there are other motives for action than the quest for profit and emphasise the idea of giving.[28]

7.2.2 Relationships at the Heart of Politics (and of a Politics of Life)

The *Accelerationist Manifesto* draws on the power of networks, as does the *Commons Manifesto*. Both place great importance on technical infrastructure in the deployment of socio-political systems (the authors of both texts emphasise how technical infrastructure can serve the needs of capitalism, or, on the contrary, serve a post-capitalist agenda). On the other hand, the accelerationists call for particularly strong vertical political power, to counter the broadness and horizontality of the networks, which the authors of the *Commons Manifesto* do not, allowing the networks around the commons to take shape, be organised and led autonomously, without interference from vertical political power. Here, they rely on the autonomous and self-regulating power of the commons, based on the way Wikipedia has supplanted the best encyclopaedias, how open-source software has been able to replace Microsoft, or how WikiLeaks has not been brought down by State apparatus. Their vision entails the marrying of the commons (based on initiative and production by peers) with cooperativism (which is closely linked to convivialism). The relational style in reference to which we are to conceive of and prepare for the future has much in common with the *Convivialist Manifesto*. On the other hand, the latter remains

[27] The *Anthropocene Manifesto* also mentions the importance of the commons (which is a central theme in other works by Andreas Weber). On the other hand, the approach outlined in this manifesto is decidedly romantic, in contrast to the technicality of the *Commons Manifesto*.

[28] The figure of the entredonneur in the *Commons Manifesto* (as distinct from that of the entrepreneur) is semantically close to the MAUSS emphasis on donation.

more radical, because of its incisive denunciation of *hybris* (which the *Commons Manifesto* does not share).[29]

The relational style of the *Accelerationist Manifesto*, on the other hand, is far from convivial autonomy. In the call for centralised power, we perceive the germ of a totalitarian component that opposes the anarchic forms of the *Commons Manifesto*, convivialism or the poetics of enlivenment of the *Manifesto for the Anthropocene*. Ultimately, the staging of the force of accelerationism is not radical – the anthropological root of *hybris* remains unchanged. The seminal authors referred to in the *Accelerationist Manifesto* are Karl Marx and Nick Land, but the authors of the manifesto do not hesitate to also employ the thinking of Lenin, whom they quote at length: 'Indeed, as even Lenin wrote in the 1918 text "left-wing" Childishness': 'Socialism is inconceivable without large-scale capitalist engineering based on the latest discoveries of modern science. It is inconceivable without planned state organisation which keeps tens of millions of people to the strictest observance of a unified standard in production and distribution. We Marxists have always spoken of this, and it is not [worthwhile] wasting two seconds talking to people who do not understand even this (anarchists and a good half of the Left Socialist-Revolutionaries)' (2013, paragraph 2.6). The tone here is the opposite of that of the *Convivialist Manifesto*, postulating that it is possible to disagree without slaughtering each other and that altering dialogue is not only possible, but absolutely necessary. *Convivialism* claims to be resolutely democratic (which *accelerationism* is not): 'The growing impotence of political parties and institutions to confront the problems of our time and to win, or keep, the confidence of the greatest number of people, is explained by the inability to reformulate the democratic ideal – the only acceptable one because it is the only one that allows for opposition and conflict' (Les convivialistes, 2013, p. 19).

The *Manifesto for the Anthropocene*, with its emphasis on co-creativity (Weber & Kurt, 2015, p. 19) at the foundation of a politics of life is an extension of this 'democratic inventiveness' (Caillé, 2011) of convivialism, from this relational space. The aim of Weber and Kurt's idea of Enlivenment is radical. Here we find the same radicality as among the convivialists or in Maurice Bellet's thinking: 'Enlivenment politics seeks alternatives to the dogma of growth and consumerism. It does not rely on technological control, but invigorates life. It facilitates material productivity through ecological stability, and does so through meaningful action'[30] (Weber & Kurt, 2015, p. 19). Andreas Weber and Hildegard Kurt are aware of the German translation of the *Convivialist Manifesto* and, in their conclusion, partly refer their own manifesto to Convivialist thinking. They define a goal of this politics of life as 'Coexistence with other beings according to the South American ethic of

[29] Thus, while the *Commons Manifesto* offers an interesting critique of contemporary capitalist lifestyles, it is far less radical than the *Convivialist Manifesto* in its denunciation of problematic anthropological foundations.

[30] 'Eine Politik des Lebens sucht Alternativen zum Wachstumsdogma und zur Konsumsucht. Sie setzt nicht auf technische Kontrolle, sondern macht Lebendigkeit erfahrbar. Sie ermöglicht materielle Produktivität durch ökologische Stabilität, und diese durch sinnhaftes Handeln' (Weber & Kurt, 2015, p. 19).

"Buen Vivir" or the newer ethics of "conviviality" recently presented by Frank Adloff and others aiming for the creation of a community and solidarity among all creatures'[31] (Weber & Kurt, 2015, p. 20). We perceive that their response to convivialism comes through the prism of their thinking, which aims for solidarity throughout the living world, because in the 2013 *Convivialist Manifesto*, this dimension of conviviality with the whole living fabric is not mentioned as such.

7.2.3 The Radicality of an Anthropology of Immersion in Nature

Ecomodernists and accelerationists (and, to a lesser extent, the appalled economists) advocate an anthropology that no longer views nature as a resource to be exploited.[32] The anthropology of the authors of other manifestos is more marked by an immersion in nature. This is particularly strong in the *Manifesto for the Anthropocene*, where the human adventure is immersed in the whole fabric of solidarity that is life, but also participates in geophysical flows, and in the *Animalist Manifesto*, where humans and animals belong to the same category of sentient beings. The anthropological conceptions at work in these two manifestos provide a solid foundation for thinking, living and acting in the Anthropocene. The age of the living evoked in the *Animalist Manifesto* and the supportive and resilient power of the vitality of organic tissues in the *Manifesto for the Anthropocene* demarcate a sort of path to follow, some way removed from any ecomodernist conception.[33] We are a species that needs others: mosquitoes at the base of the food chain, bees that pollinate our vegetables, earthworms that aerate our soils, and the bacteria in our intestinal flora. These manifestos outline a new anthropology where humans are not viewed as an entirely separate and independent entity, but as one entity among myriad others that are immersed in the living world. Moreover, there is a reversal of perspectives: it could be that our observation of other living beings could provide elements for reflection about human beings. Thus, trees and animals can allow us to think differently and to make changes in ourselves in order to survive in the Anthropocene: 'What we have failed to understood, the tree knows. Why does it stop growing? It leaves room for other trees; it does not want to be the whole forest, nor does it want to invade the sky. Man wanted to be the master, not just the inhabitant' (Notre affaire à tous, 2019, p. 12). The *Convivialist Manifesto* does not explicitly speak of such an anthropology; on the other hand, the centrality of relationships and

[31] 'Ein Zusammenleben mit andere Wesen nach Maßgabe des südamerikanischen Schöpfungsethos des "Buen Vivir" oder der jüngst von Frank Adloff und anderen präsentierten "Konvivialität", der Gemeinschaftlichkeit und Daseinssolidarität aller Geschöpfe' (Weber & Kurt, 2015, p. 20).

[32] Ecomodernists point out that treating one part of nature as a resource allows another part of nature to remain untouched, which they say is essential to spiritual and psychological well-being.

[33] For example, the type of agriculture favoured by the ecomodernists is industrial agriculture, which they encourage to continue developing. This is a far cry from permaculture or the kind of relationship to nature described in the *Manifesto for the Anthropocene*.

sharing of existence completely opens up the possibility of this shared conviviality with all living things – which is, moreover, made explicit in the new version of the manifesto with the idea of 'common naturality'.

The anthropology of most of the manifestos studied is at odds with this dominant anthropology of modernity of being separate from nature (the author of the *Integral Ecology* manifesto writes, for example, that it is a question of 'making our belonging to Nature the new driving force in history' (Batho, 2019, p. 13)). Anthropological conceptions of immersion in nature have direct economic implications. Thus, the economic proposals of the *Manifesto for the Anthropocene* centre upon an economy of the commons, conceived on the basis of belonging to the biosphere, and radically break away from a capitalist system. This is not at all the case, for example, in the *Manifesto of the Appalled Economists* or the *New Manifesto of the Appalled Economists*. While attentive to the centrality of the ecological component in the organisation of human societies, this second manifesto is not based on an anthropology of immersion in nature, and does not propose a break with the contemporary capitalist system. The anthropology of immersion in nature also has a set of social implications. This can be seen in the *Manifesto for the Anthropocene* and the *Animalist Manifesto,* with understanding of the social and political body expanded to the adventure of life as a whole, but also in the *Manifesto for Climate Justice* and the *Integral Ecology* Manifesto, which thinks of social matters in relation to environmental issues.[34]

The *Animalist Manifesto*, for its part, is not content with defining a new morality, but contributes to a new anthropological conception based on a new age that Corine Pelluchon calls the age of the living. Thus, this manifesto is at the heart of a renewed philosophy of the subject. The entry into the Anthropocene requires a broadening of humanism, which can no longer be thought of as an entity 'in itself' based on a form of 'human exceptionalism' in the history of life, but, on the contrary, based on immersion in life and participation in the adventure of life within the earth's biosphere. Thus, it is impossible to think of humanity without the cosmic space it inhabits – the Earth – and its coexistence with living beings in all their forms, including animals: those beings with whom we share sentience. The movement to redefine a political anthropology in the Anthropocene is comparable to that outlined in this book.

7.2.4 Convivialist Radicality

In reading these different manifestos, we see a dual interest in convivialism that deserves to be noted. The first lies in its capacity to serve as a common denominator for a set of alternative thoughts currently emerging around the entry into the

[34] For example, Delphine Batho stresses the fact that today, eight individuals hold as much money as 3.6 billion people who are the first victims of climate change and systemic changes in the functioning of the biosphere (for which the richest are most responsible).

Anthropocene. Convivialism allows us to make links with Weber and Kurt's politics of life, Pelluchon's animalism or the commons of Bauwens, Kostakis and Pazaitis.

The second interest of convivialism is that this capacity to gather alternative thoughts does not in any way remove their radicality (from the Latin *radix*, meaning root). On the contrary, convivialism makes it possible to grasp this destructive human *hybris*, which must be identified, mastered, and overcome through an anthropological shift. Pascal Glémain speaks of this need for a new radicality for the contemporary period, which could take the form of humanist radicalism (or convivialist radicalism),[35] such as what he identifies within the social and solidarity economy on which he works (2017). Humanism, which has carried the idea of human exceptionalism and may have been blind to humanity's domination of nature, may be an equivocal notion. The idea of convivialist radicality seems more interesting because, whilst it must obviously be able to be lived on a daily basis among humans, it must also be able to be experienced with non-human living entities, the plants and animals with which we coexist. Indeed, this is one of the challenges of the Anthropocene. To survive and thrive in this new epoch, we need a firm – and flexible – grip. Firmly anchoring ourselves in the powerful solidarity of living things could provide this life-saving support.

Moreover, convivialist radicalism claims to be non-violent – or rather, it proposes to mobilise forces of opposition without actual destruction. Convivialism, with its 'friendly' and seemingly harmless name, mobilises radical alternative political imaginaries. Being together, in addition to being possible, can become the basis of our shared existence: we coexist more than we exist, with other humans and sharing possible conviviality with non-human life. The radical component of convivialism also lies in the interweaving of these three functions of utopia, critique and resistance: it is possible to live together (utopia), but *hybris* is at the heart of human beings, and has become the norm in how our organisations and institutions function (critique), which is why it is necessary to stand up against the destructive economic hegemony (resistance). Here, the radical alternative to capitalist hegemony is not to transfer the force of the power of capital – and of the market – today to a strong State, as the accelerationists propose. Convivialist radicality resides precisely in divestment of this force, leaving room for initiative as to what happens 'among us'.

After reading these different manifestos, whose goal is the preparation of the future, it is important to note that each of them is founded upon an act of faith. This act of faith is as much about the anthropological resources of the human adventure as about the future as such. The convivialist act of faith appears to be radical, non-dangerous (the vision sketched out does not involve a systemic danger to the human adventure), intellectually honest (we have all already experienced sharing existence with others), and interesting from an educational perspective (educability being the thread underlying the convivialist anthropological conceptions).

[35] A notable difference between this convivialist radicalism and the radicalism of the Third Republic is that it is not primarily the progress of science that can bring about a new social order, but progress in relational style.

The main limitation of convivialism is that, for now, it is a poorly theorised political idea. The manifesto gives general guidelines and some authors have just begun a work of theoretical consolidation.[36] It is quite simple to say what should be thought or done, but to a large extent, the analysis of the paradoxes, tensions and contradictions of this proposed political orientation remains to be carried out.

7.3 An Earthling Anthropology

The discussion now continues in search of an earthling anthropology. Until now, human history and natural history have obeyed independent logics. Humanity's status as a geological agent brings about a paradigmatic shift.[37] The entry into the Anthropocene reconfigures the opposition in thought between nature and society: 'In this age-old affair, the Anthropocene poses a disarmingly simple question. If natural and human processes are connected through and through, if the natural curves of the Great Acceleration mimic the social curves, does it still make sense to pretend that nature and society are autonomous concepts?' (Federau, 2017, p. 311–312). What the Anthropocene means is the association, if not the 'fusion', of human history with natural history (Parizeau, 2016, p. 23), rendered possible by action through technology. Nature has been experimenting with a wide variety of life forms for 3.7 billion years, and as an extension of that series of experiments, *Homo sapiens* appeared only 195,000 years ago: we can say that 'we are *of* and *in* nature' (Deléage & Coutellec, 2015, p. 60). In her essay, *Homo natura*, French jurist Valérie Cabanes highlights our immersion in nature and the cosmos as well as our dependence on it: 'The essential element of organic compounds is carbon, which has been present on Earth since its formation: it was produced by nucleosynthesis at the heart of the stars that exploded before the formation of the solar system. Therefore, we really are stardust' (Cabanes, 2017, p. 40).

This perception of the problem posed by the distinction between humanity and nature predates the concept of the Anthropocene. Indeed, the French anthropologist Claude Lévi-Strauss had clearly identified the problematic severance of humanity and nature forced by modernity: 'We began by cutting man off from nature, and by constituting him as a sovereign kingdom; we believed that by doing so, we were erasing his most irrefutable character – namely that he is, first and foremost, a living being. And by remaining blind to this common property, we have given free rein to all types of abuse. Never better than after the last four centuries of history can

[36] Of note here is the work of Alain Caillé (2011, 2015, 2016), Francesco Fistetti (2016), Pascal Glémain (2017), Marc Humbert (2017), and the collective work we have initiated (Wallenhorst et al., 2020).

[37] Paradoxically, the Anthropocene reverses the analyses initiated by the famous Polish scientist Nicolas Copernicus in the sixteenth century and the famous English naturalist Charles Darwin in the mid-nineteenth century on the non-centrality of humanity within its cosmic environment and among the living!

western man understand that by appropriating the right to radically separate humanity from the animal kingdom, by granting to the one all that he denied to the other, he began a vicious cycle, and that the same boundary, constantly drawn back, would serve to keep men away from other men, and to claim, for the benefit of ever smaller minorities, the privilege of corrupt humanism immediately created by having borrowed the principle and notion from self-love' (Lévi-Strauss, 1973, p. 53). The French anthropologist François Flahault goes even further, showing in *L'homme, une espèce déboussolée – Anthropologie générale à l'âge de l'écologie* (2018) that human beings have a mode of existence that is ecological and cosmological *by nature*. Humans cannot be thought of without the full range of interactions with their peers and their environment. Also in 2018, Dominique Bourg published an essay entitled *Une nouvelle terre* (A New Earth) in which he proposes to go beyond the dualism of modernity to propose another figure of modernity – a monist one, 'conscious of the irreducibility of its spiritual foundations, having renounced the myth of infinite growth, concerned with the contradictions between the market and freedoms, having relativised the notion of risk, reinterpreting human rights by turning its back on both anthropocentrism and rugged individualism, rediscovering the speculative nature of knowledge, and discerning in technology more an accompaniment of nature than a domination-destruction' (2018, pp. 16–17).

The Anthropocene, which constitutes a new geological era, also constitutes a break in humanity's conception of itself. It thus seems necessary to rethink the human condition. Indeed, awareness of the entry into the Anthropocene results in a difficulty in thinking of humanity in an essentialised way against the background of an anthropology of humanity's separation from nature based on the idea of nature and, to a lesser extent, of condition. The geological power of human beings and the advent of the Anthropocene are impacting the definition we have held of ourselves since Aristotle introduced the concept of the *zôon politikon*. It is now necessary to think within the framework of an anthropology of immersion in nature (Papaux & Frigerio, 2015), since to know the human being is 'first of all to situate him in the universe, not to cut him off from it' (Morin, 2000, p. 49). This was identified by Hannah Arendt in 1958/1983, who wrote in the prologue to *The Human Condition* that 'The earth is the very quintessence of the human condition' (p. 8). As we have seen, there are different interpretations of the Anthropocene polarised around the control of planetary cycles by a form of *Homo deus* and the possible end of humanity due to human *hybris*. Bruno Latour proposed a categorisation that materialises this polarisation. The first pole is that of humans who go to the extremity of the modern logic of uprooting themselves from nature and the second is that of earthlings immersed in nature on which they are dependent (Latour, 2013).[38] Humans, who mostly conceive of themselves within the framework of an anthropology of separation from their condition with nature, must become earthlings by becoming fully aware of their belonging to the Earth.

---

[38] This is also the same tension that Alexander Federau addressed in a chapter entitled '*Martiens et Terriens: quelle anthropologie pour l'Anthropocène?* (Martians and Earthlings: what anthropology is needed for the Anthropocene?) (2019).

Thinking of man as part of nature, as a product of the Earth (Flahault, 2013), leads us to think of the issues of environmental sustainability and anthropological sustainability as linked.[39] For Aristotle, the idea of human nature refers to the idea of nature, with which it is partly linked. In this extension, Hans Jonas, with the *Imperative of Responsibility*, developed an ecological philosophy in which man is linked to nature. In fact, Jonas develops a valuable way of thinking about humanity and its environment together. He allows us to 'break with the abstract opposition of a nature that would have value in itself independently of man, and a humanity that would be the exclusive centre of all values. It is the relationship of one to the other that is the only real one, or to be more precise, the relationships of the living to their living environments' (Pommier, 2013, p. 495). Jonas particularly clearly perceived two great threats: that of the destruction of the environment (an ecological threat) and that of the destruction of humanity by posthumanism (an anthropological threat). For Jonas, the first imperative is the long-term continued existence of human beings, who must be preserved and protected from the anthropological transformation generated by technoscience. Only humans are the guarantors of their own survival. This is the foundation of the principle of responsibility.

If we are to think of an earthling anthropology, it must be distinguished from geological omnipotence. Before being the mark of a power, the Anthropocene is also the mark of humanity's immense vulnerability. Awareness of this entry into the Anthropocene structures a new relationship with the future. This is a strong characteristic of the contemporary period; it marks a break with the relentless pursuit of progress, allowing for an improvement in daily life. We have left the ascending linearity of western modernity. The entry into the Anthropocene leads us into a logic that is opposed to that of modernity: the further we advance, the more constricted the possibilities are. Beyond the doubt about the improvement of quality of life in the future, it is the very existence of a possible future that is the subject of uncertainty at present.

For the American historian of science Naomi Oreskes (2015), the Earth system sciences and the analytical models developed by their researchers must increasingly take social factors into account, and work in concert with social scientists if they are to produce accurate analyses. This author takes as an example the question of forecasting greenhouse gas emissions, which is based on political orientations, the choice of techniques used and the behaviour of citizens. Social issues thus appear to be directly linked to scientific forecasts. Behind her analyses of a range of scientific situations and predictive models (hydrology, climate sciences, seismology), Oreskes shows that the relationship between technology and the social has changed. While social considerations have always been part of Earth science analyses, they have been secondary, and now they come to the fore. It is thus in line with the analyses of

[39] The notion of sustainability incorporates human finitude. Recent decades have been marked by a series of transgressions: 'Modernity has indeed dedicated itself to a constant intolerance to the idea of limits, in many fields, whether it be technology, aesthetics, sport, or even ethics, etc., with, of course, first and foremost, the postulate of infinite economic growth' (Bourg, 2012, p. 9). Sustainability implies respecting limits and refraining from transgressing them.

the historian Chakrabarty, for whom the social and the natural can no longer be analysed separately (2009). Oreskes (2015) makes clear the epistemological problem this raises, and her arguments are in line with Federau (2017). Indeed, natural science models tend to proposing predictive analyses and do not hesitate to do so concerning human behaviour. However, history and the social sciences have constantly shown the unpredictability of human behaviour.[40]

7.3.1 A Move Towards Surviving an Anthropological Crisis

In *The human condition*, Arendt links the destiny of humanity to the Earth, in acute awareness that it is not entirely impossible for humans to manage to emancipate themselves from the Earth that birthed them. This is a transversal theme of this book, present from the prologue to the end. It shows the gradual distancing from nature in which humans are immersed: 'Before we knew how to go around the Earth, to circumscribe the sphere of human habitat in days and hours, we had put the globe in the living room to feel it and make it rotate before our eyes. [...] The decisive shrinking of the Earth followed the invention of the aeroplane –a means of actually leaving the surface of the Earth. This fact is like a symbol of the general phenomenon: one can decrease the terrestrial distance only on condition of putting a decisive distance between man and the Earth, only on condition of alienating man from his immediate terrestrial environment' (Arendt, 1983, p. 282). Yet in addition to trying to tear themselves away from the Earth,[41] humans have acquired the ability to destroy it and the biosphere. This conviction, that we 'shall probably be able one day to destroy even the earth itself' (Arendt, 1983, p. 301), is one of the foundations

[40] In the wake of Part I of this book, we may legitimately wonder whether the appellation *Homo faber*, to define human beings, would not be wiser (*sapiens*...) than *Homo sapiens* (whose etymology refers to knowledge that is linked to wisdom and intelligence). (It is fitting, though, to recall the reason for the appellation *Homo sapiens* following *Homo faber*: it is as a result of their productions that human beings have been able to begin to elaborate ideas. Action precedes reflexivity and the development of thought). This is what Henri Bergson mentioned in 1907, saying: 'If we could rid ourselves of all pride, if, to define our species, we kept strictly to what the historic and the prehistoric periods show us to be the constant characteristic of man and of intelligence, we should say not *Homo sapiens*, but *Homo faber* (p. 140). A few years later, Vladimir Vernadsky emphasised the anthropological importance of *Homo faber* in our relationship to matter: 'Man has introduced a new form of action of living matter with raw material. It is no longer only the elements necessary for the production and formation of living matter that come into play here and change its molecular edifices. They are elements necessary for technology and the creation of civilised forms of life. Man acts here not as *Homo sapiens*, but as *Homo faber*' (Vernadsky, 1924, p. 342). Alain Papaux, for his part, makes explicit in 2015 (p. 539) the disproportion between our capacity to do and to think of the consequences of our actions, following Arendt's intuition, who mentions that the importance attached to practical knowledge hinders the development of our thinking (Arendt, 1983, p. 9–10).

[41] Arendt, in *The Human Condition,* analyses the modern alienation of the flight from the Earth to explore the universe.

of *The Human Condition*, on the basis of which it is possible to follow in the footsteps of Arendt's thinking, and consider the anthropological conditions of an authentically human life, possibly preserved from destruction. This does not happen without a crisis.[42] In addition to becoming more collectively earthbound, one of the important elements of the anthropological consolidation of the political identified in this second part is to weather the storm of the contemporary anthropological crisis. This is why the next chapter is devoted to a study of the work of the philosopher Maurice Bellet. His entire work is concerned precisely with identifying the ways in which the contemporary *krisis* can be survived, starting with an anthropological shift.

Bibliographical References

Arendt, H. (1983). *Condition de l'homme moderne. Calmann-Lévy* (original edition 1958, French translation 1961), French translation.

Arnsperger, C. (2010). Changer d'existence économique: enjeux anthropologiques de la transition du capitalisme au post-capitalisme. *Revue d'éthique et de théologie morale, 258*, 23–50.

Arnsperger, C. (2016). Progrès et conscience, Eléments pour une anthropologie économique non réductionniste de la durabilité. In G. Hess & D. Bourg (Eds.), *Science, conscience et environnement, Penser le monde complexe* (pp. 179–206). PUF.

Batho, D. (2019). *Ecologie intégrale – Le manifeste*. Les éditions du Rocher.

Bellet, M. (1993). *La seconde Humanité. De l'impasse majeure de ce que nous appelons l'économie*. Desclée de Brouwer.

Bellet, M. (2013). *L'avenir du communisme*. Bayard.

Bourg, D. (2012). Transition écologique, plutôt que développement durable. *Vraiment durable, 1*, 77–96.

Bourg, D. (2018). *Une nouvelle Terre*. Desclée de Brouwer.

Brassier, R. (2016). Accélérer la raison. In L. de Sutter (Ed.), *Accélération !* (pp. 157–181). PUF. French translation.

Cabanes, V. (2017). *Homo natura – En harmonie avec le vivant*. Buchet Chastel.

Caillé, A. (2011). Au-delà du libéralisme, le convivialisme. *Revue Projet, 324–325*, 94–97.

Caillé, A. (2015). *Le convivialisme en dix questions. Un nouvel imaginaire politique*. Le Bord de l'Eau.

Caillé, A. (2016). *Les Convivialistes, Eléments d'une politique convivialiste*. Le Bord de l'eau.

Chakrabarty, D. (2009). The climate of history: Four theses. *Critical Inquiry, 35*(Winter), 197–222.

Citton, Y. (2016). Accélérer l'écologie. In L. de Sutter (Ed.), *Accélération !* (pp. 205–223). PUF.

Cuboniks, L. (2016). Accélérer le féminisme. In L. de Sutter (Ed.), *Accélération !* (pp. 253–273). PUF. French translation.

de Sutter, L. (Ed.). (2016a). *Accélération !* PUF.

de Sutter, L. (2016b). Introduction. In L. de Sutter (Ed.), *Accélération !* (pp. 7–25). PUF.

Deléage, J.-P., & Coutellec, L. (2015). L'écologie scientifique, une science impliquée? *Ecologie et politique, 51*, 55–64.

Dupuy, J.-P. (2007). Some Pitfalls in Philosophical Foundations of Nanoethics. *Journal of Medicine and Philosophy, 32*, 237–261.

Dupuy, J.-P. (2009). Le Futur bifurque-t-il? Vers une nouvelle science du futur. In M. Grossetti, M. Bessin, & C. Bidart (Eds.), *Bifurcations* (pp. 373–386). La Découverte.

[42] Crisis is another key theme in Arendt's work.

Federau, A. (2017). *Pour une philosophie de l'Anthropocène*. PUF.
Federau, A. (2019). Martiens et terriens: quelle anthropologie pour l'Anthropocène? In N. Wallenhorst & J.-P. Pierron (Eds.), *Eduquer en Anthropocène*. Le Bord de l'eau.
Fistetti, F. (2016). Le convivialisme, 'contre-mouvement' du 21ème siècle. *Revue du MAUSS, 48*, 247–258.
Flahault, F. (2013). L'homme fait-il partie de la nature? *Revue du MAUSS, 42*, 125–128.
Flahault, F. (2018). *L'homme, une espèce déboussolée – Anthropologie générale à l'âge de l'écologie*. Fayard.
Glémain, P. (2017). Penser le convivialisme en économie sociale contemporaine. *RECMA, 4*(346), 27–41.
Humbert, M. (Ed.). (2017). *Reconstruction de la société – Analyses convivialistes*. Presses Universitaires de Rennes.
Illich, I. (1971a). *Une société sans école*. Seuil.
Illich, I. (1971b). *Libérer l'avenir*. Seuil. French translation.
Illich, I. (1973a). *La convivialité*. Seuil. French translation.
Illich, I. (1973b). *Tools for conviviality*. Harper and Row.
Latour, B. (2013). *Facing Gaia Six lectures on the political theology of nature*. Being the Gifford Lectures on Natural Religion Edinburgh, 18th–28th of February 2013. https://macaulay.cuny.edu/eportfolios/wakefield15/files/2015/01/LATOUR-GIFFORD-SIX-LECTURES_1.pdf, consulté le 25 October 2017.
Les convivialistes. (2013). *Manifeste convivialiste*. Le bord de l'eau.
Lévi-Strauss, C. (1973). *Anthropologie structurale II*. Plon. (original edition 1962).
Morin, E. (2000). *Les sept savoirs nécessaires à l'éducation du futur*. Seuil.
Negri, A. (2016). Accélérer la politique. In L. de Sutter (Ed.), *Accélération !* (pp. 49–67). PUF. French translation.
Notre affaire à tous. (2019). *Comment nous allons sauver le monde – Manifeste pour une justice climatique*. Massot éditions.
Oreskes, N. (2015). How earth science has become a social science. *Historical Social Research, 40*, 246–270.
Papaux, A. (2015). Homo faber. In D. Bourg & A. Papaux (Eds.), *Dictionnaire de la pensée écologique* (pp. 536–540). PUF.
Papaux, A., & Frigerio, V. (2015). Droit de l'homme et écologie. In D. Bourg & A. Papaux (Eds.), *Dictionnaire de la pensée écologique* (pp. 292–296). PUF.
Parizeau, M.-H. (2016). De l'Apocalypse à l'Anthropocène: parcours éthique des changements climatiques. *Revue de métaphysique et de morale, 89*, 23–38.
Pelluchon, C. (2017a). L'éthique des vertus: une condition pour opérer la transition environnementale. *lapenseeecologique.com, PUF, 1*(1), 1–18.
Pelluchon, C. (2017b). *Manifeste animaliste – Politiser la cause animale*. Alma.
Pommier, E. (2013). Le sens de la vie chez Hans Jonas. *Etudes, 418*, 485–495.
Rosa, H., & Wallenhorst, N. (interview). (2017). Apprendre à écouter le monde. *Chemins de formation, 21*, 19–31.
Vernadsky, V. (1924). *La géochimie*. Librairie Félix Alcan. French translation.
Wallenhorst, N., Mellot, S., & Theviot, A. (2020). *Interconnectés? Numérique et convivialisme*. Le bord de l'eau. (*forthcoming*).
Weber, A., & Kurt, H. (2015). *Lebendigkeit sei! Für eine Politik des Lebens. Ein Manifest für das Anthropozän*. Klein Jasedow, Think Oya.

Chapter 8
Weathering the Storm of the Contemporary Anthropological *krisis*

Abstract The notion of crisis discussed in this chapter refers to the Great Acceleration curves, which highlight a civilisation-wide systemic problem. It is becoming apparent, through this research, that the means of getting through this civilisational crisis are anthropological, so this chapter questions the anthropological strategies for weathering the coming storm.

Keywords Anthropocene · Maurice Bellet · Anthropological plasticity · Anthropological change · Anthropological *krisis*

For Daniel Curnier, for example, the challenge of the Anthropocene, which consists of 'reintegrating social systems into ecological systems […] makes it necessary to solve an anthropological problem' (2017, p. 59). The concept of crisis,[1] or *krisis*, is not suitable to speak of the entry into the Anthropocene, because we are dealing with a genuine tipping point, having durably altered the conditions which make the biosphere habitable. On the other hand, this environmental situation is forcing us to make anthropological shifts, as has begun to be sketched out through the various manifestos. This moment of transition leads the human adventure to the heart of an unprecedented *krisis* and, to use the words of the French philosopher and theologian Maurice Bellet, must humanity not be born anew in order to ensure its continued survival? The entry into this new geological epoch of anthropic origin encourages us to think about possible anthropological shifts to correct the course we are on. To do so, it is interesting to draw upon the thinking of Maurice Bellet – a contemporary author to whom reference has already been made on several occasions in this work. In his work, Bellet weaves together three themes: *krisis*, humanity threatened by its own actions, and the birth of humanity. In Bellet's thinking, the political is frequently worked on the basis of an existential reflection on possible collective action

[1] Beyond the contemporary economic crisis, several authors evoke a crisis of civilisation. Examples include Raymond Aron and Edgar Morin, the latter of whom states unambiguously that 'the time has come to change civilisation' (Morin & Lafay, 2017).

© The Author(s), under exclusive license to Springer Nature Switzerland AG 2023

N. Wallenhorst, *A Critical Theory for the Anthropocene*, Anthropocene – Humanities and Social Sciences, https://doi.org/10.1007/978-3-031-37738-9_8

(Bellet, 1993, 1996, 1998, 2013). The interconnection of the three themes mentioned based on existential reflection is an interesting resource for thinking prospectively about the contemporary period, marked by the threat to life posed by the entry into the Anthropocene. In this chapter, we begin to sketch out the act of anthropological uprising or consolidation at the heart of this contemporary *krisis*. What will gradually emerge as the basis of the 'between-us' will be explored in greater depth in the following chapters; then we shall examine the underlying theory, through Arendt's work; and finally, present a model designed to also integrate non-humans into the 'between-us' at the root of an anthropological shift that will consolidate the human adventure's existence on Earth.

Maurice Bellet, in his works, does not show real concern for ecological matters. His anthropology remains anthropocentric or theocentric. On the other hand, what makes Bellet worthy of our attention is the anthropological proximity of the shift he calls for to the type of anthropological shift that can be brought about by reconnecting with the Earth. Moreover, his thinking on 'between-us' as the basis of what constitutes humanity and makes humans human is also a strong point of convergence with the political thinking of Hannah Arendt, and with the anthropology of Hartmut Rosa or Andreas Weber.[2]

Maurice Bellet passed away on 4 April 2018 in Paris. In the context of this work, we met this author four times in 2014, and conducted recorded interviews on his intellectual influences and the notion of anthropological shift to consolidate the political.[3] The intellectual and poetic gesture[4] peculiar to Maurice Bellet's work is indeed that of the critical theorists mentioned by Hartmut Rosa in the previous chapter. Maurice Bellet is relatively unknown in the field of human and social sciences, so this chapter begins with a quick introduction to the defining features of this author of more than 60 books, before developing a response to his work around the anthropological shift to consolidate politics in the Anthropocene. The anthropological themes worked on in relation to Bellet's idea of *krisis* resonate strongly with Arendt's thinking, which is sometimes understood as relating to crisis (Revault d'Allonnes, 2011; Ehrwein Nihan, 2013).[5]

---

[2] On the other hand, a limitation of Maurice Bellet's work for thinking in the present time is that his notion of 'between us' is exclusively human. The Earth, animals and things are not included in this 'between us that makes us human'. The same is true of Arendt's thinking.

[3] We have reworked these interviews into a dialogue book (Bellet & Wallenhorst, 2022).

[4] What is noteworthy about Maurice Bellet's romantic work and thought (which is not, strictly speaking, a theoretical work) is that he bases his 'revolutionary' approach to changing the 'world and life' on a 'science of the human'.

[5] The question of crisis is at the heart of Arendt's thinking, as it is at the heart of the approach to critical theory, of which crisis is both the presupposition and the mode of theoretical elaboration (Assoun, 2016, p. 47).

8.1 Thinking About a Humanity in Motion with Maurice Bellet

8.1.1 *Some Biographical Details*

After publishing a first philosophical work under the direction of the French philosopher Henri Gouhier, entitled *La liberté selon l'Action 1893 de Maurice Blondel*, Maurice Bellet studied for a doctorate in philosophy, with his thesis *Vocation and Liberty* under the direction of the French philosopher and academician Jean Guitton. Bellet then wrote a *doctorat d'état* in philosophy, *La fonction critique dans la certitude religieuse*, under the direction of the French philosopher Paul Ricœur. His viva panel consisted of the philosopher Emmanuel Lévinas, the philosopher Henry Duméry (friend of and commentator upon Blondel), the historian of philosophy Clémence Ramnoux (who co-founded the University of Nanterre with Paul Ricœur), and the philosopher Olivier Lacombe, member of the Académie des Sciences morales et politiques. He also defended an additional thesis entitled *La difficulté de la parole et du sens* under the direction of Clémence Ramnoux. Bellet taught at the Institut Catholique de Paris and collaborated with the Jesuits on the journal *Christus*. Maurice Bellet was ordained a priest in 1949 at the age of 26. He was then appointed to teach philosophy at Saint Louis in the minor seminary of Bourges in the wooded region of Sologne, where he remained for 15 years. On reading his book, *La longue veille* (The Long Vigil), published in 2002 with a biographical dimension, one understands that Maurice Bellet had a particularly strong experience of the 'perverse god' of certain ecclesial traditions, which he denounced (Bellet, 1979).[6] The first years of his ministry were plainly difficult. The discovery of psychoanalysis[7] was important, and allowed him to remain in the Catholic Church while developing his thinking. He then spent several decades serving as a listening ear for others.

Maurice Bellet was marked by Marc Sangnier's movement 'Le Sillon' (The Furrow), his own parents having met as members of the movement. He regularly points to the way in which this movement, which was one of the first attempts to think about a Christian democracy (Lambert, 2001, p. 63), marked his relationship to Catholicism and to social commitment. For Bellet, who was not directly involved in the Sillon (which ceased operating in the early twentieth century), the Sillon constituted a form of bringing Catholicism and democracy together – an aspect that was particularly innovative for the time, if not revolutionary.[8] While developing a critical analysis of the Sillon – and in particular, of the excessive attachment to its

[6] In this sense, he is one of the first thinkers to have identified the structural perversion of ecclesiastical bodies as theological and spiritual foundations, which is particularly perceptible through the various ecclesiastical scandals of the present period (paedophilia, sexual abuse of nuns, the prevalence of homosexuality among Vatican prelates in spite of homophobic discourse, etc.).

[7] Maurice Bellet's psychoanalyst was Robert Gessain; his listening skills had a particular impact on Bellet.

[8] Bellet mentioned these aspects in an interview with him (Bellet & Wallenhorst, 2022).

founder Marc Sangnier – Bellet states that 'In its own way, it touched upon the crucial thing. I have spent my life trying to say it. But it goes infinitely far beyond all its categories and quarrels where they were, where we were' (Bellet, 2002, p. 57). He regularly mentioned how much the spirit of Le Sillon had marked his life.[9]

8.1.2 General Presentation of Maurice Bellet's Work

Maurice Bellet's work crosses disciplinary boundaries. It is marked by the articulation of the critical function of three disciplinary fields: philosophy, theology and psychoanalysis. To date, few studies have been carried out on Bellet's work. We have identified a thesis in twentieth-century literature by the Iranian Moussavi Chirazi at Nancy 2 University, *L'évolution de l'écriture de Maurice Bellet* (1999, 2007); another thesis in Theology, *L'expérience du point-lieu chez Maurice Bellet – Etude d'un parcours discursif*, written by Daniel Lagacé-Roy and defended in 2003 at the University of Montreal; a few articles (Maldamé, 1988; Lefebvre, 2005); and numerous reviews of his works. His work is interesting when thinking about the complexities of the contemporary period, in the way it intertwines various disciplinary fields and in its fight against any form of reductionism. Bellet produced an abundant body of work, which he describes as "unclassifiable",[10] particularly consequential, whose object of study is 'the humanity of humanity' – that is, fundamental anthropological themes. He works on questions of meaning. While producing philosophical works, Bellet sometimes ventures into prospective hypotheses in publications that do not conform to academic standards resembling essays. In his work, he frequently addresses the question of preparation for the future on the basis of the peculiarities of the contemporary period.

While some of his books are theological, the majority of his work is intended to be relevant to readers outside the circle of believers. Bellet is characterised by pragmatism and a critical conception of truth. This brings theology into contact with Human and Social Sciences (Donegani, 2011). As Frédéric Lambert reminds us, with *Gaudium et Spes*, 'The Roman Catholic Church now admits that it is no longer

---

[9] Maurice Bellet was born in 1923 to a working-class father. Although he did not experience destitution, he did experience poverty (his grandfather was a carriage driver). Bellet was marked by the war: his parents spoke to him regularly about the First World War, and his brother Jean died in combat in 1940. In *La longue veille*, Bellet recounts an interaction with his father that particularly marked him and had an impact on his work: 'In *La Voie*, I spoke of the 'little flame of eternal gaiety'. I heard that phrase from my father towards the end of his life, about his wife, my mother: 'What can I say? She's a sunny person'. It was perhaps *through this remark* of my father's that I caught a glimpse of this prodigious truth: the profound nature of man, above all, below the abyss, is light'" (Bellet, 2002, p. 64).

[10] Bellet put a great deal of energy into thinking outside the box, and in this sense his work is deeply alternative and original. This is what he explicitly seeks: the freedom to operate outside of the norms: 'Here, we are outside the established cleavages – free, all free, in the largest possible space' (1992, p. 14).

the sole custodian of the truth and that there is a range of ways to salvation as well as an autonomy of earthly realities' (Lambert, 2012, p. 177). Bellet belongs to the category of pragmatic 'postmodern theologians' (Donegani, 2011, p. 14) who unreservedly subscribe to this perspective, and we can say that he 'takes up the intellectual and anthropological challenge of modernity' (Donegani, 2011, p. 14). His work is marked more by the conviction of urgency (to save humanity from despair and chaos) than by the discussion of a revealed truth.

8.1.3 Maurice Bellet's Manifesto for an Anthropological Shift: **Incipit, ou, le Commencement** *(1992)*

In 1992, Maurice Bellet wrote a short text that resembles a manifesto: *Incipit, ou, le Commencement* (The Beginning). This text, initially written in a poetic register and which could be read without any political reception, is ultimately a form of political manifesto for a revolution without violence based on an anthropological mutation.[11] It lays the groundwork for a series of politico-social works that Bellet would later write, such as *La seconde humanité* (1993), *La longue veille* (2002), *Le paradoxe infini* (2004), and *L'avenir du communisme* (2013).

Incipit is a poetic text, written in one go by Bellet, and he declares that he would not change a single line(!).[12] The theologian Myriam Tonus, his close collaborator, author of *Ouvrir l'espace du christianisme – Introduction à la pensée pionnière de Maurice Bellet* (2019 – Opening up Christianity – An introduction to Maurice Bellet's pioneering thinking), views *Incipit* as his most important book, in which he formulates his aim most clearly. Bellet's life was marked by waiting[13]: 'Long have I waited, and long have I hoped. Something had to emerge, someone would speak, we would be carried along by the current again' (1992, p. 7). He then expresses what is at the very heart of his thinking[14]: 'What is left of us? What is left when there is nothing left? This: that we show humanity towards humans, that between us there remains that between-us which makes us human. If this were lacking, we would fall into the abyss, not of the bestial, but of the inhuman or the de-human – the monstrous chaos of terror and violence where everything comes undone'[15] (1992, p. 8). This sacralisation of the 'between-us' at the root of humanity is characteristic of our

[11] We will refer regularly to this text in this chapter.

[12] We gained this information through one of our interviews with him.

[13] He would continue to build upon this idea of 'expectation' 10 years later, in *La longue veille* (2002).

[14] 'I am finally hearing what I have been trying to say for 30 years, for as long as I can remember' (1992, p. 7).

[15] These few sentences at the beginning of *Incipit* are, along with the closing sentences of the book, among the most important in his work: 'No man is condemned' (1992, p. 77).

author's thoughts.[16] Indeed, this 'between-us' is 'in advance of the beginnings', 'even before "I think; therefore I am", as that which separates from madness in the "I think" could coalesce, to the point of thinking it is the only thing' (1992, p. 12). Here we perceive that, in Bellet's mind, there is no doubt that we *co*exist more than we exist, and our humanity is derived from such coexistence. The 'between-us' is also: 'upstream of the political, which, without this reference, descends into slaughter' (1992, p. 13).[17] Here, this coexistence can be understood to be of absolutely prime importance, in the organisation of social life of self-managed groups or organisations where humanizing solidarity reigns. The idea is even more explicit in the following passage: 'What sets the human apart from the inhuman is *that without which there is no humanity*. [...] And in principle, that is devoid of any specific support. It is not contained within anything – anything tangible that can be grasped, be it doctrine, method, institution. Not even the Divine Order, to which all things are subject, and which takes a different shape according to the ages and places of humanity" (1992, pp. 11–12). Here, Bellet rejects everything that is other than what is among us and that would come from an institution. He then continues with this same logic: 'It is upstream of that which we call morality or ethics, because more than being an obligation, it is a gift' (1992, p. 13).

The basis of Bellet's thinking is intersubjectivity, in keeping with the heritage of contemporary philosophy: 'it is therefore not the isolated man, the self, the individual, the subject as solitude. Man's essence is in his primordial relations' (Bellet, 1992, p. 26). What brings everything into being (starting with the fact that humans are human) is the relationship between humans.[18] Bellet views the autonomy of groups or organisations, with primacy given to intersubjectivity, as the very core of social and political organisation, but stresses that such communities are not isolated from one another, but rather, are interconnected. This conception of the 'between-us' is unequivocally political, with unconditional welcome being extended to those who are outsiders.[19]

---

[16] This 'between-us', understood as the origin of humanity, is the site of love. However, such love is lacking: 'Whilst this very necessary [love] is always there, since we are human, lack thereof and deviance therefrom lead us to atrocity. If, with this glimpse of light, we look at the world, it is mud and ashes' (1992, p. 15). It is this love of each other that allows us to look critically at the world and at what is dysfunctional at the root. Bellet continuously expounded on that theme, denouncing the destructiveness of *hybris* (a form of counterpart to love) and especially of the combination of economics and technology in what he calls 'the great mixing' (in *L'avenir du communisme*, 2013).

[17] For our part, we would agree with Hannah Arendt that, on the contrary, the 'between-us' is the very foundation of the political, from which the political emerges.

[18] Thus, intersubjectivity – or interpersonal relations – is the seat of truth: 'Its place of truth is at once much humbler and much more deeply out of control: it is the relationship between humans, an unforgiving place, an exile of all pretence' (Bellet, 1992, p. 49).

[19] 'Everything that has emerged as a result of the irrepressible presence of love – above all, love that makes us human – is not separate and does not set us apart. It can only be that which makes every man close. Not an overshadowing, triumphant universal – that is more ferociously exclusive than anything else – but instead infinite hospitality, love for one's enemies, and openness to strangers' (Bellet, 1992, p. 57).

The whole of *Incipit* goes towards identifying the between-us that cannot be affected by external principles and organisations. It becomes the principle at the foundation of everything, and the basis for rethinking and rebuilding everything: 'We can deploy this power [of the "between-us"] *ad infinitum*, in all fields; we can modify the available fields, shake up the distributions; we can, by rediscovering anew the powerful flavour of nascent humanity, continuously evolve, with a bone-deep shift which is far more radical than a revolution. We can, and we must' (Bellet, 1992, pp. 65–66). In this passage, we perceive Bellet's idea of anthropological shift, his revolutionary anarchist radicality.[20] It is a question of changing everything – without losing our humanity, but instead discovering it more profoundly.

8.1.4 Experiencing Humanity as an Adventure

One of Bellet's significant contributions is the way in which he makes it possible to think of humanity as an adventure. In his thinking, nothing is certain, everything can be undone at any moment, except forward motion – which in no way guarantees humanity immunity from destruction. Bellet is, first and foremost, a literary artist – he works with language and has a singular style: the main tool with which he causes the reader to experience humanity as an adventure, in the grip of peril and with an uncertain outcome. The following paragraphs highlight the singularity of Bellet's writing, which allows us to think of humanity as an adventure, not reducible to what we already know about it, and which opens up the possibilities – thus allowing for prospective anthropological thinking.

Indeed, Bellet's writing style strikes the reader straight away. Bellet uses a narrative, poetic and performative style. Bellet's relationship to language is sometimes more akin to that of a poet to words than that of an academic to concepts. The performativity of Bellet's writing is an essential characteristic of the author's work. Bellet writes to *do* something to the reader[21] – in particular, to make him experience his own humanity. Thought must act on reality: it 'has the function of drawing this genesis out of a lethal obscuration or impediment' (Bellet, 2004, p. 164). This style of writing joins his interest in action as an extension of his entry into academic work with his dissertation on *L'Action 1893*. Bellet's writing is both provocative and open, sometimes tinged with exaggeration or caricature (with heavy use of

[20] In *Incipit,* Bellet sets out the fundamental themes of anarchist thought: the fact that there is a plurality of possible paths, that autonomous communities are capable of living and organising themselves, mutual aid, respect: 'Each of these paths (or others) encourages community, being together, with unique characteristics every time' (1992, p. 72); 'Of the essential community I will say this: one is included in it by a single principle: that of unfailing friendship' (p. 72); 'It is mutual aid: we will not leave anyone in need. It is respect for everyone, non-judgment, discretion, respect for secrecy and privacy' (p. 72). To enable these alternative communities to emerge, Bellet explicitly says: 'It is not a matter of things or rules; what counts is the depth and truth of the bond. But things and ways of living matter. We can hope for a life together' (Bellet, 1992, p. 74).

[21] This observation was confirmed to us by Maurice Bellet during our interview on 20 January 2014.

superlatives).[22] His provocations, his impertinence, and the openness left invite readers to question and to develop their own ideas: 'I sometimes say the same thing in several ways: this is so that, in this discrepancy, you can find your own views' (Bellet, 1990, p. 14). Similarly: 'Evocative writing goes where it goes, on the wings of a mockingbird sailing on the wind. It does not care if everything is in its proper place, each object (each word) arranged as in the bourgeois house of a well-bred homemaker. It says without saying. It repeats freely and without concern. It takes no consideration of objections – what people may think, what others will say. It delights in mixing genres. Nothing is more pleasing to such a writer as a sidebar, a tangent, a digression. For a fussy essay examiner, it is pointless, careless and poorly put-together excursuses (the cursus is in the textbooks). Ellipsis and allusion are natural in such writing." (Bellet, 1976, p. 100).

Bellet has a unique relationship with bibliographic sources. His writing differs from that of academics who make reference to the authors on whom they draw. Bellet does not quote; there are no footnotes in his works. On this point, our author compares himself to an artist: a sculptor creating a work of art does not cite the origins of his inspiration at every stroke of the chisel.[23] Here is what he writes about his relationship to his predecessors: "In truth, one may wonder whether the great philosophers of modernity and of the west did not themselves practise this type of suspense. There are no notes in Spinoza, Descartes, Kant or Hegel! Yes, you may say, but they still acknowledge their references; they recognise their work's position within a certain tradition; they admit their relationship to those who have gone before them; but to what extent? Their relationship to religion is particularly questionable. The philosopher, in the modern fashion, tends to keep *a certain silence* about his origins – about the ultimate reference which created him. That is to say, the suspense is there, but not recognised – leaving philosophy in that hinterland that must not be spoken of and that will resurface, obliquely, through psychoanalysis and other returns of the unrecognised. I recognise that speaking in this way is extremely irreverent (there is, in the intellectual world, an almost religious respect for "the great authors"). I recognise that we could debate this point endlessly. But I am not defending a thesis; I am trying to open a door" (2004, p. 191).

Bellet's thinking is synthetic and complex and does not lend itself to thematic analysis. Everything appears to be interrelated with everything else, and one can only grasp a theme through its interaction with at least one other. Bellet's thought is particularly marked by the articulation of paradoxes. His points of support are never presented in a clear-cut manner, but as suggestion or hypothesis. His thinking is off the beaten track. He leads the reader towards a foreignness or a different place from which he can gain a different perspective on the initial object of his research.

There is, in Bellet's work, a divestment of what he writes, by going beyond it – a systematic criticism of his own words: 'And yet everything I have just said can be

[22] One of the limitations of Bellet's writing is perhaps a tendency towards allegorical developments marked by darkness (Le Corre, 2006, p. 135).

[23] See interview with Maurice Bellet (Bellet & Wallenhorst, 2022).

contested, refuted, with the best arguments. For, in a sense, it is not true' (1993, p. 112). In *L'avenir du communisme*, for example, he writes that '"the Master's Master is desire-envy', and then a little further on, on page 118, that he is no longer certain that this is the case. Even in the writing of a novel like *L'insurrection* (1997), he finds a way to criticise his novel by going beyond it with, in the continuation of the novel (after the final page of the story), the writing of a second text differentiating itself, and then of a third text. Bellet is a thought experimenter. He writes to see what happens. There is an epistemological element structuring his work on this point: 'I name the major possibilities. The risk is total; it is up to each person here to hear their own words; we are removed from the safety of knowledge' (1993, p. 122).

We are dealing with itinerancy. It is a train of thought that wants to be in constant motion, and is structured in a moving way: what is said is regularly undone or overtaken for elsewhere or otherwise, which the reader sometimes struggles to discern: 'the point of support is no longer here, or behind, but forward, in the depths of the unknown. And we have faith in the Unknown' (Bellet, 2002, p. 42). Movement is an invariant or structuring element in Bellet's – even his conception of reality is shifting and can only be grasped by a movement of thought. Reality is like the Big Bang, an explosion; it appears only in crisis (Bellet, 1993, p. 102). For Bellet, the point of support for the elaboration of thought is systematically a critical point of support, but it is a point of support that displaces thought at the same time as it displaces itself. Bellet does not elaborate arguments comparable to pyramids as René Descartes proposes in the second part of the *Discourse on Method*, with his third precept 'to conduct my thoughts in order, beginning with the simplest and easiest-to-know objects, and then to ascend little by little, as if by degrees, to the knowledge of the most complex' (1637, pp. 137–138). Bellet does not climb: he moves. He arrives at one point and then jumps to another. Bellet's thinking is like a fast, profound, forward movement trying to identify what is 'essential'. Moreover, we can say of Maurice Bellet something similar to what the jurist and literary scholar Pierre-Louis Mathieu says of the thinking of Teilhard de Chardin who 'had a very clear awareness of being only a moment of human thought in the making' (Mathieu, 1969, p. 16).

Maurice Bellet seems to be guided by an urgency, by the idea that 'that the way out is in movement, in stepping forward' (Cupillard, 2001, p. 565). Far from developing a thought marked by unity and a conceptual architecture reinforced in the course of more than half a century of research, we can say of Bellet, as does French Jesuit theologian Dominique Bertrand, that he 'bursts with perpetual invention [...] that he undermines certainties [and] questions everything' (Bertrand, 2005, p. 272). Bellet offers stimuli or provocations for thinking. This is one of the benefits of his critical thinking in motion. Bellet's provocative thinking is not without links to MAUSS. Indeed, Alain Caillé, Editor of the *Revue du MAUSS*, in a review of Bellet's *Plaidoyer pour la gratuité et l'abstinence*, mentions that his thinking could lend elements to Serge Latouche's movement of convivial degrowth (Caillé, 2003, p. 434).

8.1.5 *Thinking About Politics from the Perspective of a 'Between-Us'*

Bellet is close to other authors who think of politics in terms of the social link with a metaphysical foundation, denouncing the liberalism of modernity.[24] Bellet envisages human action resisting against economic domination and the threats facing humanity, thus following in the footsteps of Hannah Arendt, the German philosopher Jürgen Habermas, or the French philosopher Claude Lefort. Bellet, who is not a Marxist, takes up Marxism's ambition of marrying an understanding of the world with a transformation of it. It is as a metaphysician that Bellet analyses the social; not as a sociologist. His approach is similar to that of Hans Jonas, whose primary strength is that of his metaphysics.

Bellet is on the side of the insane. He is the spokesman for people who have been annihilated, crushed by the weight of life – who are not necessarily in a situation of desocialisation. However, he thinks of the universality of human beings, on the basis of the singularity of the stories heard in his capacity as an analyst (in his texts, he constantly weaves together particularity and universality).[25] Bellet does not think from a social point of view, but from the standpoint of the 'between-us' – the founding basis of his thinking. Here, he is in keeping with the worldview of Arendt or even Levinas – a member of his viva panel – and his notion of 'between-us' (1991) or of the face.

8.2 *Krisis* and Criticism

8.2.1 *Krisis*

The concept of crisis, often used to refer to the contemporary environmental situation, is not appropriate. There is no environmental crisis. On the other hand, contemporary human civilisation is facing a crisis that must be overcome. The concept of *krisis* is operative here, in thinking about this decisive transition.

Bellet is a dissenting thinker – though he denied it during our interview, preferring the term 'critical'.[26] For him, criticism is 'the astonishment of Socrates, the doubt of Descartes, the power and the right, jointly, to question thoroughly, without

---

[24] Moussavi Chirazi, in his study of Bellet's discourse, notes the presence of threefold non-religious discourse: philosophical, psychoanalytical and political (2007). Bellet's political language attempts, in some directly political books, to say what is most essential or most necessary for humans to live (Bellet, 1993, 1996, 2013).

[25] On the other hand, he does not relay any cultural or identity-based claims of constituted minority groups. There is no claim to cultural rights in Bellet's thinking, as may be the case with the French sociologist Alain Touraine, for example (2010, 2013). There is no communitarianism in Bellet.

[26] See interview with Maurice Bellet.

reserve. That is the point: *without reserve*' (Bellet, 2002, p. 93). Bellet's contestative thinking is marked by what he identifies as the radical critique inherent in the Gospels, which feeds his desire to see another type of world and humanity come into being.[27] It is possible to identify a two-pronged evangelical critique marking Bellet's thinking. Firstly, he thinks in reference to a figure of the human presented in the Gospels through the person of Christ, manifesting what humanity could be. He develops his thinking in this gap between what he observes or hears[28] and what he identifies as an achievable goal for humanity. It is in relation to this possible gap that he opposes, contests and denounces. Secondly, Bellet's contestation is grounding in his reading of 'the death of God' – a moment of radical criticism, as he develops in a publication for *Etudes* in 2008, with the particularly evocative title: '*La contestation évangélique*' (Bellet, 2008).[29] An interesting element for the contemporary period in Bellet's anthropology is that he does not conceive of 'life' without *krisis*. *Krisis* is not to be avoided, it is not indicative of an end or a failure; it is by getting through a *krisis* that one is able to start afresh. *Krisis* allows for a birth, and in particular, the birth of humanity. The concept of humanity and that of *krisis* are at the heart of Bellet's work. He does not think of human beings without *krisis*, without passing through the crisis, through that 'zero point' which becomes the origin.

The notion of *krisis* is at the origin and heart of Bellet's research.[30] His apprehension of *krisis* as something that forces us to think is very Arendtian: for Arendt, the event is unprecedented, and 'breaks and resists [...] its integration and thus its dissolution in a causal series' (Revault d'Allonnes, 2011, p. 198); it is 'a crisis of reality in the sense that, in accordance with the original meaning of the word (*krisis*), it marks a rupture within a process' (Revault d'Allonnes, 2011, p. 199). For Arendt, crisis is a revelation: the emergence of an event is accompanied by a novelty to be considered and forces one to identify the essence of what is happening. The concept of crisis runs through Arendt's work, notably with totalitarianism being identified as a crisis of the political, referring to this fundamental crisis of the human condition. As human life is hosted by the world, living in a world in crisis connotes the way in which each person is returned to their human condition.

[27] It might be possible to classify Bellet's work among the anarcho-Christianisms, though we perceive the ambivalence of the author's fascination with strong powers. This is perceptible, for example, in *L'avenir du communisme* (2014) and especially in the conclusion, where Bellet holds out the hope of something from the current Chinese power.

[28] Maurice Bellet worked as a psychoanalyst.

[29] An expression that recurs in the writings of Maurice Bellet is 'evangelical radicality', which is a point of reference in his thinking. This radicality can be understood as a discourse on the roots of humanity.

[30] This development of the idea of *krisis* also makes Maurice Bellet a thinker of critique, allowing him to develop epistemological conceptions. Bellet's thought is a thought of deconstruction, from a perspective that is related to that of Derrida, where deconstruction 'is a deconstruction of criticism [… and where] deconstruction tries to think in history on the authority of the critical instance' (Poché, 2007, p. 44). Bellet places particular emphasis on critiquing critique. For Derrida, as for Bellet, deconstruction also refers to resistance – to refusal to give in to any form of hegemony (Poché, 2007, p. 55).

Crisis is one of the most widely used conceptual tools in the political and economic analysis of the contemporary period. We are in a period of crisis; the crisis has become structural and seems to have lost all the singularity of its temporality. Yet this contemporary global political and economic crisis is also an anthropological crisis. In the eyes of the Swiss ethicist Céline Ehrwein Nihan, the crisis 'constitutes our "living together"' and our being in the world' (2013, p. 45). Crisis must be understood in its complexity: it is both what can be salvific and constructive as well as what can represent the danger of destruction: 'there are situations where the crisis we experience no longer has anything to do with the instability constitutive of all human experience, but represents a state of *destabilisation* that results in the destruction of freedom and community bonds' (2013, p. 56).

8.2.2 Ethics of Resistance and Critique of Technology

Bellet's works challenge the economic hegemony and its liberal individualism. Bellet's work is a critique of total capitalism. Our author denounces the fact that society is founded upon the economy – that is, that utility is the reason for societies' existence, and that contemporary individual only exist by virtue of participating in consumption: 'I consume; therefore I exist', as Flahault puts it (2013, p. 781). Bellet's critique presents a resistance. Before stating precisely what needs to be done, Bellet states what needs to be fought against: what humans are not and what causes alienation. The reign of economics and its support of technology is one of the first obstacles identified here. Bellet's thinking is marked by the main criticisms of modernity formulated at the end of the nineteenth century and during the twentieth century, on which he relies in the development of his critical thinking of the contemporary period. He is close to Arendt's denunciation of industrial modernity with its alteration of the human condition. Technology and the inclusion of an individual's action in a chain of activities reshape the notion of responsibility. How can one perceive oneself as being responsible for one's actions when the ultimate consequence is located at the end of a chain of activities of which one is not aware? For Arendt, modern man has become increasingly incapable of thinking about what he does. Thus, technology brings with it a significant loss of responsibility, and thinking about responsibility involves rethinking our relationship to the world (Poizat, 2009, p. 180).

Thus, based on the idea of humanity in the process of emerging – in the process of being born – he is critical of technology, which, though created by man, could escape him and cause alienation of him.[31] In the wake of a certain number of twentieth-century authors, starting with the great German philosopher Martin

[31] For Bellet, modernity may have taken us away, somewhat, from what is at the heart of the human adventure: 'We come from this modernity that strove to emerge from the dark ages of humanity; its work has been prodigious, we are part of it, and we live from it. But what happens, in the course of the adventure, to that tiny thing without which we cannot exist?' (1992, p. 17).

Heidegger, Man needs, to some degree, to be defended from technology. In Bellet's work, the human and the technical could sometimes appear as two oppositional categories[32] – which Bourg contests, believing that there is no humanity without technique (Bourg, 1996, 2012, p. 10). Bellet's fear, along with Arendt and Jonas, is that human existence is viewed from the perspective of technology, causing dehumanisation. In the face of this concern, Jonas's response is to develop an ethics of responsibility for life and in relation to the future; and Arendt's is to develop politics on the basis of human action, distinct from the vital cycle (Pommier, 2013, p. 272). Bellet's response is a consideration of the preservation of humanity through the renewal of its origin – i.e. the 'between-us'. He feels the need to reflect on the conditions of emergence of humanity. For Bellet, as for Jonas, the threat which technology poses to humanity relates to the very *raison d'être* of humanity *as* humanity, making us responsible to future generations as well as to the ontological idea of humanity. In Jonasian thought, technology does not consist, primarily, of the mastery of know-how, but of a threat: 'the promise of modern technology has become a threat' (Jonas, 1990, p. 15). The technology developed by humans has the potential to end human existence. One of Jonas' contributions is his emphasis on the gap between the development of technology and the lack of knowledge as to its long-term effects. He had clearly perceived the entry into a new period of an anthropic nature, which threatens the survival of humanity. Indeed, one of the contemporary problems is that humanity has 'considerable technological means of destruction and, on the other hand, technologies of remediation that do not exist on a global scale' (Bourg, 2012, p.11). The concern about the dehumanisation, if not the destruction, caused by technology has been growing during the twentieth century, with the realisation of the effects of the Industrial Revolution, then the horror of the Second World War and the Holocaust. It has culminated, in the twenty-first century, with transhumanism, obliging us to make a fundamental choice: 'Transhumanist fantasies of technological omnipotence and claimed immortality force us to choose quickly between these two forms of modernity: that of *hybris* and all-out excess, or that of self-limitation and the patient elaboration of collective decisions' (Bourg, 2012, p. 12).

Bellet's thinking also includes links with Christian thinkers of the twentieth century such as the French theologian Jacques Ellul who, as early as 1954, saw in technology (*La Technique ou l'Enjeu du siècle*) an alteration of the humanity of human beings that would destroy inner life; or the French philosopher Emmanuel Mounier who, as early as 1950, saw the machine as a potent force for depersonalisation (Mounier, 2001). In Bellet's thinking, technology brings profound anthropological reconfigurations.

[32] In an interview, Maurice Bellet weighs up that statement.

8.3 Anthropological Plasticity

8.3.1 The Idea of Humanity in Maurice Bellet's Thinking

Maurice Bellet's work was shaped by his dialogues with Jean Trouillard, a Sulpician priest specialising in Plotinus and Blondel, who was particularly marked by negative theology: 'I have certainly kept from these encounters a resolute and radical apophatism, as far as God is concerned: any discourse on God makes me suspicious. God is not an object. Even Blondel's "subject of subjects" leaves me reticent' (Bellet, 2002, p. 89). 'The void, the void, and the void of the void. Empty space. Take away from God all the clutter: the very word God, the fear of God and the question of God, and piety – the padding, pretentious, nauseating piety. So what about atheism? Oh no, much harder – much more radical and merciless. Atheism is still a theory, a stance in relation to God… Which one? The atheist depends on the theist, and the theist is often sure to be a mediocre thinker. Farewell, mediocre atheist' (Bellet, 2002, p. 90). It is against a background of negative anthropology, in the tradition of negative theology, that Bellet approaches the idea of humanity. In contrast to all positivism, humanity is not directly defined but is understood on the basis of the mystery of its origin and the mystery of the meaning of its adventure.

In his work, Bellet regularly addresses the idea of 'the humanity of humanity' (most notably in *Le paradoxe infini*). With this expression that emphasises the idea of humanity, he affirms the need to rethink it, and the urgency of doing so if there is to be a possible future. The idea of humans or of humanity is directly linked, in Bellet's eyes, to the question of the future and preparation for it: 'Is there not something constant in man, from the moment when he becomes human? Is it not the illusion of all illusions to believe that all of our transformations and inventions have altered the human condition? The conditions, no doubt; the means, the environment, the artifices – but the substance? Yes, the essence of man, that mixture of anguish and fury, traversed by the light of thought itself, and the need for love and history and science itself? For if there is becoming, the becoming is still human. It is part of makes the humanity of humans' (Bellet, 2004, p. 28).

For Bellet, the human is that which is undefinable, is not objective and is always more. On the other hand, the human is found in boundaries, and in particular, in the boundary demarcating the *in*human – integrating the paradox that what is inhuman is an integral part of humanity (Bellet, 2004). The navigation of *krisis* allows for a distinction, and even a separation (the Greek word *krinein* means to separate, to decide), between the human and the inhuman. It is a question of thinking about the way in which the contemporary journey which humanity is obliged to undertake as a result of the entry into the Anthropocene could involve drawing a distinction from the inhuman element which is linked to human nature. The interconnection he proposes between *krisis* and the emergence of humanity is particularly interesting, and offers hope, when thinking about getting through the civilisational *krisis* highlighted by the Anthropocene, which could represent an 'opportunity' to become more human.

Human beings are marked by movement, by effusiveness, by the genesis of humanity. For Bellet, humans cannot be thought of without outside of the context of their emergence. He shows that, in the unknown of the mystery of humanity, there is a solid point upon which to base our thinking. In contrast to reifying theory about humanity, Bellet thinks of humanity from the point of view of the mystery it holds – the meaning of its presence in the world – and refuses to say *anything about* the human.[33] He works on the basis of questions rather than assertions, which is one of the strongest contributions of his thinking.[34]

Bellet's political work is based upon his unique idea of humanity and his negative anthropology. This idea of humanity as the basis for politics is reminiscent of the work of French thinker Edgar Morin, formally presented in *Introduction à une politique de l'homme* (1999) but perceptible throughout all of his work, as the French political scientist Gil Delannoi notably pointed out in 1982. Moreover, there are several similarities between Edgar Morin and Maurice Bellet's ideas, with the use of identical expressions such as 'human emergency' or 'humanity's humanity' in Morin (2001); and the use of the concept of *krisis* (Morin, 2012).

What are the notional similarities between human beings and subjects, in the philosophical sense? There are two immediately perceptible differences between humans and subjects: the mention of the species in the name, and the reference to ontology rather than to the Human and Social Sciences: the human is a human being. Is Bellet's human a 'human subject'? It seems to include characteristics of the Tourainian subject, for example, to which are added the belonging to a species and the ontological dimension. As Donegani states, 'The notion of subject thus seems to its proponents to better account for the human vocation in that, unlike the individual, the subject possesses a linguistic and vocative dimension, its choices being subject to the approval of others and its identity established on the basis of a relationship that is no longer self-reflexive but complementary with others' (Donegani, 2013, p. 355). The notion of the human and the universalism it covers in Bellet's work is not immune from question. Indeed, it could refer to the idea of a single human nature without the essential social, cultural and political differentiations that emerge when talking about the Anthropocene – as we saw in Part I of this book.

In his understanding of the idea of humanity, Bellet's thinking has another limitation for a work on the Anthropocene. The anthropology underlying Bellet's thinking does not at any point integrate the questions thrown up by the Anthropocene, and remains marked by the separation of humanity from nature (Bellet does not think of humanity's being immersed in nature). The 'between-us', the place of

[33] Indeed, Bellet's work is strongly marked by apophaticism. One of the complexities of discussing Bellet's work is finding a way to say something on the basis of a text that says what it is not.

[34] Rejecting the positivist hegemony in the field of human and social sciences, one contribution of Bellet's work lies in his negative anthropology. Bellet, by rejecting any knowledge about human beings that would allow us to grasp what humanity is – by trying to think of humanity through differentiation from what they are not – is an interesting epistemologist to refer to in contemporary sciences.

genesis of humanity, is exclusively 'between-humans': it does not include animals, plants, or inorganic matter, which are nevertheless what allows us to live in the primary sense of the term. We live only through exchanges with these entities, as Andreas Weber clearly shows. Bellet's poetic and intellectual argument, which is particularly inspiring when thinking about the sustainability of the human adventure in the Anthropocene, runs the risk of being sterile if not viewed in combination with contemporary thinkers like Weber, Pelluchon or Flahault. That is what we shall do in the rest of this book.

8.3.2 From a Real Man to a Possible Man

One of the characteristics of Bellet's work is that he thinks about human existence against the background of a foundation of humanity that 'makes us human, and human to each other' (Bellet, 2013, p. 51). To use Bellet's ideas in developing a critical theory for the Anthropocene is to think about a different way of living together, and a different conception of what a human is. The author of *Le Sauvage indigné* is indeed 'a good guide' (Arnsperger, 2010a, b) for those who wish to return to the existential anthropological foundations in order to think about politics in the contemporary postmodern period.[35] Bellet's existential anthropology allows us to comprehend human beings in light of their finitude, facing the prospect of the end of their adventure on Earth. Bellet's work deals human beings from an anthropological perspective that is both political,[36] with the 'between-us' as the unit on the basis of which he thinks, and existential, questioning the meaning of our presence in the world and the possibilities of present and future action to defer our own destruction. By questioning the meaning of our presence in the world, the possibility of collective action and a collective adventure, this type of thinking raises political questions.[37]

The human being as perceived by science is a real-world human, who is observable, and whose behaviour it is possible to analyse. However, the Swiss theologian Maurice Zundel speaks of a distinction between the 'real man' and the 'possible man' (1990). In actuality, it is the same distinction that Bellet continually makes, when he speaks of his desire to see the advent of a second humanity that he

[35] In *The Second Humanity*, Maurice Bellet addresses the problem of economics, rather than an economic problem. He puts the meaning of economics to work: 'Why does the capitalist love money? Why do people want to improve their standard of living? Why do people have to work beyond what is strictly necessary?' (Bellet, 1993, p. 24). These questions, which are essential in human life, can have a bearing on economic science, as can be seen in Arnsperger's proposal of existential economics.

[36] Most of the reviews of Bellet's works have been published in the journal *Etudes*, whose subject matter is partly political.

[37] Consideration of these existential themes is particularly important in the contemporary period marked by capitalism (Arnsperger and Parijs, 2003; Arnsperger, 2002, 2005, 2009, 2011, 2013) and Prometheism.

considers possible. To put it in theological terms, it is a question of leading a change from psychosomatic man to pneumatic man (Arnsperger, 2010a, b, p. 41). The possible man or the empirical man refers, for Arnsperger, to the *homo oeconomicus*, who is tortured by the fear of death and who pursues his quest for the absolute through materialism (capitalism). For Arnsperger, the radical anthropological critique of the Gospels reveals a flaw or a void at the very heart of real man or *homo oeconomicus*, and entails the proposal of a way of overcoming this alienated human condition. Far from proposing a religious affiliation, the Gospels 'outline a critical method' (2010a, b, p. 41). They tell of how to elevate the human condition from real man to possible man. Bellet's thinking is permeated with this dynamic.

8.4 Anthropological Change

8.4.1 Dealing with the Threat

The capacity that humanity has acquired to destroy itself with atomic weapons has been particularly thoroughly discussed in the twentieth century by thinkers like the German philosophers Karl Jaspers and Günther Anders. Then, work began to focus on the environmental risk, following directly in the footsteps of these thinkers, notably by Hans Jonas. The road travelled over the last 40 years, and the report *The limits to Growth* published in 1972 with the support of the Club of Rome by the American ecologists Donella Meadows and Dennis Meadows, the Norwegian climatologist Jorgen Randers and William Behrens is not that of remedial political action, but the transition from the hypothesis of a global environmental problem to the absolute certainty of it. Dennis Meadows wrote, in 2010, that 'forty years ago, it was still theoretically possible to slow things down and achieve a balance.' Then he immediately clarifies, 'It no longer is' (2010, p. 197). The entry into the Anthropocene is a threat that needs to be thought through, as do the reasons for our lack of response to the scale of structural change underway. As Jean-Pierre Dupuy keeps reminding us, 'Even when we know that a catastrophe is about to happen, we do not believe it: we do not believe what we know' (2009, p. 374). Dupuy continues: 'No one can tell which way it will go. The judgment can only be retrospective. However, *it is possible to anticipate, not the judgment itself, but the fact that it must depend on what will be known once the "veil of ignorance" cloaking the future is lifted.* Thus, there is still time to insure our descendants will never be able to say "too late!"–a "too late" that would mean that they find themselves in a situation where no human life worthy of the name is possible' (2009, p. 376). Indeed, the trajectory of the human adventure on Earth is 'literally suicidal' in the view of authors like the French physicist Jean-Paul Deléage (Deléage, 2010, p. 24), which can refer to anxieties similar to those of a nuclear war. Our anxiety is, quite simply, the fear of 'the death of the species' (Deléage, 2010, p. 29). The contemporary situation was anticipated by a number of authors, such as the ecologist Garrett Hardin who, as early as 1968, announced that humanity was at risk of self-destruction if we continued to consider the commons as being unlimited. For Dupuy, 'We are living under a new regime of evil—an evil without harmful intent' (Dupuy, 2012, p. 731).[38]

Yves Cochet, former French Minister of the Environment, identifies three particularly different worldviews that are currently in conflict in our industrial societies. The first isa productivist model presupposing continuity between the past, present and future, based on growth. This is the dominant model in political discourse and is strongly relayed, without reflexivity, by the media. Cochet helpfully quotes the conclusions of a European Council report of 28–29 June 2012: 'Strong, smart, sustainable and inclusive growth, based on sound public finances, structural reforms and investments to boost competitiveness, remains our top priority' (Cochet, 2013, p. 50). The second is an 'Augustinian' model, in reference to St Augustine's statement: 'The world is like a man: he is born, he grows and he dies' (St Augustine, Sermon 81, paragraph 8, December 410). This means that a series of phenomena will come to an end during the twenty-first century, such as oil extraction, the mining of certain metals, or sea fishing. The third is a discontinuous model in which a large-scale change is foreseen without it being possible to anticipate it precisely, and in which the possibility of social and political chaos cannot be ruled out.

The threat we are facing is of unprecedented magnitude, as the French philosopher Frédéric Neyrat forcefully points out. However, in the case of sudden disasters, whether linked to CO_2 emissions alone or to those in combination with the *techno-fixes* implemented by geo-engineers, we could imagine this strange scene: an Earth that is indelibly anthropogenised, but in the absence of human beings – or sparsely populated, stripped of their former "geological power". Like an Anthropocene without the *anthropos*' (Neyrat, 2014, p. 47). We are facing a threat of which we – or at least some of us – are the cause: 'it is too late to avoid the collapse, and there will be no soft landing' (Cochet, 2013, p. 71).

Bellet's work is not based on awareness of this ecological threat. The threat he perceives is techno-scientific and economic. It refers precisely to all the factors that have propelled the Earth system into the Anthropocene. This is why Bellet's reading is interesting in order to identify the anthropological shift that is needed in the Anthropocene. In *L'avenir du communisme*, Bellet criticises the supreme authority of economics and denounces the 'great mixing': this combination of technology and science in service of the economy, and therefore of desire, which, through money, becomes master of events. Bellet's ideas are similar to those of the German philosopher, who emigrated to France, Eric Weil. One of his contributions is the awareness that he is inhabited by something other than human, which threatens his humanity. For Weil, a new 'sacred' factor appeared with the rationalisation of work during the Industrial Revolution: techno-scientific power gaining dominion over nature (Guibal, 1995, p. 501). This process of mixing is particularly violent: it is

[38] Today, the number of people displaced by natural disasters is between 20 and 30 million per year, not including those who have migrated in response to slow changes in their environment. This figure is higher than the number of political refugees recorded by the United Nations High Commissioner for Refugees (16.7 million in 2013). (Gemenne, 2015, p. 234).

characterised by the exploitation of humans by other humans, and by exclusion. Humanity is dealing with violence that 'eradicates the humanity from humanity' (Bellet, 2013, p. 64). Such naked violence 'is wordless and thoughtless [...] The violence of which I speak is silent and invisible. It is so in the very people who carry it out and can also suffer it. It is part of those processes, which we have described, that take away from the human being the distance, the inner distance, the duration, the meaning of relationships, the basic wisdom that shields him from madness. It is a destruction of what is human in man [...] The heart of social life is that there should be this connection between humans, which endows them with humanity. This link can be severed or perverted. Here, the link has disappeared. The violent process of mixing, pushed to the limit, comes to this' (Bellet, 2013, pp. 65–66).

Bellet attempts to name a threat of the order of a breakdown or deflagration of what is human in each of us generated, by a great 'techno-scientific-economic mix': 'Yet the world is suffering from the same disease. [...] The decomposition of – of what? I've spent much of my life trying to figure it out. What I am sure of is that it defies our categorisation. It is a human necessity, a primordial necessity, in which the Order of which I have spoken so ill finds a pretext and justification for its worst abuses. It is a primitive order of things that makes it possible for us humans to exist. Human history, for the most part, may be said to be made up of the avatars of this thing. The latest is what we call 'economics', in its overwhelming and universal domination, as it is today. However, this avatar is a failure' (Bellet, 2002, p. 86). Bellet's thinking has elements in common with Mounier's rejection of liberalism or 'politics first' (Lambert, 2001, p. 64). Bellet echoes Mounier's critique of capitalism and its dehumanising effects. Like Mounier, Bellet does not have words harsh enough to describe money and the perversion it brings. Following in Mounier's footsteps, for Bellet, capitalism – and particularly, the hegemony of economics – is akin to an alienating dictatorship from which humanity must free itself.

8.4.2 *From Humanity Under Threat to the Rebirth of Humanity*

The thematic universe of threatened humanity is always articulated with that of the rebirth of humanity. This articulation is at the root of the way in which the Anthropocene can present the opportunity for an anthropological shift, or an anthropological consolidation of politics.[39] For our author, the way in which the contemporary world functions is part of a humanity on the verge of collapse. Bellet does not write outside of this threat of humanity towards itself in political and anthropological

[39] For some authors, the contemporary period is characterised by an 'anthropological revolution'. Thus, for the French economist Bernard Perret, the consideration of environmental issues will lead to an 'anthropological revolution that will move us from an enclosed, hermetic self, solely concerned with its own autonomy, to a more vivid perception of the web of relationships in which we are embroiled' (Perret, 2014, p. 38).

resonance – which we cannot help but take seriously after the massive destruction of the twentieth century. We find, in Bellet's work, the same characteristic as in the Critical Theoreticians and, more generally, the German thinkers who fled Nazism, for whom politics is the horizon of thought and for whom humanity represents the greatest threat to itself.[40] For Bellet, on the one hand, the economy mistreats humans and can lead them to their own destruction, and on the other hand, humans have a never-ending job to do in order to distinguish the human from the inhumanity inside of them. Bellet notes a gap between the rationality of contemporary thought and the irrationality of the foundations on which the contemporary economy seems to be built. This observation generates disquiet; it is 'as if this highly rationalised world rests on decisions, or perhaps even uncontrolled impulses, that make it fundamentally crazy' (Bellet, 2013, p. 12).[41] For the French theologian François Euvé, Bellet's work constitutes a call 'for man to free himself from the inhumanity that threatens him' (Euvé, 2013, p. 713). In order for threatened humanity to be reborn and thus be preserved from destruction, it must go through *krisis*. Bellet's work is constructed as a call for collective action based on a proposal to reshape humanity on the basis of relationship and the 'between-us': it is on this point that it is especially relevant to consult Bellet's thought when aiming to strengthen the political in the Anthropocene. Only a 'base of humanity' guards humans from possible destruction by the economic hegemony and contemporary technoscience. It is from the point of view of a humanity under threat that Bellet envisages its future, and the horror of the Holocaust is viewed as a possibility from which the future is not exempt. In the name of humanity, he calls for a radical anthropological change and for political control of the human species' anthropological future.

For the author of *La seconde humanité*, the contemporary period is marked, in a manner of speaking, by the end of politics, resulting from the end of the ambition to change the world and life (1993, p. 17).[42] Bellet, who notes the 'sad state of humanity' (1993, p. 43), does not cease to believe in the possibility for human beings to live differently and to free themselves from the madness intrinsic to the hegemony of economics. This imperative to liberate man from the grip of economic domination becomes the object of politics for Bellet, who thinks of collective action of a revolutionary nature[43] preserving the most precious common good: the primordial

---

[40] Having seen the horror of the Holocaust, Jonas, while he may have difficulty defining what is human in a clear way, he clearly identifies what is *not* human – the inhumanity which we need to combat (Saudan, 2008, p. 114).

[41] Bellet thinks only in terms of a threat whose looming presence he perceives but which he struggles to identify explicitly. He fears that we may 1 day descend into murder and madness and lose what makes us human, capable of love and care for others. The threat is perceptible when we look at the world, but it is far more consequential than its manifestation through acts of deviance or atrocity. It underlies our existence like magma, which is there but has not yet emerged: 'In this world of science and reason, troubles come to the surface that were thought to have been forgotten. It is as if, underneath what we have built, the strange tremors could be heard; as if the abyss threatened to open up under the ground where we still walk' (Bellet, 1992, p. 18).

[42] After writing this in 1993, Bellet went on to develop the idea extensively in 2013, in *L'avenir du communisme*.

[43] Bellet imagines a revolution without violence, but marked by strong resistance and existential collective action.

'between-us'. Thus, he tries to think of a collective action that serves this excessive ambition to change the world, life, and coexistence, while trying not to fall into the trap of violence – after having seen, during the twentieth century, the horrors that man is capable of through ambitious collective actions. Mankind has no guarantee that its adventure will continue. There is neither an established institution nor a 'great power' that would 'ensure its rebirth' (Bellet, 1993, p. 161) and protect it from the threat it poses to itself. It is up to humanity to take charge of its own survival. This is where the responsibility of politics comes in: to protect humanity from the inhuman swerves of which it is capable (Bellet, 1993, p. 48). Bellet's analysis is unambiguous: 'The modern explosion presupposes and anticipates this second coming of humanity; it demands it. We must change humanity' (Bellet, 1993, p. 141).[44] Modernity alters humanity, which must think politically and take charge of its own anthropological future. Indeed, Bellet continues, 'We must take power, take ownership of the frightening and marvellous process that puts genesis before us and within us' (Bellet, 1993, p. 141).[45] For this to happen, it is necessary for people

[44] In Bellet's thinking, science has a special place in bringing about this 'anthropological shift', but it is a 'human science' as opposed to technoscience. As early as 1992, Maurice Bellet laid the foundations of a human science, the basis of which would be to enhance humans' humanity – a theme to which he then devoted a book in 2004, *Le paradoxe infini*. In *Incipit* (1992), he writes: 'It is through this that we will be able to create human science that truly looks at humanity as a whole, with all their constitutive relations, and which is nevertheless no longer a totalising pretention' (1992, p. 68). We can see here that listening, inspired by psychoanalysis, occupies an important place in this human science: 'when the research in which each person is most radically involved does not isolate him or her, but rather invites him or her to listen to the stranger who has drawn near' (p. 68). Here, the human science is the means of bringing about an anthropological mutation whose purpose is to allow us to live together. He writes, for example: 'This human science is required by the very movement of the sciences, which disperses, and which at the same time increases the demand for unity. It is an impossible project, but in the attempt, we will reveal the possibilities, and thus be able to direct our energies: in this way, the true face of humanity comes to light – the difference which gives rise to our capacity to be' (Bellet, 1992, p. 69). It is on that basis that politics takes shape or must take shape ('Through this, a politics freed from totalitarian folly will be possible, and yet not resigned to the day-to-day nor subjected to imperatives that are never critical nor confronted. The politics that is now necessary goes beyond politics. It requires a recovery, a recasting of culture, a rediscovery of the human, the boldness of a second humanity' (Bellet, 1992, p. 69). One of the parts of the *Manifesto for the Anthropocene* is entitled 'Science as compassionate practice' ('Wissenschaft als mitfühlende Praxis') (Weber & Kurt, 2015, p. 16). It is astonishing to note the strength of the parallels with Bellet's thinking, especially as outlined in *Le paradoxe infini*. For Weber and Kurt, the function of science is to serve the development of the living and to orient the *anthropos*' practice of vitality, as a practice of solidarity. The function of science is to directly guide human societies, and even the orientation of the living world and its vitality. Thus: 'The sciences, instead of simply making knowledge available, must develop knowledge of orientation to become a practice of investigating vitality' ('Wissenschaften müssen, anstatt Verfügungswissen zu produzieren, Orientierungswissen erarbeiten zu einer forschenden Praxis der Lebendigkeit werden') (Weber & Kurt, 2015, p. 17). The world is, for Weber and Kurt, a relational fabric based on the power of vitality.

[45] In the face of the latent threat, the challenge is that 'we may recover our forgotten genesis', 'that we may relate to each other, giving each other peace and freedom to live' (Bellet, 1992, p. 19). Here, we can see how much importance Maurice Bellet attaches to autonomous organisations, to social modes similar to communities that live in reference to one another. It is important 'that we be that primordial tenderness for one another' (1992, p. 20).

to make use of that part of themselves that 'dares to carry the weight of the world' (Bellet, 1993, p. 201). The courage and the firm will to participate (Tillich, 1999) in this collective adventure of humanity is therefore particularly necessary: 'I persist and I sign: everyone can take part, here and now, in the great work of humanity. There is no such thing as a hopeless existence' (Bellet, 1993, p. 202).

Bellet approaches this movement, which he readily describes as a second birth, from a unique perspective. His questioning integrates a lively critique with a subversive faith. This articulation gives rise, in his understanding of the contemporary world, to the idea of protest and hope centred on possible action, and in particular common, shared action. Bellet's thinking, with this critical vivacity, remains structured around an 'even though'. The hope of seeing another world and another type of human come into being remains, in spite of everything: 'And in this appears the great "even though": even though we may languish at the bottom of the abyss, an incomprehensible dawn comes. If the crux of our life is to be born – that is to say, to overcome the evidence of non-life and the omnipotence of death, here is the *absolute* birth – that which transcends absolute death' (Bellet, 2002, p. 114). Even though humanity is under threat, a collective future remains possible. Even though disappointment is present, the hope of change is unaffected. In this 'even though', it is possible to perceive the power of revolutionary energy. For this author, 'the prototype of this revolutionary energy that demands change, liberation from oppression, the fulfilment of hope' (Bellet, 2013, p. 127) was born in Israel and is told of in the Bible.

Bellet's idea of the birth of the human is comparable to Arendt's concept of natality referring to "the possibility for humans to assert themselves in a singular way within the world, precisely because they bring new promise and unpredictability" (Pommier, 2013, p. 275). Natality means the capacity of every woman and man to generate something new – to initiate new actions. For Arendt (1983) newborns carry within them a power to reshape the world that is not the mere continuation of the past. Indeed, if tradition – and therefore the past – and its logic of transmission from one generation to the next are fundamental to the reception of newborns into the world, the appropriation of tradition necessarily takes place in an act of creation and modification by the younger generations (Chalier, 2008; Prouteau, 2006).

8.4.3 *Towards an Alteration of the Desire of* homo oeconomicus

Desire is the anthropological root giving economics a fulcrum to exercise its destructive logic on humanity (Wallenhorst, 2014). Bellet hopes for a change in humans, and tries to outline what such a change might look like. He thinks of humanity only in terms of evolution to become more human. The anthropological shift he calls for is an alter of the desire of *homo oeconomicus*: he thinks of humans as having thrown off the yoke of economics, which binds them by the enjoyment of consumption and profit, in order to become more capable of both attention toothers and attention to the future.[46]

Bellet imagines a democracy at the service of communion, allowing humans to be born anew, combatting destructive violence, and thus working for a world freed from alienating economics. One of the political implications of the rejection of violence manifests in the concept of communion: 'What is at stake here, from the social point of view, is what consensus is possible, on which society can build. Thus, for example, democracy is that of a people where, if not all, at least the great majority has this sense of human communion which makes democratic power serve that communion. "Communion", this beautiful word, which has come from so far away, is perhaps the word that is closest to what we are talking about here. It suggests a shared mode of human life, which is, in reality, the strongest principle against destructive violence' (Bellet, 2013, p. 80).

For our author, 'it is appropriate to want more' (Bellet, 2013, p. 17) than the hoped-for great revolution. In *L'avenir du communisme*, it is not communism *per se* that Bellet deals with, but communion, in an anthropological and political approach. One of the strong points of communism is the alliance of theory and action enabling another type of humanity to come into being (Bellet, 2013, p. 30). The aim of *L'avenir du communisme* is to revive 'the communist ambition but differently' (Bellet, 2013, p. 35), in which he takes up Marx's project and his 'desire to end the inhuman' (Bellet, 2013, p. 35). In Bellet's view, one of the strengths of communism is that it incorporated both a vision and a political will.

The means of action is disconcertingly simple and stands in opposition to the hegemony of economics: the quality of human presence in the relationship – what is at stake there is priceless. It is from the perspective of the 'between-us' that he hopes for and envisages a change in society, because it is there that humanity is born. There is an existential dimension to Bellet's response to societal and political questions – an existential dimension that is disarming: 'Everything rests on this point, on this infinitesimal and total giving, which makes it possible for a human to know himself as human, and for humanity to be as one, instead of a magma of violence where everything is destroyed' (Bellet, 2013, p. 79). To change humanity is to allow communion in relationship. Here, the reader is entitled to question the scope of this ambition. The author, acutely aware of the risks of possible 'madness' and 'savagery' entailed by the enterprise, cannot help but move in this direction because it is 'humanity that is threatened, not stock market prices!' (p. 85). It should be mentioned here the extent to which Bellet's work is opposed to transhumanist thinking, which reinvents humanity but, in so doing, destroys it. Coexistence is the foundation of Bellet's thinking. Changing humanity is indeed the theoretical and political ambition of Bellet's work, 'but not at all as a fabrication of a new human, for all that remains is the presence and the urgency of retaining within ourselves that which allows us to be there' (Bellet, 2013, p. 85).

---

[46] This aim is directly related to that of the convivialists.

8.5 Action

Bellet, whose thinking is particularly in tune with postmodernity, never ceases to question the possibility of action: 'What can we do in this time of crisis?' In dealing with this question, the author implicitly draws on this primordial – and unanswered – question: 'What are we doing here?' Bellet thinks of humanity as a political adventure with an existential foundation. Possible action is thought of as collective action or as a common enterprise resulting from plurality. In this way, Bellet's ideas are part of a conceptual extension of Arendt's Action, in which the 'between-us' is seen as the fulcrum of politics. On the other hand, he places more emphasis than does Arendt on the idea of communion; for him, communion is a starting point for thinking about the political, which he formally expresses in *L'avenir du communisme* (Bellet, 2013). Bellet's communionist thinking is close to the convivialist thinking of Alain Caillé and MAUSS (Caillé & Chanial, 2014; Caillé, 2015; Caillé & Les convivialistes, 2016). Faced with humanity's capacity to destroy itself, which is Bellet's intellectual spur, his argument rests on the fragility of the 'between-us' and the importance of communion among humans. As we journey through the Anthropocene, finitude seems to be the only thing left to humanity. This political thought based solely on human fragility is an interesting resource.

8.5.1 Thinking of Action from the Perspective of Revolution

One of the strengths of Bellet's thinking lies in the articulation of paradoxes. This communionist thinking, in which the 'between-us' is the origin of humanity, is characterised by a form of non-violence and non-voluntarism, while being structured around the concept of revolution. Here, again, there is a proximity to Arendt's model of collective action or concerted action, constructed on the basis of her analysis of revolutions (especially the American War of Independence). Bellet's conception of revolution is similar to that of Lévinas, who considers that a revolution takes place when human beings are emancipated – i.e. when they wrench themselves free of economic determinism. Lévinas' revolutionary incitement encourages the development of ways of life that are not dependent on the economy, and organisation of society contingent upon 'truly caring for one's fellow man' (Poché, 2003, p. 101). These themes are particularly close to what Bellet sets out in *La seconde humanité*. Moreover, he thinks of revolution in terms of the idea of conversion. In this sense, his work fits in with that of Mounier, for whom 'The essential thing remains preparation for the spiritual revolution, which entails a slow personal and moral conversion' (Lambert, 2001, p. 65). Bellet is not a revolutionary in the Marxist sense; he is more of a resistance fighter, or an *indignant savage*: 'I am, I want to be the indignant savage' (Bellet, 2002). Indeed, in Bellet's view, the 'greatest power' is not political but relational. Thus, the power of the psychoanalyst is much more consequential than political power in Bellet's eyes, facilitating a much more profound change.

8.5.2 *The Existential Foundations of Action*[47]

As we have seen, for Bellet, humanity is threatened by the contemporary way of life, marked by techno-economic hegemony. Thus threatened, humanity must be born again – that is to say, evolve, and this implies a revolution without violence. Bellet's approach is political thinking with existential foundations, upon which we shall draw in this work. What matters to him is the action of transforming the world and humanity. Bellet does not so much seek to determine what the exemplary City and its means of organisation should be as what humanity should be, delivered from murder, madness, and monetary monopoly. He thinks of possible actions that humans can take on humanity, on the basis of the possible presence of human-to-human relations, and possible actions that humans can undertake with other humans. The goal is to allow humanity to be reborn. This is what is at the root of the social, and therefore of the political (Aron, 1962, p. 9). Here, we see similarities with the Ricœurian and Arendtian conception of politics, marked by conviviality.

It is as a theologian that Bellet thinks about the world, humanity and possible actions. 'Bellet's entire work is permeated by the radical challenge of "defining the place of *theos*" outside the mortifying cleavages to which it has been confined by classical theology and institutional religion' (Arnsperger, 2010a, b, p. 35–36). Bellet's theology is similar to fundamental anthropology (Arnsperger, 2010a, b, p. 35). Indeed, it boils down to a 'discipline of thought and action that focuses on the set of questions to which human beings need answered – without which it is impossible to live. The *theos* of *theo*-logy is not a being – it is a gesture in a place: a gesture of uplift in the very place where humans' humanity is determined. A gesture or a word, moreover – a word so powerful that it is like a hand holding us back from the edge of the abyss. A very great theologian, Maurice Bellet, spent his whole life saying just that, in a thousand-and-one ways – for it is the essential of the essential, without which the human condition dissolves into nothingness, even if everything seems fine on the surface: democracy, wealth, freedom, etc.' (Arnsperger, 2010a, b, p. 35).

8.5.2.1 A Post-Promethean 'Between-Us' to Get Us Through the Anthropological *krisis*

The notion of 'between-us', borrowed from Lévinas, is important in Bellet's thinking, because it is at the root of what makes humans human. The next chapter deals with this notion, understood in oppositional complementarity to that of the

[47] Action can, in fact, quite often refer to existential foundations in an extension of Blondel's thinking: 'it is action which finally reveals the transcendence present in the heart of man, who cannot be satisfied with being a simple anonymous representative of the species. It is action that both leads to a transformation of the world, and does so within the context of personal faith that addresses all the challenges of the times' (Breuvart, 2009, p. 189). Philosopher Eric Weil's category of action opens onto an absolute, and is also connected with metaphysical foundations. For Weil, the driving force behind action is the quest for a form of profound serenity and freedom (1982).

individual, before it is extended, through Andreas Weber's work, to the nonhuman living world and the inorganic. The normative and speculative reflection undertaken in Part II of this book is to be distinguished from the body of theories and research highlighting the need for humanity to evolve toward progressive hybridisation with machines and computer circuits (Alexandre, 2011; Hottois et al., 2015). In the anthropological shift we are outlining here, it is a post-Promethean type of 'between-us' that is suggested as contributing to the consolidation of the political.

Bibliographical References

Alexandre, L. (2011). *La mort de la mort*. Lattès.
Arendt, H. (1983). *Condition de l'homme moderne*. Calmann-Lévy (ed. or. 1958, tr. fr. 1961), tr. fr.
Arnsperger, C. (2002). Justice et économie. Latitudes d'égalisation et obstacles existentiels. *Revue de métaphysique et de morale, 33*, 7–26.
Arnsperger, C. (2005). *Critique de l'existence capitaliste – Pour une éthique existentielle de l'économie*. Cerf.
Arnsperger, C. (2009). *Éthique de l'existence post-capitaliste – Pour un militantisme existentiel*. Cerf.
Arnsperger, C. (2010a). Changer d'existence économique: enjeux anthropologiques de la transition du capitalisme au post-capitalisme. *Revue d'éthique et de théologie morale, 258*, 23–50.
Arnsperger, C. (2010b). Monnaie, dette et croissance sans prospérité: portée et limites du 'tournant' jacksonnien. *Etopia, 8*, 109–116.
Arnsperger, C. (2011). *L'homme économique et le sens de la vie*. Textuel.
Arnsperger, C. (2013). Fonder l'économie écologique. *Revue d'éthique et de théologie morale, 276*, 93–120.
Arnsperger, C., & Parijs, P. (2003). *Éthique économique et sociale*. Éditions La Découverte.
Aron, R. (1962). A propos de la théorie politique. *Revue française de science politique, 1*, 5–26.
Assoun, P.-L. (2016). *L'Ecole de Francfort*. PUF. (first edition 1997).
Bellet, M. (1976). *Le lieu du combat*. Desclée de Brouwer.
Bellet, M. (1979). *Le Dieu pervers*. Desclée de Brouwer.
Bellet, M. (1990). *Dire ou la vérité improvisée*. Desclée de Brouwer.
Bellet, M. (1992). *Incipit*. Desclée de Brouwer.
Bellet, M. (1993). *La seconde Humanité. De l'impasse majeure de ce que nous appelons l'économie*. Desclée de Brouwer.
Bellet, M. (1996). *L'Europe "Au-delà" d'elle-même*. Desclée de Brouwer.
Bellet, M. (1998). *Le sauvage indigné*. Desclée de Brouwer.
Bellet, M. (2002). *La longue veille 1934–2002*. Desclée de Brouwer.
Bellet, M. (2004). *Le paradoxe infini*. Desclée de Brouwer.
Bellet, M. (2008). La contestation évangélique. *Etudes, 408*, 636–648.
Bellet, M. (2013). *L'avenir du communisme*. Bayard.
Bellet, M., & Wallenhorst, N. (interviews). (2022). *Sapere. Saveurs et savoirs d'un monde qui nous parle*. Le Pommier.
Bertrand, D. (2005, February). Maurice Giuliani, Maurice Bellet. *Etudes*, 271–273.
Bourg, D. (1996). *L'Homme artifice*. Gallimard.
Bourg, D. (2012). Transition écologique, plutôt que développement durable. *Vraiment durable, 1*, 77–96.
Breuvart, J.-M. (2009). Eric Weil et la question de Dieu selon H. Bouillard. *Recherches de Science Religieuse, 97*, 185–193.
Caillé, A. (2003). Maurice Bellet: Invitation. Plaidoyer pour la gratuité et l'abstinence. *Revue du MAUSS, 22*, 434.

Caillé, A. (2015). *Le convivialisme en dix questions. Un nouvel imaginaire politique*. Le Bord de l'Eau.

Caillé, A. (2016). *Les Convivialistes, Eléments d'une politique convivialiste*. Bord de l'eau.

Caillé, A., & Chanial, P. (Eds.). (2014). *Du convivialisme comme volonté et comme espérance* (Vol. 43). Revue du MAUSS.

Chalier, C. (2008). *Transmettre de génération en génération*. Buchet Castel.

Cochet, Y. (2013). Trois modèles du monde. In A. Sinaï (Ed.), *Penser la décroissance* (pp. 49–72). Presses de Science Po.

Cupillard, D. (2001, April). Maurice Bellet: L'amour déchiré. *Etudes*, 565–566.

Curnier, D. (2017) *Quel rôle pour l'école dans la transition écologique? Esquisse d'une sociologie politique, environnementale et prospective du curriculum prescrit*. Doctoral thesis in Environmental Sciences from the University of Lausanne, supervised by Dominique Bourg and Philippe Hertig.

Deléage, J.-P. (2010). En quoi consiste l'écologie politique? *Ecologie & Politique, 40*, 21–30.

Donegani, J.-M. (2011). L'autocompréhension du catholicisme, entre critique et attestation. *Raisons Politiques, 4*, 5–14.

Donegani, J.-M. (2013). Crise de l'Occident, crise du christianisme, crise de la différence. *Recherches de Science Religieuse, 101*, 351–376.

Dupuy, J.-P. (2009). Le Futur bifurque-t-il? Vers une nouvelle science du futur. In M. Grossetti, M. Bessin, & C. Bidart (Eds.), *Bifurcations* (pp. 373–386). La Découverte.

Dupuy, J.-P. (2012). On peut ruser avec le destin catastrophique. *Critique, 783–784*, 729–737.

Ehrwein Nihan, C. (2013). Penser la crise ou plaidoyer pour une réflexion critique sur la crise de la crise à partir de l'œuvre de Hannah Arendt. *Revue d'éthique et de théologie morale, 276*, 43–60.

Euvé, F. (2013, December). Recension de Maurice Bellet, *L'avenir du communisme*, Paris, Bayard, 2013. *Etudes*, 713.

Flahault, F. (2013). Pour une conception renouvelée du bien commun. *Études, 418*, 773–783.

Gemenne, F. (2015). L'Anthropocène et ses victimes: une réflexion terminologique. In F. Gemenne (Ed.), *L'enjeu mondial* (pp. 233–240). Presses de Sciences Po, "Annuels".

Guibal, F. (1995, November). Eric Weil. Le défi de la violence. *Etudes*, 495–504.

Hottois, G., Missa, J.-N., & Perbal, L. (Eds.). (2015). *L'humain et ses préfixes – Une encyclopédie du transhumanisme et du posthumanisme*. Paris.

Jonas, H. (1990). *Le principe responsabilité*. Cerf. French translation 1990 (or. 1979).

Lambert, F. (2001). *Introduction à l'histoire des idées politiques*. Armand Colin.

Lambert, F. (2012). Le 'silence' des intellectuels catholiques français. In F. Hourmant & A. Leclerc (Eds.), *Les intellectuels et le pouvoir*. PUR.

Le Corre, F. (2006, January). Maurice Bellet: La traversée de l'en-bas. *Etudes*, 135–136.

Lefebvre, P. (2005). Une conversion "hors les murs": *Les Allées du Luxembourg* de Maurice Bellet. In N. Brucker (Ed.), *La conversion. Expérience spirituelle, expression littéraire* (Vol. 8, pp. 573–582). Actes du colloque de Metz (5–7 June 2003).

Lévinas, E. (1991). *Entre nous – Essais sur le penser-à-l'autre*. Grasset.

Maldamé, J.-M. (1988). L'itinéraire de Maurice Bellet. Une phénoménologie de la création. *Revue Thomiste, 88*(2), 299–314.

Mathieu, P. L. (1969). *La pensée politique et économique de Teilhard de Chardin*. Seuil.

Morin, E. (1999). *Introduction à une politique de l'homme*. Seuil.

Morin, E. (2001). *L'identité humaine*. Seuil.

Morin, E. (2012). Krisis. In E. Morin (Ed.), *Journal 1962–1987* (pp. 1123–1180). Seuil.

Morin, E., & Lafay, D. (interview). (2017). *Le temps est venu de changement de civilisation*. L'Aube.

Mounier, E. (2001). *Le personnalisme*. PUF (ed. or. 1950).

Moussavi Chirazi, S. D. (1999). *L'évolution de l'écriture de Maurice Bellet*. Doctoral thesis in 20th-century, supervised by Philippe Lefebvre, Université de Nancy 2.

Moussavi Chirazi, S. D. (2007, July). Un style moderne pour le langage religieux: l'écriture de Maurice Bellet. *Archive of Scientifique Information Database*. http://www.sid.ir/en/VEWSSID/J_pdf/1046320100106.pdf, consulted online on 3 October 2013.
Neyrat, F. (2014). Critique du géo-constructivisme. Anthropocène & géo-ingénierie. *Multitudes, 56*, 37–47.
Perret, B. (2014). Transition écologique ou choc de la finitude? *Revue du MAUSS, 43*, 35–40.
Poché, F. (2003). *Penser avec Arendt et Lévinas*. Lyon.
Poché, F. (2007). *Penser avec Derrida*. Chronique Sociale.
Poizat, J.-C. (2009). *Assumer l'humanité, Hannah Arendt: la responsabilité face à la pluralité* de Gérôme Truc. *Le Philosophoire, 31*, 177–188.
Pommier, E. (2013). Ethique et politique chez Hans Jonas et Hannah Arendt. *Revue de métaphysique et de morale, 78*, 271–286.
Prouteau, F. (2006). *Former, oui... mais dans quel sens?* L'Harmattan.
Revault d'Allonnes, M. (2011). Hannah Arendt penseur de la crise. *Etudes, 415*, 197–206.
Saudan, A. (2008). Hans Jonas: la tentative ou la tentation d'un fondement ontologique du devoir. *Le Philosophoire, 30*, 111–125.
Tillich, P. (1999). *Le courage d'être*. Cerf. French translation.
Touraine, A. (2010). *Après la crise*. Seuil.
Touraine, A. (2013). *La fin des sociétés*. Seuil.
Wallenhorst, N. (2014, July). Recension de L'avenir du communisme de M. Bellet. *Esprit*, 142–143.
Weber, A., & Kurt, H. (2015). *Lebendigkeit sei! Für eine Politik des Lebens. Ein Manifest für das Anthropozän*. Klein Jasedow, Think Oya.
Weil, E. (1982). *Philosophie et réalité, Derniers essais et conférences*. Beauschene.
Zundel, M. (1990). *Recherches sur la Personne*. Desclée de Brouwer.

Chapter 9
From the (Augmented) Individual to a Post-Promethean 'Between-Us'

Abstract The idea of an anthropological shift has been worked on for several decades in the United States, and for about 10 years in a set of works that have attracted media coverage and gained public recognition: the transhumanist theories and research, mentioned previously in this book. These are linked to the theories of a good Anthropocene because of the underlying techno-modernism and the refusal to understand finitude (environmental or human) as a limit.

Keywords Anthropocene · Transhumanism · Donna Haraway · Modernity · Education

Just as it would be possible to increase or improve nature through technological mastery, it would be possible to enhance human beings as well.[1] The economic and medical sectors liberally propose structuring directions for the future of humanity. While the Anthropocene has brought human mortality into sharp focus, this set of theories and transhumanist research offers the possibility of negating individual death.[2] What, though, is the political anthropology underlying these perspectives? In this chapter, transhumanism is viewed as an economic project of bringing about an anthropological shift based on the accomplishment of the Promethean project,

[1] Finitude is one of the greatest contemporary anthropological problems. Humanity has been brought face to face with this reality as never before: 'we are confronted in a very concrete way with the finitude and fragility of the world, and with the limited character of a number of vital resources […] it is as if humanity were discovering a new facet of its mortality' (Perret, 2014a, b, p. 36). (In *Tristes Tropiques*, the French anthropologist Claude Lévi-Strauss had mentioned that the world would end without humanity, the same way it had begun).

[2] 'Not to accept our finitude is once again to exclude ourselves from the natural circle of life. Our fantasy of omnipotence pushes us towards a hope that is incompatible with reality, especially with the finitude of the earth's resources themselves. Thus, ironically, we mortgage our own survival as a species, destroying the conditions of life itself' (Cabanes, 2017, p. 41).

© The Author(s), under exclusive license to Springer Nature Switzerland AG 2023

N. Wallenhorst, *A Critical Theory for the Anthropocene*, Anthropocene – Humanities and Social Sciences, https://doi.org/10.1007/978-3-031-37738-9_9

whereby the political is ousted. While in transhumanist theories, the emphasis is placed on the shift of the individual, thought of outside of any relational context, we on the other hand emphasise the 'between-us' as a political space for human genesis. For the French pedagogue François Prouteau, it is a matter of the future of humanity to 'newly centre teaching on this human "We" linked to the cosmos, inescapable and vital' (2018, p. 115). He then goes on to say, 'This "We" configures a world – that is, a habitable space where intersubjective relations are woven, without which a human subject cannot come into being' (p. 115). The first section of this chapter presents our understanding of transhumanism as an absence of politics; the second section is spotlights the achievement of the Promethean project as the substance of transhumanist theories. Then, inspired by the reading of Donna Haraway's *A Cyborg manifesto* (2007),[3] we will propose a critical counterpoint so as not to hastily engage in too simplistic a criticism of transhumanism. Then the dynamics of augmentation are questioned, with regard to educational authority in a fourth section. As the achievement of the Promethean project is identified in this chapter as possible but not desirable, the fifth section proposes fundamental elements for a political thought of education, aiming to bring about the emergence of a post-Promethean 'between-us'. The French philosopher Jean-Marc Liautaud shows that the transition from the question 'Is it possible?' to 'Is it desirable? is crucially important, and refers to the emergence of a 'post-Promethean subject capable of setting limits to his or her action' (2018, p.75). Transhumanist theories and research appear, in this chapter, as the exact opposite to the anthropological shift outlined in this work, consisting of a consolidation of the political.

9.1 Transhumanism in Questions

9.1.1 Some Technological Developments in the Contemporary Period

An important development in the last two decades has been the progressive technological convergence known as 'NBIC': nanotechnologies, biotechnologies, information technologies and cognitive sciences (primarily neurosciences) are coming together in their capacity to make nanometric adjustments to the fabric of life. As a result, a whole range of anthropic techniques have become possible and developed – i.e. modifications made to human beings for non-medical purposes. These capabilities raise a series of anthropological and political questions: having learned to heal

[3] Donna Haraway's thinking has already been touched upon briefly in this book, with her conception of the Anthropocene as the Chthulucene.

humans, should we improve them? Should we augment them? What would be the purpose?[4] Is a perfect, sanitised and controlled individual really what we need?[5]

Technology now has a paradigmatic function in our relationship to the world. It has become a way of existing, and therefore of orienting ourselves – that is, ultimately, a way of making decisions (Capelle-Dumont, 2019).[6] Technological innovations have been the carriers of ideologies with differing conceptions of what human beings are, and what their relationship to the world should be. This means that technology is not axiologically neutral – it influences the directions our societies take: 'Put more bluntly' argues Marie-Hélène Parizeau, President of COMEST (UNESCO's World Commission on the Ethics of Scientific Knowledge and Technology), argues: 'Technology is part of a system of domination – economic, political or cultural' (2019). Parizeau thus offers a retrospective on the last 40 years of technological innovation, and shows that four increasingly short cycles have followed one another: biotechnologies, nanotechnologies, the NBIC convergence with the idea of the augmented human, and finally Artificial Intelligence (AI). Looking at the succession of cycles of technological innovation from this standpoint, we understand that what is currently at stake is quite simply human freedom – no more, no less – individual, certainly, but above all, collective freedom.[7]

Consider, for example, the use of drugs, in a non-therapeutic but ameliorative function: in today's world, certain substances are the means of choice to improve performance (cognitive, physical or emotional).[8] There are a number of psychostimulants that can be used for purposes other than therapeutic: modafinil increases

[4] French physicist Thierry Magnin questions the simplified conception of life underlying NBICs, which are commonly viewed as a driving force for improvement (2019). On the contrary, the scientific work of recent decades shows the complexity of life and its permeability to the environment. Thus, epigenetics highlights the way in which our genetic heritage is not only received, but is shaped by our behaviour and our environment, and work on neuroplasticity demonstrates that psychological experience alters the biological functioning of the brain. In light of these findings, Thierry Magnin points out a paradox: while biology is gradually breaking away from a functionalist paradigm, biotechnologies could become locked into one. He emphasises that there is a dynamic relationship between vulnerability and robustness which gives life its astonishing plasticity. Magnin proposes a form of reversal by considering the vulnerability inherent to living beings as an ethical spur in the contemporary period, marked by the quest for power through technological developments. This makes it possible to break with a utilitarian and dominance-based view of living beings, in favour of relational ethics.

[5] This is a question that is delicately posed in François-Régis de Guényveau's novel (2017).

[6] Technology requires an ethics of decision-making. This is sought by Philippe Capelle-Dumont, based on an understanding of technology as an intermediary between humans and nature. Capelle-Dumont discusses the illusion of neutrality and the danger of Martin Heidegger's Enframing, Michel Foucault's technology as a creative force, and Hans Jonas' heuristics of fear and responsibility. Capelle-Dumont then highlights three aporias and proposes to think of this necessary ethics of decision-making for the contemporary period on the basis of the principle of alliance.

[7] 'We have less to fear from artificial intelligence than from human stupidity', says French scientist and futurist Joël de Rosnay (2018, p. 234), noting the importance of education.

[8] However, it is not only explicitly transhumanist theories and research. This quest for increased human performance is currently one of the marked trends in medicine (Le Dévédec & Collin, 2018).

periods of attentive wakefulness, ritalin increases concentration capacity, pheniso-propamine (an amphetamine) improves cognitive performance.[9] Such neuropharmacology, whose aim is to improve brain function, raises a number of questions.[10]

This question of the brain and how to increase its capacities is currently the subject of a great deal of attention and prospective research. Thus, the objective of Neuralink, a company founded in 2016 by young South African billionaire Elon Musk, is to interface the brain with electronic components. Similarly, the goal of CALICO (*California Life Company*), founded in 2013 by Google, is to fight aging and 'defeat death'. While these projects attract investment from private companies, they also benefit from private funds, as is the case with the *Human Brain Project*, which has received one billion euros of funding from the European Union over 10 years to simulate a functioning human brain on a computer by 2024.

The current work on understanding the function of the brain and its interfacing with computer chips brings up the question of the body, which is sometimes understood to be a transport vehicle for the brain. Proposals have been made to improve the body's performance,[11] or actually change the body when it becomes too old. Italian neurosurgeon Sergio Canavero, from Turin, announced in November 2017 that he had performed the first head transplant on a cadaver. He performed the procedure in China, with the help of Chinese surgeon Ren Xiaoping and his team, who say they have performed head transplants on mice and monkeys (Barthélémy, 2018).

Among other contemporary technological developments, cryonics has become a reality. This is a technique for preserving bodies at extremely low temperatures (−196 °C), which is a rapidly expanding sector in the United States. At present, this technique comes under funeral law, but the objective is for revival of the subject in the future, with the cryogenically preserved person being 'woken up' once medicine has found a cure for their condition (Caire, 2018).

9.1.2 *Transhumanism as the Absence of Politics*

The term 'transhumanism' was coined by Julian Huxley in 1957 in the book *New Bottles for New Wine*, meaning a form of 'transcendentalisation' –, the ability to become 'more than human' through a lasting modification of human nature.

[9] It might also be possible to develop oxytocin sprays in the near future to combat shyness and aggression, or to use MDMA (ecstasy) to increase cooperation and empathy within groups.

[10] Is brain enhancement really only a matter of individual freedom, when it brings about an imbalance in social relationships? (Larrieu, 2018.

[11] One avenue being explored is robotics (a rapidly expanding sector), with the creation of exoskeletons – a form of armour or artefact external to the human body that improves its performance. Note that the French physician Vincent Rialle proposes the expression 'humanitarian robotics' to identify particularly positive initiatives to mobilise contemporary technoscientific research, which he differentiates from 'suicidal robotics' seeking only to increase human power and combat individual death (2018) (through this expression, Vincent Rialle identifies the potential of 'civilisation' brought about by technoscientific prowess).

Currently, in the use of the term, there is the idea of the transition of humanity to a more advanced stage of 'evolution' which could be considered posthumanity.[12] The prefix 'trans' also refers, for authors like the German philosopher Jürgen Habermas of the second generation of the Frankfurt School, to a major transition between external and internal nature (2002, p. 41). Transhumanist theories and research lend themselves to partisanism as opposed to neutrality, as they touch on completely fundamental aspects of the human adventure. The term 'transhumanist', which was still unknown to the general public only a few years ago, is now universally understood.

Today, it is possible to identify two main schools within the set of transhumanist theories and research: a Californian school made up of Silicon Valley web companies around the *Singularity University* led by American engineer Raymond Kurzweil and Greek-American engineer Peter Diamandis, subsidised by Google; and a more academic, Oxfordian school, in the shape of the *Future of Humanity Institute* at the University of Oxford, led by the Swedish philosopher Nick Bostrom, who has been heavily involved in shaping the intellectual content of transhumanism (Bostrom, 2002, 2003, 2014; Savulescu & Bostrom, 2008).[13] In addition to these two main transhumanist currents, there is a set of associations and movements. Transhumanism consists of different areas of interest: the extropians, defined by the British philosopher Max More (More, 1993, 2003, 2010), are particularly neoliberal, with an emphasis on individual responsibility and non-intervention by the State. The main feature of the work of Raymond Kurzweil, who was hired by Google a few years ago, is the search for amortality based on artificial intelligence, as well as that of Bruce Klein, who set up the *Immortality Institute*. Nick Bostrom and David Pearce, on the other hand, who founded the *World Transhumanist Association* in 1998, which has since changed its name to *Humanity+*, are more democratic. In this sector of research integrating a democratic component, we can also include American sociologist James Hughes, who asserts a concern for the fair redistribution of wealth, as set out in his book *Citizen Cyborg* (2004). (Also see Goffi, 2011, p. 22; and Frippiat, 2011, p. 36).

When transhumanism is discussed in the French context, the ethical perspective often comes to the forefront, and the political component appears to be greatly underestimated in the analyses. However, the aim of transhumanism with the pursuit of economic hegemony in our globalised societies is to extend the private sphere until the public sphere disappears entirely. Transhumanism is eminently problematic

[12] This argument holds in the French-speaking context. In the English-speaking world, 'posthumanism' as an umbrella term covers all perspectives that differ from conventional humanism, some of which are distant from transhumanism.

[13] Surprisingly, in a statement in Paris in June 2017 at the *Unexpected Sources of Inspiration* (USI) conference on June 19–20, Nick Bostrom mentioned that he did not define himself as a transhumanist: 'I don't tend to define myself as a transhumanist, because this term is used by very different people with widely different meanings, which creates confusion' (Benoît Georges, *Les Echos*, June 26). We see, through this statement, that there is no unified transhumanist movement, but a set of very different trends. That is why, as opposed to the term 'transhumanism', we prefer the expression 'transhumanist theories and research' which incorporates the idea of a broad raft of visions.

from a political perspective.[14] We can define transhumanism as a hegemonic economic set of ideas concerning the alteration of humanity, carried by a type of capital from technological mastery. The common aim of the different groups of transhumanists consists of the preparation of a new type of human, a form of '*homo sapiens augmentens*' (Blin, 2015, p. 88). A researcher from the Coventry University in the UK, cyberneticist Kevin Warwick, for example, presents himself as the first cyborg in history since the linking of computer components into his nervous system (Maestrutti, 2011, p. 61). The alteration of humanity that transhumanists are working towards is corporeal: it is about increasing sensory, but above all cognitive, capacities and postponing death in order to ultimately eliminate it.[15] This is an economic project of defining the future (by suppressing the future) which is not carried by out citizens but by shareholders of extremely new and powerful multinationals. The companies performing this transhumanist project on behalf of all humanity – first and foremost Google, Amazon, Facebook and Apple (but also the Asian group of companies BATX: Baidu, Alibaba, Tencent and Xiaomi) – have in common that, in the space of around 20 years, they have acquired a power for action perhaps greater than that of a State – without any restrictions other than the budget having to be voted by a handful of retired shareholders on the west coast of the United States.

Max More (2003) describes three of the seven 'extropian principles' of transhumanism as follows. These present the vast ambition of the hoped-for anthropological change (some of the principles not noted in the following incorporate a component that is intended to be more democratic and social):

> Perpetual Progress – Seeking more intelligence, wisdom, and effectiveness, an indefinite lifespan, and the removal of political, cultural, biological, and psychological limits to self-actualization and self-realization. Perpetually overcoming constraints on our progress and possibilities. Expanding into the universe and advancing without end.
>
> Self-transformation – Affirming continual moral, intellectual, and physical self-improvement, through critical and creative thinking, personal responsibility, and experimentation. Seeking biological and neurological augmentation along with emotional and psychological refinement. [...]
>
> Intelligent technology – Applying science and technology creatively to transcend "natural" limits imposed by our biological heritage, culture, and environment. Seeing technology not as an end in itself but as an effective means towards the improvement of life. [...]

[14] In line with French sociologist Thierry Blin (2015, p. 100), French historian Franck Damour (2017, p. 61) and the German political scientist working in France Klaus-Gerd Giesen (2018), we were tempted to call the transhumanist cause a political one. Upon reflection, though, it seems to us that this is not the case and that transhumanism consists, on the contrary, of suppressing the political, understood in an Arendtian sense as the possibility of collective action emerging from human plurality. Moreover, as Arendt shows in *The Origins of Totalitarianism*, transforming humans can only be part of a project aiming to achieve total domination. A project of human transformation goes beyond politics, and is therefore no longer political as such. Indeed, Arendt has shown that exceeding the political amounts to destruction of the political (Tassin, 2017, p. 161). It is precisely this excessive, totalitarian aspect that is problematic.

[15] A defining element of transhumanism is its anthropotechnics, which consists of the modification of human beings through interventions that have no medical purpose.

The above principles are relatively typical of transhumanism and are shared between the different ideological sensibilities.[16] It is possible to identify three central characteristics of transhumanist theories and research: to push back the limits of death and to reach amortality[17]; to push back the limits of knowledge and to reach omniscience; and finally, to acquire unlimited power (in particular, with a view to intergalactic travel and colonisation). This enterprise is to be achieved, it is proposed, by interfacing the brain with computer media. One of the fundamental elements on which transhumanism is based is that the current anthropological limits must be exceeded to allow humanity to solve its major problems (such as global warming). We can see the degree to which it is a question of bringing the Promethean project of modernity to its conclusion in surpassing limits, as in the demiurgic access to omnipotence; and some may find the radicality of this prospect frightening. Rather than bringing the Prometheanism of modernity (which is at the root of transhumanism) to an end, this chapter proposes to try to go beyond it.[18]

The promise of amortality that transhumanism offers has earned it the status of a new religion (Luzi, 2018) – a form of ultramodern religion with capitalist foundations and ecocidal implications. What is truly questionable within transhumanist theories and research is the claim that a techno-scientific solution can be found for all the problems we face. The Anthropocene raises sociopolitical problems with anthropological roots, to which an appropriate response must be provided that a single techno-scientific paradigm is unlikely to be able to provide.

9.2 The Fulfilment of the Promethean Goal of Modernity

9.2.1 The Promethean Individual

Prometheus is a mythological figure who allows us to think of human power in the context of a transgression in men's relationship with the gods. The myth of Prometheus probably originated with the Caucasians, with whom the Greeks were in regular contact. There is a set of variants of the Prometheus myth based on two main versions: the *Theogony* of the Greek poet Hesiod, written in the eighth century BCE, and *Prometheus Bound*, by Greek tragedian Aeschylus in the sixth–fifth century BC. The narrative elements that structure these two stories are quite similar:

[16] In this transhumanist aim of surpassing humanity, man may seem superfluous. This is a characteristic of totalitarianism highlighted by Arendt.

[17] Arendt believes that this quest for immortality is indicative of the sin of vanity (1983, p. 67). We prefer to use the prefix *a-* rather than *im-* to signify that the attempt to avoid death is made by beings who are born, as opposed to the immortality of gods who are never born.

[18] The secularisation of modernity has both made men mortal and given them the hope of emancipation from that condition. Therein lies the intrinsic ambiguity of modernity and the transition from religion to technoscience.

Prometheus is the Titan who stole fire from Zeus to give to mankind; he then incurs the wrath, and retribution, of Zeus. Fire symbolises a collection of attributes such as intelligence, the beginning of civilisation and technical prowess. On the other hand, these two authors' interpretations of the tale diverge: for Hesiod, Prometheus' imprudence and transgression are responsible for all human suffering, whereas for Aeschylus, Prometheus is responsible for the transition from savagery to civilisation. He is a true hero to men, with whom he joins forces against the gods (Haudry, 2016, pp. 327–346).

The figure of Prometheus is that of the man of technique (Prometheus teaches men arts and techniques, including metallurgy), boundless avarice, transgression and excess (*hybris*), and refers to contemporary individuals' endless search for emancipation and rejection of the ties imposed by nature (Flahault, 2008a, p. 35). This figure of Prometheus 'not only expresses a desire for emancipation, greatness and power; it also constitutes a model, a stimulus and a justification. [...] What is special about the Promethean ideal, what made it strong, is the intimate mixture of a realistic program of knowledge and action with a figure that captures the imagination and arouses a desire for identification' (Flahault, 2008a, pp. 14–15). Descartes' proposal that we become 'masters and possessors of nature' (1637, Tome 1, 6th part) is the first explicit formulation and modern scientific manifestation of this Promethean ideal. It has been directly linked to the idea of progress, from the Latin *progressus*, meaning 'to advance' and 'to go forward', with the presupposition that doing so will allow us to improve. However, as French lawyer Valérie Cabanes, in an extension of this etymological reflection, asks, does progress really improve us? (2017, p. 36). Indeed, Prometheism is related to *hybris*, which can be understood as excess and limitlessness: it is 'the claim that there is no reason to assume a predetermined limit to what we can accomplish or the ways in which we can transform ourselves or our world' (Brassier, 2016, p. 160). Since the Enlightenment, western individuals 'aspire to enjoy a sense of absolute existence, forever' (Flahault, 2008a, p. 281). There are several particularly visible manifestations of Prometheism in the contemporary period, such as technoscience and capitalist neoliberalism, which come together here in transhumanist theories.[19] However, it can also be seen in the ideals of progress, freedom and emancipation. Prometheism is also a characteristic of humans that animals lack. As Cabanes reminds us: 'Whilst wolves and lions, for example, are powerful predators, whilst they rise to the top of the food chain as we do, they do not massacre, they do not enslave' (2017, p. 21). We could add that they

[19] For the critical theorist Herbert Marcuse, the world is chosified – we have subjugated it and imposed control upon it. As a result, it is no longer capable of a 'warm', responsive relationship with us, and in Marcuse's eyes, it is necessary to develop an 'erotic' relationship with the world. Marcuse constantly denounced Prometheism and proposed alternative myths (such as those of Orpheus, Narcissus and Epimetheus). In *Eros and Civilisation*, Marcuse seeks to free us from this Promethean figure of modernity where only productivity matters.

do not feel this need to exceed limits, to reach record levels of omnipotence, to gain ever-increasing mastery over others and their environment.[20]

It is astonishing to note how Hannah Arendt perceived that the Promethean project of modernity had in its germ the prospect of reaching its fulfilment in amortality: 'the wish to escape the human condition, I suspect, also underlies the hope to extend man's life-span far beyond the hundred-year limit. This future man, whom the scientists tell us they will produce in no more than a hundred years, seems to be possessed by a rebellion against human existence as it has been given, a free gift from nowhere (secularly speaking), which he wishes to exchange, as it were, for something he has made himself. There is no reason to doubt our abilities to accomplish such an exchange, just as there is no reason to doubt our present ability to destroy all organic life on earth. The question is only whether we wish to use our new scientific and technical knowledge in this direction, and this question cannot be decided by scientific means; it is a political question of the first order and therefore can hardly be left to the decision of professional scientists or professional politicians' (Arendt, 1983, pp. 8–9). This primordial political question is indeed the responsibility of the philosopher, the citizen and the educator. It is the question that is put to work in this critical theory for the Anthropocene.

In her *Journal de pensée* (*Journal of Thought*), Arendt goes so far as to imagine that humans could conquer death, and that this discovery would have extremely damaging consequences for thought: 'The greatest and most atrocious danger to human thought would consist of the fact that what has once been thought should be nullified by the discovery of some fact which had hitherto remained unknown; for example, it might be possible 1 day to make men immortal, and everything that has been thought about death and its depth would then become simply laughable. It would be possible to say that this price is too high in return for the overcoming of death' (Arendt, 2005). Let us continue with Arendt who, as early as 1958, had finely perceived the way in which machines could not only alter the human condition but also risk replacing humans on Earth: 'But it could be that we, earthly creatures who have begun to act as inhabitants of the universe, will never again be able to understand, i.e. to think and express, the things that we are nevertheless capable of doing. In this case, everything would happen as if our brain, which is the material, physical condition of our thoughts, could no longer follow what we do, so that from now on we would really need machines to think and speak for us. If it turns out that knowledge (in the modern sense of know-how) and thought have become separated for good, then we will be the playthings and slaves not so much of our machines as of our practical knowledge – mindless creatures at the mercy of every technically

[20] As was mentioned in the interview with Hartmut Rosa, this question of imposing control on the world was identified and denounced by Theodor W. Adorno and Max Horkheimer (along with its corollary of the domination generated by rationalism). The same is true of the thinking of Günther Anders, showing that, with the division of labour (and the experience of the concentration camps), we were in the process of transforming humans, like the world, into machines.

possible device, however deadly (Arendt, 1983, pp. 9–10).[21] Arendt then expounds on the importance of speech in the human condition as a political condition – that is, a space for welcoming a wide range of people that is not subject to the tyranny of sameness: 'The reason why it may be wise to distrust the political judgment of scientists qua scientists is not primarily their lack of "character" – that they did not refuse to develop atomic weapons – or their naïveté – that they did not understand that once these weapons were developed they would be the last to be consulted about their use – but precisely the fact that they move in a world where speech has lost its power. And whatever men do or know or experience can make sense only to the extent that it can be spoken about (Arendt, 1983, p. 10). Here, Arendt poses the importance of the space between humans as a relational space, one of the challenges of the contemporary period being to ensure that space is safe from the Promethean project.[22] The accomplishment of the Promethean project proposed by transhumanism would have an inevitable impact on the world and on others – those two structuring components of all political thought.

9.2.2 Towards the Disappearance of the World

In transhumanist theories, it is striking that the human community is no longer conceived of from a regulated public space of expression of plurality. This political function of regulation seems to have disappeared, in favour of blind confidence in the technical progress of which humans are capable. The proper function of political thought is to organise the sharing of human existence in the world. How is this organisation thought of in transhumanist theories? The uploading of our minds onto computer media is presented by several authors as particularly interesting because of the possibility of saving one's mind, acquiring any type of information, and moving – as informational data – at the speed of light (Goffi, 2011, p. 26). That this eventuality is unlikely is not what matters to us here. Instead, what is problematic is

[21] A number of Arendt's analyses have this astonishing topicality and are particularly relevant for understanding the contemporary period. This does not mean that Arendt was a visionary, but simply that she was fully contemporary with her time, and that she knew how to analyse the present time with finesse and acuity. So it is with *Obsolescence of Man*; its author, Günther Anders, upon rereading it 25 years after the first edition, wrote: 'Not only does this volume, which I completed more than a quarter of a century ago, does not seem outdated to me, but now more topical than then. […] These observations were not prognoses but diagnoses' (Anders, 2012, p. II).

[22] Two of Hannah Arendt's contemporaries – Hans Jonas and Günther Anders – also examined this tension between Prometheism and post-Prometheism. Jonas' *imperative of responsibility* lays post-Promethean foundations in the face of the threat to humanity posed by the Promethean excess of technical promise. Being responsible means accepting to be bound by what is fragile and whose existence is threatened. For Jonas, man appears to have been overtaken by his productions. His Promethean critique is similar to that of Günther Anders, who analyses a Promethean shame in which man is ashamed before the power of his technical artefacts: 'the shame that seizes man before the humiliating quality of the things he himself has made' (Anders, 2002, p. 37). This is the shame of 'having become' rather than of 'having been made' (2002, p. 38).

the way of organisation of the world underlying this thinking, to ensure the sharing of human existence. If the world is defined, in line with Arendt's thought, as the space between men, or the 'tent' sheltering humanity in all its diversity, we note that the world is absent from these theories.[23] Inclusion in the world no longer comes through the body's physical existence: 'The mode of existence that transhumanists aim to achieve, in effect, is to circulate as an informational flow in computer networks, freed from all gravity and adherence: not in the world, let alone of the world; but quite forever outside the world' (Goffi, 2011, p. 27). The world no longer appears here as the space of hospitality (Theobald, 2008) accommodating humanity in its plurality, where, as the French philosopher Renaud Hétier encourages us to do, everyone learns to feel one another (2018). Here, the world is seen as a limit to individuals' power.

9.2.3 *An Absence of the Other*

Another important dimension of a political thought is the way in which people entering the world are received. Transhumanist theories are predominantly libertarian, with a partial revival of the thought of Russian-born American philosopher Ayn Rand[24] and her denunciation of altruism, which has been interpreted as an encouragement to selfishness (Goffi, 2011, p. 31).[25] Some of the transhumanist theories may indeed be content with a minimal liberal ethic, in line with the thinking of the French philosopher Ruwen Ogien and his principle of non-harm: 'Since the individual choice to "augment" oneself is not imposed on others, there should be no principle to condemn it. In other words, private choices of self-augmentation, of bio-technological intervention, will engage only the sphere of individual freedoms over which nothing has authority: no collective truth, no institution, no book, no thing other than the property of oneself' (Blin, 2015, p. 95). Indeed, in the search for amortality as a central point, transhumanist theories centre on individuals who do not wish to leave the world, rather than the reception of ones coming into the world – within which we perceive this exaltation of the individual specific to the

---

[23] The analysis of the Ivorian philosopher Augustin Dibi Kouadio fits with our own. While inhabiting space in a given time is one of the characteristics of the human adventure, Kouadio shows that contemporary technological developments, in their assertion of power, are giving rise to a distanceless universe (2020). How, then, can others become our neighbours? What do we seek to increase with the power of our technological artefacts? What kind of human do we want to become – if not to produce? Does the instantaneous information of a disaster occurring thousands of miles away make us attentive to our neighbours, understood as a face?

[24] Ayn Rand, born in 1905 into a Jewish family, was a contemporary of the Critical Theoreticians and Hannah Arendt. On the other hand, she worked on a very different political philosophy; hers was libertarian, with a form of theorisation of capitalism based on the rational egoism of individuals.

[25] Note, however, that the libertarian influence is not the only one within transhumanist theories and currents that increasingly incorporate logics of framing individual responsibility and freedom (Hottois, 2018).

Promethean project (Flahault, 2008a, b). Hospitality seems to be absent from transhumanist theories. The power sought does not seem in any way oriented towards welcoming others, but appears to serve the self. The welcoming of others who come into the world is replaced by the individual who does not leave, as we can see in the proposal to issue procreation permits as a remedy to the overpopulation generated by the increased life expectancy. Since the world is not renewed by the birth rate (Arendt, 1972a), here again, we see it disappearing from view. Thus, in addition to granting little place to the world, transhumanist theories also grant minimal roles to others, aiming only for hegemony of the identical (Blin, 2015, p. 97). The ousting of others refers to the ousting of neighbours, which directly impacts the ousting of the individuals ability to be a neighbour to others. We can say that we are not dealing with thoughts that assume the responsibility of preparing the world for the generations to come, but with a suppression of morality for the benefit of an all-powerful neoliberal individual. There is no longer any other: no otherness, no neighbour. This generates an identity configuration quite different from that described by Paul Ricœur in *Oneself as Another* (1990) with the eviction of ipseity from the identity dynamic. The incompleteness of the human, thought of until now in the philosophical tradition as that which allows one to continue becoming human, is replaced by a completion that generates obsolescence. The articulation of sameness and strangeness inherent to the dynamics of the self seems to be reached: this reign of the identical causes damage to the self. While we could believe that transhumanist research and theories lay the groundwork for the future, it would seem that they draw a future that is the extension of the present, populated by the same individuals. The permanence of the identical, by extending the present indefinitely, hinders the future and, in a way, stops time.

We are dealing with a way of thinking in which the world, the other, the self and time appear to be affected by the permanence of the identical, reminiscent of a form of totalitarianism that can even lead to a renunciation of reality.[26] Indeed: 'One renounces one's own capacities since technology can do everything and so much more efficiently than any human will. We even renounce our memory since Google offers it in a permanent snapshot. And finally, we renounce the real. Because the real and the fictitious are no longer distinguishable. The real has become fictitious through its all-out digitization: it is put into numbers, into photos, and thus simplified, it is put into screens, but is it still the real?" (Thiel, 2015, p. 147). This regime of the same crowds out the plurality from which democratic convictions, disagreements and debates have hitherto emerged. This regime appears totalitarian insofar

[26] Arendt's understanding of modernity involves an understanding of totalitarianism and its extermination of politics associated with the development of the power of technosciences. *The Human Condition* shows that the logic of capitalism, which overvalues labour, not only fails to create a common world, but also contributes to the destruction of public space. This capitalist logic, the driving force behind transhumanist research, has strong analogies with that of totalitarianism: 'It must be admitted that if totalitarianism is a form of domination unknown before the modern era, it is also because the constitutive elements of modernity are linked – in a way that needs to be analysed – to this new power structure' (Tassin, 2017, p. 135).

as it is no longer capable of producing a plurality of convictions (Arendt, 2002, p. 215).[27] It is important here to remember that in Arendt's thinking, plurality is the condition of Action, which manifests the political. Without plurality, there is no possibility for the political to find a space of materialisation. Without plurality, we are faced with totalitarianism. Transhumanist theories constitute a form of paroxysm of this hypermodern and Promethean individual that develops in the contemporary globalised neoliberal period around the surpassing of limits, the quest for power, and the exaltation of the individual. The Promethean project refers to a primacy given to things – and to the individual as an object – over others. This preference of the relationship to things over the relationship to others means that the individual is primary over society (Flahault, 2008a, p. 66). One of the direct extensions of this idea is the certainty that 'the human being can wipe out the society in which he was born without himself being destroyed, so that it is possible to rebuild a new society. Better still, it is possible to destroy the old man and make him a new man, since, when he is reduced to nothing, the human being is not annihilated: he retains the power to regenerate' (Flahault, 2008a, pp. 66–80). In the face of this destructive illusion, working through education to create a 'between-us' form of interrelationship that is safe from this Promethean project is one of the challenges of the contemporary period – education being understood here as fundamentally cosmopolitical (Lamarre, 2012), based on receiving of those who come into the world with hospitality.

9.3 The *Cyborg manifesto* (1985) by Donna Haraway as a Counterpoint

Before continuing to set out some elements of oppositional political anthropology to the Promethean individual of modernity in order to allow for an anthropological shift facilitating the progressive emergence of post-Promethean social spaces, we propose a counterpoint. Another manifesto, Donna Haraway's *Cyborg Manifesto*, is a valuable resource to look at this issue of transhumanism from another point of view and enrich the political anthropology at work. Indeed, an important element for understanding one of the logics of a transhumanist anthropology is that there is no fundamental difference between naturally embodied, culturally embodied and technologically embodied tools, humans being 'natural-born cyborgs', to use philosopher Andy Clark's expression. Thus, there is no moral problem in modifying a human nature that is not sacred (Baetschi, 2018). This element is a critique of the naturalistic foundations of the way in which humanity is understood. In other words,

[27] Arendt, in *The Origins of Totalitarianism*, shows how the mindset that 'everything is possible' leads to totalitarianism. It is indeed this liberal 'everything is possible' that raises questions in the theories and transhumanist research.

transhumanism can also be an interesting critique of humanism. This is what Donna Haraway does in a particularly interesting way.

In 1985, the American feminist thinker Donna Haraway published a long article entitled 'A Cyborg Manifesto: Science, Technology and Socialist-Feminism in the Late Twentieth Century' in the magazine *Socialist Review*. This manifesto was a great success.[28] Donna Haraway's work has great density and richness; in particular, it shows the work of modernity in anthropological transformation and of the social world through techno-scientific power.[29] Haraway's political ideology is firmly rooted in the living world,[30] which she shows to be more complex than it appears at first glance, with the way in which the technosciences blur the boundaries between the living and the non-living. Hence, the boundaries between nature and culture, man and woman, human and animal are also blurred. This manifesto, which is at the crossroads of allegory and essay[31] helps us to break out of certain Manichaean dualisms or certain conservate ways of thinking within which we have willingly restricted ourselves.

At the very beginning of her manifesto, Donna Haraway presents her intellectual approach: 'This chapter is an effort to build an ironic political myth faithful to feminism, socialism, and materialism' (2007, p. 29). However, she immediately makes it clear what kind of fidelity she is talking about. 'Perhaps more faithful as blasphemy is faithful, than as reverent worship and identification. Blasphemy has always seemed to require taking things very seriously' (2007, p. 29). Thus, she attempts to take a position from within a moral majority and utter a blasphemy. What is it? Is it the desacralisation of humanity to present a possible end to it, or its possible overcoming in the form of the cyborg? Is it the Promethean individual who no longer needs links to others? Is it the end of sexual differentiation? 'At the centre of my ironic faith, my blasphemy, is the image of the cyborg' (Haraway, 2007, p. 30).[32]

The cyborg myth is subversive. Haraway's aim, through this partly fictional account, is ultimately to alter the power structures in place: 'Gender, race, or class consciousness is an achievement forced on us by the terrible historical experience of the contradictory social realities of patriarchy, colonialism, and capitalism' (Haraway, 2015, p. 39). Donna Haraway's stance is that of socialist feminism.

---

[28] This is an incredibly seminal text, which is still regularly cited nearly 40 years on.

[29] In addition to using a range of images and metaphors, Haraway is very well informed about a number of scientific fields: biology, computer science, the potential of technical production, and more.

[30] Donna Haraway originally trained as a biologist.

[31] The style of the *Cyborg Manifesto* also uses the register of the science-fiction short story, with a constant search for performativity and interpellation of the reader. The last sentence is an illustration of this: 'I would rather be a cyborg than a goddess'. The image of the cyborg is used to transgress the usual idea of humanity.

[32] However, cyborgs have a major problem, as 'they are the illegitimate offspring of militarism and patriarchal capitalism, not to mention state socialism' (Haraway, 2007, p. 33).

She keeps coming back to this positioning during her *Cyborg Manifesto*.[33] Among the intellectual enemies that Donna Haraway attacks in her *Cyborg Manifesto*, the idea of nature is perhaps the idea against which the attacks are harshest. Donna Haraway is decidedly non-essentialist.[34] The world is what we make of it, Haraway seems to write.[35]

9.3.1 Technosciences as Political Provocation and Expansion of the Field of Thought

What is Donna Haraway's position in relation to technosciences? One of her translators, Delphine Gardey in the foreword to the French translation of the *Cyborg Manifesto*, argues that it is first and foremost a passionate relationship. This passion is expressed by the fascination with the creations made possible by technologies. While she is attentive to the opportunities for the emancipation of nature allowed by technology, she is also vigilant to ensure that technosciences do not generate a new set of dominations. Finally, Donna Haraway seeks to challenge public opinion and provoke critical democratic debates on the evolution of technosciences. Thus, Haraway's political view of technosciences is partly dissonant with ours, since they do not necessarily lead to domination. It is necessary here to specify that the writing of the *Cyborg Manifesto* in 1985 shifts the point of view adopted up until this point on transhumanism. Until now, we have been speaking from the perspective of the anthropological alteration projects of the web giants that emerged in the 2000s (or the academic theories of transhumanism emerging in the years 2000–2010). Donna Haraway wrote her manifesto over 35 years ago about how human/machine fusion gives us food for thought. It is not primarily an analysis of the actual economic development projects of large corporations, but the identification of how new technical potentials could open up the space for political thought.

What seems to be at the heart of Donna Haraway's thinking and the *Cyborg Manifesto* is feminist socialist thought. Haraway defines the cyborg as 'a cybernetic organism, a hybrid of machine and organism, a creature of social reality as well as

---

[33] We can even say that this manifesto is a tool to assert this intellectual and political position, as she mentions through the expression 'cyborg feminists': 'Cyborg feminists have to argue that "we" do not want any more natural matrix of unity and that no construction is whole' (Haraway, 2007, p. 43).

[34] On several occasions, Donna Haraway cites Michel Foucault – one of her contemporaries. In some respects, her efforts are in line with Foucault's in identifying the ways in which power takes a hold on individuals. On the other hand, the cyborg tries to find an 'elsewhere', away from the subjugation to biopolitics.

[35] The complexity and intellectual finesse of the *Cyborg Manifesto* make us cautious in our reading and analysis of this seminal text (which is partly, but only partly, about what technosciences do to the human adventure and how they call our common world into question). The images regularly evoked by Donna Haraway make it difficult to read the *Cyborg Manifesto* in isolation from the rest of her work (as we do here).

a creature of fiction. Social reality is lived social relations, our most important political construction, a world-changing fiction' (Haraway, 2007, p. 30). The cyborg is a figure which helps think about a set of transcendencies. Indeed, it is situated at the intersection of nature and culture or biology and technology, and goes beyond sexual reproduction. It refers to other bodily forms and to a different experience of sexuality. This is what interests Donna Haraway in her thinking and activism. Thus, the cyborg allows a critique of sexual norms and sexually constructed forms of social domination (Gardey, 2007). The technosciences are presented as a space from which to draw sources of knowledge, from which it is possible to act newly in the City.

Through this metaphor of the cyborg, is Donna Haraway one of the first posthumanists? Nothing could be further from the truth.[36] Donna Haraway's aim is not to augment humans or to endow them with superpowers. She does not seek to reach an 'elsewhere'. Her questioning is different. The author appears to ask: How can we think outside of the dominations that seem structurally integral in our society, and which interfere with, if not actually destroy, our humanity? The cyborg, then, is a metaphor which opens the door to other accounts of humanity, nature, culture and sexuality which would not be possible without this medium. This figure of the cyborg is not Donna Haraway's own; on the other hand, it is she who invests in it to develop a radical political thought on that basis.

This intellectual exercise is an undeniably important one. We need to think afresh, and radically; to think resolutely outside of a set of shackles that have led to the contemporary disaster and to identify what lies at the root of it. We need absolute creativity rather than epistemological rigor (which can sterilise thought) in the contemporary period. Through the political figure of the cyborg she presents, Donna Haraway is the precursor of a school of thought denouncing the artificial nature of the boundaries between humans and the rest of the living world, between human and animal, between nature and culture, which is perceptible both in the entry into the Anthropocene and in the techno-scientific feats made possible by the NBIC convergence.

The *Cyborg Manifesto* is in a completely different register to the manifestos studied previously, which seek both to sketch out a possible new world and to facilitate its concrete enactment. The *Cyborg Manifesto* is not a guide to the implementation of concrete actions (so does not represent a 'post' anything). On the other hand, everything is written so that the reader is taken to a completely different place, but an elsewhere that is not pure fantasy since it is made possible by the contemporary development of science – the essential place of the here and now, of what is. We are dealing here with the description of something *else*.[37] For our part, it is not the technosciences that are the inspiration for thinking about a radical other or a radical

[36] One of her translators, Laurence Allard, states that, contrary to a number of posthumanist readings of Haraway's work, she 'has never been a posthumanist' (Allard, 2007, p. 19).

[37] The image used by one of her translators, Laurence Allard (2007), to describe Haraway's intellectual stance is not that of reflection, but of diffraction, whose function is not to reproduce an identical copy.

elsewhere – here, we draw more heavily on the poetic approach of Maurice Bellet or Andreas Weber. On the other hand, the benefit of Donna Haraway's work is to broaden the representations around possible common worlds between humans and animals, this interdependent fabric of the living, and the great and powerful animated rock that is the Earth, without forgetting the technical artefacts whose complexity is constantly increasing and becoming more and more organic (with the possibility of creating living matter as well as animated objects).

9.3.2 The Cyborg Myth and the Question of Limits

Through the figure of the cyborg, what Haraway seems to be saying is not an apology for the Promethean contemporary individual, whose power makes him independent of all relationships. On the contrary, it is rather an ode to deep connections, to encounters, to relationships, to mergers. There remains here the problem of boundaries, which is posed by Haraway. What does Haraway say about boundaries? What are they for? Are they necessary? Are they, so to speak, merely there to be broken? It seems possible to read the *Cyborg Manifesto* as a metaphor for a politics of coexistence that could, paradoxically, nourish convivialist ideology – while displacing it through openness to artefacts and the profound alteration of nature.

Among the things which Donna Haraway denounces are limits and boundaries. Indeed: 'In the traditions of "Western" science and politics – the tradition of racist, male-dominant capitalism; the tradition of progress; the tradition of the appropriation of nature as resource for the productions of culture; the tradition of reproduction of the self from the reflection of the other – the relation between organism and machine has been a border war. The stakes in the border war have been the territories of production, reproduction and imagination'. In response to these boundaries, Haraway thinks in terms of a metaphor that is the separation of these boundaries, and a form of fusion between the different parts separated by them. Thus, she describes her manifesto as a twofold argument: 'for pleasure in the confusion of boundaries and for responsibility in their construction' (2007, p. 31). One of Donna Haraway's struggles is to give humanity the face of a woman and not just of a man. Indeed, she is fighting against this male domination. The purpose of this manifesto is to nourish a socialist feminist thought that no longer refers to the idea of essentialised nature.

Haraway mentions (as early as 2007) that the divide between animal and human has almost completely disappeared.[38] Today, it is possible to distinguish humans from animals, but this is done almost without a border, based on differences of degree rather than of nature. Yet there is still a border that is gradually disappearing – that which differentiates the organism from machine. Our machines are gradually becoming alive, says Donna Haraway, a few decades before artificial intelligence

[38] She points out that it is not about tools, or language, or social behaviour, or neural functioning.

took off, and before the intensification of robotisation. It is precisely through the erasure of these borders that the cyborg emerges. Who is it? Animal? Human? Organic? Machine? It is becoming more and more difficult to tell what is human and what is animal and what is nature or culture.[39]

The cyborg has no gender, and is 'resolutely committed to partiality, irony, intimacy and perversity. It is oppositional, utopian, and completely without innocence. No longer structured by the polarity of public and private, the cyborg defines a technological polis based partly on a revolution of social relations in the *oikos*, the household. Nature and culture are reworked; the one can no longer be the resource for appropriation or incorporation by the other' (Haraway, 2007, p. 32). In a way, the cyborg represents the end of politics in its Arendtian sense, and the final victory of the *oikos*. There is no longer any possible political space – that is, one marked by the circulation of speech and the emergence of unplanned and possibly subversive action within which individuals emerge as political actors.

Through the figure of the cyborg, Haraway wonders 'if cyborgs can subvert the apocalypse of returning to nuclear dust' (the Cold War). Indeed, as she mentions, they have no attachment to the earth, from whence they do not come and 'cannot dream of returning to dust'. Would cyborgs not be more detached and therefore less likely to enter into this type of violent conflict?

9.3.3 A Political Anthropology Beyond Dualisms and Domination

In the organisation of collective life and the development of thought, we have needed to categorise, differentiate, draw boundaries and create polarities or oppositions. These allow us to conceive and build a common world. On the other hand, they generate a set of particularly problematic forms of domination.[40] Whilst they help to alleviate confusion, they can also nourish and legitimise the violence of domination: 'Feminisms and Marxisms have run aground on Western

[39] Donna Haraway had identified this blurring, which in fact has not been removed.

[40] In the *Cyborg Manifesto*, the State is perceived negatively. It is too permeable to the capitalism that erodes it, and too centred around military and control functions. In an extension of this argument, the school is identified as a problematic institution. The school is also too porous to the logic of the market. It produces in children what the ruling (i.e. technocratic and militarised) social classes want: the continuation of their domination over the masses. The school works to produce a two-tier society between highly educated elites and the rest of the population. Haraway is aware of her position: 'I am conscious of the odd perspective provided by my historical position: a PhD in biology for an Irish Catholic girl was made possibly by Sputnik's impact on US national science-education policy. I have a body and mind as much constructed by the post-Second World War arms race and cold war as by the women's movements. There are more grounds for hope in focusing on the contradictory effects of politics designed to produce loyal American technocrats, which also produced large numbers of dissidents, than in focusing on the present defeats' (Haraway, 2007, p. 68).

epistemological imperatives to construct a revolutionary subject from the perspective of a hierarchy of oppressions and/or a latent position of moral superiority, innocence, and greater closeness to nature' (Haraway, 2007, p. 73).

The myth of the cyborg allows us to overcome these dualisms, as Haraway says in the last sentences of here manifesto: 'Cyborg imagery can suggest a way out of the maze of dualisms in which we have explained our bodies and our tools to ourselves' (Haraway, 2007, pp. 81–82). She then concludes with these words, measuring the extent of the provocation of certain right wingers, by deconstructing what is at the heart of her: 'This is a dream not of a common language, but of a powerful infidel heteroglossia. It is an imagination of a feminist speaking in tongues to strike fear into the circuits of the supersavers of the new right. It means both building and destroying machines, identities, categories, relationships, space stories. Though both are bound in the spiral dance, I would rather be a cyborg than a goddess' (Haraway, 2007, p. 82).[41]

Through this questioning of limits, Haraway criticises dualisms because of the way they inevitably lead to domination of one of the two terms. This is the case with masculine/feminine, which leads to the domination of women. She identifies a range of other dualisms that are seen as 'troubling': self/other, mind/body, culture/nature, but also right/wrong, truth/illusion, total/partial, God/man, maker/made, reality/appearance, civilised/primitive. This is why the figure of the cyborg allows us to conceive of an elsewhere. By thinking of an elsewhere free of these dualisms, domination fades away. Basically, this is Haraway's aim, rather than a stance in favour of technology or fusion with machines. In fact, it seems that Donna Haraway is not a transhumanist. She tries to think of humanity fundamentally freed from the violence of domination.[42]

Unlike other manifestos, the *Cyborg Manifesto* does not end with a set of proposals or recommendations for implementing the described vision of a common world. It is not a manual for action. This manifesto denounces and allows us to think differently about socio-political situations.[43]

Haraway's description of the cyborg is reminiscent of the progressive rise to hegemony of artificial intelligence in the contemporary period – or at least, the pervasive presence of algorithms: 'The ubiquity and invisibility of cyborgs is precisely why these sunshine-belt machines are so deadly. They are as hard to see politically as materially. They are about consciousness – or its simulation'. This ties in with what was discussed in the second part of this chapter, with the way in which

---

[41] This last sentence condenses the intellectual effort which Donna Haraway outlines in the *Cyborg Manifesto*. Indeed, the figure of the goddess embodies two dualisms: the man/woman and the god/human duality. The author tells us that she would prefer to break free from such dualisms.

[42] Donna Haraway's feminism and her denunciation of the essentialism of the notion of culture appear to be more the gateway to understanding the cyborg myth than technophilia.

[43] It provokes thought and brings together in a single text contemporary dominations, cutting-edge research in 1985, and the promises of science, while mobilizing the resources of science fiction and evocation (and not just rationalized explicitness).

something escapes the political and, in escaping it, annihilates it.[44] As early as 1985, Donna Haraway perceived the way in which new technologies are reorganising the public space and impacting democracy. She wrote: 'The new communications technologies are fundamental to the eradication of "public life" for everyone (2007, p. 60).

In a way, through the fiction of the cyborg, Haraway evokes our dual kinship: with animals and with machines. Humans do not exist outside of the living and the animal kingdom, and are profoundly impacted by their technical artefacts. Our tools alter us and cause us to mutate, but always in a continuation of life and in a necessary solidarity with the rest of the living world – without which we would be lost.

We perceive that with the *Cyborg Manifesto*, we have another political point of view on the question of technosciences and their potentials. On the other hand, this point of view is speculative: it is not actually concrete, and was expressed prior to the massive investment in technosciences by neoliberal capitalism in the early twenty-first century.

9.4 To Augment or to Educate?

After the counterpoint of Donna Haraway's apprehension of the potential of technosciences to devise an anthropological shift allowing us to imagine (and not to implement) a common world without domination, this fourth part of the chapter deals with the question of the augmentation of the 'augmented human', with regard to education, the etymological foundations of which are partly common.

Etymologically, 'augment' comes from the Latin *augere*, which itself has the Indo-European root *aug*. This etymological root *aug* and the Latin *augere* are particularly interesting because of their multiple meanings. They have given rise to a set of terms that are partly inseparable from each other: first of all, 'to increase' (carrying the triple idea of greatness, power and intensity today); but also 'authority' – that which develops and makes one grow; then 'actor', that which acts or spurs one to action. For the French linguist Émile Benveniste (1969), *augere* means first of all the fact of producing outside one's own heart, even before the fact of increasing. *Augere* first means a creative act and refers to the idea of beginning or origin. The French linguist Bernard Jacquinod recognises Benveniste's interpretation, but weights it with the two main meanings of *augere*: to increase and to create. He considers that the transition from 'increase' to 'create' was probably made by designating an increase from an initial quantity that could be zero (1988, p. 316). For Jacquinod, the root *aug* refers to 'the appearance of something that begins to grow' (1988, p. 318), and *augere* means both a simple increase and an origin or birth. Thus, the etymological stem *aug* has budded a fourth term, 'augur' or

[44] Here education appears to be the key – and, as briefly mentioned in Chaps. 6 and 7, it seems important that teachers have room to manoeuvre in the formulation of their thoughts. As we enter the Anthropocene, it is important to think about education that is free from the constraints of self-righteousness and axiological neutrality. The political function of educators and teachers must be fully assumed.

'inaugurate': the fact of adding something new, or of starting something. These four terms (increase, authority, actor, augur) refer strongly to the power and influence we have over our environment and the people around us. Basically, this is what is at stake: the ability to act. The derivative *auctoritas* thus means the power of initiative (Benveniste, 1969, p. 150). The act of adding something is an extension, but also an origin. *Augere* is related to *origo* (origin). For the French historian of law Pierre Legendre, the subject is founded at the junction between origin and the authentication of what legitimises its existence – that is, at the crossroads between *origo* and institutionalised *auctoritas* (1998). These etymological elements give indications as to the anthropological depth of the questions surrounding augmentation.

Authority was particularly studied during the twentieth century by Hannah Arendt, and especially in an article published in the United States in 1954, '*What is authority?*', then published in 1961 in *Between Past and Future*. In this text, authority is rooted in the past through tradition: 'The authoritarian relation between the one who commands and the one who obeys rests neither on common reason nor on the power of the one who commands; what they have in common is the hierarchy itself, whose rightness and legitimacy both recognize, and where both have their predetermined stable place' (Arendt, 1972b, p. 123). Authority is identified as being in crisis and directly linked to the crisis of modernity. Even at this point, Arendt links authority, augmentation and origin (through the concept of foundation): 'It is in [the context of the Greeks] that word and concept of authority originally appeared. The word *auctoritas* derives from the verb *augere*, "augment," and what authority or those in authority constantly augment is the foundation' (1972b, p. 160). By augmenting the foundation, authority increases the possibility of creation. This political scientist's analysis includes the four-pole dynamic with the emergence of the author[45]: 'In order to understand more concretely what it meant to be in authority, it may be useful to notice that the word *auctores* can be used as the very opposite of *artifices*, the actual builders and makers, and this precisely when the word *auctor* means the same thing as our "author". Who, asks Pliny at the occasion of a new [theatre[, should be more admired, the maker or the author, the inventor or the invention?–meaning, of course, the latter in both instances. The author in this case is not the builder but the one who inspired the whole enterprise and whose spirit, therefore, much more than the spirit of the actual builder, is represented in the building itself. In distinction to the *artifex*, who only made it, he is the actual "author" of the building, namely its founder; with it he has become an "augmenter" of the city' (1972b, p. 161). The dynamic specific to augmentation, becoming the origin of an actor through authority, if it is at the heart of the modalities of sharing existence in the world is difficult to apprehend. Let us continue with Arendt in 'What is authority?' 'However, the relation between *auctor* and *artifex* is by no means the (Platonic) relation between the master who gives orders and the servant who executes them.

[45] The importance of the emergence of the one who acts was worked on in particular depth by Arendt in *The Human Condition*, published in the United States in 1958, in which she shows that action (distinct from the making or construction of a work, as mentioned in the example above) reveals the one who acts – who did not exist as such before the engagement in a concerted action. This aspect will be worked on, in particular, in the next chapter.

The most conspicuous characteristic of those in authority is that they do not have power. *Cum potestas in populo auctoritas in senatu sit*, "while power resides in the people, authority rests with the Senate". Because the "authority," the augmentation which the Senate must add to political decisions, is not power, it seems to us curiously elusive and intangible' (1972b, p. 160).

In '*The Crisis in Education*', also published in the United States in 1954, Arendt associates authority with responsibility, since the authority of an adult over a child (or of a teacher over a pupil) amounts to assuming responsibility for the world over those who enter the world. The crisis of authority that she identifies reconfigures the world (which is the space of the political): 'Authority has been discarded by the adults, and this can mean only one thing: that the adults refuse to assume responsibility for the world into which they have brought the children' (1972c, p. 244). In Arendt's eyes, the crisis of authority, understood as a crisis of responsibility for the world on behalf of the generations entering it amounts to a form of implicit discourse by adults, saying, 'In this world even we are not very securely at home; how to move about in it, what to know, what skills to master, are mysteries to us too. You must try to make out as best you can; in any case, you are not entitled to call us to account. We are innocent, we wash our hands of you' (1972c, p. 245).

'Authority has vanished from the modern world' (1972b, p. 121). Authority does not need to be restored, as we sometimes hear when we observe incivility among pupils or children. In line with Arendt's thought, what we need to bring about is something quite different: it is the possibility of political action – conceived as concerted action at the heart of human plurality – that reveals the actor. This dynamic of augmentation with authority at its heart refers directly to the notional relations among education, politics and natality – understood as the possibility of the emergence of the new – in Arendt's thinking.[46] Authority is linked to the properly political vocation of each individual. The question of authority in education relates to the means of empowering each person to be an actor in the world. Does the techno-scientific and digital revolution, with its exponential possibilities of augmenting human beings, favour this political vocation – this capacity to be an actor in our environment? What is currently being constructed, with the augmentation of man caused by the intensification of production of his artifices? This is a troubling question of the present time. The terminological offshoots from the root *augere* show the dynamic between authority, augmentation, origin and emergence of actors. Authority is possible by augmentation, and its purpose is to augment others. The exercise of authority goes far beyond mere augmentation, but results in the emergence of actors among those present. This means that a novelty occurs through the exercise of authority, and that augmentation is necessary to allow actors to emerge.

Humans derive some of their authority from augmentation as a result of their artifices (this is noticeable, for example, when only one person in a group has a smartphone, and the group needs to check directions or information). What is the

[46] As mentioned in the introduction, Arendt has been wrongly accused of conservatism: the only thing she tries to conserve is the possibility, through birth, for radical novelty to emerge. Her thinking goes against the usual dynamics of conservatism.

specific mission of education in this context? The emergence actors is one of the aims of education in its pre-political component – the actor being understood here as one who acts in concert with others, appearing in conjunction with the political. Educating and augmentation share this same aim of helping actors to emerge and, in so doing, break with what was – a break that becomes the spark for a renewal. There is no need to be afraid of the human augmentation that is developing in the contemporary period, but to observe, think and educate the actors emerging from these new forms of augmentation. How are we to participate in the revelation of the other as an actor, no longer based on the heritage of a tradition but on artefacts produced by human hands that allow us to augment ourselves? The techno-scientific and digital revolution and its ambition to augment man can be the occasion for the emergence of a radical novelty in the world so imbued with the past, provided that it does not aim to replace the actor but to reveal him – which is not the case in most transhumanist theories and research, as we mentioned earlier. This is a criterion of discernment for the educator.

9.5 Educating for Post-Promethean Interleaving

As the myth of the creation of a new human being is characteristic of a totalitarian system, as we have seen in history with Nazism and Stalinist communism, we shall try to think differently about human beings without falling into the trap of totalitarianism, by thinking about the conditions for the emergence of a post-Promethean 'between-us'. Understanding the 'between-us' rather than the human also avoids over-essentialising humanity. It is also a necessary exercise in a debate with transhumanist theories that have the peculiarity of being solely focused on the human as an individual, rather than as a relational space. We apprehend the 'between-us' – which is to be distinguished from the 'inter-se' by its integration of plurality, therefore of the other, and is not part of the regime of sameness – as what shapes humanity from day to day. To think about what is 'between us' is to think the common good and the precedence of coexistence over existence. The 'between-us' is as much the space of emergence of humanity as it is the arena of the political, and refers to what Hannah Arendt calls '*inter homines esse*': the common being of men (1972, p. 16).[47] What is at stake in the emergence of a post-Promethean 'between-us' is humanity's long-term survival – that is, its non-obsolescence. Gunther Anders had perceived this issue in 1956 when he identified the alienation of humans from their productions and that this alienation would end with the obsolescence of the human being (Anders, 1956). In *The Obsolescence of Man*, he criticises the myth of Prometheus, upon which the anthropology of modern man is founded. He calls for the figure of

[47] The world, understood as that which is 'between', is Arendt's real concern. 'The world lies in between people, and this in-between – much more than (as is often thought) men or even man – is today the object of greatest concern and the most obvious upheaval in almost all the countries of the globe' (Arendt, 1974, p. 12).

a human being who distances himself from *hybris* and integrates limits. To think of a post-Promethean interrelationship is to understand the individual in relation to and dependent on others as well as on his environment: 'We must eventually recognise that in order to humanise ourselves and to be ourselves – or rather to become ourselves – the human being depends on that which is not himself' (Flahault, 2005, p. 44). The post-Promethean 'between-us' outlined in this contribution represents distancing from this hegemonic *homo oeconomicus* in the contemporary period, which transhumanism proposes to overcome by full emancipation.[48]

The quest to overcoming Prometheism in education in the politics of the augmented human should in no way be confused with a rejection of technology. In the famous words of the French philosopher Maurice Merleau-Ponty, 'Everything is both manufactured and natural in man as it were in the sense that there is not a word, not a form of [behaviour] which does not owe something to purely biological being and which at the same time does not elude the simplicity of animal life' (1976, p. 220–221). Technology contributes to our anthropisation and then our humanisation, in its association with the symbol. The human does not exist in isolation; it emerges as a continuation of the anthropisation of the environment made possible by technology, with the appearance of social bodies in which the symbolic component is particularly prevalent (Bourg, 1996). We can distinguish the three processes of hominisation with the progressive physical transformation of animals into hominids, then that of anthropisation with the transformation of objects through techniques, and finally that of humanisation with the objective transformation of objects by symbols (Leroi-Gourhan, 1992).

9.5.1 From Prometheus to Soteria, Aidos or Epimetheus

This reflection on post-Promethean interleaving converges on an anthropological characteristic that we could call Soterian. Soteria is the allegorical goddess of safety, preservation and deliverance from evil (Bonneuil and Fressoz, 2013), who had temples in Patras and Aegium. The myth of Soteria, etymologically meaning 'salvation', is based on observance of boundaries, where Prometheus' is their transgression. We can identify another oppositional category to the Promethean approach through the figure of the Greek goddess Aidos (Baskin, 2014), goddess of shame, modesty, and humility. Thinking of a Soterian 'between-us' involves a reduction in risk-taking and an ethics of virtues such as moderation, material restraint (Parizeau, 2016, p. 36) or the prudence of what is desirable for all. French sociologist Alain Ehrenberg, in *La fatigue d'être soi* (1998), highlighted the problematic passage of

[48] This reflection on education in politics takes the 'between-us' (a category opposed to the individualism of transhumanism) as a starting point. Whilst it is *necessary*, in a political text, to mention the 'between-us', it is not *sufficient*; the idea must be examined in connection with that of a capable community. As the philosopher Eric Weil points out, the political is also characterised by the organisation of a historical community that makes it capable of making decisions.

modernity from a regime polarised between the authorised and the forbidden to one polarised between the possible and the impossible, of which we currently perceive a form of paroxysm in transhumanist theories.[49]

Ivan Illich, in the closing lines of *Deschooling Society* (1971a, b), is also looking for a different name by which to represent a 'post-Promethean' relational style. He finds it in Epimetheus (which, etymologically, refers to 'hindsight'), the brother of Prometheus (which means 'foresight'). Indeed, for Illich, Epimetheus, who married the ancient goddess of fertility Pandora, represents the figure of a carer, attentive to life and guardian of hope, far removed from the Promethean productivism of modernity: 'We now need a name for those who value hope above expectations. We need a name for those who love people more than products [...]. We need a name for those who love the earth on which each can meet the other [...]. We need a name for those who collaborate with their Promethean brother in the lighting of the fire and the shaping of iron, but who do so to enhance their ability to tend and care and wait upon the other [...]. I suggest that these hopeful brothers and sisters be called Epimethean men' (Illich, 1971a, b, pp. 187–188). This figure of Epimetheus was later taken up by Hans Jonas in *The Imperative of Responsibility*.

In the wake of the proposals for anthropological displacements in the chapters of this second part (The idea of anthropological shift; Weathering the storm of the contemporary anthropological *krisis*; From the (augmented) individual to a post-Promethean 'between-us'), the next chapter, which views humanity from the perspective of the human adventure, proposes an anthropological model for the Anthropocene on the basis of which to think about convivialist education.

Bibliographical References

Allard, L. (2007). A propos du *Manifeste cyborg*, d'*Ecce Homo* et de *La promesse des monstres* ou comment Haraway n'a jamais été posthumaniste. In D. Haraway (Ed.), *Manifeste cyborg: science, technologie et féminisme socialiste à la fin du 20ème siècle* (pp. 19–27). Exils éditeur. (original edition 1985), French translation.

Anders, G. (2002). *L'obsolescence de l'homme, Sur l'âme à l'époque de la deuxième révolution industrielle*. Ivrea (ed. or. 1956), tr. fr.

Anders, G. (2012). *L'Obsolescence de l'homme – Sur la destruction de la vie à l'époque de la troisième révolution industrielle*. Editions Fario (ed. or. 1979), tr. fr.

Arendt, H. (1972a). *La crise de la culture*. Gallimard (ed. or. 1961), tr. fr.

Arendt, H. (1972b). « Qu'est-ce que l'autorité? ». In H. Arendt (Ed.), *La crise de la culture*. Gallimard (pp. 121–185) (ed. or. 1961b), tr. fr.

Arendt, H. (1972c). « La crise de l'éducation ». In H. Arendt (Ed.), *La crise de la culture*. Gallimard (pp. 223–252) (ed. or. 1961c), tr. fr.

Arendt, H. (1974). *Vies politiques*. Gallimard (ed. or. 1970), tr. fr.

Arendt, H. (1983). *Condition de l'homme moderne*. Calmann-Lévy (ed. or. 1958, tr. fr. 1961), tr. fr.

---

[49] This is an element highlighted by the French psychosociologist Jean-Yves Robin in a text entitled '*Conditions anthropologiques de l'émergence d'un homme sans limite*' (Anthropological conditions for the emergence of a man without limits) (2018).

Arendt, H. (2002). *Les origines du totalitarisme III. Le système totalitaire*. Seuil (ed. or. 1948), tr. fr.

Arendt, H. (2005). *Journal de pensée, 1950–1973 – 2 tomes*. Seuil. French translation (original edition 2002).

Baetschi, B. (2018). Plus vite, plus haut, plus fort, plus… pour aller où? Transhumanisme et aspiration au bonheur. *ESKA Journal International de bioéthique et d'éthique des sciences, 29*(3), 170–188.

Barthélémy, P. (2018, December 23). Sergio Canavero, l'homme qui veut greffer des têtes. *Le Monde*.

Baskin, J. (2014). The ideology of the Anthropocene? *MSSI Research Paper (Melbourne Sustainable Society Institute), 3*, 3–19.

Benveniste, E. (1969). *Le vocabulaire des institutions indo-européennes, Tome II Pouvoir, Droit, Religion*. Les Editions de Minuit.

Blin, T. (2015). Et si on liquidait l'homme! Sur le bonheur transhumaniste qui nous menace. *Le Débat, 183*, 87–100.

Bonneuil, C., & Fressoz, J.-B. (2013). *L'événement Anthropocène*. Seuil.

Bostrom, N. (2002). *Anthropic Bias, Observation Selection Effects in Science and Philosophy*. Routledge.

Bostrom, N. (2003). Are you living in a computer simulation? *Philosophical Quarterly, 53*(211), 243–255.

Bostrom, N. (2014). *Superintelligence: Paths, Dangers, Strategies*. Oxford University Press.

Bourg, D. (1996). *L'Homme artifice*. Gallimard.

Brassier, R. (2016). Accélérer la raison. In L. de Sutter (Ed.), *Accélération !* (pp. 157–181). PUF. French translation.

Cabanes, V. (2017). *Homo natura – En harmonie avec le vivant*. Buchet Chastel.

Caire, A.-B. (2018). La cryogénisation. *ESKA Journal international de bioéthique et d'éthique des sciences, 29*(3), 54–70.

Capelle-Dumont, P. (2019). Technique, finitude et éthique de la décision. In P. Capelle-Dumont & C. Roche (Eds.), *Puissance technologique et éthique de la décision*. Parole et silence.

Damour, F. (2017). Le transhumanisme, une idée chrétienne devenue folle? *Etudes, 7*(July-August), 51–62.

De Rosnay, J. (2018). Transhumanisme ou hyperhumanisme? L'avenir de l'humanité. *ESKA Journal international de bioéthique et d'éthique des sciences, 29*(3), 233–240.

Ehrenberg, A. (1998). *La fatigue d'être soi. Dépression et société*. Odile Jacob.

Flahault, F. (2005). Vers une nouvelle pensée sociale. *Revue du MAUSS, 26*, 377–382.

Flahault, F. (2008a). *Le crépuscule de Prométhée – Contribution à une histoire de la démesure humaine*. Mille et une nuit.

Flahault, F. (2008b). Comment l'homme peut-il être à la fois égoïste, bon et méchant? *Revue du MAUSS, 31*, 307–317.

Frippiat, L. (2011). L'amélioration technique de l'être humain: introduction aux différents courants du débat. *Journal International de Bioéthique, 22*, 33–50.

Gardey, D. (2007). Avant-propos. Deux ou trois choses que je dirais d'elle. In D. Haraway (Ed.), *Manifeste cyborg: Science, technologie et féminisme socialiste à la fin du 20ème siècle* (pp. 9–16). Exils éditeur. (original edition 1985), French translation.

Giesen, K.-G. (2018). Le transhumanisme comme idéologie dominante de la quatrième révolution industrielle. *ESKA Journal international de bioéthique et d'éthique des sciences, 29*(3), 189–203.

Goffi, J.-Y. (2011). Nature humaine et amélioration de l'être humain à la lumière du programme transhumaniste. *Journal International de bioéthique, 22*, 18–22.

Guényveau, F.-R. (2017). *Un dissident*. Albin Michel.

Habermas, J. (2002). *L'avenir de la nature humaine. Vers un eugénisme libéral?* Gallimard. (original edition 2001), French translation.

Haraway, D. (2007). *Manifeste cyborg: science, technologie et féminisme socialiste à la fin du 20ème siècle*. Exils éditeur (ed. or. 1985), tr. fr.

Haraway, D. (2015). « Anthropocene, Capitalocene, Plantationocene, Chthulucene : making kin ». *Environmental Humanities, 6*, 159–165.

Haudry, J. (2016). *Le feu dans la tradition indo-européenne*. Archè.

Hétier, R. (2018). Augmenter le pouvoir de sentir à l'heure de l'homme augmenté. In N. Wallenhorst, D. Coatanéa, & F. Prouteau (Eds.), *Eduquer l'homme augmenté – Pour une société postprométhéenne* (pp. 119–131). Le Bord de l'eau.

Hottois, G. (2018). L'anthropologie philosophique technicienne du transhumanisme. *ESKA Journal international de bioéthique et d'éthique des sciences, 29*(3), 135–153.

Hughes, J. (2004). *Citizen Cyborg: Why democratic societies must respond to the redesigned human of the future*. Basic Books.

Huxley, J. (1957). *New bottles for new wine*. Chatto and Windus.

Illich, I. (1971a). *Une société sans école*. Seuil.

Illich, I. (1971b). *Libérer l'avenir*. Seuil. French translation.

Jacquinod, B. (1988). Etude de vocabulaire grec: αὔξω et ἄωτος. *Revue des études anciennes, 90*(3–4), 315–323.

Lamarre, J.-M. (2012). L'éducation cosmopolitique: apprendre le propre, apprendre l'étranger. *Le Télémaque, 41*, 31–46.

Larrieu, P. (2018). La neuro-amélioration des sujets "sains": enjeux anthropologiques, sociologiques et juridiques. *ESKA Journal international de bioéthique et d'éthique des sciences, 29*(3), 71–91.

Le Dévédec, N., & Collin, J. (2018). Le médicament augmenté: l'usage du médicament dans les discours transhumanistes et ses significations sociales. *ESKA Journal international de bioéthique et d'éthique des sciences, 29*(3), 93–108.

Legendre, P. (1998). La totémisation de la société. Remarques sur les montages canoniques et la question du sujet. In A. van Debeek & K. van der Toorn (Eds.), *Canonization und Decanonization* (pp. 425–434). Brill.

Leroi-Gourhan, A. (1992). Le geste et la parole, I. Technique et langage. Albin-Michel (ed. or 1964).

Liautaud, J.-M. (2018). La Conscience contre la Condition. Penser l'éducation de l'homme augmenté avec Eric Weil. In N. Wallenhorst, D. Coatanéa, & F. Prouteau (Eds.), *Eduquer l'homme augmenté – Pour une société postprométhéenne* (pp. 57–77). Le Bord de l'eau.

Luzi, J. (2018). Le capitalisme transhumaniste et la mort. *ESKA Journal international de bioéthique et d'éthique des sciences, 29*(3), 33–62.

Maestrutti, M. (2011). Humain, transhumain, posthumain. Représentations du corps entre incomplétude et amélioration. *Journal International de Bioéthique, 22*, 51–66.

More, M. (1993). Technological self-transformation: Expanding personal extropy. *Extropy, 10*(4/2), 15–24.

More, M. (2003). *Principles of extropy*. http://editions-hache.com/essais/more/more1.html

More, M. (2010). The overhuman in the Transhuman. *Journal of Evolution and Technology, 21*(1), 1–4.

Parizeau, M.-H. (2016). De l'Apocalypse à l'Anthropocène: parcours éthique des changements climatiques. *Revue de métaphysique et de morale, 89*, 23–38.

Perret, B. (2014a). Transition écologique ou choc de la finitude? *Revue du MAUSS, 43*, 35–40.

Perret, B. (2014b). De la difficulté de penser un avenir sous contrainte écologique. *Transversalités, 130*, 69–81.

Prouteau, F. (2018). Tous connectés, ou comment on devient un homme numérique. In N. Wallenhorst, D. Coatanéa, & F. Prouteau (Eds.), *Eduquer l'homme augmenté – Pour une société postprométhéenne* (pp. 15–34). Le Bord de l'eau.

Rialle, V. (2018). Robotique humanitaire *versus* robotique suicidaire: ou comment ré-enchanter la 'silver économie'. *ESKA Journal international de bioéthique et d'éthique des sciences, 29*(3), 17–25.

Ricœur, P. (1990). *Soi-même comme un autre*. Seuil.

Robin, J.-Y. (2018). Conditions anthropologiques de l'émergence d'un homme sans limite. In N. Wallenhorst, D. Coatanéa, & F. Prouteau (Eds.), *Eduquer l'homme augmenté – Pour une société postprométhéenne* (pp. 79–95). Le Bord de l'eau.
Savulescu, J., & Bostrom, N. (2008). *Human enhancement*. Oxford University Press.
Tassin, E. (2017). *Le trésor perdu – Hannah Arendt, l'intelligence de l'action politique*. Klincksieck.
Theobald, C. (2008). Le christianisme comme style. Entrer dans une manière d'habiter le monde. *Revue d'éthique et de théologie morale, 251*(HS), 235–248.
Thiel, M.-J. (2015). L'homme augmenté aux limites de la condition humaine. *Revue d'éthique et de théologie morale, 286*, 141–161.

Chapter 10
From the Human Condition to the Human Adventure

Abstract In the proposal of a critical theory for the Anthropocene, this chapter has a special place because it shows the development of an anthropological model, to use to think about convivialism as a paradigm of education for politics in the Anthropocene. This chapter consists of an appropriation of the Arendtian concept of action and the proposal of an underlying anthropological model, on the basis of which it will be possible to identify what kind of learning should be favoured to consolidate the political, in order to allow us to navigate the civilisational *krisis* with which the Anthropocene confronts us. This will make it possible to propose a shift in the apprehension of humanity from the notion of the human condition to that of the human adventure (the groundwork for which is prepared by the reading of the various political manifestos, the work of Bellet and the idea of getting through the *krisis* and the oppositional proposal to the transhumanist anthropological mutation), but which has yet to be deeply rooted in the whole adventure of the living within the earthly realm.

Keywords Anthropocene · Hannah Arendt · Human adventure · Human condition

Hannah Arendt wrote *The Human Condition* following her three volumes on *The Origins of Totalitarianism*. After addressing the enigma of evil, Arendt wrote a work of fundamental anthropology in which she attempted to identify the core of the human condition.[1] In this work, several oppositional categories are presented. The first is the *vita activa* (active life), which is the title of the German edition of

[1] Ricœur, in the French preface to *The Human Condition*, shows that this work is a stand against totalitarianism. After writing *The Origins of Totalitarianism*, Arendt worked on the fundamental anthropological condition on which terror can and should be based. She explores the conditions for the emergence of concerted action as a bulwark against violence. Politics and its power are understood as dissipating when people no longer act together in concert. This is why there is a kind of temptation to substitute violence for the power of politics.

© The Author(s), under exclusive license to Springer Nature Switzerland AG 2023

N. Wallenhorst, *A Critical Theory for the Anthropocene*, Anthropocene – Humanities and Social Sciences, https://doi.org/10.1007/978-3-031-37738-9_10

The Human Condition –Arendt's personal favourite of all the translations produced.[2] The human condition is marked by *vita activa*, which differs from *vita contemplativa* (contemplative life)[3] to which she devoted her last unfinished work, *The Life of the Mind*.[4]

The Human Condition is organised around two other oppositional but complementary categories: the private (economic) sphere and the public (political) sphere. It is in the work of these two concepts that Arendt's critique of modernity lies. In particular, she criticises modernity marked by a form of domination of the private sphere over the public sphere, which refers, among other things, to the domination of self-interest over selflessness. Within this tension between the private and public spheres, Arendt develops a three-faceted anthropological model of man's relationship to the world: the labour of the *animal laborans*, the work of *homo faber*, and the action of the *zoon politikon*.

One of Arendt's criticisms of modernity is that 'We have changed work into labour' (1983, p. 142). As Ricœur notes in the preface to the French edition of *The Human Condition*, the difference between labour and work lies in the capacity to endure, and, in the process, to create a common world among men. Arendt denounces the way in which, instead of using *works*, we consume the product of *labour*. That is, we deplete our environment to the point of exhaustion instead of building a world. The world is that which bears witness to the presence of other human beings. Human life is only possible in a world (Arendt, 1983, p. 31). The world 'transcends our lifespan into past and future alike; it was there before we came and will outlast

---

[2] We worked with the German and French versions of Arendt's work which, admittedly, are not identical in every respect. Arendt returned to this work several times on the occasion of the publication of the translations in which she was particularly invested.

[3] The *vita activa* can only be practiced with others, while the *vita contemplativa* is a solitary activity. Using a twelfth-century author from the, Hugh of Saint Victor, Hannah Arendt contrasts the *vita activa* with the *vita contemplativa*: the former is laborious, while the latter is 'pure tranquility'; "the *vita activa* takes place in public, the *vita contemplativa* 'in the desert'" (1978, p. 23). The *vita contemplativa* is 'the point where mental activity comes to rest' (p. 24).

[4] *The Life of the Mind* is about human activities that are not integrated into the three activities of the *vita activa* (labour, work and action) that are thinking, willing and judging. Arendt died 2 days after completing the second part, 'Willing'. A posthumous book was produced from a set of lectures on Kant's political philosophy, on the basis of which Arendt had planned to write the third part, which would have been entitled 'Judging', and which is entitled *Lectures on Kant's Political Philosophy* (1982). In a letter to Martin Heidegger on 20 March 1971, she had informed him of her project of writing *The Life of the Mind* and of her intention to dedicate it to him (Arendt & Heidegger, 2001, p. 203). Reading the exchanges between Arendt and Heidegger, we understand that it was with Heidegger that Arendt discovered this passion for thinking that never left her. Arendt's writing is, among other things, a dialogue with Heidegger (this is particularly the case with *The Human Condition*). As the French philosopher Étienne Tassin, a specialist on Arendt, rightly points out: 'to the being-towards-death of Dasein (mortality) she opposes being-towards-birth (natality), to the solitary action of thought the effective plural action of the actors, to the tyranny of the [impersonal] 'One' the manifest plurality of actors, to the inauthenticity of the public sphere the revealing power of action, to the authenticity of being with oneself the distinction in acting with others, to chatter the acting word, to the glorious heroism of the solitary thinker the ordinary glories of civic citizenship, etc.' (2017, pp. 9–10).

our brief sojourn in it. It is what we have in common not only with those who live with us, but also with those who were here before and with those who will come after us. But such a common world can survive the coming and going of the generations only to the extent that it appears in public' (Arendt, 1983, p. 66). The products of work ensure permanence and durability of the world, which characterises the world: 'The world, the man-made home erected on earth and made of the material that earthly nature delivers to human hands, consists not of things that are consumed but of things that are used. If nature and the earth generally constitute the condition of human life, then the world and the things of the world constitute the condition under which this specifically human life can be at home on earth' (Arendt, 1983, p. 151). The world needs the earth, but is not identical with the earth. It is established on Earth by humanity: 'It is related [...] to the human artifact, the fabrication of human hands, as well as to affairs which go on among those who inhabit the man-made world together. To live together in the world means essentially that a world of things is between those who have it in common, as a table is located between those who sit around it; the world, like every in-between, relates and separates men at the same time' (Arendt, 1983, p. 63). The use of objects produced by *homo faber* that are not consumed allow for the building of the world through the solidity that allows for the accommodation of men on Earth (Arendt, 1983, p. 153).

Arendt's critique of modernity is particularly interesting in order to identify the way in which unregulated economic hegemony has led to this problematic entry into the Anthropocene.[5] The third category developed by Arendt – action – is enigmatic. In Arendt's thinking, there is no greater human activity than the action of sharing speech.[6] Action and speech are the two faces of political activity par excellence.[7] However, to make the political happen, an additional condition is necessary: that this happens in a public arena that Arendt calls 'the space of appearance'.[8]

[5] Whilst Arendt presents a critique of a certain type of modernity, it can by no means be said that she is anti-modern. As mentioned in the introduction, Arendt's thinking has been interpreted in different ways, and her work has been classified in several of the usual categories of political theory. She appears fundamentally republican (with a kind of cult of tradition in her ideas). On the other hand, her relationship to Greek philosophy has sometimes led to her being classified as a conservative. However, the communitarians have even held her up as a champion of communitarianism because of her thinking of the '*inter homines esse*'. What is very clear in Arendt's arguments is her criticism of liberalism (and of a certain modernity). This is our reading of Arendt's work, which has marked our own work in critiquing the hegemony of liberal capitalism that the Anthropocene unmasks.

[6] Arendt has a very high opinion of action, as opposed to thought, which is usually valued more highly in the universe of philosophers or theoreticians. For example, she states: 'It is not ideas, but events that change the world' (Arendt, 1983, p. 307).

[7] Though the 'between-us' on which we rely in conceiving of the political may appear a restrictive framework, it is nevertheless the space in which the political emerges, and in this sense, is original: 'Action and speech, directed towards humans, occur between humans' (Arendt, 1983, p. 205).

[8] The concept of appearance has particular importance in Arendt's work, since Arendt gives ontological status to the phenomenological nature of appearance: 'In this world into which we enter, appearing from nowhere, and from whence we return to nowhere, *Being and Appearing coincide*' (1978, p. 37).

Actions and speech must be shown. If the world is what accommodates human life, it is 'by word and deed that we insert ourselves into the human world' (Arendt, 1983).[9] Action and speech are impossible for isolated individuals; they are made possible by the interconnection of human acts and words (Arendt, 1983, p. 212). Action does not take place without speech; it needs speech. Action and the speech associated with it help give meaning to existence: 'Denied action and word, deprived of the articulation of natality, we would be condemned to revolve endlessly in the eternal cycle of becoming; but denied the faculty of undoing what we have done, of controlling at least in part the processes we have set in motion' (Arendt, 1983, pp. 276–277).

Action is characterised by novelty and unpredictability is always a 'miracle' (Arendt, 1983, p. 200). This political activity *par excellence*, which is action,[10] is made possible by birth, meaning that a beginning or novelty is possible in the world: 'To act, in the most general sense, means to take an initiative, to undertake (as indicated by the Greek *archein*, 'to begin', 'to guide' and eventually 'to govern'), to set in motion (which is the original meaning of the Latin *agere*). Because they are *initium*, newcomers and innovators by virtue of their birth, men take initiative, they are moved to action' (Arendt, 1983, p. 199). In Arendt, action is directly related to natality: 'If action as beginning corresponds to the fact of birth, if it is the actualization of the human condition of natality, speech corresponds to the fact of individuality, it is the actualization of the human condition of plurality, which is to live as a distinct and unique being among equals' (Arendt, 1983, p. 200).

Arendt has a very high opinion of the possible action of men.[11] Action appears as a partial remedy to this domination of the private sphere. In particular, Arendt seems

[9] The space between us, the space in which action and speech occur, is what confers humanity on people. The space of politics is a space of humanisation.

[10] Action is distinguished from production: 'Distinct from both consumer goods and utilitarian objects, finally, there are the "products" of action and speech, which together form the fabric of human relations and affairs. Left to themselves, objects not only lack tangibility: they are even less durable, more futile than what we produce for consumption. Their reality depends entirely on human plurality, on the constant presence of others who can see, hear, and thus attest to their existence' (Arendt, 1983, p. 107).

[11] Arendt had various exchanges of correspondence that are particularly interesting to read because they extend over long periods. Among others, we can note the correspondence with her great friend Kurt Blumenfeld – an important Zionist activist and intellectual, between 1933 and 1963; that with her thesis director and friend Karl Jaspers in Heidelberg covers 1926–1969; and that with Martin Heidegger, her philosophy professor in Marburg and briefly her lover, began in 1925 and ended with Arendt's death in 1975 (with a hiatus between 1933 and 1950 in connection with Heidegger's membership of the Nazi party). Each of these sets of correspondence is a form of ode to friendship. (There is also a collection of her correspondence with her husband Heinrich Blücher from 1936 to 1968, correspondence with her friend Mary McCarthy from 1949 to 1975, and a collection with her friend the philosopher Gershom Scholem, which began in 1939 and ended in 1963, with a rupture following the publication of *Eichmann in Jerusalem*, whose thesis was unbearable for Scholem). In Arendt's epistolary exchanges with her friends, we discover her aesthetic and poetic sensibility (she both read and wrote poetry). In these exchanges, we identify the articulations of Arendt's intellectual work with the exchanges and moments shared with her dear friends, which

to have created the category of action on the basis of the American War of Independence, democratic and without violence.[12] Arendt's action is light-bringing, gives hope and appears as the counterpart to the darkness of totalitarianism.[13] It is, to use her lexicon, a miracle. Action is as powerful as it is vulnerable. It is as much about long-term changes through the lasting imprint it leaves on the heart of the world, as it is ephemeral. Its power lies in its lightning speed. Action only lasts when people act together.[14] Action, which is the category of politics, is not expressed through the institutionalised forms of social life, but it contributes to their institutionalisation.[15] Action is instituting. Action is linked to natality, which carries within it the power to renew the world. However, it is as fragile as a newborn baby.

Action is an emergence in the public sphere, born of the sharing of existence at the heart of human plurality.[16] It is non-Promethean and non-programmatic in essence. This Arendtian action can easily evoke utopianism. It can be relatively easily dismissed. Action is the counterpart of economic hegemony. It is opposed to techno-scientific Prometheism. Its emergence is opposed to transhumanist planning in pursuit of a good Anthropocene. The development of action directly consolidates the political. It is what has been lacking in the course of modernity (and postmoder-

were enormously important for her. The reading of these letters allows us to identify the experiential and existential importance that Hannah Arendt attached to the exercise of freedom inherent in the relationships, the debate and the interactions at the foundation of her political thought. The correspondence with Kurt Blumenfeld elucidates Hannah Arendt's experience of concerted action, of shared struggle with peers, as a source of meaning in existence.

[12] The examples of revolutions are of great importance in the development of Arendt's political thinking. She discusses concerted action using the example of the American War of Independence, which culminated in 1776 in Philadelphia, the French Revolution in Paris in 1789, and the Hungarian Revolution in Budapest in 1956.

[13] In *The Origins of Totalitarianism*, Arendt highlights the intrinsic desire of totalitarianism to eradicate the political component that is linked to freedom. Action is opposed to this annihilation. The veiling of reality by a 'permanent revolution' is an important Arendtian insight in the conceptualisation of totalitarianism as a common feature of the Nazis and the Bolsheviks (Bruneteau, 2014, p. 21). Hannah Arendt sought to identify 'the essence of totalitarianism' which is especially within the camps that did not merely aim to eliminate a portion of the population, but represent an ideal of social domination whose aim is a tragic 'mutation of human nature' (Bruneteau, 2014, p. 25).

[14] What is at stake in action is power and might. When men speak and act together, they have this power, which dissipates as soon as their group is divided.

[15] Action with others, or concerted action, has three effects: it reveals actors who did not exist before the action; it allows actors to create relationships among themselves that did not exist before the collective action, and institutes a political community; and it establishes a space of appearance that can be institutionalised into a public-political space taking over community or social relationships (Tassin, 2017, pp. 15–16). The 'between-us' is thus the place of genesis of the political which may even go so far as to take institutional form.

[16] 'It is not man, but men who populate our planet. Plurality is the law of the land' (Arendt, 1978, p. 38). For Arendt, Plurality is a fundamental condition of both action and speech and has as its specificity 'the double character of equality and distinction'; human plurality 'is the paradoxical plurality of unique beings' (1983, pp. 197–198). Speech and action are what allow men to distinguish themselves in their uniqueness and not to be merely distinct.

nity), that has led us straight into the jaws of the Anthropocene.[17] Because action, and the natality associated with it, is a category bringing hope in the face of the threat to human life posed by the Anthropocene, we will take some time to study it.[18]

10.1 The Three Dimensions of the Human Adventure: *hybris*, the World and Coexistence

10.1.1 Human Nature, Human Condition or Human Adventure?

The characteristic feature of naturalism is that it does not consider individuals with free will, but a unified species with a simplified history. The simplicity of the notion of 'human nature' seems to be a permanent background to research in the Earth System Sciences. The contingent humanity component of the notion of the human condition is all the more relevant as we enter the Anthropocene. Thinking of humanity without reference to the idea of human nature poses a problem because it removes humanity's reference to an external counterpart (nature). For our part, we choose to understand humanity in reference to the living, and more specifically, to the interdependent fabric that is the living universe.[19]

It would seem interesting to introduce another nuance with the notion of human adventure, placing emphasis on the collective component, the uncertainty and the

[17] The contemporary period is marked by the collapse of the metanarratives (Lyotard, 1979), which marks a turning point from modernity to postmodernity. This postmodern period refers, in particular, to a disenchantment of the world (Weber, 1904–1905; Gauchet, 1985). Social actors no longer mobilise in the same way, and action, both individual and collective, is reconfigured. In the 1970s, action appeared as a sociological object, with the actor and the system no longer being two sides of the same coin (Dubet, 1994, 2002). Action raises questions insofar as it is no longer an extension of the system. The individual thus emerges, if not, for some contemporary sociologists, the subject (Touraine, 1984; Dubet & Wieviorka, 1995) and the contemporary period is regularly described as being marked by growing individualism, which is not necessarily antagonistic to the social bond (de Singly, 2003). For the managerial scientist Jean-Pierre Bréchet and the economist Lionel Prouteau, collective action is based on the recognition of 'a rich anthropological action, that is to say a projective action, both creative and anticipatory' (Bréchet & Prouteau, 2010, p. 110). The French sociologist Jean-Pierre Boutinet and Jean-Pierre Bréchet mention the modalities of a collective human action to lay the foundations for the future. In their view, it is through hopeful action that the human condition is founded and legitimised, and not through action that explores results. For these authors, collective action is our most precious common good, because it generates relational solidarity and allows for a preparation of the future enabling humanity to come to its senses (Boutinet & Bréchet, 2014).

[18] It is indeed important, in the Anthropocene, more than at any other time in human history, to consider all forms of hope and not to underestimate the mobilising component of the ideal. It seems to us that an ideal, even a misguided one, is worth pursuing if it is mobilising.

[19] This is a choice of how to read the living world, referring to the readings of Andreas Weber, François Flahault, Corine Pelluchon and Pablo Servigne.

future trajectory of humanity. The human condition is that contingency common to all human beings; it is shared by all. The notion of the human adventure brings with it the sharing of a common fate related to the uncertainty of the future. The idea of adventure is representative of the threat we face. This threat weighs on the whole of humanity and thus we all share in the same potential fate. The idea of the human adventure – unlike that of the human species – also integrates our social, cultural and political differentiations.

The experience of the camps, and the project to transform human nature, made Arendt particularly cautious about the notion of human nature, preferring the contingent dimension of the concept of condition.[20] While the concept of nature can be adapted to grasp objects, it stumbles over the question 'who are we?', for which there is no answer about nature without reference to a deity. The human condition, on the other hand, is subject to a set of contingencies, starting with the fact of being earthlings. Arendt, in *The Human Condition* published in 1958, raises the issue of the human condition of space research and men's attempt to escape the Earth: 'The most radical change we can imagine for the human condition would be emigration to another planet. Such an event, which is no longer quite impossible, would mean that man would have to live in fabricated conditions radically different from those offered to him by the Earth […] These hypothetical travellers who escaped from the Earth would still be human: but all we could say about their "'nature" is that they would still be conditioned beings, although their condition would then be, to a considerable extent, self-made' (1983, p. 18). Today we are those travellers of whom Arendt once spoke. We are both the travellers of Space X and Elon Musk's project to colonise Mars within the next decade, but also those of Google or Apple, whose genetic characteristics will have been modified, but above all those of the Anthropocene, which confronts us with an uncertainty unequalled to date in human history. Our adventure remains conditioned both by elements partly external to ourselves, such as the Earth system in which we participate, and by elements produced by the hands of certain men who have an impact on all. Writing in 1958, Arendt identified the Industrial Revolution and the Atomic Revolution as two important changes in the world but which do not profoundly alter human life: 'For it is still probable that the enormous changes of the industrial revolution behind us and the even greater changes of the atomic revolution ahead of us will remain changes of the world, and not changes in the basic condition of human life on earth' (Arendt, 1983,

[20] The desire to transform human nature in order to perfect it is indeed at the heart of the totalitarian enterprise; it is even an indicator of totalitarianism. As our political scientist mentions in the third volume of *The Origins of Totalitarianism*: 'The purpose of totalitarian ideologies is therefore not to transform the external world, nor to carry out a revolutionary transmutation of society, but to transform human nature itself. The concentration camps are the laboratories in the experiment of total domination, for human nature being what it is, this goal can be achieved only under the extreme circumstances of a human-made hell, and their infamy is therefore not only the business of their inmates and those who administer them according to strictly "scientific" criteria; it is the business of all men' (Arendt, 2002, p. 277). The utmost vigilance must be exercised with regard to enterprises that transform human beings. As Tassin notes, Arendt's analysis of the concentration camps shows that any quest to transform human nature leads to its destruction (2017, p. 153).

pp. 136–137). The dual change of the entry into the Anthropocene and the techno-scientific revolution of NBIC, on the other hand, will have a lasting impact on the human condition and dispatch us on an adventure into the unknown.

Humanity is not only defined intrinsically and statically, but also in terms of what it can and will make of itself within its host environment. The word 'adventure' comes from the Latin *adventura*, that which is to come, and *advenire*, to happen, to occur. The derivatives of *advenire* include *advena*, foreigner, and *adventicius*, that which comes from outside, that which occurs. In the current use of the term 'adventure', there is the idea of uncertainty and peril in relation to the environment in which it takes place. Adventure contains a sequence of events plot: something happens whose outcome is unknown. In fact, today, we do not know the outcome for humanity to which the entry into the Anthropocene will lead. Adventure is an experience – a test (experience comes from the Greek *experior*, to test). The second term, human, has the same etymological root as *humus*, earth. *Humus* is what gives *humanitas*, human nature. Human nature comes in part from the soil, humans being the ones who tread the earth of the soil. Biologically, we come from the earth and are part of a flow of matter with it that constitutes us. Human beings are partly defined by the relationship they have with the earth, with *humus*. The word 'humus' today defines the layer of natural compost on top of the earth resulting from organic decomposition. It is the place that welcomes death (decomposition) to give life (germination). Thus, humans are linked to the humility that comes with the finitude of their condition. *Humus* and *humilitas* define *humanitas*. 'Humans' and 'earthlings' have two different Latin etymological origins with the same meaning: Earth. Humanity must learn to mobilise its power in a different way, starting from an integration of finitude, both anthropological and environmental. The notion of the human adventure integrates what we are becoming.[21]

We therefore propose to think of humanity in the Anthropocene on the basis of the idea of an adventure rather than of a condition. The notion of human nature refers to an essence, that of condition to the contingency of the present, that of adventure to the uncertainty of the future and to the possibility of altering contingent characteristics (including, in particular, modifications of the systemic characteristics of the Earth, or the possibilities of modification of the biological substratum of humans).[22]

Dominique Bourg (2013) proposes the notion of 'transcendental damage' to think in a new and broader way about the notion of risk related to an individual and

[21] This notion of the human adventure has anthropological foundations that are close to the analyses of the anthropologist François Flahault, who breaks with the modern essentialisation of the individual: 'Whereas, in the conventional western mode of thought, the subject and the world are distinct and being belongs to that which exists by itself, the contemporary paradigm emphasises, on the contrary, that the individual is not external to his or her living environment: something exists in contact with something else, a being actualises and sustains itself through interactions with someone or something else' (2018, p. 38). Later on, he goes so far as to write: 'Let us stop opposing being and becoming: being is not a state – it is an activity' (p. 38).

[22] In Jonas' writings, it is possible to find the expression 'human adventure', which he articulates with the idea of collective action and the ethics of responsibility.

monetary dimension. The notion of risk refers to the hazards encountered by the contingency of the lives of singular individuals. However, as far as environmental characteristics are concerned, this is limited: 'Risk can only concern damage affecting a limited number of individuals, literally or figuratively' (Bourg, 2013, p. 110). The notion of transcendental damage refers to the endangerment of our 'natural conditions of existence, by definition collective, whose alteration could even jeopardise the very deployment of human societies, currencies included. In this case, we would be dealing with a *transcendental* form of damage, on a *meta-level*, conditioning our daily life in its material and relational aspects' (p. 109). This notion highlights the collective destiny of humanity, which seems to be subject to the same, uncertain fate. While some are responsible for the environmental and anthropological uncertainties of the future, all are affected by this destiny. The notion of 'transcendental damage' is important in the understanding of humanity as an adventure.

The Anthropocene brings something new to the human condition by jeopardising the sustainability of the world, so important to Arendt in the understanding of the human condition. In Arendt's thinking, the world is a fundamental element of the human condition: it allows for human productions that are works of art, thus giving men the possibility of inscribing the narrow temporality of their existence in the temporality of the world; the world is also the shared space of action. The world is what precedes birth and succeeds death. It is perennial. To evoke the world, Arendt uses the metaphor of a tent set up on the earth to accommodate the whole of humanity.[23] If we extend this metaphor, the Earth being in movement, it is undeniable that the tent has been shaken. The Anthropocene is changing the human condition.

10.1.2 Anthropocenic Criticism of Arendtian Anthropology

Figure 10.1 (mentioned in the following pages) presents the three dimensions of the human adventure that we propose to identify. These do not correspond to the Arendtian categories discussed in *The Human Condition* for several reasons. Firstly, the last few decades have seen a body of work that mentions the hegemony of *homo oeconomicus*. The category of *homo oeconomicus* has become a standard category on the basis of which a collective of authors work. *Homo oeconomicus* is an extension of Arendt's *animal laborans*, but with emphasis on consumption rather than the production of labour. Arendt clearly saw that the productions of all labour were immediately destined to be consumed and thus marked by complete lack of durability. The 60 years that have passed since the publication of *The Human Condition* have considerably hardened the process of consumption, and *homo oeconomicus*, the consumer, has taken precedence over the *animal laborans*.

Secondly, the Arendtian category of *homo faber* also deserves to be re-examined. One of the benefits of Arendtian anthropology is that it integrates the space of

[23] The geographer Michel Lussault shows that the world 'is now capable of overflowing its terrestrial location' (2013, p. 21) with all the satellites observing humanity from outside.

Sphere	Private sphere Economic	Public sphere Policy	
Dimensions of the human adventure	***Hybris***	**World**	**Coexistence**
Logic driving the activity	Logic of profit	Logic of responsibility	Logic of hospitality
Pole	Individual	Plurality	Others
Anthropological modelling	*Homo oeconomicus*	*Homo collectivus*	*Homo religatus*
Biosphere	Anthromes	Agora	Milieu

Fig. 10.1 The three dimensions of the human adventure. These categories are connected and stem from one another. Chemical exchanges enable the political world, which in turn can foster coexistence between humans, as well as between humans and non-humans, which enables the fluidity of 'living' relationships (and allows everyone to live without the threat of the domination of one species over the others). These categories are thought of in a perpetual cycle, or in a form of conviviality

articulation between the private and the public sphere with the category of the creation of *homo faber*. *Homo faber*, through his works, productions that are not destined for consumption, participates in the construction of a common world among men. In the contemporary world, which has been marked by a considerable socio-economic acceleration since the 1950s, as shown by the 'Great Acceleration' curves or the analyses of German sociologist Hartmut Rosa the works produced by men are largely part of the globalisation of a consumer society. This is the case with NBICs in the service of web companies, progress in aerospace and the human desire to colonise Mars, not driven by an association of states but by recent multi-billion dollar companies, or even artistic productions intended for consumption (Menger, 2003). Can we still say that *homo faber* really participates in the construction of a common world through his works that are not directly and immediately intended for consumption?

Thirdly, we wish in this work to delve into the Arendtian category of action of the *zoon politikon*. After reading the fifth chapter of *The Human Condition*, it seemed important to us to distinguish two components which allow for action: the primacy given to collective interest over the maximisation of individual interests, on the one hand, and the sharing of human existence, on the other. The category of coexistence that we propose is at the interface of space and time, of Arendtian life and the world, of the private and the public spheres. The space of the political is at the juncture of plurality and the sharing of existence. For Arendt, the world and life are two distinct – if not oppositional – categories, both of which refer to fundamental components of the human condition. Life is the interval between birth and death, between appearing in the world and disappearing from the world: 'To be alive means to occupy a world that preceded your arrival and will survive your departure' (Arendt, 2016, p. 39). Life refers to the economic sphere[24] and to that movement which needs to be maintained continuously by consumption; it consists of the activities related to

[24] For Arendt, the economy is confined to the private sphere, close to its etymological origin: the management (*nomos*) of the house (*oikos*). As Ricœur notes in his preface to *The Human Condition*, in this respect she disagrees with Karl Marx, for whom the economy is political and belongs to the public space.

the maintenance of life: work, consumption and leisure (Foray, 2001, p. 82). The sphere of life is individual and refers to *homo oeconomicus*. The world, on the other hand, is what precedes and follows life. It refers to a form of permanence and is part of long timescales.

Fourthly, the oppositional categories of 'life' and 'world' deserve to be rethought in the light of the Anthropocene. The anthropological shift at work in this research attempts to consolidate 'the world' (this political space in which concerted action emerges). On the other hand, it should not be thought of in opposition to 'life', because it is indeed this interdependent fabric that is the living, which is the medium in which the body politic (and the world) exist. The political world exists only because living beings interact organically, chemically, with each other. The political world is only the result of sharing with the living – and even with the non-living, the non-organic.

The 'politics of life' of Andreas Weber (discussed in Chaps. 6 and 7) provides an interesting link between biological work on the living and Hannah Arendt's political thought, which opposes two spheres – the sphere of the world and the sphere of life. One of the strengths of Weber's thought in Arendt's reading of politics is the profound link between the paradigm of the world and that of life. The world is characterised by its vitality.[25] This means that reality is not only objective, but above all creative and expressive (and therefore uncontrollable) because it is alive. This is an overlook component in the Anthropocene analyses that Andreas Weber compensates for. Weber's thesis is that the world in which we are immersed is not based on a dissociation between ideas and things, resources and users, nature and culture, but on the contrary, on continual creative relationships and transformations.[26] At the heart of the Anthropocene a movement, and therefore a hope, is possible because the main feature of the living reality that constitutes us and in which we are immersed is vitality. Weberian vitality integrates the Arendtian political component based on relations within plurality and the emergence of novelty that natality allows. Its lexicon field is different: Weber evokes sharing rather than plurality; creativity rather than natality; but the proposed dynamic includes strong elements of proximity, without opposing life to the world. This vitality is central to Weber's thinking and his view of the world, and is characterized by its expansiveness. On the other hand, its expansiveness is not that of insatiable capitalism characterised by the constant need for more. The paradigm of life, for example, is clearly positioned differently than in reference to that of consumption (thus breaking with Arendtian modelling): 'The vitality to which we aspire is not simply a private mental state that can be secured by way of consumption' (Weber, 2017, p. 17).[27] Expansiveness is that of life, not that of the individual. Indeed, Weber apprehends the human subject as an inter-subject. This breaks with an individualistic conception of life and founds a thought based on coexistence (a term he does not use himself). Thus, it is not the

[25] This is in line with the approach of Corine Pelluchon, who shows how the world is based on life.

[26] Weber's approach is decidedly poetic. As he sees it, 'Any practice of vitality can only be a poetic practice'. 'Jede Praxis der Lebendigkeit kann nur eine poetische Praxis sein'. (Weber, 2016a, b, p. 15).

[27] 'Die Lebendigkeit, die wir ersehnen, ist nicht nur ein privater seelischer Zustand, den man sich womöglich auf dem Weg des Konsums sichert' (Weber, 2017, p. 17).

individual who has the possibility of developing and expanding, but life, through sharing it. Here, coexistence precedes existence: 'I can only be because you are'[28] (Weber, 2017, p. 10). For Andreas Weber, sharing does not reduce us, does not cut us off from something, but augments us.[29] In the development of this thesis, his starting point is phenomenological: Andreas Weber shows that breathing is a type of sharing, inhabiting one's own body is also based on sharing with the elements around us (2017, p. 9). This sharing is understood from the point of view of sharing with the biosphere – ecosystems rely only on exchange. Andreas Weber's work consists of an overcoming of modern forms of understanding of life. Existence (*Dasein*) is understood biologically, in its participation and continuous interaction with the environment that surrounds it: 'Without reference to the body, our language could not develop. Our ways of thinking do not come from "pure reason" but are correlated with the living body' (Weber, 2007, p. 142).[30] Sensibility here is one of the defining elements of his anthropology. He develops what he calls a 'biopoetics' or a 'creative ecology of the living'[31]: 'Our interiority is a phenomenon of matter. It is a dimension of the body (*Leib*), not a disembodied spirit (*Körper*)'[32] (Weber, 2007, p. 111).

Here, we propose to distance ourselves from the Arendtian categorisations insofar as the sphere of politics acquires materiality only through the tissues of the living. The differentiation between life and the world loses its relevance in the contemporary period marked by the Anthropocene, where we need to root ourselves more deeply in solidarity with the living world, while at the same time extracting ourselves from the destructive capitalist hegemony. It is a question of conceiving of politics in opposition to economic hegemony while taking root in the vitality of the living. The category of life is of limited utility in expressing this economic hegemony, and it is necessary to propose another category signifying this destructive domination, because it is limitless, of economic logics. We propose here to identify it within *hybris*. The choice to name this oppositional component to politics in this way is not insignificant. It means that destruction is a fundamental anthropological component that must be kept under control or else the continuity of the human adventure will be endangered. It also means that *hybris* will not easily be eradicated from the human adventure, of which it is a component.

Hannah Arendt does not have a developed set of ecological ideas. To understand this dimension of her thought, it is possible to formulate different hypotheses. This seems to be due, in particular, to her grounding in Greek political philosophy with

---

[28] 'Ich kann, nur sein, weil du bist' (Weber, 2017, p. 10).

[29] This is the subject of the book *Sein und Teilen* (Being and Sharing) (Weber, 2017).

[30] 'Ohne Bezug auf den Körper könnte sich unsere Sprache nicht entwickeln. Unsere Denkwege entspringen nicht der "reinen Vernunft", sondern der Ratio des lebendigen Körpers' (Weber, 2007, p. 142).

[31] 'eine schöpferische Ökologie des Lebendigen'; *Lebendigen* could be translated as living, lively.

[32] 'Unsere Innerlichkeit ist ein Phänomen der Materie. Sie ist eine Dimension des Leibes, nicht die eines körperlosen Geistes' (Weber, 2007, p. 111). In German, there are two different terms for the body: *Leib* and *Körper*.

the differentiation between humans and nature, the difference between *polis* (the city-state of humans) and *cosmos* (the good order of the world). Moreover, we may wonder whether ecological thinking might have been a painful reminder of the biological for Jewish thinkers, scarred by the Nazi obsession with land and nature. Finally, if a biological-political conception is problematic for Arendt, it is also likely to be related to her reliance on the thinking of Heidegger, whose philosophy does not include a biologising conception of the political.[33] For example, Arendt's and Heidegger's critiques of technology are not ecological critiques of technology. We do not find, in Arendt's or Heidegger's work, the idea that the living world has rights. Rather, we are dealing with a Cartesian perspective vis-à-vis nature, with humans being conceived as 'masters and possessors of nature' (*The Discourse of Method*, 1637, Part Six).

10.1.3 The Biosphere as an Anthrome, an Agora, and A Milieu

Once *hybris* is accepted as one of the three components of the human adventure, the category of life fits into the biosphere, which gives the three dimensions materiality and power. Life, understood as the biosphere, refers to this interdependent fabric open to the cosmos (which receives the sun's rays). The biosphere that gives the human adventure its materiality is made up of three facets. Firstly, the anthromes, described in the introduction, means the way in which humans appropriate and profoundly modify the earth's surface to cater for their own needs (which leads to the destruction of animal habitats). Today, the biosphere is largely anthropised, which poses risks of systemic shifts.

The second facet is the agora[34] that place of gathering and dialogue to bring about action. If we view the biosphere as an agora, it is the place where the various political subjects of the earth's surface (humans, animals, plants, minerals, etc.) come together and conduct exchanges. As Andreas Weber forcefully reminds us, the biosphere is a place of exchange and sharing. Through respiration and food, chemical elements are exchanged between the atmosphere and the various living organisms; this exchange is what allows the various organisms to live. Life is never individual; without others, life gives way to death. Life is always exchange, circulation, and sharing (life is understood here as conviviality, sharing of life-giving sustenance). Understanding the biosphere as an agora means that these exchanges that constitute life must also be the subject of speech. This is the responsibility of humans.

[33] It is different in terms of its relationship to Nazism.

[34] The agora is the meeting place in Greek cities, treated and used as a social space, political forum and marketplace, at need.

The milieu is the third dimension of the biosphere. In the notion of milieu, emphasis is placed on the way in which spaces are imbued with representations and relationships from which they are inseparable. The term milieu (of which mesology is the science)[35] has been employed once again by the French geographer and philosopher Augustin Berque in his work, as a translation of the German term *Umweltlehre* and the Japanese term *fûdogaku* (Berque, 2016, p. 88). The specificity of mesology is the perception of the relationship with the milieu from the point of view of being. Mesology, therefore, incorporates the existential and symbolic component that the milieu has for the being. In other words 'for the same *Umgebung*,[36] there may be different corresponding *Umwelten*'[37] (Berque, 2016, p. 91). This term, milieu, is equivocal because it means as much what is in the middle as what is around. We will use the term milieu to define what is around, the space of a certain place, what Berque calls 'the *chôra* of a certain *topos*' (Berque, 2010, p. 20). Berque's philosophical notion of milieu integrates being and is even 'necessary for the concretisation of being' (2010, p. 20): the milieu allows the being to grow; concret- is derived from the Latin *concretus*, meaning 'to grow together' (Berque, 2010, p. 21).

10.1.4 Homo Oeconomicus, Homo Collectivus and Homo Religatus

We propose to identify the human adventure at the juncture between three poles of intelligibility of the contradictory behaviours of actors. The first pole is the individual marked by the profit-driven logic of the neoliberal *homo oeconomicus*, seeking to maximise his own individual interests. This pole of the individual is in tension with two others: that of plurality, marked by the responsibility logic *homo collectivus*, driven by a commitment to the welfare of the collective, but also with the pole of others, marked by the sharing of existence (coexistence[38]) and the logic of hospitality of a *homo religatus* who knows how to be neighbourly to his *alter ego* (linked to the Ricœurian *socius*). This existential component manifests an understanding of humanity in all its finitude, facing the prospect of its own inevitable death.

Homo oeconomicus is an integral part of the human adventure. In a way, it is what allows for the human adventure to be made up of singular individuals with

[35] On 7 June 1848, the term 'mesology' was coined by the French physician Charles Robin at the Société de biologie. It means the science of milieux, or environments. This term predates 'ecology', proposed by the German biologist Ernst Haeckel in 1866, but which then supplanted the term 'mesology', which has long since fallen out of use.

[36] *Umgebung* can be translated as milieu in the sense of surroundings.

[37] *Umwelt* also means environment/milieu, but with a symbolic and existential singularity. It is the environment of individuals.

[38] Arendt recalls that for the Romans, 'to live' is synonymous with 'to be among men' (1983, p. 16), reflecting the importance of coexistence.

their own interests that diverge from those of their alter egos. What is currently problematic is the hegemony of *homo oeconomicus*. First of all, as Alain Caillé (2008) points out, the logic driving *homo oeconomicus*' activity does not explain the diversity of human behaviour. To understand the logics of human activity in the contemporary period, we must add anthropological components that have oppositional complementarity with *homo oeconomicus* to the model.[39] By identifying these oppositional anthropological complementarities, we shall be able to treat them with the appropriate degree of importance in educational thinking and pedagogical practices. Basically, the proposed model opposes *homo oeconomicus* with *post-Promethean political interleaving*. This is what gives structure to the three-dimensional development presented here. The individualistic component of *homo oeconomicus* is not opposed to an individual whose activities are underpinned by other logics, but to a void. This void is a relational space that does not accommodate anyone.[40] Two anthropological components are necessary to facilitate the emergence of this post-Promethean political interweaving: the *homo collectivus* of plurality and the *homo religatus* capable of hospitality towards others (which is manifest in the dual figure of the alter ego and neighbour). The post-Promethean political 'between-us' here represents Arendt's category of action: it emerges only when human beings act together.

These contradictory driving logics of action between *homo oeconomicus*, *homo collectivus*, and *homo religatus* refer to tensions between commercial goods (*homo oeconomicus*), collective goods (*homo collectivus*) and the common good (*homo religatus*), in the model developed by Flahault (2013a, b). This proposal has points of similarity with the theoretical architecture of Dubet's sociology of experience, in the wake of Touraine's sociology of action. Dubet analyses social experience at the intersection of relations of tension between logics of strategy (interest in the situation) that can, in some respects, be those of *homo oeconomicus*; logics of subjectivation action (historical figure of the subject) that can, among others, be those of *homo collectivus* (Touraine indeed has a very lofty representation of a Subject mobilised by the collective); and logics of socialisation (logic of integration) that can. in part, be those of *homo religatus*.

[39] This proposal for theoretical modelling of the human adventure has some similarities with the anti-utilitarian theory of action developed by Alain Caillé (2009), who does not deny the motivations of individual interest maximisation, but highlights the plurality of motivations of social actors.

[40] Alain Caillé develops a three-faceted typology of action, with self-interest, interest in others or love (which is similar to *agape*), obligation and freedom (2009). Alain Caillé's typology (20096) is a particularly interesting critique of *homo oeconomicus*, but it has the peculiarity of being an anthropology of the acting subject. The ambition of our model of the human adventure is to place the 'between-us', rather than the subject, at the heart of action. The 'between-us' has to do with interaction, but cannot be reduced to the interaction between two or more individuals. It signifies that empty space that is between people and that can accommodate projects, exchange, *agape*, hospitality, conciliation of interests, or listening. The centre of this anthropological model is 'a certain void' between people that can be creative and mobilising. This is a non-anthropocentric anthropological model.

One of the challenges of the twenty-first century revealed by the Anthropocene is the need to move away from the capitalist and commercial economy and to integrate a new conception of the human being,[41] as Flahault has stressed on several occasions. In *L'homme: une espèce déboussolée – Anthropologie générale à l'âge de l'écologie* (2018), Flahault's highlighting of the ecological and cosmological dimension of the modalities of existence of human beings[42] gives credence to the concept of the between-us developed in this work (oppositional to that of individual or subject) and to the resulting thesis relating to convivialism as an educative style, which can help imbue humanity with an anthropological power.

10.2 *Hybris*: The Profit-Driven Logic of *homo oeconomicus*

Since the Industrial Revolution, capitalism and consumption have gone hand in hand with increasing individualism. The contemporary individual becomes a consumer and seeks to maximise his own individual interests. The capitalist economy becomes all-consuming and hegemonic (Peyrelevade, 2005) and the contemporary individual is becoming confused with *homo oeconomicus*, defined by a logic of profit-making and pure consumption. *Homo oeconomicus* is an effect of the Industrial Revolution, but also the manifestation of the utilitarian paradigm, whose limits have been demonstrated by a number of anti-utilitarian authors. *Homo oeconomicus* is the self-interested individual so well described by eighteenth-century Scottish economist Adam Smith in *The Wealth of Nations*, with his famous 'invisible hand'. Smith's idea is that the pursuit of individual interests benefits the community. This utilitarian logic was developed and expanded by the British economists Jeremy Bentham (1748–1832) and John Stuart Mill (1806–1873), and then by a group of liberal economists pursuing so-called 'classical' economics.

The limitations of *homo oeconomicus* are considerable and regularly denounced in the literature; indeed, it 'knows neither "altruism" nor "responsibility" for others and future generations' (Faber et al., 2002, p. 324). In some respects, *homo oeconomicus* is an extension of the Cartesian subject, external to his environment, which he tries to grasp or exploit to his advantage (Flahault, 2005, p. 378).[43] However, the rationality of *homo oeconomicus* is relative, and for Arnsperger, he is incapable of

[41] Flahault makes this interesting remark about human nature in his latest book: 'Can we say that "there is no human nature"? Yes, if we mean that there is no essence of man that makes him good or bad, selfish or altruistic by nature. No, if we mean that man has the Promethean power to make himself. For there is indeed a human nature in the sense that man necessarily exists in and through a living environment (beginning with his own body)' (2018, p. 39).

[42] 'We are spontaneously aware of ourselves as having an existence distinct from what we are aware of: there is *I*; and there is *what* I perceive, *what* I feel, *what* I think. But we can also perceive ourselves quite differently: by experiencing the feeling that our *mind*, our 'ego' exists insofar as it maintains contact with what gives it life'.

[43] Jean-Philippe Pierron evokes, for example, the figure of the trader as 'the flesh-and-blood version of *homo oeconomicus*' (2014, p. 36) who 'embodies the modern economic animal' (p. 40).

reflexivity on what influences his rationality and is 'deaf and blind to his own existential concerns and anxieties' (Arnsperger, 2011). Thus, the dominant contemporary anthropology based on *homo oeconomicus* is that of 'economic growth rooted in a psychology of lack' (Arnsperger, 2010a, b, p. 25). The logic of accumulation at the heart of capitalism develops an illusory sense of immortality for those who have too much. The individual is afraid of the future, which will inevitably contain suffering, fragility and ultimately death, and he struggles against this 'horizon of existential finitude' (Arnsperger, 2010a, b, p. 24) through capitalism. We can even go so far as to say, as Arnsperger does, that the contemporary economic crisis is rooted in the anguish of fragility and death, which *homo oeconomicus* cannot know.

Arendt shows that in contemporary society, manufacturing and consumption have taken on the role that ought to be played by political action. Thus, we are seeing the economic win out over the political, which is the emancipation of labour[44] with the occupation of the public space by the *animal laborans* (which is, in the final analysis, nothing more than a private space exposed to the light of day) (Arendt, 1983, p. 150). Objects intended for use, products of the craft industry, have become consumer goods. Thus, a dress or a chair is consumed almost as quickly as food (Arendt, 1983, p. 140). *Homo oeconomicus* occupies the space between people.

10.3 World: The Responsibility-Driven Logic of *homo collectivus*

Despite this hegemonic tendency of the logics driving *homo oeconomicus*, they are nevertheless in conflict with another anthropological component marked by collective interest. The academic literature shows us traces of the anthropological model underlying this maximisation of collective interests, under the term *homo politicus*.[45] For the German economists Malte Faber and Johannes Schiller and German philosopher Thomas Petersen (2002), the concept of *homo politicus* is an alternative conception of human behaviour, marking its differentiation from *homo oeconomicus*. *Homo politicus* is 'grounded in political philosophy and focused on the human interest in justice in the welfare of the community' (Faber et al., 2002, p. 324). *Homo politicus* refers to the sense of responsibility that stands again contemporary individualism.[46] It is the *homo* of responsibility and of the civic logic

[44] Labour obeys the regime of necessity. Arendt recalls that the Greeks refused to countenance calling slaves 'men', although they were members of the human species, because they were subject to necessity (1983, p. 96). The characteristic of labour is that 'nothing is left behind to indicate result of the effort, almost as soon as it is spent' (1983, p. 99).

[45] As early as the sixteenth century, the word *politicus* appears in the Robert Estienne's *Dictionarium Latinogallicum*, and refers to 'civilian' or '*civilis*' (Demonet, 2005).

[46] It is not uncommon for the logic of responsibility of *homo collectivus* to emerge as the result of horror at the vulnerability of others, as we saw in Abbé Pierre's appeal for help for the homeless during the winter of 1954, or in the revolt and then the collective commitment of the indignant during the Arab Spring in 2011 (Pierron, 2012).

of maximizing 'social welfare' (Nyborg, 2000, p. 306). The logic of responsibility of *homo collectivus* also integrates the project dynamic with its projective component and its aim to foment harmony and consensus. Faber, Petersen and Schiller, following in the footsteps of the Norwegian economist Karine Nyborg and the American economist Kenneth Arrow (1963), define *homo politicus* as 'human behaviour that considers what is best for society' (Faber et al., 2002, p. 328). For the British historian John Greville Agard Pocock, *homo politicus* refers to Aristotle's *zôon politikon* affirming his being and virtue, and signifying that man is part of the City (in the sense of the broader community) and cannot be otherwise. *Homo politicus* is the *homo* of the world and its exteriority, by differentiation from *hybris*.[47]

Thus, *Homo politicus* is important in this apprehension of the human adventure, which it helps to grasp beyond a logic of essentialization: what constitutes and defines man actually comes from outside himself. However, the concept of *homo politicus* may be difficult to grasp, because it is linked to *homo religatus* – the third anthropological component that we wish to study. It has no existence outside of the connection with other human beings. Indeed, 'Politics is based on a fact: human plurality [...]. It is because philosophy or theology are always concerned with man, because all their statements would be correct even if there were only one man or only two men or only identical men, that they have never found any philosophically valid answer to the question: what is politics? [...] as if there were something political *in* man that is part of his essence. This is precisely where the difficulty lies: man is a-political. Politics originates in *the space-which-is-between-humans* – hence, in something fundamentally *external to man*' (Arendt, 1995, pp. 31–32).

Two conditions appear to be required for the emergence of the political in this 'space-which-is-between-humans': the presence of a pluralist collective and the sharing of human existence. The term *politicus*, referring mainly to the first of these two conditions, is here equivocal in relation to the politics we wish to grasp. The main component of *homo politicus* as defined by Faber, Petersen and Schiller, and by Arrow, is the capacity to focus on the interests of the collective in a very broad sense, not reducible to the group to which the individual belongs. Thus, we understand this anthropological component as *homo collectivus*. The term *collectivus* does not fit into a Marxist communist or totalitarian collectivist tradition, with the removal of the pluralist component of the collective. The term *homo collectivus* signifies the capacity of individuals to apprehend the interests of a vast pluralist whole, based on the logic of responsibility.

Homo collectivus constitutes a form of resistance to the hegemony of *homo oeconomicus*, which is embodied, for example, in Jonas and Arendt's ideas of resistance to the mindset that 'everything is possible' (Frogneux, 1996, p. 685). Faber, Petersen and Schiller go so far as to conceive of *homo politicus*, whose main characteristics are included here under the umbrella of *homo collectivus*, as a new model of human behaviour. The link to the environment of *homo collectivus* is that of

[47] In Arendt's thinking, the world has categorical primacy over life; this marks a real difference with Jonas' ethics of responsibility, in which life has primacy. The model that we propose in this work associates the world with life without the one having primacy over the other.

sustainability and durability, but against a background of solidarity with the rest of the biosphere, which constitutes a space for political coexistence.[48] *Homo collectivus* is not only responsible for other humans, but for everything in the biosphere, which then becomes an agora.[49]

10.4 Coexistence: The Logic of Hospitality of *homo religatus*

10.4.1 The Primacy of Coexistence

Homo religatus is the second anthropological component of Arendt's category of action explored in this chapter. *Homo religatus* refers to an individual who does not exist in himself in an essentialised form, but in relationship with others. It is based on the same idea relationship to something outside oneself as does *homo collectivus*: the world viewed on the basis of a human plurality, for *homo collectivus*, and the singularity of the other in the case of *homo religatus*. *Homo religatus* is the man of relationship, linked to others, whose coexistence (i.e. shared existence) precedes existence. The notion of existence is not synonymous with life or vitality – it is situated at the meeting point of life and the world, of the poles of the individual and the plurality. The word 'existence' derives from the Latin *ex stance*, standing outside. It refers to the idea that we live only by being in the world (Boutinet, 1990). Existence is this condition shared among the whole of the human adventure, and we coexist more than we exist (Flahault).

Since the time of Aristotle, it has been understood that the fact of being in society makes man a human being. For Flahault, it is only in the context of socialised coexistence that a newborn can find his place as a human being. The fact of being a member of a group of many, of coexisting, precedes self-existence' (Flahault, 2013a, p. 775). Thus, one of the fundamental paradigms for understanding human beings is that humans are relational: our existence is preceded by that of others. To exist is to exist with others and 'self-interest is inseparable from interest (positive or negative) in others' (Flahault, 2008, p. 315). This logic of coexistence presupposes belonging to a clearly delimited framework in which everyone has a place. Having

[48] These modes of existence are only possible if there is justice among the people living on Earth, and among present and future generations as well. Sustainability 'requires both intra- and intergenerational justice' (Faber et al., 2002, p. 329). *Homo collectivus* is related to the *homo sustinens* described by German economist Bernd Siebenhüner (2000). The anthropological components of *homo psychologicus* and *homo reciprocus* are also fundamental for sustainability (Faber et al., 2002, p. 332).

[49] Such an assertion follows various writings regularly referred to in the course of this work, such as *Homo natura – En harmonie avec le vivant*, written by the lawyer Valérie Cabanes (2017), *Lebendigkeit sei! Für eine Politik des Lebens. Ein Manifest für das Anthropozän*, by Andreas Weber and Hildegard Kurt (2015), the *Manifeste animaliste – Politiser la cause animale*, by Corine Pelluchon (2017a, b), and *Mutual aid: the Other Law of the Jungle*, written by Pablo Servigne and Gauthier Chapelle (2017).

coexistence precede existence presupposes the integration of certain barriers against the current of unlimited desire and the response provided by the economy. Indeed, the contemporary period is indeed marked by a denial of the finitude of our environment, which compels man to deny his own finitude: 'the denial [of natural finitude] is inseparable from the affirmation of the omnipotence of technology, of a malleability of the world without any limits' (Bourg, 2009, p. 69).

In his latest book, *L'homme, une espèce déboussolée – Anthropologie générale à l'âge de l'écologie* (2018), French anthropologist François Flahault continues to emphasise that the individual does not precede society. This element is of prime importance because its corollary is that the economy is at the foundation of societies: "Today, we know that the process of hominisation was preceded and made possible thanks to the millions of years of social life that preceded it – the very one that primates observe and whose complexity, already astonishing in chimpanzees, has developed in the successive representatives of the genus *Homo*. Languages themselves, which are specific to *Homo sapiens*, could not have developed in the absence of a living environment that required individuals to interact' (p. 29). Based on a body of contemporary scientific work, particularly in biology and the principle of 'coevolution', François Flahault works on a post-Promethean anthropology that runs counter to that of an author such as the Israeli historian Yuval Noah Harari in the globally successful *Homo deus* (2015). Indeed, Flahault is at pains to show that man, and that each individual, did not make himself. Against the current of a modern utilitarian ideology centred on the individual, Flahault shows the importance of the relationship and the need to think together about the living world and its environment, the organism and its biotope.[50] This is a paradigmatic break with modern conceptions of the individual and the subject.[51]

The paradigm of coexistence is not unrelated to Mauss' idea of the gift where, in an exchange, the individuals themselves have an interest in each other, in addition to the object exchanged. The human being in Maurice Bellet's thinking is a man connected to others who becomes truly human only through the relationship. What is in the relationship is more essential than man as such. Indeed, Bellet thinks of the human being only in relation, enabling him to be human (with the proviso that only relations with other humans humanise).[52] This author's hope – to see another type of man and world come into being – refers to the underlying evocation of a *homo collectivus* in interaction with the *homo religatus*. Bellet's *homo* requires courage and

---

[50] Hominisation cannot take place without interactions with 'that essential part of the living environment that are the other members of the group' (Flahault, 2018, p. 30). The underpinnings of this work by Flahault include proximities with agronomist Isabelle Delannoy's recent work, *The Symbiotic Economy* (2017). In this economic essay, which puts to work a relational anthropology that breaks with modern individualism, Isabelle Delannoy emphasises our relations with the environment and the possibility of living in harmony with ecosystems.

[51] 'The etymology of the word "individual" says it well: a being, not a composite, but one being. A core, a self, a me, around which the world extends: the not-me. The inner clearly distinct from the exterior' (Flahault, 2018, p. 31).

[52] The world-famous book by German author Peter Wohlleben, *Das geheime Leben der Bäume. Was sie fühlen, wie sie kommunizieren – die Entdeckung einer verborgenen Welt*, published in 2015, shows how trees and the care they receive are part of humanisation.

is committed to a future shared with others. In this respect, he is like Arnsperger's *homo*, captured in the figure of the existential activist, who is one of the forms of postmodern *homo collectivus* in his link with *homo religatus*, confronted with the fall of the Grand Narratives, integrating his finitude and sharing it with others. The conception of a human becoming such through relationship is in line with the observations of the French ethnologist Maurice Leenhardt in New Caledonia where 'in the Houaïlou language, *do kamo* means "the true man", truly human, who defines himself solely through his relationships' (Donegani, 2013, p. 358).

An authentic man is a connected man because it is in 'connection' that he becomes human. For the Belgian philosopher Jacques Taminiaux, the Cartesian 'I am' does not fit this man, connected to others (2007, p. 22). For Heidegger, it is better to say: 'I am One, one is, one undertakes this or that, one sees things in such-and-such a way. It is this One that is the true modality of everydayness, of being one-with-another, concrete, average. It is from the perspective of this One that the kind and manner in which man first and most often sees the world, in which the world approaches men, in which man expresses the world, arises…' (Heidegger, *Gesamtausgabe*, 18, pp. 63–64, quoted by Taminiaux, 2007, p. 22). Man cannot be thought in separation from a 'We' or a community component.[53] *Homo religatus* also refers to the figure of the man defined by French philosopher Emmanuel Mounier, and to that defining characteristic of relationship and communication in which man finds a form of fulfilment (Lambert, 2001, p. 65). In Mounier's view, availability to others is one of the characteristics of the person: 'The person is an inside that needs the outside' (Mounier, 2001, p. 56).[54]

As mentioned earlier, *homo religatus* is also a symbolic being linked to the environment[55] –in relation to trees, rivers, the wind or animals, for example. It is here that conviviality takes on its full meaning in the Anthropocene and acquires all its strength of resistance to the hegemony of capitalism. Thus, for example, forests cannot be considered wood factories owned by large companies with a view to making a profit, when what is at stake is the common good with air quality, atmospheric CO_2 absorption, the renewal of the nutritive quality of the soil, water quality, the accommodation of various forms of terrestrial and aerial life, etc. In the same way, is it possible to consider animals mere producers of meat for human consumption?

[53] In this sense, he is a symbolic being referring to Jonas' *homo pictor*.

[54] *Homo religatus* is linked to *homo religiosus*, whose connection with others is expressed through the sharing of a religious practice. *Homo religiosus* can be understood here as a manifestation of *homo religatus*. For the French political scientists Jean-Marie Donegani and Marc Sadoun, religious sentiment is a 'fundamental anthropological structure' (Portier, 2009, p. 817). Arnsperger, for his part, considers the human condition to be a religious thing: *religare* is 'so constitutive of the human condition that the *religio* in question is not an institution, but a mode of being, an anthropological necessity' (Arnsperger, 2010a, b, p. 35).

[55] By inhabiting the Earth, man participates in the creation of the ecumene, the habitable space of Earth. In line with the thinking of Augustin Berque, Jean-Philippe Pierron encourages us to think of humans and the milieu as 'partners in interactions and relationships' (2013). This relationship to the milieu is anthropologically fundamental because it allows humans to have a sensory experience, which will serve as the basis for a transcendental relationship or spiritual experience (Pierron, 2016) that orients the desire for limitlessness in a non-Promethean way.

10.4.2 Homo Religatus, Socius and Hospitality

One of the manifestations of *homo religatus* is the *socius* developed by Ricœur. This is a man who behaves in a neighbourly fashion towards others and becomes capable of hospitality, enabling him to move on from simply meeting others to having relationships with them (Boudou, 2012, p. 272), and allowing us to grasp what is happening in the 'together' (Bergier, 2014). When using the concept of *socius*, it is possible to integrate the notion of hospitality, in its symbolic sense, to think about political issues based on the anthropological capacity to welcome others and accept their differences.[56] The hospitality of *homo religatus* allows for dialogue, and thus the emergence of speech, giving rise to the individual as a citizen.[57]

This notion of *socius* is similar to the way in which Lévinas, in his thinking, gives priority to the other through the notion of the face, which makes it possible to move from staring to considering in relationships. Another way of formulating this is perhaps to consider 'the stranger and the poor as always higher than oneself' (Poché, 2003, p. 90). Lévinasian thought apprehends social life in terms of singular faces that become singular utterances. This makes it possible to grasp social reality as inhabited spaces where the reception of the other, who is different, and constitutes otherness. Social relations in the contemporary democratic space are thus thought of in terms of hospitality, which can be what gives meaning to the existence of social actors by focusing on others[58] and by sharing their frailties.[59] Hospitality refers to the capacity for engagement, and to the fact of 'accepting to get wet, and therefore [to] risk being vulnerable' (Poché, 2013, p. 60). This relationship between hospitality and alteration structures the dynamics of politics.

Socius integrates this logic of *agapè*, originally meaning the representation of an ideal (unconditional) love of God for humans, which has social effects. *Agapè* is linked to grace, which consists of fervently welcoming as well as being fervently welcomed. Grace, like hosting, has a dual meaning: active and passive. This conception of grace has aspects in common with the notion of hospitality. The first sign of hospitality is trust of another's word (Cornu, 2007, p. 27). The host who

---

[56] In this sense, Boudou proposes a support on an anthropology of hospitality for Political Theory (Boudou, 2012).

[57] In Arendt's thinking, it is a question of learning to inhabit the world and to inhabit it with others in the awareness of transience. Thus, learning hospitality is fundamental.

[58] The Irish philosopher Richard Kearney uses the term 'the Other' 'to refer to an otherness worthy of respect, consideration and welcome (hospitality)' (Kearney, 2001, p. 95) and the term 'the Stranger' 'to refer to the experience of an otherness linked rather to *selection* (as in immigration policies and all that differentiates nationals from non-nationals, allies from enemies) or to *distrust* (as in the case of UFOs and other unwelcome intruders)." (Kearney, 2001, pp. 95–96).

[59] *Socius*, which refers to an idea of selflessness or concern for others, can also be one of the modalities of contemporary self-realisation (Poché, 2003).

welcomes another into their home receives, the hospitality given from their home in addition (Derrida, 1997).[60]

10.5 Homo Collectivus, Oeconomicus, and Religatus

10.5.1 Homo Collectivus and Homo Religatus

Homo religatus developed in the course of modernity by gaining importance over the *homo politicus* of the Ancients, thus accompanying the history of liberalism from a 'freedom-participation' of the Ancients to a 'freedom-autonomy' of the Moderns (Baudouin, 2002, p. 130). Since Aristotle's time, it has been recognised that it is the fact of being in society that imbues man with humanity. This is the idea which Flahault then developed with the notion of coexistence: 'It is only in the context of socialised coexistence that the newborn can find his place as a human being. The fact of being among many, of coexisting, precedes self-existence' (Flahault, 2013a, p. 775). Indeed, for Flahault, the ultimate purpose of politics is to 'foster human relationships' (Flahault, 2013a, p. 775). Insofar as relationships are what allows one to become human, the object of politics is a work of humanisation. The articulation of *homo collectivus* and *homo religatus* is indeed at the foundation of a conception of politics.

The political, in Arendt's work, refers to spatiality: it is in the space between human beings that it is born – the space where a world emerges. The political, situated in this space between men, is thus 'horizontal before it is vertical' (Goetz & Younès, 2009, p. 32). The Arendtian image of the materialisation of the world is that the space between men is both what separates and what connects, like the table between men. Sometimes, though, this common world is interrupted by deserts: when action is no longer possible or speech is no longer shared. The image of the desert signifies this absence from the between-us. The desert is not the end: 'one can travel through the desert, cross many deserts, but one must know where the oases are' (Goetz & Younès, 2009, p. 30): those of art, thought, culture, love. Here, the oasis is thought of as a private space for rest when it is taking too long to cross the desert. However, it must not become a refuge, otherwise there is a risk that it will be altered by the desert, and ultimately become a desert itself. It is interesting to cross-reference Bellet's thinking on the 'between-us' as the birthplace of humanity and that of Arendt as the place where the world, and therefore politics, emerges. Politics and humanisation go hand in hand. *Homo religatus* is inseparable from *homo collectivus*. Their articulation makes the emergence of

[60] From an urban perspective, hospitality is a 'gift of space to be inhabited, crossed or contemplated' (Gotman, 1997, p. 16). The hospitality in question is not only intimate or intrapsychic – it has an economic and material support in the money in circulation. Circulating money, which is opposed to accumulated money, refers to the pole of the other and can be the economic support for hospitality (Arnsperger, 2004).

action possible. *Homo collectivus* and *homo religatus* refer to the human of the political, by differentiating from *hybris*, which refers to the pole of the individual thought of as an "in-itself".

Etymologically, the term 'world' means cosmic order, integrating the heavens, centred on the city (referring to the political component of the collective). For Berque, the world is 'the set of holds we have on the Earth (nature)' (2009a, b, p. 77). The world is political and is grasped, in particular, through action. The terms *monde* (French), *mundo* (Spanish, Portuguese) and *mondo* (Italian), meaning world, come from the Latin *mundus*, which means 'a general setting in order, from bodily care to the world in the sense of universe, passing through the domestic household, but all from that navel represented by the *mundus*, a sacred pit in the heart of the city, serving as a conduit between the world of the dead and that of the living, and which was opened three times a year to offer the dead the fruits of the earth in tribute' (Berque, 2015, p. 353). Today, the term has two etymological roots: *mundus* and *kosmos*. For traditional societies, the idea of *kosmos* encapsulates the question of the immensity of nature, the moral rules of civilisations, and of adornments and ornaments. The *kosmos*, which we can translate as 'world', integrates the idea of order and a moral order. The order of nature, the order of moral rules and the order of adornment is good. Today, because of the unsustainability of our lifestyles on an ecological and ethical level, we no longer have a *kosmos* that brings together our conduct and knowledge: 'We moderns have lost this capacity from the moment when, for us, things have become morally neutral objects, ontologically distinct from the moral subjects that we are' (Berque, 2004, p. 11). The modern dualism on which science is based has led to a decosmisation, a 'loss of *kosmos* as an order linking the being of things and our own' (Berque, 2004, p. 11).

10.5.2 Homo Oeconomicus and Homo Collectivus

In the contemporary individual, there is a tension between *homo collectivus*, attempting to act as a citizen concerned with collective wellbeing and preparation for the future, and *homo oeconomicus*, seeking to maximise his personal wellbeing. One of the keys to contemporary anthropological tension can be summarised by the tension between 'individual preferences' and 'interest in justice and sustainability' (Faber et al., 2002, p. 326).

For Boutinet and Bréchet, the contemporary period is marked by predominance of the logic of profit over the logic of the project. Given this observation, it is essential to rebalance these logics in their 'oppositional complementarity' (Boutinet & Bréchet, 2014, p. 46). One of the challenges is to combat the alienating presentist timescales of profit (marked by urgency and ephemerality) to develop the emancipating timescales of the project. In a fluid and evanescent present (Baumann, 2013) where actors struggle to anchor their existence, integration over a period of time involved in the project is particularly structuring and necessary. Boutinet and

Bréchet's analysis highlights the contemporary tension between *homo oeconomicus* and *homo collectivus* – though they do not state it in these terms. In proximity to the oppositional complementarity of these binary logics, Nyborg constructed an anthropological and economic model interpreting the individual as experiencing tension between 'personal welfare functions' that 'are applied in contexts where the individual sees himself as a consumer' and 'subjective social welfare functions' that 'are used when the role of the citizen is seen as most relevant' (Nyborg, 2000, p. 305).

10.5.3 Homo Oeconomicus and Homo Religatus

Aristotle articulates the *zôon politikon* with the *zôon koinonikon* (community animal) and the *zôon oikounomikon* (family animal) (Aristotle, *Politics* I, 2[1242a]). In some respects, *homo religatus* and *socius* refer to the way in which man is able to carry on close relationships in the community and the family. It is a relational modality of the *zôon koinonikon* and the *zôon oikounomikon*. It is important to note this historical singularity of the totalising dimension of *homo oeconomicus* in the contemporary period (which does not appear in this form among the ancients[61]). In fact, as we have already mentioned, it seems necessary, in the contemporary period, to undertake a 'profound existential reflection on the human condition' (Arnsperger, 2010a, b, p. 40). The conflicting logics between *homo oeconomicus* and *homo religatus* refer, in part, to the tension between self-interest and selflessness. In his economic work, Amartya Sen has, for example, extended the *homo oeconomicus* model by integrating moral questioning and feelings into the model explaining actors' activity. A challenge at present is to think about what sustains contemporary individuals in their sense of existence. *Homo religatus* allows us to identify some of the limits of the contemporary *homo oeconomicus*, which presents itself as a single contiguous entity, removed from the relational sphere.

10.5.3.1 Educating for the 'Birth of New Men' (Arendt, 1983, p. 278)

In Arendt's thinking, the foundation of action is natality, that ceaseless renewal of the possibility of a beginning, of a newness in our common world: 'The miracle that saves the world, the realm of human affairs, from normal, 'natural' ruin, is ultimately the fact of natality, in which the faculty of action is ontologically rooted. In other words: it is the birth of new men, the fact that they begin anew, the action of which they are capable by birthright' (Arendt, 1983, p. 278). It is the consequence of this birthright that is explored in this third section, which focuses on what can be

[61] The logic of *hybris*, on the other hand, was already well identified.

learned in order to strengthen the political in the Anthropocene and enable the human adventure to survive in this new geological epoch. Arendt goes on to state that 'Only the full experience of this capacity [natality] can grant human affairs faith and hope' (Arendt, 1983, p. 278). In reaction to the darkening of the future generated by the awareness of the entry into the Anthropocene, it is advisable to explore without reservation the paths to hope, among which are education and pedagogy, in connection with the miracle of natality. The control of *hybris* appears, here, as one of the fundamental political issues of education. It is mainly to this that Part III of this book is devoted.

Bibliographical References

Arendt, H. (1983). *Condition de l'homme moderne*. Calmann-Lévy (original edition 1958, French translation 1961), French translation.

Arendt, H. (1995). *Qu'est-ce que la politique?* Seuil (original edition 1993), French translation.

Arendt, H. (2002). *Les origines du totalitarisme III. Le système totalitaire*. Seuil (original edition 1948), French translation.

Arendt, H. (2016). *La vie de l'esprit*. PUF (original edition 1978), French translation.

Arendt, H., & Heidegger, M. (2001). *Lettres et autres documents 1925–1975*. Gallimard (original edition 1998), French translation.

Arnsperger, C. (2004). L'inestimable valeur de l'autre: argent, altérité et socialité. In M. Drach (Ed.), *L'argent* (pp. 33–59). La découverte.

Arnsperger, C. (2010a). Changer d'existence économique: enjeux anthropologiques de la transition du capitalisme au post-capitalisme. *Revue d'éthique et de théologie morale, 258*, 23–50.

Arnsperger, C. (2010b). Monnaie, dette et croissance sans prospérité: portée et limites du 'tournant' jacksonnien. *Etopia, 8*, 109–116.

Arnsperger, C. (2011). Dépasser le capitalisme, mais par étapes. *Revue Projet, 324–325*, 73–81.

Baudouin, J. (2002). *Les idées politiques contemporaines*. PUR.

Baumann, Z. (2013). *La vie liquide*. Fayard. French translation.

Bergier, B. (2014). *Comment vivre ensemble?* Chronique Sociale.

Berque, A. (2004). Ce qui fonde l'éthique environnementale. *Diogène, 207*, 3–14.

Berque, A. (2009a). Les travaux et les jours. Histoire naturelle et histoire humaine. *L'Espace géographique, 38*, 73–82.

Berque, A. (2009b). *Ecoumène – Introduction à l'étude des milieux humains*. Belin.

Berque, A. (2010). Logique des lieux de l'écoumène. *Communications, 87*, 17–26.

Berque, A. (2015). Mythologie de l'urbain diffus. *Annales de géographie, 704*, 351–365.

Berque, A. (2016). La relation perceptive en mésologie: du cercle fonctionnel d'Uexküll à la trajection paysagère. *Revue du MAUSS, 47*, 87–104.

Boudou, B. (2012). Eléments pour une anthropologie politique de l'hospitalité. *Revue du MAUSS, 40*, 267–284.

Bourg, D. (2009, December). L'impératif écologique. *Esprit*, 59–71.

Bourg, D. (2013). Dommages transcendantaux. In D. Bourg, P.-B. Joly, & A. Kaufmann (Eds.), *Du risque à la menace, Penser la catastrophe* (pp. 109–126). PUF.

Boutinet, J.-P. (1990). *Anthropologie du projet*. PUF.

Boutinet, J.-P., & Bréchet, J.-P. (2014). *Logiques de projet, logiques de profit*. Lyon.

Bréchet, J.-P., & Prouteau, L. (2010). A la recherche de l'entrepreneur. Au-delà du modèle du choix rationnel: une figure de l'agir projectif. *Revue française de socio-économie, 6*, 109–130.

Bruneteau, B. (2014). *Les totalitarismes*. Armand Colin.

Cabanes, V. (2017). *Homo natura – En harmonie avec le vivant*. Buchet Chastel.

Caillé, A. (2008). *Théorie anti-utilitariste de l'action. Fragments d'une sociologie générale*. La Découverte.
Cornu, L. (2007). Confiance, étrangeté et hospitalité. *Diogène, 220*, 15–29.
de Singly, F. (2003). *Les uns avec les autres. Quand l'individualisme crée du lien*. Armand Colin.
Delannoy, I. (2017). *L'économie symbiotique – Régénérer la planète, l'économie et la société*. Actes sud.
Demonet, M.-L. (2005). Quelques avatars du mot 'politique' (XIVème-XVIIème siècles). *Langage et société, 113*, 33–61.
Derrida, J. (1997). *Adieu (à Emmanuel Lévinas)*. Galilée.
Donegani, J.-M. (2013). Crise de l'Occident, crise du christianisme, crise de la différence. *Recherches de Science Religieuse, 101*, 351–376.
Dubet, F. (1994). *Sociologie de l'expérience*. Seuil.
Dubet, F. (2002). *Le déclin de l'institution*. Seuil.
Dubet, F., & Wieviorka, M. (Eds.). (1995). *Penser le sujet autour d'Alain Touraine*. Paris.
Faber, M., Petersen, T., & Schiller, J. (2002). Homo oeconomicus and homo politicus in Ecological Economics. *Ecological Economics, 40*, 323–333.
Flahault, F. (2005). Vers une nouvelle pensée sociale. *Revue du MAUSS, 26*, 377–382.
Flahault, F. (2008). Comment l'homme peut-il être à la fois égoïste, bon et méchant? *Revue du MAUSS, 31*, 307–317.
Flahault, F. (2013a). Pour une conception renouvelée du bien commun. *Études, 418*, 773–783.
Flahault, F. (2013b). L'homme fait-il partie de la nature? *Revue du MAUSS, 42*, 125–128.
Flahault, F. (2018). *L'homme, une espèce déboussolée – Anthropologie générale à l'âge de l'écologie*. Fayard.
Foray, P. (2001). Hannah Arendt, l'éducation et la question du monde. *Le Télémaque, 19*, 79–101.
Frogneux, N. (1996). Hans Jonas développe-t-il une anthropologie arendtienne? *Revue philosophique de Louvain, Tome, 94*(4), 677–686.
Gauchet, M. (1985). *Le désenchantement du monde*. Gallimard.
Goetz, B., & Younès, C. (2009). Hannah Arendt: Monde – Déserts – Oasis. In T. Paquot & C. Younès (Eds.), *Le territoire des philosophes* (pp. 29–46). La Découverte.
Gotman, A. (1997). La question de l'hospitalité aujourd'hui. *Communications, 65*, 5–19.
Kearney, R. (2001). L'autre et l'étranger: entre Derrida et Ricœur. In J.-F. Mattéi (Ed.), *Philosopher en français* (pp. 95–111). PUF.
Lambert, F. (2001). *Introduction à l'histoire des idées politiques*. Armand Colin.
Lussault, M. (2013). *L'avènement du Monde – Essai sur l'habitation humaine de la Terre*. Seuil.
Lyotard, J.-F. (1979). *La condition postmoderne*. Editions de minuit.
Menger, P.-M. (2003). *Portrait de l'artiste en travailleur, Métamorphoses du capitalisme*. Seuil.
Mounier, E. (2001). *Le personnalisme*. PUF. (original edition 1950).
Nyborg, K. (2000). Homo Economicus and Homo Politicus: interpretation and aggregation of environmental values. *Journal of Economic Behavior & Organization, 42*, 305–322.
Pelluchon, C. (2017a). L'éthique des vertus: une condition pour opérer la transition environnementale. *lapenseeecologique.com, PUF, 1*(1), 1–18.
Pelluchon, C. (2017b). *Manifeste animaliste – Politiser la cause animale*. Alma.
Peyrelevade, J. (2005). *Le capitalisme total*. Seuil.
Pierron, J.-P. (2012). L'indignation. *Etudes, 416*, 57–66.
Pierron, J.-P. (2013). Au-delà de l'anthropocentrisme: la nature comme partenaire. *Revue du MAUSS, 42*, 41–48.
Pierron, J.-P. (2014). *Mythopées*. Vrin.
Pierron, J.-P. (2016). L'écologie, lieu d'une renaissance spirituelle. *Revue Projet, 354*, 84–89.
Poché, F. (2003). *Penser avec Arendt et Lévinas*. Lyon.
Poché, F. (2013). (entretien avec Piolot, F.), *A-t-on encore le droit d'être fragile?* Chronique Sociale.
Portier, P. (2009). La modernité libérale à l'épreuve de la théorie politique. *Revue française de science politique, 59*, 813–820.

Servigne, P., & Chapelle, G. (2017). *L'entraide – L'autre loi de la jungle*. Les Liens qui Libèrent.
Siebenhüner, B. (2000). Homo sustinens als Menschenbild für eine nachhaltige Ökonomie. *sowi-online, 1*, 1–13. www.sowi-onlinejournal.de/nachhaltigkeit/siebenhuener.htm
Taminiaux, J. (2007). La déconstruction arendtienne des vues politiques de Heidegger. *Cahiers Philosophiques, 111*, 16–30.
Tassin, E. (2017). *Le trésor perdu – Hannah Arendt, l'intelligence de l'action politique*. Klincksieck.
Touraine, A. (1984). *Le retour de l'acteur*. Fayard.
Weber, A. (2007). *Alles fühlt: Mensch, Natur und die Revolution der Lebenswissenschaften*. Berlin Verlag.
Weber, A. (2016a). *Enlivenment. Eine Kultur des Lebens. Versuch einer Poetik für das Anthropozän*. Matthes und Seitz.
Weber, A. (2016b). *The biology of Wonder – Aliveness, Feeling, and the Metamorphoses of Science*. New Society Publishers.
Weber, A. (2017). *Sein und Teilen – Eine Praxis schöpferischer Existenz*. Transcript Verlag.
Weber, A., & Kurt, H. (2015). *Lebendigkeit sei! Für eine Politik des Lebens. Ein Manifest für das Anthropozän*. Klein Jasedow, Think Oya.

Part III
Conviviality as a Paradigm of Political Education

Chapter 11
Learning Convivial Citizenship in the Anthropocene

Abstract The Anthropocene is characterised by the weaving together of certainties and uncertainties, in relation to both geoscientific and political/social data. The reliable geoscientific data pertain to what we identify as the long term (i.e. at least the next few millennia): the climate will be warmer; the atmosphere denser in CO_2 and methane; the oceans more acidic and covering a greater surface area; and consequently, there will be less land, meaning the ecumene (the space on Earth habitable to humans) will be reduced; biodiversity noticeably diminished; heatwaves so severe that conditions become inhospitable to human life will be more frequent and more intense; and so on. The still uncertain data pertain to a change in the global state of the biosphere and a break in its systemic organisation, leading to the reorganisation of lifeforms; global food production potentially in jeopardy; the extent of the destruction of biodiversity; the level at which Earth's surface temperature will eventually stabilise; and so forth.

Keywords Anthropocene · Convivial citizenship · Education in the Anthropocene · Political education

11.1 Education in the Anthropocene: Between Changing Nothing and Changing Everything

11.1.1 An Interleaving of Certainties and Uncertainties

Barnosky et al. (2012) highlight the percentage of ecosystems which have been directly transformed by human action, and the possible effect of tipping the biosphere from one state to another when a certain threshold is crossed. This threshold is estimated at between 50% and 90% of direct transformation of ecosystems, taking account of the observations that can be made in the context of changes in the state of local ecosystems. Thus, these authors identify a potential critical transition

© The Author(s), under exclusive license to Springer Nature Switzerland AG 2023

N. Wallenhorst, *A Critical Theory for the Anthropocene*, Anthropocene – Humanities and Social Sciences, https://doi.org/10.1007/978-3-031-37738-9_11

on a global scale.[1] There are many uncertainties surrounding the other type of state towards which the biosphere could shift. Barnosky et al. estimate that the probability of a new global state of Earth's biosphere is high (2012, p. 55). Our proximity to this tipping point is still uncertain, and these authors call for further research to identify indicators for this global state change. To date, 43% of terrestrial ecosystems have already undergone major transformations, and in the view of Barnosky et al., it is 'highly plausible' that in decades or centuries to come, 'if it has not already begun', we shall see a global state change of the biosphere (2012, p. 57).

In relation to political/social matters and the organisation of human societies, again, there is much that is known and much that is not. We can say for certain that the dominant economic model of globalised postmodern societies is not sustainable beyond a few decades, and that the few elements of collapse that were experienced at the beginning of the twenty-first century, with the subprime mortgage crisis and the collapse of the American bank Lehman Brothers, are bound to worsen in decades to come. We also know that the twenty-first century will be marked by a series of population migrations as it becomes impossible to remain living in certain geographical areas. It is also a given that access to drinking water will be a problem for many human beings. What we have not yet fully identified is the extent of the political and social changes: will 75% of the human population be exposed to life-threatening heat levels (Pal & Eltahir, 2015)? Will global food security be jeopardised? Will we experience the widespread collapse of our economic system and a series of wars for survival across the globe?

As has been stated many times since the beginning of this book, the Anthropocene links the long-term history of the Earth with the short-term history of the human adventure. When we look at these geoscientific and politico-social issues from an educational perspective, we can identify two main orientations that coexist. The first is the idea that we should not change much and simply go on living. Faced with the possibility of a disaster of unprecedented proportions, is the most important thing not for parents to continue playing with their children – whether it is at handball, Mario Kart, or Monopoly? Faced with an uncertain future, should teachers' priority not be to ensure that 'little Leo' and 'little Lea' learn at least to read and write? What can we expect from a teacher and researcher – in psychosociology, fluid mechanics or medicine – who is aware of the seriousness of the catastrophe we face in the Anthropocene? Is it not to continue teaching so that students obtain a qualification and fit into our globalised societies, even if those societies are fundamentally marked by destructive economic *hybris*? What is more important for an adult educator than to provide help and guidance to one of their trainees in their search for a vocation, thus focusing on others while at the same time finding fulfilment in their work? The four figures of the educator mentioned here allow life to be lived in a human and dignified way. We perceive that the Anthropocene does not alter the function of the

[1] Humans have altered landscapes and even 'natural' areas are being affected. Forty percent of the earth's ecosystems have been transformed into agricultural land. Today, there are no longer any areas on Earth not impacted by human activity, as the chemical make-up of both the oceans and the atmosphere have been altered. Even ecosystems that are not directly affected by human activity are being impacted indirectly.

educator – fortunately. There are things that must continue: playing with one's parents; developing independence for one's future journeys in the world through reading, 'riting and 'rithmetic; developing skills and finding a place on Earth among humans; constructing meaning in one's existence throughout one's life.

The second conception of the way forward is radical – but can coexist with the first. The Anthropocene is characterised by a break with the previous geological epoch, the Holocene, in terms of systemic elements that are perceptible in every corner of the globe (alteration of the bio-geochemical cycles of carbon, nitrogen and phosphorus; modification of the chemical composition of the atmosphere; modification of the climate; unprecedented increase in the rate of extinction of living species; etc.). Just as the Anthropocene is characterised by this systemic change from the Holocene, so education in the Anthropocene must be characterised by radical changes from education as it was in the Holocene. These changes may be of the same magnitude as those that characterised the changes between the modes of knowledge transfer in effect during the Pleistocene (2.58 million years to 11,700 years before present) and the modes of knowledge transfer that developed during the Holocene (11,700 years before present to the present day), which were characterised by the emergence of formal education. Part III of this work deals with the characteristics of education in the Anthropocene, consisting of guiding the necessary anthropological shift outlined in Part II.[2]

11.1.2 Moving Beyond the Paradigm of Education for Sustainable Development

The Centre for Educational Research and Innovation (CERI), one of the bodies of the Organisation for Economic Co-operation and Development (OECD), invested in environmental education in the 1970s in a rather liberal way as a field of educational innovation. In 1972, UNESCO developed the 'Man and the Biosphere' programme, in particular in conjunction with the International Union for Conservation of Nature

[2] As mentioned in the introduction, Part III puts to work an anthropology of educability, in opposition to the anthropology of ineluctability which can sometimes be seen behind the words of Earth System Science researchers. The conception of education employed in our research has the direct function of enabling humans to become more human (Lamarre, 2002b). This understanding of education is in line with the thinking of two German authors, steeped in the notional universe of German thinking and its humanist enlightenment, aimed at fully realising man's humanity: Fichte, with his emphasis on education, and Arendt, for whom man is defined by politics. Arendt's thinking, however, is at odds with German humanist idealism. It is through education and politics that we become human, it being understood that the objective of education is political, involving participation in the world and assumption of responsibility for the world. The human being is educable and of education – he is shifting, and comes into being through the changes wrought by education; he is *homo educandus*. (In the Greek idea of *paedeia*, education has the function of enabling man to become human: 'You have been instructed in these activities not for the acquisition of a technique but for your education, as befits a private individual and a free man' (Dialogue from Socrates to Hippocrates, described in *Protagoras* (Plato, −399/−390, p. 72).)

(IUCN), which had previously developed expertise in environmental education. In 1972, at the United Nations Conference on the Human Environment held in Stockholm, environmental education was held up as an international priority (Sigaut, 2010–2011, pp. 60–64).

Education for sustainable development, education for the environment and environmental education are of considerable importance for education, because they propose a different way of thinking about education which, in addition to the objective of ecological preservation, aims to enrich life and give it meaning (Brière et al., 2010–2011, p. 251). Environmental education also has the benefit of including a political component, as it contributes to the shaping of responsible citizens (Sauvé, 2009).[3] In the *Dictionnaire de la pensée écologique* (Bourg & Papaux, 2015a), education is identified as one of the themes of ecological thinking, with two of the 350 entries being devoted to it: 'Environmental education' and 'Education for sustainable development'. These view environmental education as a matter of 'learning to "live *here* together"' (Sauvé, 2015, p. 376). These two notices set forth the fundamental educational issues, but they do not mention the paradigm shift that results, for education, from the planet's entry into the Anthropocene. Yet the Anthropocene is shaking up educational paradigms that need to be recast. By working differently on a problematisation of the relationship between education and the environment, based on education in the Anthropocene, we propose to refocus the debate and move past the tensions around education for sustainable development (integrating economic and sustainable development issues), education for the environment, or environmental education (integrating more educational and social issues). Indeed, the Anthropocene undermines some of the fundamental paradigms on which western education has been based since the Enlightenment. The entry into the Anthropocene can be viewed as a revelation of our educational foundations and allow us to think differently about education. What may have been valuable a few decades ago may become obsolete in the light of the Anthropocene.

Major changes are needed in the way we think about education. The Australian physicist David Shearman and the Australian philosopher Joseph Wayne Smith in their educational and formative thinking incorporate the depth of the long-term structural problems, but in a way that raises questions. They believe that citizens' ignorance is a problem that cannot be solved without the emergence of a new kind of elite, particularly cognisant of environmental issues. They therefore propose to create a new type of university – 'real universities' – focused on the future and preparing for it by training 'eco-elites'. They follow the eco-authoritarian tradition, integrating the institution of the university into it, and believe that the salvation of humanity will come from science rather than from politics (Eckersley, 2017, p. 6). The approach worked on in this section rests on entirely different foundations to Shearman and Smith's model. It is more in line with the approach worked on by Arnsperger in his ethics of a post-capitalist existence. Arnsperger identifies capitalism – which, in this work, is seen as shouldering much of the responsibility for the

[3] The great merit of developmental education, as described by Lucie Sauvé Professor of Environmental Education at the University of Quebec in Montreal, or eco-training (Pineau, 1992; Taleb, 2016) is that its explicit goal is to contribute to the re-enchantment of the world.

entry into the Anthropocene – as a historically constructed 'culture' from which it is possible to break away without resorting to totalitarianism (Arnsperger, 2003, 2005a, c, p. 481). Rosa, for whom 'Capitalism is not a given of nature!' and who reminds us that there have been non-capitalist civilisations (Rosa & Wallenhorst, 2017, p. 5), corroborates Arnsperger's analysis. This cultural component of capitalism highlighted by Arnsperger is an important fact to consider from an educational perspective. Since capitalism boils down to cultural practices based on the idea that individual freedom is sacrosanct – an idea that can be discarded – it must be possible to put these practices to the test through democratic political experiments that allow new social norms to emerge (Arnsperger, 2005b: 480, 2006b). In this way, educational work can contribute to the '*political* reflection on the *democratic* construction of *social* norms', allowing us 'to discover what kind of human beings we are capable of being' (Arnsperger, 2005b, pp. 486–487). How can educational action contribute to the sharing of existence within a common world? How can education play a part in an anthropological consolidation of politics?[4]

The data from the Earth System Sciences, presented in Part I of this book, brought to bear anthropological notions such as that of boundaries and Prometheism – understood as the quest for power through transgression (and in particular, transgression of the boundaries which create a safe space for the human adventure to exist). These elements led to the question of a different type of power, rooted in finitude in relation to the sharing of existence with others (i.e. coexistence). Part I, which highlighted the centrality of individualism in how contemporary globalised societies function, ended with a questioning of the political, that space which exists between humans. This first part highlighted the need to think about post-Promethean ways of sharing existence within the earth, our common home. The space of the 'between-us' (distinct from the '*inter se*') was worked on in Part II. The proposed break with Prometheism is not a break with technological development, nor even with the idea of the augmented human, but with this drive to obtain power by transgressing boundaries. What lessons must be learned, given the entry into the Anthropocene, to allow the human adventure to endure? What kind of education for politics is needed?' These are the questions around which Part III of this book is organised. With this in mind, we offer a critical discussion of Daniel Curnier's thesis in Environmental Sciences, defended in 2017 and already mentioned in this work: *Quel role pour l'école dans la transition écologique? Esquisse d'une sociologie politique, environnementale et prospective du curriculum prescrit*. From there, we put to work the notion of citizenship upon which Curnier's work hinges. What type of citizenship does we need to promote in the Anthropocene? Here, we propose a notional dialogue between the French universe of citizenship (*citoyenneté*) and the German one of *Bildung*, drawing on an 'existential' register.[5] This will support the proposal of the notion of convivial citizenship as one of the necessary lessons in the Anthropocene.

[4] While this normative reflection is necessary, it is by no means sufficient for the efforts of which it is part. A range of other means of action are needed in terms of international relations, a change in national constitutions, European treaties binding on companies, etc.

[5] The adjective 'existential' here is part of the German intellectual tradition – not of Sartrean existentialism.

11.2 Curnier's Work on the Role of the School in the Ecological Transition

11.2.1 Curnier's Prescribed Curriculum

Curnier's educational thinking is part of the trend towards education for sustainable development based on strong sustainability. In his prospective thinking, he often refers explicitly to the notion of the Anthropocene. In his work, Curnier weaves together the Anthropocene, prospective educational thinking and the conception of politics. As this type of interweaving is particularly rare, it is interesting to begin this section with a critical dialogue with that author's ideas. Through the discussion of Curnier's prospective proposal, we can identify the points of convergence and divergence with the conceptual modelling presented in *this* book. With other theoretical references, Curnier employs the same understanding of politics as ours. Indeed, he considers that one of the missions of the school is to promote the emergence of a society based on a principle of strong sustainability,[6] through educating its students for citizenship.

Curnier's research is highly stimulating. It proposes knowledge to be transmitted in the Anthropocene in order to allow for the transition to a type of society based on strong sustainability, through the modelling of a prescribed curriculum. This is defined as the educational policies 'that set the framework and structures within which the student's experience will unfold throughout his or her journey' (2017, p. 20),[7] enabling the school to contribute to the ecological transition. The Anthropocene brings with it a set of challenges that necessitate a radical transformation in the organisation of contemporary societies. Curnier starts from the observation of the entry into the Anthropocene, and discusses the way in which the schooling system can be a key actor with a view to bringing about sustainable development with reference to the principles of strong sustainability – i.e. taking note of the extent of the systemic changes of the Earth related to the entry into the Anthropocene. In his view, it is a question of thinking about a new model of society with the education system as part of its foundations: 'This transformation of the system of representations and values that dominates contemporary societies cannot be carried out without considering the role that education plays in it' (p. 23). Curnier lays down the foundations for an alternative educational system in line with the epistemology of the Environmental Humanities.[8]

---

[6] For Curnier, strong sustainability 'requires respect of ecological limits to the development of human activities' (2017, p. 192), in line with what was proposed in our Chap. 3. This choice of strong sustainability 'therefore means the rejection of productivism, whether of the liberal or planned type' (p. 192).

[7] The prescribed curriculum differs from the actual curriculum, which is the way in which the actors actually put the prescribed curriculum into practice or not.

[8] Daniel Curnier, in his prospective work of defining an educational system, identifies the paradigm of modernity from which we need to distance ourselves by the following set of characteristics: 'rationalism, reductionism, anthropocentrism, the ideology of progress, the specialisation of knowledge, the subject-object duality and the domination of technology' (2017, p. 112).

Curnier's prospective work is deliberately radical (p. 189). He has a high opinion of the possible utility of school in the transition necessary to guarantee the human adventure's future survival in the Anthropocene: one of the functions of school is to ensure a future for our societies on the basis of the learning instilled in the pupils. After a prospective reflection, Curnier analyses data from an empirical field: the Swiss educational system of the canton of Vaud, for pupils aged 12–15.

11.2.2 A Critique of the Neoliberal Dimension

Curnier identifies a failure of sustainable development since the Rio Conference in June 1992: 'This failure can be attributed, in particular, to the contradiction between the declared objectives and the political and economic context of recent decades, which is strongly marked by the values of neoliberal ideology. The contemporary societal project is therefore neither favourable to the ecological transition nor to the implementation of ESD [education for sustainable development]' (2017, p. 369). It has become a necessity to leave the neoliberal paradigm behind: 'A transformation of the school system contributing to the ecological transition can therefore only be envisaged on the basis of a paradigm shift affecting the whole of society' (p. 388). Curnier's educational conceptions, anchored in a model of strong sustainability, are at odds with the usual understanding of sustainable development, which he identifies as a form of legitimisation of the paradigm of modernity, marked by a Cartesian vision of the world and the productivist conception of development' (p. 97). Curnier states bluntly that the role of the school 'will be to contribute to a change of paradigm' (p. 125). Here, we see the importance of the political purpose of education. Education is a political means of choice for bringing about societal change.

Curnier, before thinking about the question of content (prescribed curriculum), thinks about the attitudes and behaviours necessary to base our societies on strong sustainability. Curnier's reflection on the prescribed curriculum has political and existential depth that goes beyond classical disciplinary knowledge. Curnier's approach stands in opposition to a conception of education whose goal is to make individuals ready to enter the job market – i.e. preparing them for integration into the contemporary neoliberal market. In this respect, Curnier shows how institutions such as the European Commission and the OECD, in their policy texts, employ educational conceptions 'explicitly guided by the imperatives of global competitiveness rather than by the goals of responsible citizenship' (p. 276). Here, we see how reference to the PISA study produced by the OECD is problematic. This standardisation of education systems on the basis of skills acquired at a certain age (15 in this case) for the purpose of optimising professional integration in a global job market is particularly problematic from the point of view of the development of another type of fundamental skills in the Anthropocene, as opposed to the contemporary neoliberal job market. These orientations are partly contradictory to those of

UNESCO.[9] Here, Curnier highlights the quality of a set of texts from UN organisations such as UNESCO from the point of view of the learning required to address the challenges we face in the Anthropocene. These assessment practices of the OECD, with their focus on the utilitarian component of schooling and competition between countries, greatly hamper the school's ability to contribute to the (absolutely necessary) ecological transition. It is regrettable that there is not yet a counterpart to the PISA study, offered by UNESCO and based on foundations other than the race for economic competitiveness, such as the emergence of responsible and hospitable citizenship.

11.2.3 Citizenship Learning as a Goal of Education in the Anthropocene

Given the seriousness of the issues of sustainability and the challenges posed by the Anthropocene, learning about citizenship throughout schooling and up to graduation appears to be an important goal. For Curnier, education for citizenship 'should be the goal of all disciplines' (p. 295). Learning for citizenship is thought of in terms of three forms of education: education for citizenship, education for global citizenship and education for sustainable development, against a background of strong sustainability. Thus, Curnier develops a political conception of education for sustainable development, far from the simple standardised learning of eco-actions, but subversive and at the heart of the advent of a different way of living together in the world[10]: 'In the framework of ESD, in concrete terms, it is a debate between a transformation of behaviours integrated in a mechanical way and the goal of an emancipating citizenship based on democratic principles' (Curnier, 2017, p. 161).

Curnier expounds upon four central concepts in the definition of school content: the Anthropocene, development, the biosphere and socio-economic systems (while also ensuring that solutions for living in the Anthropocene are transmitted).[11]

---

[9] Curnier highlights the illusory nature of the hypothetical axiological neutrality of educational systems. To the question of whether it is possible to have educational concepts that are not axiologically neutral, Curnier replies that it is 'misguided to claim that education is politically neutral. Nevertheless, an educational project can take place with respect for religious, moral, and political convictions, as long as its political intentions are not glossed over and space for citizen debate takes the place of the imposition of a predefined order' (2017, pp. 291–292).

[10] In his work, Curnier signifies the need to identify 'measures of consensus' (2017, p. 27). This search for consensus in Anthropocene is particularly interesting. Indeed, there is strong temptation, perceptible among Earth system researchers, to decide for others in the name of scientific expertise, thus negating any political component. While the search for consensus is important, it is particularly demanding because a future seems difficult to achieve without a certain degree of radicalism in changing the organisation of contemporary societies, and the actors must have possession of the structuring knowledge of the long term when making the decisions.

[11] Through the transmission of this type of knowledge, the aim is to acquaint students touch with the question of the future, both from the point of view of the challenges to be met and the solutions that could be implemented.

However, beyond the knowledge to be transmitted, Curnier addresses the end goal of the relationship to knowledge, which is primarily political, with the learning of citizenship[12]: 'The learning of critical, complex, prospective and transformative thinking, as well as ethics and the concrete implementation of actions, are in fact part of the civic and humanist conception of education, rather than tools for specifically grasping the challenges of sustainability' (p. 217).[13] As Curnier mentions, 'the challenges of the Anthropocene require a collective political effort that goes far beyond the incorporation of behavioural changes at individual level' (p. 353). Indeed, it is precisely for such collective political mobilisation that education in the Anthropocene must be put to work. It must be attentive to 'the individualistic vision of self-fulfilment' present in personal development activities that allow each person to better seek individual happiness, which is found 'in certain pedagogical discourses centred on the needs of the student' (p. 353). This component is very much present and recurs regularly in citizenship education.[14]

Curnier's educational ideas are close to those developed in our work. He writes: 'Acting on attitudes from a citizen's perspective does not mean moulding minds and ramping up the practice of eco-actions in schools, but rather enabling students to understand the world and the challenges facing humanity in the 21st century. It also means encouraging them to reflect on the ethical and philosophical dimensions of the issues at stake and developing their capacity to decide and act' (p. 162). He then goes on to say: 'There are two opposing trends in the work of defining ESD. The first considers that the student must be 'shaped' to become a docile social agent. The second postulates that the acquisition of skills such as critical thinking and taking action will lead to the emergence of socially and environmentally responsible citizen-actors'. The substance of Curnier's approach is reformative and transformative, with a certain radicalism. The shift in education that he proposes breaks with the usual conformist approaches, through his critique of disciplinary knowledge and his proposal of education based on interdisciplinary knowledge whose aim is to help bring about a societal transformation. For Curnier, part of the knowledge worked on at school should concern socially relevant issues – i.e. those elements that are debated in the public arena and that affect political issues. In this context, socially relevant issues are presented in opposition to the decontextualised discipline-based segmentation of knowledge usually conveyed at school.

[12] Based on his work on the prescribed curriculum, with a strong interdisciplinary and contextualised component, Curnier identifies the emergence of the political vocation of citizenship as an aim of schooling.

[13] His understanding of knowledge is interesting, because it is not seen as the only medium for students' relationship with the world, but also includes experiential, emotional and cultural components (p. 203).

[14] He even – albeit rarely – addresses the collaborative component, which we understand as one of the keystones in our work: 'In the context of a paradigm shift driven by the challenges of the Anthropocene, however, learning to collaborate is essential for collective decision-making' (p. 216). In contrast, Curnier does not really make explicit the need to learn political thinking.

11.2.4 Extending the Paradigmatic Breaks Outlined

Curnier, in his normative and prospective approach, focused on the knowledge to be inculcated, whereas we reflect on the learning to be achieved. While we have asked the question 'What must be learned in the Anthropocene?' to enable the human adventure to endure, Curnier asks the question 'What must we teach in the Anthropocene?'

Curnier conceives of education as having a political purpose, contributing to the advent of a society based on foundations other than the logic of maximising individual interests. Curnier's transformative approach aims at emancipatory socialisation for students, defined from the perspective of learning citizenship, with ethical debate as a favoured educational tool (p. 178). In order to conceive of education for sustainable development based on strong sustainability, Curnier organises the prescribed curriculum around the objective of acquiring six operative capacities that he takes from Pierre Varcher: critical thinking, complex thinking, prospective thinking, transformative thinking, ethics and the transition to action. The six proposed learning skills are interesting, clear and important. However, it seems to us that a certain perspective is missing, manifesting the break with the dominant developmental paradigm in education, which extends economically into the unlimited quest to maximise individual interests and the pursuit of growth. In our research, the lessons identified as referring to this other type of paradigm are the learning of hospitality and the learning of responsibility.[15] These two types of learning can be grouped together under the term 'post-Promethean'. insofar as they do not pursue a search for power based on transgression (of the link to others, to nature, to the community, etc.). The characteristic of these two types of learning is that they do not add something to the learner, but, in a certain way, take away something which allows them to relate to others outside of themselves, allowing the other (rather than the learner) to have access to speech, to exist in the world and to act in it.[16]

Curnier's schooling system is not thought of as a sanctuary of the Republic, sheltered from social division, but instead as 'a place where academic knowledge, personal convictions, and public controversy inevitably come together' (p. 172). Thus, our author mentions that, 'rather than shutting down controversial issues (a posture of absolute neutrality), the teacher should engage students in debate by adopting a posture of engaged impartiality' (p. 172). On the other hand, the aim is 'that students learn to form a personal constructed opinion'. Though we are not opposed to this aim, it seems to us, on the other hand, that this should be done against a background of primacy given to learning to listen to, and be hospitable, to others.

[15] Learning to listen is particularly important. It allows for educational hospitality and is at the heart of the ability to exercise responsibility suited to the situation. The idea will be discussed in depth in this Part III.

[16] Curnier, who regularly draws upon Edgar Morin's *Enseigner à vivre – Manifeste pour changer l'éducation* (2014), does not seem to perceive the equivocal component of teaching how to live from the standpoint of the political education worked on in his thesis (teaching how to live can take the form of teaching how to be oneself).

Hospitality toward others' ideas as well as toward their existence allows for the emergence of what Arendt calls 'speech' – which is political in the true sense. As put forward in Chap. 10, the two lessons of hospitality and responsibility that can be put to work in the light of debates are oppositional to *homo oeconomicus*, and, in this sense, crucially important in the Anthropocene.

Curnier proposes to rethink the whole form of the school – namely 'the material and organisational conditions in which school learning takes place' (2017, p. 237) (disciplinary decompartmentalisation, school hours, class beyond the four walls of the classroom, organisation of classroom furniture, etc.). Though the stance is resolutely reformative and transformative, Curnier's approach, while political, remains marked by the dominant developmental paradigm of Educational Sciences, focused on the subject of learning (i.e. the learner) and not on the figure of the other or on the 'between-us'. It does not push to its conclusion the break with this developmental paradigm in order to develop a different paradigm on which to base educational conceptions in the Anthropocene.

One of the findings of both our work and Curnier's is that it is no longer citizenship that is central to participation in society, but consumption. The logic of profit maximisation, without consideration of the environmental context, is self-destructive and threatens to drive human beings to extinction (Curnier, 2017, p. 61). Hence, for Curnier, the challenge with which the Anthropocene confronts us 'consists of nothing less than redefining the human being on the one hand, and his relationship to the world on the other' (p. 60). The anthropological challenges of the Anthropocene that Curnier identifies as referring to a necessary anthropological metamorphosis constitute, in his view, 'a paradigm shift at least as important as the beginning of sedentarisation some 11,000 years ago following the spread of agriculture' (p. 66).[17] Though Curnier evokes the necessity of an anthropological change, his work does not go into detail about this dimension, contrary to what we do in this book. In this regard, the work is based on a set of authors who present political anthropologies characterised by their radicalism (Rosa, Arnsperger, Abram, Bellet and Weber), taking seriously the need for an anthropological shift. This research focuses on the anthropological changes necessary to make human existence sustainable in the Anthropocene.

To the question 'What kind of education is needed in the Anthropocene?', Curnier seems to answer that it is education for citizenship or education for sustainable development, based on the concept of strong sustainability. Curnier's reflection focuses on the prescribed curriculum – his work is about educational policies. After having discussed education for sustainable development based on strong sustainability with the question of citizenship at its heart, he questions the knowledge traditionally taught at school and, in particular, the disciplinary segmentation of that knowledge. On that basis, he develops a pointed criticism of the schooling system, which could have a fundamental role to play in the ecological transition. We agree with this; however, we believe it possible to go further and think of things in a more radical and subversive

[17] Among the fundamental anthropological principles on the basis of which to think, live and act in the Anthropocene, we note the acceptance of the impossibility of gaining total mastery of the world (Curnier, 2017, p. 116) – thus, there is a limit to our power.

way. It seems necessary to go all the way to the paradigmatic ruptures for education in the Anthropocene, such as learning that the world can speak; that it is possible to learn together and that the act of learning is not only an individual appropriation; that it is possible to think of education differently than in terms of emancipation; or that it is possible to experience conviviality with non-humans. Despite the strong similarities with Curnier's theorisation, the fundamental differences between his work and ours are as follows: taking seriously the need for an anthropological shift[18]; bringing the question of the relationship with the Earth and with the non-human into a necessary activity of restoring the humanity to humanity; basing an educational approach on the 'We' and thinking about the possibility of 'learning together' (which is not found in education); and the development of a political anthropology of education as opposed to a political sociology of education.

11.3 Citizenship and the Anthropocene

11.3.1 What Kind of Citizen Do We Want?

11.3.1.1 The 'Facing' Adult as a Figure of the Anthropocene Postmodern Adult

The Anthropocene calls into question some of the paradigms underlying conceptions of education, but also the figure of the adult in reference to which the act of education is thought of. The term 'adult' comes from the Latin *adultus*, meaning 'one who has grown up' or 'one who has completed his or her growth'. In recent decades, the field of education and training sciences has been marked by a body of work that rightly emphasises the educability of adults (Avanzini, 1996). Some works are critical of the concept of an adult and its etymological roots and lead to its withdrawal from certain educational and formative thoughts (Lapassade, 1997; Colin & Le Grand, 2008). Yet does the entry into the Anthropocene not remind us of the adult in education, in the extension of its etymology: one who has completed their biological growth and who can now focus on something other than themselves? Is the adult not thus defined by their responsibility towards those who have not yet experienced this phase of growth? Does not the Anthropocene strengthen adults' responsibility for the youngest and for future generations?

In the contemporary context of the decline of the institution (Dubet, 2002), the postmodern adult appears as one who is exposed and who must *face*[19] whatever

[18] As has been pointed out, however, Curnier does raise this question of an anthropological shift, but it is not what he focuses on in his work.

[19] This 'facing' that characterises postmodern adults has all the limitations of an expression that refers to a fictitious external entity. As mentioned, the adult is immersed in nature. What the expression 'facing' conveys is the singularity of what the adult can see *before* him or her: a decline in the conditions of habitability of the biosphere, which alters the human condition.

problem is at hand; but on the contrary, the figure of the adult has been particularly invested in other types of educational and formative thinking. Examples include Boutinet's works questioning the figures of adult life, such as *L'immaturité de la vie adulte* (1998), *Psychologie de la vie adulte* (2005), *Penser l'accompagnement adulte* (Boutinet et al., 2007), *Où sont passés les adultes?* (Boutinet & Dominicé, 2009). The adult *must* indeed face up to the threat – the 'mother of all threats' (Les convivialistes): the capacity we have to destroy ourselves – which is now taking the place of risk (Bourg, 2013) as well as the terrifying possibility of *the end of the world and of humanity* (Afeissa, 2014). The contemporary adult is returned to the limits of *homo faber*, who still remains *homo oeconomicus crescens* (Arnsperger, 2016), fascinated by his own achievements. Now, though, he finds himself forced to invent new ways of living and new ways of inhabiting the Earth together. In a context of institutional decline, and having 'completed his growth', the adult has this impossible responsibility to address a problem whilst standing within it (Martuccelli, 2002). Basically, what is in front of adults? More than ever before in history, they are facing an uncertain collective future. If the adult is in the process of transformation (Boutinet, 2006), the main changes are still to come and will be mainly due to the entry into the Anthropocene.

There is some debate as to how the contemporary period should be qualified. Is it a postmodern period (Lyotard, 1979; Boutinet, 2004), late modernity (Rosa, 2013), second modernity (Giddens, 1994), hypermodernity (Aubert, 2005)? For some Promethean authors, the Anthropocene propels the human adventure into an age of hypermodernity marked by an exacerbation of the characteristics of modernity, and in particular the linear temporalities of progress where tomorrow will be better than yesterday. On the contrary, it seems that the contemporary adult is no longer the modern adult born in the wake of the Enlightenment and pursuing progress to improve daily life. We have departed from the ascending linearity of western modernity. Beyond the doubt about the improvement of the quality of life in the future, it is the very existence of a possible future that contemporary uncertainty concerns.

The postmodern condition is marked by the collapse of the grand narratives, as highlighted by the French philosopher Jean-François Lyotard. The apprehension of the human adventure on the basis of the three components of *hybris*, the world and coexistence, also highlights the way in which the postmodern condition of the adult is marked by paradoxes within oppositional complementary facets. Time in postmodernity is thus understood between acceleration and the blocked time of delay[20]; space is characterised as being between globalisation and territorialisation; and finally, coexistence is under tension between the poles of individualisation and networking. It is these three structural transformations that the postmodern adult must face: his or her

[20] On the other hand, as the French political scientists Luc Semal and Bruno Villalba suggest, note the paradox between the omnipresence of the sense of urgency and short-term thinking in the political world, and the political playing down of the time left to reorganise human societies and attempt to ensure humanity's survival on Earth (2013).

relationship to space, time[21] and the sharing of existence. Edgar Morin has proposed complexity as a framework for thinking about the postmodern period (1990). It is no longer a question of thinking about the simple, but the complex; thus, he proposes a break with modern Cartesian thinking.[22] Today, we have difficulty thinking about complexity. Often, in trying to think of complexity we think of confusion – hence our difficulty in setting priorities. One of the challenges of thinking in the contemporary period appears to be identifying priorities for action in a complex world. It is a question of learning to think about the possibility of concerted action.

11.3.1.2 The Figure of the Citizen in Corine Pelluchon's Ethics of Virtues

To bring about other lifestyles and other ways of sharing human existence, Corine Pelluchon identifies the moral dispositions that make this possible through the notion of ethics of virtues – of which the virtue of consideration is the keystone – in an essay, *Ethique de la considération* (2018) and in an article entitled *L'éthique des vertus: une condition pour opérer la transition environnementale* (2017). What can enable citizens to change their behaviours by giving more space to future generations, other species and nature can be qualified as ethics of virtues, designating 'an approach to morality that is characterised by the fact that one is not exclusively interested in the norms and principles of action, but that the emphasis is placed on the concrete motivations of agents, on the set of representations, emotions and affects that determine their way of being' (Pelluchon, 2017, p. 3). The strength of Pelluchon's approach is that it looks at the ways in which citizens adhere to the changes necessary to enable the environmental transition; thus, she identifies the 'moral traits' favouring such a transition. In this philosophical enterprise centred on the modalities of passage from thought to action, we perceive political intentionality rooted in an existential approach. Thus, environmental issues are understood as an extension of a 'crisis of subjectivity' (p. 4).

Among these moral traits, the intellectual virtue of prudence occupies an important place. It is the foundation of moral virtues such as 'courage, temperance, liberality, magnificence, magnanimity, a just desire for glory, gentleness and friendliness, truthfulness, a sense of jest, reserve or modesty and justice' (p. 7). Prudence refers

---

[21] Postmodern adults, in the course of their lives, are regularly faced with biographical turning points which bring about moments of deceleration, some of which are chosen (return to education, travel, change of career path, or retirement), and others suffered (illness, unemployment, compulsory retirement). Here, the control of speed appears as one of the characteristics of the postmodern adult. Contemporary social acceleration may give him a feeling of control over his existence, but sometimes control escapes him. Thinking of the adult in reference to the figure of the subject, taking back control of his or her destiny is in line with the Tourainian subject (1997, 2010, 2015) or the emerging subject in biographical approaches (Delory-Momberger, 2003, 2004, 2005; Dominicé, 2007, 2016).

[22] As Descartes explicitly mentions in Part II of the *Discourse on Method*, 'The Rules of Method': 'to divide each of the difficulties I examine into as many parts as may be necessary to solve them better' (1637, pp. 137–138).

to the adequacy of the means to the end. Secondly, the implementation of an environmental transition requires the virtue of courage to see things as they are by taking the measure of the environmental damage done. For Pelluchon, this environmental transition also requires an ethic of consideration, integrating solicitude towards others and compassion towards one's neighbour (p. 8).

The virtue of moderation, referring to the avoidance of excess and deprivations, is what can allow the environmental transition to take root in a sustainable way. It goes hand in hand with sobriety, which integrates pleasure and can only be happy (p. 8). At the heart of this constellation of virtues, humility has a special place. Humility is linked to listening and its primacy over speaking. It is what allows the construction of a quality 'between-us' that favours the emergence of a common world. This constellation of virtues favours the learning of a healthy relationship with oneself, with others and with the city, underpinned by magnanimity (p. 11).

An interesting foundation of Pelluchon's work that has a bearing on the style of political education in the Anthropocene worked on in this research is that 'that there is not, on the one hand, an ethics of environmental virtues and, on the other hand, an ethics proper to the relations between human persons' (p. 9). This reinforces the importance of educational work on the relationship between us in the contemporary period marked by the Anthropocene. The ethics of virtues approach, despite misleading appearances, is not individual but collective for political purposes. The environmental transition we must achieve is not only based on an individual ethics of virtues, but on the coming together of citizens to become a political force. This presupposes a 'sense of obligation' (p. 5) in respect of pluralism that could also be defined as an ethics of responsibility.

11.3.1.3 The Figure of the Existential Citizen Described by Christian Arnsperger

Arnsperger calls for a new type of citizen, which he describes as the 'entrepreneurial citizen', whose main characteristic is to be marked by existential citizenship. This 'entrepreneurial citizen' is a 'restorer of life principles' (2011a, p. 75). The exercise of reflexivity and work on oneself – otherwise labelled anthropological experimentation by Arnsperger – can be one of the foundations of political engagement (2011b). Arnsperger's figure of the existential citizen is particularly close to Touraine's figure of the subject (1997, 2010, 2015), especially in its capacity to be involved in collective struggles. It particularly embodies the *homo collectivus* presented against the background of the existential reflexivity of *homo religatus*.

The principles of life at the basis of Arnsperger's existential citizen[23] can be divided into two categories: principles on the personal level of motivation and

[23] In this work, we will not evoke, as Arnsperger does, an existential citizen, so as not to run the risk of excessive essentialisation. Instead, we prefer to examine existential citizenship as a process. It is therefore the notion of convivial citizenship upon which we shall draw, because it is open to non-human entities as well as humans.

principles on the collective level of convention. On the motivational register, the defined life principles are: 'a principle of detachment and letting go, a principle of shared finitude, a principle of accepted dependence, a principle of seeking ontological truth (i.e. the truth of being) and a principle of care and "helping to be"' (2011a, p. 83). On the collective level of conventions, Arnsperger defines the following principles: 'a principle of broadening and deepening (which fully values the fundamental notion of growth but radically shifts its focus), a principle of fundamental creativity, a principle of mutuality, and a principle of voluntary simplicity' (2011a, p. 84). Fundamentally, Arnsperger shifts the locus of growth: from economic growth within corporations to growth within every human existence. A two-faceted condition is required to make this shift: 'enough goods and services, but not too many' (2011a, p. 86). Frugality is necessary – that is, etymologically, the 'full and fruitful use of resources', not to be confused with the depletion and exhaustion of natural resources, both anthropological and environmental. It is a matter of 'deliberately hatching a new form of citizenship within new economic rules' (2011c).

The concrete political modalities developed by Arnsperger to enable an ecological and economic transition link the issue of lifelong education with that of economic alternatives. In Arnsperger's thinking, citizens must be educated in 'convivial sobriety and egalitarian ethos' (2008, p. 99) in order to cope with knowledge of their existential precarity (2001). British economist Tim Jackson's popular book *Prosperity without Growth? The Transition to a Sustainable Economy* (2010) marked a turning point in the dissemination of ecological and economic transition thinking. However, for Arnsperger, his thinking remains unsatisfactory, as it is too social-democratic and does not sufficiently assume a complete structural overhaul of the financial system (Arnsperger, 2010). Existential citizens have analysed the contemporary economic system in which they no longer wish to participate and are experimenting with other, simpler and more cooperative ways of living (2011a, p. 87).

For Arnsperger, the acceptance of the precarity of one's life is one of the factors that hinders proper implementation of distributive economic justice. He distinguishes between the 'apparent individual freedom' that can be seen in decision-making and the 'fundamental existential unfreedom that leads people to obey a set of unconscious compulsions linked to the denial of death' (2002, p. 24). He thinks of distributive justice as including two additional components: 'the bases of self-respect not reducible to the possession of goods or income' and 'the existential bases of an assumption of self-as-mortal, not reducible to social recognition or self-respect mediated by language, codes of conduct, political correctness' (2002, pp. 13–14). That is, he adds to the conception of distributive justice the question of symbolic goods and the ways in which economic officials make decisions that promote distributive justice.

Arnsperger's approach emphasises the importance of existential learning in civic life; it is the foundation on which to build a viable future for the generations to come, and is seen as an essential part of university education. Thus – and this is of particular interest to us as we begin Part III of this work – an educational paradigm must be conceived that integrates the learning of 'the harsh existential reality that wealth, economic power and symbolic prestige are fundamentally the outlets for a vital anguish and the denial of a fragile physicality that must be assumed in a different way so as not to make others, less endowed with talents and/or initial wealth, bear the material burden. Everyone has seen, at one time or another in his or her life, how hard such a renunciation can be, if one is not endowed with the psychological capacities (which many 'powerful' people lack, and this is no coincidence) and existential resources to make this renunciation' (Arnsperger, 2002, p. 25).

Arnsperger's citizen is alternative, based on a new way of living one's human condition in rupture with the way capitalist societies live it (2011a, p. 79). The existential citizen's imperative is to break away from capitalist alienation. It is fundamentally missionary and militant: 'the pioneers of frugality will emigrate first – and some are already emigrating – to the new "frugal frontier". They will dare to initiate new forms of life, new forms of economy. [...] And in this sense, they have a very important social role: that of *initially* paving the way for others' (2006a, p. 38).

11.3.2 What Type of Citizenship Is Necessary in the Anthropocene?

Following on from the reflection on the educational policies that need to be developed in the Anthropocene and the three figures of the citizen in reference to which citizenship can be thought of, we now reflect on the type of citizenship that needs to be learned to exercise in the Anthropocene, allowing for the anthropological shift identified in Part II of this work.

For the French political scientist François Gemenne, one of the main issues of this new geological era is cosmopolitanism, insofar as the concept of the Anthropocene generates this 'false impression of a unified humanity, where all humans are equally responsible for the planet's transformations'. Yet the inequalities in the Anthropocene are glaring: 'the actions of some are responsible for the suffering of others' (2015, pp. 238–239). This observation of the extent of inequality in the Anthropocene is also made by French physicist Jean-Paul Deléage, who considers it fundamentally important in the Anthropocene to rethink the social contract within societies as well as between the countries of the Global North and the Global South (2010, p. 28). Moreover, the Anthropocene leads us to think outside the usual growthist paradigm – even if it is green – which carries with it this Promethean excess and 'disregard for the quality of relations between human beings' (Sinaï, 2013, p. 40).

It is a matter of learning a new relational *habitus* marked by responsibility and hospitality. The cosmopolitical component is at the heart of an education in politics in the Anthropocene, with the dual meaning of cosmopolitics: the welcoming of human plurality as the physical (cosmic) manifestation of the exercise of politics. It is to citizens[24] or to future citizens that education for politics is addressed. For an author like Eckersley, only a judicious practice of citizenship[25] will enable democracy to survive the challenges posed by the Anthropocene (2017, p. 11). Conversely, the French political scientist and educator Pierre Statius believes that the profound crisis currently facing democracy has anthropological foundations and must necessarily be worked on from an educational perspective (2017, pp. 10–11).

Citizenship must be examined in connection with a raft of notions. It can be linked to that of identity and its construction in relation to the institutionalised forms of social life. Citizenship can sometimes be confused with nationality when national identity is mentioned. Thinking about citizenship and the citizen is not only about the factual recognition of belonging to a people or a community (Lamarre, 2002a, p. 26). The political conception of citizenship in this work – as distinct from an ethnic approach to citizenship – is to be compared with the politicality of the French sociologist Numa Murard and the French philosopher Étienne Tassin, who note the importance of 'thinking in political terms about life experiences that are carelessly relegated to the intimate sphere of existence and thinking in existential terms about political conflicts and struggles for dignity' (2006, p. 31).[26]

In addition to being a process, citizenship is a practice; it can only be active and integrates the dimension of the relationship to oneself and the relationship to others (Balibar, 2001, p. 109). Citizenship is a process of becoming human. For the German philosopher Gunter Gebauer, the fundamental anthropological question of the contemporary period is indeed 'How does man become man' and no longer 'What is man?' It is from the world that man receives his humanity with politics (Balibar & Gebauer, 2012). Citizenship, involving the political arena that is the world, concerns humanity in what is most essential. Citizenship as understood in this research has notional proximities with the concept of 'reflexive citizenship' developed by psychologist Margarita Sanchez-Mazas of the University of Geneva and the Belgian philosopher Raphaël Gély (2005), who describe the historical singularity of the contemporary experience of citizenship: 'Today, it is as if the effective exercise of citizenship implied a perpetual questioning of one's own presuppositions and one's own modalities. This citizenship would be reflexive in the sense that it could only be exercised by returning at the same time to the very conditions of its exercise. Contemporary citizenship is not only what is exercised. It is what is exercised by reflecting on oneself' (Sanchez-Mazas & Gély, 2005, pp. 73–74).

[24] As Ricœur notes in his preface to *Condition of Modern Man*, 'the "who" that action reveals is the citizen as distinct from the worker and even from the maker of artifices made with human hands' (Ricœur, 1961, p. XX). The importance of speech and action is that they reveal the speaking and acting agent and constitute him as a citizen (Arendt, 1983, p. 205).

[25] Action reveals the citizen, but action consists mainly of setting processes in motion (Arendt, 1983, p. 346). In thinking about education, we are particularly interested in the process of citizenship in order to grasp the means to learn it.

[26] The term 'citizenship' is to be contrasted with the German term *Bürgentum*. The German concept of *Bürgentum* means citizenship as well as bourgeoisie (it is the same term). The German *Bürgentum* refers to the idea of connection between autonomy and subjection that is conceptually identified in subsidiarity (Balibar, 2001, p. 247).

11.3.2.1 Education and Politics in the Notions of *Bildung* and Citizenship

Educating for Citizenship?

To evoke the citizen is to evoke the State and a legally and socially founded community of belonging (Neveu, 2004). The first State and republican institution which a future citizen will encounter in the course of their lives is the school. School is regularly seen as the sanctuary of the republic: it is the school that establishes the citizen. The republican school has a responsibility to train citizens in the double sense of creating (instituting) the future citizen and enabling pupils to learn about citizenship (Pélabay, 2011, p. 117). Thus, everything related to citizenship is closely linked to the question of the school as an institution in its normative component. Over the centuries, the school has served an agenda of 'disciplining and moralising' which has 'in turn taken the form of a project of training the believer and then a project of training the citizen' (Monjo, 2013, p. 80). The normative component of citizenship training is a point that is seen as problematic by some authors, who dissociate the school's function of instilling basic knowledge from education for citizenship. Jonas' work on the notion of responsibility, on the other hand, leads him unambiguously to understand the citizen as the 'immanent goal of education' (1990, p. 202), both by parents and by the State. Being a citizen, being an adult and being responsible go hand in hand for Jonas who writes: 'The goal of education: to be an adult' (p. 211).

There is a paradox between learning or initiation into the world and the school, which is closed off from the world, functioning as a sanctuary. The school is not the world; it is 'the institution that stands between the world and the private domain… to facilitate the transition between the family and the world' (Foray, 2004, p. 179). The transmission of knowledge at school must enable pupils to gain an understanding of the world, but also an existential experience of the world, as highlighted by Arendt. Educating the citizen as a person implies a form of elevation to universality. The other, or the stranger, is the fulcrum of such education. It is in others that we can learn to recognise ourselves in order to become another by experiencing 'the common humanity' that links us to [the other]' (Foray, 2004, p. 190).[27]

---

[27] The notion of citizen contains the same ambivalence as that intrinsic to the notion of subject. The subject refers to a form of freedom in the face of determinism, while it can also mean subjection to other individuals or systems; for a long time, the subject meant someone who did not have his own freedom. The same is true for the citizen who is educated to become compliant, law-abiding, and suited to living with others (assimilation can lead to subjection). However, the education of the citizen also takes on another meaning when it refers to emancipation or has a subversive component with regard to the State or a defined system. This ambivalence allows for a refinement of the concept of emancipation; emancipation can only take place on the basis of a mastery of norms. The education of the citizen as we understand it includes within it the production of subjectivity that leaves subjection behind (Balibar, 2009).

Bildung*: A Political Conception of Education*

In order to go further in this exploration of the type of citizenship to be promoted in the Anthropocene, it may be interesting to compare this notion with the German notion of *Bildung,*[28] which has the characteristic of mobilising an existential component. Generally, the notion of *Bildung* is positively received by French educators for two main reasons: its reference to a global approach to the subject and the fact that it goes beyond the school framework of learning, with an emphasis on the way in which 'daily life', through the experiences it contains, enables learning (Brougère & Ulmann, 2009). Among the strengths of the notion of *Bildung* is the understanding of the subject in a form of anthropological globality with the integration of the directly existential component where *Erfahrung* (experience) is *Bildung*; the experience of existence is thought of as formative. This conception is in line with the anthropo-training theorised by Pineau (2003). In the French literature on *Bildung*, frequent mention is made of the idealist universe[29] in which it was born and the ideals it continues to convey. What is less well known in France, however, is the political dimension of this idealism. It is to the political sphere, as much as to the educational sphere, that the idealism of the notion of *Bildung* must be linked. *Bildung* is a political concept because it refers to the advent of civil society in the same movement as subjects. One of the virtues of the notion of *Bildung* is to understand the social link as a pedagogical question (Masschelein, 2002, p. 40). The notion of *Bildung* integrates the idea of the possible creation by man of a new man[30] and is in keeping with what has been worked on in this research with the idea of an anthropological shift.

This particularly interesting notional universe has two inseparable limitations: structural individualism (Dumont, 1991) and a lack of education in politics. In the universe of *Bildung*, the educational act is thought of as a political act, contributing to the emergence of a new type of human. Despite its strong idealism, one of the strengths of the notion of *Bildung* is its ambition. On the other hand, paradoxically, the notion of *Bildung* refers in a rather individualistic way to the world, since it is

[28] *Bildung* was at the heart of the creation of the University described by German pedagogue Alexander von Humboldt (whereas the contemporary French University has a (small) part of its origins in addition to the Napoleonic project or the British project of cardinals Paul Cullen and John Henry Newman).

[29] Within the German world, the pedagogue Johann Friedrich Herbart provided a different voice. While giving central importance to *Bildung* and the development – through education – of that which is human in man, he developed a realistic and empirical approach to this notion, breaking away from its idealism. On the other hand, Herbart remains steeped in the individualism underlying the notion of *Bildung*, where it is man as such who is the goal to be achieved' (Maigné, 2002, p. 51).

[30] Perfectibility is an essential concept in the educational anthropology of the German Enlightenment, which is built on the notion of *Bildung* (Ricken, 2002). The evolution of modernity then gave rise, in Germany, to a critique of this idealistic philosophical anthropology with the historical anthropology of the Berlin School and thinkers like the German educational anthropologist Christoph Wulf and the German philosopher Gunter Gebauer.

about oneself, from oneself, and for oneself that the world is thought of. The dimension of collective belonging is hardly present. The subject is inserted into a world that is apprehended as a space for self-development. However, it is from within that the subject derives his own identity. It is in the intimacy of existence that self-development takes place: the ultimate goal of *Bildung*. With *Bildung*, we are dealing with self-education, with the subject discovering and developing through contact with the world. Historically, the notion of *Bildung* emerged in a pietistic Protestant universe in the late eighteenth century in Germany. Protestantism, with its emphasis on individual faith to the detriment of institutional authority, reinforced the importance of intimacy, which became the gateway to the outside world in which the believer must prove himself. With Protestantism, politics is based on the intimacy of existence. The place of *education* is, first and foremost, the intimate (Wilhelm Graf, 2002, p. 783), as opposed to the world and confrontation with others who inhabit it. The individual is obliged to work on his or her personal realisation (because of the intimacy of salvation) rather than on the necessities of the world.

The subject is not received from the world and is not intended to be given to the world. In the spirit of *Bildung*, educating is a powerful political act, essential to humanity. On the other hand, it is not so much a citizen who is *gebildet* (educated, trained, cultivated), but a human being. The man who is formed is essentialised – it is the humanity of the man who is educated or elevated, as opposed to the citizen or the man of politics anchored in an economic-social reality. From an idealist perspective, by raising the humanity in each person, one participates in the creation of a new humanity which brings a future to the world, without having had recourse to a political education of the subject as a citizen. *Bildung* is a political conception of education, with existential foundations that do not really integrate an education in politics.[31]

The German intellectual tradition of the sciences of the mind (*Geisteswissenschaften*) provides valuable resources for thinking about the links between education and politics. The German resources of the world of *Bildung* make it possible to think about education in a global approach integrating a reflection on one's humanity. In fact, education in Germany is also based on the

[31] In Germany in the nineteenth century, the idea of the university was particularly lofty. Its function is a kind of mediation between the individual, the community and the State. Despite their underlying individualism, *Bildung* and the University in Germany have a social mission: 'The university embodies thought as action (i.e. as research and education at the same time), as aspiration to an ideal. The university and the State are linked, here. The university seeks to embody thought as action towards an ideal; the State seeks to realise action as thought' (Masschelein, 2002, p. 45). In Germany, the mission of *Bildung* is at the heart of the foundation of the Humboldtian model of higher education. This implies that the university should not only have a formative but also an educational purpose; this is an important issue for the contemporary University in France and for French higher education in general. The idea of adult education emerged during the twentieth century and has historically been institutionalised outside of universities (Manifet, 2015). We have experienced a semantic shift from the emergence of 'adult education' or 'continuing education' to 'lifelong learning'(even though the notion of 'lifelong learning' has been emerging for the last 20 years).

experience of existence. This approach has the advantage of training people who have not ignored the fundamental questions of humanity, capable of recognising others as an *alter ego*. Having been educated about my own humanity, I am able to reach out to the other and share this common humanity with them. One of the benefits of the notion of *Bildung* lies in its component of existential consolidation, which is particularly interesting for thinking about education for politics in the Anthropocene.

Citizenship: An Education in Politics

The political dimension of belonging to a collective is hardly present in the spirit of *Bildung*, and this is why it is interesting to compare this notion to that of citizenship. The concept of citizenship etymologically refers to the city and to participation in its functioning. Compared to the concept of *Bildung*, citizenship is rooted more in the exteriority of the world. The notion of *Bildung* apprehends the subject partly as an entity 'in itself', marked by a form of disconnection from the world, whereas the notion of citizenship apprehends the subject on the basis of their relationship to the world and participation in it.

Sometimes the notion of citizenship has a strong normative component, entailing learning behaviour suitable for social life. In this respect, it is far removed from the developmental logic of *Bildung*. Overall, citizenship refers to individuals' capacity to act on their environment and to the means of living together. In an extension of this idea, we can identify two main types of citizenship: civic adaptation to social life, and the struggle for equality and justice, which can result in a conflict with the State or society (Murard & Tassin, 2006, p. 24).

Whereas the notion of *Bildung* refers to the self, the notion of citizenship refers to the city, the dwelling place of others. The world between others, and between the other and the self, is the notional centre of gravity of citizenship. From a political perspective, citizenship, which is related to the other, is partly related to the idea of hospitality: the welcoming of those who come from elsewhere into the city. The etymology of hospitality refers to a certain ambivalence (perceptible in the proximity of the notions of host and hostile) and to the risk incurred in welcoming a stranger.[32] Thus, citizenship can be linked to educational hospitality. This hospitality rests on the same foundations as the political hospitality of the stranger: 'It is on condition that we welcome the strangeness of the other and our own that the world ceases to be foreign to us, and we cease to be strangers to the world, on this condition that

[32] The notion of citizenship integrates that of hospitality – which *Bildung* does not. Hospitality here does not necessarily mean the selfless welcome of others; it can refer to logics of alliance between hostilities (Cornu, 2007). The notion of hospitality is close to Mauss' notion of gift: it is given, received, returned and exchanged; it is more a continuous process than a state (Gotman, 1997). Hospitality is part of a process of hominisation that refers to Mauss' total fact. The gift and exchange of services constitute the human being: 'man has become and becomes such insofar as he practises hospitality' (Schérer, 1997, p. 60).

hospitality can be the hallmark of cosmopolitics' (Tassin, 2004, p. 280). This stranger refers to the figure of our neighbour, the one we are or could be for others. Welcoming them allows us to be part of an economy of giving or spending that is also an economy of the other's debt to us.[33] The notion of citizenship, thought of in its interaction with that of hospitality, is ambivalent because it presupposes a citizen who is both educated in conformity but also committed to action on behalf of the other, which can take a subversive turn.

As the Belgian sociologist and pedagogue Jan Masschelein (2002) points out, it was against the background of the notion of identity that humanist thought developed (based on humanity's relationship with itself), offering a crucible for educational and pedagogical thought (this is particularly noticeable in Germany with the notion of *Bildung*). The entry into the Anthropocene may allow us to think afresh about this individualistic humanist foundation. Is it possible to think in education, as in politics, on the basis of the other or otherness – a hypothesis that is the opposite of that of identity?[34] It is understood here that one of the radical figures of the other will be non-humans, which the type of citizenship to be exercised in the Anthropocene will have to integrate.

11.3.2.2 World Citizenship?

The ancients associated the territory with public life. An individual, depending on his condition, could participate in public life as long as he lived there. Now, modernity has produced the idea of the citizen, not as directly contingent on community membership but as access to universal rights (Balibar, 2012, p. 20). Indeed, the European project, like globalisation, dissociates territory and citizenship. The French philosopher Étienne Balibar identifies a 'social citizenship' specific to Europe and implemented following the Second World War by social democracy (2001). When Socrates was asked where he was from, he answered, 'from the world' and not 'from Athens'. Yet what kind of world and citizenship are we talking

---

[33] To understand hospitality from an anthropological perspective is not to reduce it to a moral or legal dimension, but to understand what it means in terms of symbolic exchange. It predates any legal component because it is *de facto* before *de jure*. Hospitality consists of taking the risk of trust and the need to form an alliance with the outside world (Cornu, 2004, p. 292 and p. 293, 2007). The work of the French educational philosopher Laurence Cornu around the development of a theory of political anthropology of trust is particularly interesting for thinking about citizenship.

[34] This comes down to thinking in the tradition of Emmanuel Lévinas and his 'Humanism of the Other' (Lévinas, 1972). In the German humanist tradition from the mid-1800s to the early 1800s, the first question is 'is what we are and what we want or need to be' (Masschelein, 2002, p. 42). Thus, 'the pedagogical/political problem is not, in the first place, a problem of the relationship with the other or others in the sense that it is not formulated on the basis of the needs or the question of the other, but the relationship with oneself. The education of man is the identification, delimitation, definition, reading and understanding of the world so that man understands himself, identifies himself, realises himself, so that he becomes human' (Masschelein, 2002, p. 42).

about? The notion of citizenship has its origin in belonging to the Athenian city. It is to a local, clearly defined, legal entity that the citizen belongs (of which he is a native). This notional extension to the world was achieved by the Stoics (Chrysippus, Zeno) who apprehended a *polis* on a global scale with the invention of cosmopolitanism (Schérer, 1997). An examination of the notion of world citizenship raises the issue of 'a double and disjointed responsibility, for the particular on the one hand and for the universal or being-in-the-world on the other' (Vincent, 2004, p. 15). However, above all, this notion of world citizenship raises the thorny question of the possibility of citizenship without a state, which would be the responsibility of a 'metapolitical body' (Foessel, 2004, p. 20). World citizenship is both what the great German philosopher Immanuel Kant identified as a 'sense of belonging to a global sphere', but also consists of the possibility 'to make legal claims against a State of which one is not a member' (Foessel, 2004, pp. 21–22): it is 'the right of a foreigner not to be treated in a hostile manner by the State whose soil he treads' (Kant, 2006, VIII, 358, p. 56). The benefit of this notion of 'world citizenship' is 'to maintain man *in the consciousness of the totality* in order to guard against any identity-based withdrawal' (Foessel, 2004, p. 31)[35]; moreover, it makes it possible to link politics to cosmopolitics through the reception of foreigners (Lamarre, 2020).

In Kant's notion of universal citizenship defined in *Perpetual Peace: a Philosophical Sketch* (2006), the citizen is first and foremost a citizen of the world before being an inhabitant of his or her own city, with the right to accept or deny the other as a citizen. The notion of world citizenship, which can be understood as human hospitality, has the primary interest of helping reshape the notion of citizenship, based on a questioning of humanity.[36] For example, according to the French philosopher Seloua Luste Boulbina, 'not everyone who wants to be is a citizen of the world, but everyone who can. He who is at home everywhere may be cosmopolitan, but he is not a citizen of the world. The citizen of the world is the one who is at home elsewhere. It is not his open-mindedness (commonplace among the poor), it is his (readiness to) availability by which he can be characterised. This is, of course, understood on all levels: geographical as well as intellectual'.[37] For the French philosopher Laurence Cornu, the figure of the citizen of the world is, for the French philosopher Laurence Cornu, 'the host of the human race' (2004, p. 285) in the dual sense of the one who receives and is received.

---

[35] The ongoing construction of the European community as a political force should make it possible to dissociate citizenship and identity – but this is not what we are seeing, with the contemporary rise of identity-based nationalist sentiment.

[36] Indeed, Cornu proposes a political anthropology approach to the concept of citizenship rather than the more common one of combining moral, legal and political components (2004, p. 285).

[37] For Arendt, the work of the German philosopher Karl Jaspers (who supervised her doctoral dissertation and with whom she kept up a lifelong correspondence) is conceived with world citizenship as its outcome. This world citizenship necessarily brings with it a relativism which it is important to be educated about. It questions conceptions of truth and reinforces the importance of communication between people of different cultural and national backgrounds.

It is necessary to redefine this notion of 'world citizenship', with which we are uneasy because of the equivocal component of the underlying mention of a global or supra-national state. Citizenship refers to belonging to a state and is therefore normally antagonistic to belonging to the world. However, it is also possible to understand world citizenship as an understanding of 'the world' within the singularity of a nation. Rather than speaking of world citizenship, it might be possible to speak of worldliness – an existential worldliness, or a 'convivial worldliness': an 'existential' or 'convivial' way of inhabiting and being present in the world – which is made up of far more than just humans. It must be understood that we do not inhabit the world as a whole, that we cannot be responsible for the world in a general way, but that we always inhabit a defined place and assume a limited and defined responsibility. While this notion of world citizenship is interesting, it does not sufficiently integrate the importance of the link to the Earth, which is one of the fundamental characteristics of the type of citizenship to be exercised in the Anthropocene.

11.3.2.3 Earthly or Cosmic Citizenship?

Environmental citizenship made its appearance in schools in the context of a 1977 circular which begins as follows: 'At a time when environmental deterioration is posing problems of choice for mankind that are decisive for their future, environmental education is clearly necessary. This education also responds to the need generally found in children and adolescents to understand nature and the world around them. It is therefore essential that schools provide pupils, from an early age and throughout their schooling, with education that enables them to grasp environmental problems intelligently and constructively' (Circular No. 77–300 of 29 August 1977, p. 1).

Since the Earth Summit in Rio in 1992, the involvement of citizens in political decisions concerning the environment has been understood as necessary, in accordance with Principle 10 of the Rio Declaration: 'Environmental issues are best handled with participation of all concerned citizens, at the relevant level'. The Anthropocene reminds us all that we are earthlings, and thus redefines citizenship and membership in nation-states and communities. Now more than ever, human beings share a common destiny: that of belonging to the Earth. Sovereignty is thus partly reconfigured – we are citizens of a State but also of the Earth, which is our master: 'By no means does being an Earthling mean renouncing national citizenship, but it does imply necessarily considering citizenship and the territoriality of democracy with a more critical and less exclusivist eye' (Eckersley, 2017, pp. 14–15). For Cabanes (2017), 'we are first of all the inhabitants of the Earth before being citizens of a country'. For the safeguarding of democracy, Pelluchon goes so far as to write that citizens must learn to recognise the beauty of nature, which is 'a relational quality and testifies to the vital link that unites us to nature and to other beings, human and non-human' (Pelluchon, 2017, p. 14).

How, though, can we conceive the place of citizens in the exercise of power, given the seriousness of the entry into the Anthropocene? Jonas, from the critique of technology and the imperative of responsibility, develops a heuristic of fear. Despite the totalitarianism of Nazism, Jonas does not fear making a political choice that is partly authoritarian in order to force individuals to take these threats into consideration. He envisages a system constraining citizens who are no longer completely sovereign in order to preserve the human adventure.[38]

11.3.2.4 Emancipating or Responsible Citizenship?

When citizenship is approached from an educational perspective, it is frequently based on the notion of emancipation, which is one of the determining and dominant notions in educational debates. On the other hand, whenever the Anthropocene is addressed in the social sciences, it is against the background of the notion of responsibility, both in terms of identifying those responsible for the entry into the Anthropocene and of the way in which we will be able to live together in this new geological epoch. What, then, should we think of citizenship, which is understood as being between the notion of emancipation and that of responsibility? The notion of responsibility, integrating a form of constraint, may be relatively distant from that of self-emancipation, which can imply a liberation from constraints.

Emancipating Citizenship?

Curnier, in his reflection on citizenship, draws upon the concept of emancipation. As one of the quests or promises of modernity – perceptible in the individualistic hypermodernity of transhumanist theories and research – has been a form of emancipation of the human adventure from the limits of the biosphere and from its mortal human condition, we have opted for caution in our use of the concept of emancipation. Is it not necessary to apprehend emancipation as a liberation at the price of a tie-up with responsibilities? Certainly, the understanding of emancipation in the concept of 'emancipatory citizenship' proposed by Curnier integrates the social component and differs from an individual understanding of emancipation: 'The emancipatory conception is based, above all, on the collective dimension of social organisation. It therefore recognises the social function of the school institution in

[38] The French engineer Jean-René Brunetière believes, in accordance with an element of Jonasian thought, that only a feeling like fear seems able to reinforce contemporary liberal power. According to Brunetière, contemporary issues of environmental sustainability and survival are one of the results of liberal ideology based on impulses that he defines in a collective interview published in the journal *Esprit* as 'the appetite for enjoyment, greed, the fear of lack, the desire for autonomy…' (Badré et al., 2011, p. 212). To counter this liberal hegemony, he believes that only an ideology based on very strong feelings can gain acceptance and that the arguments of reason alone will not lead to the necessary individual sacrifices.

the training of future citizens. Without excluding the needs of the pupil, it complements them by catering to their social needs, to which it attaches greater importance. It is therefore part of a conception of individual freedom subject to limits imposed by the collective interest' (Curnier, 2017, p. 354). He goes on to state that 'the collective dimension is central' (p. 356) to his notion of emancipation. Curnier's emancipatory conception is not individual since, in line with Varcher and some UNESCO guidelines, he defines it as a 'collective emancipation of learners so that they can become active citizens, able to face societal challenges together'. The equivocal dimension of the term 'emancipation' is that we cannot free ourselves from the nature in which we are immersed. Also, if it is possible to envisage collective emancipation, the collective must refer here to the adventure of terrestrial life as a whole (humans, animals, plants, bacteria, fungi…). An educational goal such as the emancipation of a collective runs the risk of alienating other social groups or categories of life in the emancipation process.

In education, emancipation is an important concept. It is linked to hope: I am not just an extension of my social determiners. Action is possible that is not determined by the place of my birth. Is it possible to learn to think in education without emancipation, but to keep that creativity and freedom which thwart determinism? For Andreas Weber, in a way it is not emancipation that is the goal: 'A culture of *Enlivenment* allows, in the tradition of the Enlightenment, a liberation from a higher order – a liberation, of course, which does not draw its impetus from its claim to perfect emancipation, but from the desire for solidarity'[39] (Weber, 2016, p. 86). This is also a point shared with Rosa. Is not the ideology of emancipation that accompanied the Enlightenment (of humans vis-à-vis nature, of humans vis-à-vis their painful human and social condition, etc.) a mistake? If it is indeed a mistake, are we then dealing with a dangerous ideology (Weber, 2017, p. 91)? This is an element that is put to work in this, Part III of this manuscript, but which challenges some of the axiological foundations of the educational act.

The Commitment that Compels Responsibility[40]

One form of intelligibility of commitment refers to the idea that the shortest path from self to self is through others. In other words, paradoxically, the emancipation of the self is not reduced to the self, but involves others. This idea is manifested in the legend of the founding of the companions of Emmaus. Abbé Pierre's first companion, Georges Legay, when he comes to ask for help, was invited to perform a

[39] 'Durch eine Kultur des *Enlivenment* wird, ganz in der Tradition der Aufklärung, eine Befreiung höherer Ordnung möglich – eine Befreiung freilich, die ihre Schubkraft nicht aus dem Anspruch auf vollkommene Emanzipation nimmt, sondern aus dem Begehren nach Verbundenheit' (Weber, 2016, p. 86).

[40] Elements of the following paragraphs were written with Jean-Yves Robin and Jean-Pierre Boutinet and have been published in French (Wallenhorst et al., 2017). We thank them for the exchanges and this co-editing.

selfless act. He was even invited to change his attitude. It was no longer a question of receiving, but of giving: 'I don't know what I can do for you', said Abbé Pierre, 'but I know that I need you, because an immense task lies ahead of us in the service of others'. The founding legend of the companions was born (Bergier, 1992). It stands as a stimulating ethical, practical and political model, illustrating the way in which acceptance of responsibility partly puts an end to disorientation (and its torments of meaninglessness). Georges Legay renounced the idea of suicide, following his meeting with Abbé Pierre, during which he was offered the opportunity to help others who had been downtrodden by life to escape their misery. Thus, he became the first companion of Emmaus. This man did not first find his place thanks to Abbé Pierre's attention to him, but by taking on responsibility for others. It is the acceptance of others that allowed Georges Legay to take his place in this social group, the Companions of Emmaus, and to be saved from despair and meaninglessness. From this logic of hospitality emerges something interesting for thinking about educational emancipation. Entering into this logic of welcoming others appears to be a condition for emancipation.

Educational Emancipation at the Crossroads Between Three Paradoxes

It seems possible to understand emancipation in education as being at the intersection between three paradoxes which have to do with emancipation seen simultaneously as a state and a process, emancipation of oneself through others, and emancipation as liberation at the price of anchoring to responsibilities.

Emancipation can be as much a matter of a new state achieved that needs to be maintained as of a process leading to uninterrupted transformations. These two forms of emancipation refer to the two timescales, highlighted in his time by Saint Augustine when, in his *Confessions*, he contrasted the present of eternity with the present of the future which fades into the past. These are timescales which spontaneously exclude each other but which can be reconciled, under certain conditions of cohabitation. The feeling of serenity, of tranquillity, of feeling at ease in one's place, of having regained self-confidence or of having discovered a new way of living in a group can characterise a new state in education that should be allowed to endure; against the background of this new state, a process continues to unfold: the development of a critical mind or of progress in the construction of the self. Thus, emancipation would be a privileged time that would favour the cohabitation, in each person, of two presentist temporalities; one concerns the present of a new continuity that opens up to the self thanks to a certain striking event, or a significant encounter, and allows one to settle in the long term; the other concerns the present of change, that favours the deployment of achievements or the development of skills, in the quest to build a future. Between these two forms of presentism, there is a tension with the risk, depending on the circumstances, that one of the presents will overwhelm the other: being concerned with enduring continuity can annihilate any commitment to an action of change; embarking on projects, developing initiatives entails the risk of favouring the more or less immediate future, leaving aside the present moment.

Without doubt, this mixture of states and processes to be articulated is one of the characteristics of educational emancipation, but – in order to make compatible what it still associates – it requires being constantly re-examined by the actor involved, with a view to an adequate resumption of what constitutes the duration of his or her present with what orients his or her future prospects.

In explaining the emancipation of oneself through others, it is worthwhile taking a tour through etymology and history. Let us look at the Latin equivalent of our verb *emancipate*: *emancipare*, and its semantic opposite, *manucapere.* In the original Latin, *manucapere* meant the act of taking a slave by the hand: it was the physical act of seizing the hand when a slave was purchased. *Emancipare*, from which we get the noun emancipation, evokes the opposite logic: it is a legal act that marks the elimination of the right to sell a slave, whom no citizen is entitled to take by the hand. Emancipation thus refers to the idea of freedom from guardianship or domination and the entry into liberation – a guarantee of subsequent autonomy. The etymological origins of the term reveal the power and density of this act founded in law: to free one's slave; historically, to emancipate (*emancipare*) is therefore an act performed by another on an enslaved individual, invited to enter into a new social existence that heralds a radically different mode of being in the world. Thus, emancipation refers to the passage from a state of enslavement to another state of liberation, through the intermediary of another: I do not emancipate *myself*; I am emancipated by another, another person, the situation, an event, an encounter, etc.

Emancipation at the cost of responsibilities is the third paradox of emancipation. From the moment the subject is emancipated, he becomes, by his acquired or conquered autonomy, the author of what he does, having to answer for his actions. This is what Génard (1999) calls the empowering interpretation of action. This interpretation continually confronts the author, because of his emancipation, with being blamed by others for actions taken, with undesirable effects; the author must always justify himself for what he does or for what he has done. In this respect, it is interesting to underline the historical conjunction revealed by the Society of the Enlightenment: a conjunction between the quest for emancipating autonomy embodied by Kant (1991) in his famous *Sapere Aude*[41] and the emergence of concerns about responsibility for the action taken to *dare* to exercise one's autonomy: political responsibility, civil responsibility, and criminal responsibility first emerged in France within a few years of each other in the late 18th and early 19th centuries, at the time when the king's subject was emancipated by the Revolution to become a citizen. This political emancipation, seen as a transformation, was taken up by Marx (2014), but he preferred to substitute emancipation-dialectic for emancipation-responsibility – i.e. a societal emancipation that does not rely on the logic of actors and their responsible involvement.

[41] The English translation of this Latin expression is *Dare to use your reasoning* – i.e.: *be autonomous.*

Educational Emancipation Based on the Consideration of Constraints

The three paradoxes highlighted above can help to understand educational emancipation, defined as an educational act with a political aim. Such an act carries within it the 'essence of education' proposed by Arendt in her definition of education as preparing for the political vocation of the individual (1972)[42] – a vocation marked by its imperfection and its need to be constantly reworked. The expressions 'self-emancipation' and 'collective emancipation' posited as unconditional human imperatives through modernity from Kant to Marx, today appear quite problematic in our postmodern vulnerabilities and the complexity generated by the entry into the Anthropocene; they are, to say the least, surrounded by a certain naivety. We cannot dissociate the two major aspects of responsibility, as they have become established in past centuries: responsibility for one's own actions, and one's responsibility towards others, neighbours, and henceforth towards the immensity of the earth's expanse and the wealth of the biosphere.

Far from the emancipation generated by the philosophical and industrial modernity of the 18th and 19th centuries, it is a question here of thinking of a paradoxical emancipation in our environments which have been weakened by the Anthropocene. It is a question of learning the art of the possible through the management of constraints. Learning responsibility, which is fundamental to living and acting together in the Anthropocene, is not self-evident in education, insofar as it can sometimes be understood as being at a distance from the notional universe of self-emancipation.

11.4 Learning Convivial Citizenship Together

11.4.1 Thinking of a Convivial Citizenship Against the Background of Existential Citizenship

As mentioned, in order to think about education in the Anthropocene, it is interesting to link the spirit of *Bildung* with the more directly political notion of citizenship. This notional articulation makes it possible to view the political responsibility of the world towards others with an understanding of the intimacy of its existence. This intercultural exploration allows us to associate the adjective 'existential' with the noun 'citizenship', signifying the German richness of the notional universe of *Bildung* and the importance of this existential component – the poor relation of education in a strictly French perspective. The richness of the notion of existential citizenship is that it apprehends the existential component within the framework of

[42] The contemporary act of emancipation closest to its historical definition of the emancipation of slaves seems to be that of coming of age – a passage from minority to majority that is legally founded and modifies the individual's relationship to the world (responsibility for one's actions, legal emancipation from parental authority).

political education and orders it for participation in the world. From this perspective, the existence of the subject is understood as the point of support in the opening on the world. The idea of humanity is not integrated into the concept of citizenship (with its individualistic quest for self-understanding); hence, it is a concept that differs from that of *Bildung*. Existential citizenship and *Bildung* are not two notional universes in tension with each other. Human existence is their common foundation, but they have two separate purposes: the self (or one's own emancipation), on the one hand, and the world, on the other. What the adjective 'existential' attached to 'citizenship' means is the connection with others and the collective dimension of the human adventure.

The notion of existential citizenship was proposed by Christian Arnsperger in *L'homme économique et le sens de la vie* (The Economic Man and the Meaning of Life – 2011) and in some ways an existential citizenship education underlies many of his works. His notion of existential citizenship is not unrelated to the notion of *Bildung*, as Arnsperger himself acknowledges: 'we should bring back to the forefront the dimension of *human existence as experimentation, as a quest for self, for a freedom*' (Arnsperger, 2011b, p. 76). We find here in this German author the characteristic features of the idea of value inherent in *Bildung* and its experimentation with existence. Existence is thought of as the predominant experience on which to base one's actions. Existential citizenship, on the other hand, has a completely different purpose than experiencing one's existence and finding one's own freedom.

The adjective 'existential' recalls the main characteristic of existence, which is finitude, and this questioning of meaning and the world in the face of the prospect of one's own death. The challenge of existential citizenship is to assume one's share of responsibility for the world by integrating the shared finitude of one's existence. With Arnsperger, this existential component marked by finitude, the anguish of death and the search for meaning can appear fundamental to understanding the life of contemporary societies[43] in the Anthropocene. Arnsperger's existential citizenship is both a pronounced reflexive capacity of individuals on the contemporary period (and particularly the alienating hegemony of the capitalist system) as well as an awareness of the existential component of the human adventure. Existential citizenship is situated at this juncture between the positioning of the individual in the contemporary world and his or her responsibility towards preparation for the future, and questions of the meaning of existence: 'If the citizen of tomorrow does not become more aware of the existential roots of our system, of his or her relationship to the anguish of death, and to the haunting of misery, we will continue to idle [...] within the increasingly narrow margins left to us by social-democratic capitalism' (Arnsperger, 2011b, p. 75).[44]

[43] Arnsperger describes these societies as consumerist, productivist and growthist (Arnsperger, 2013).

[44] Arnsperger's existential citizen is militant. This conception of citizenship as perceptible in the militant act is in line with the notion of '*insurgent citizenship*' developed by American political scientist James Holston, on the basis of his 20 years of observation in Brazil of the struggles to appropriate space to build houses (2009). This ties in with the notion of democratic citizenship developed by the French philosopher Étienne Balibar (2009).

This existential citizenship – sometimes at odds with the logic of subjection to sovereignty – is akin to Arendt's 'freedom of the mind' which, when the world descends into madness, can 'convince people to remain human in a world that has become inhuman' (Ehrwein Nihan, 2013, p. 51). Existential citizenship presupposes the exercise of one's reflexivity on the alienations of the contemporary period and a mobilisation aimed at collective emancipation taking account of the social, economic and environmental constraints of the future; it is related to the Tourainian subject. The proposed notion of existential citizenship is in line with Australian politician Robyn Eckersley's Anthropocene reflexive democracy (2017). Indeed, existential citizenship, through its participation in the strengthening of *homo collectivus* and *homo religatus*, is directly reflexive because it consists of distancing oneself from the hegemony of *homo oeconomicus*: 'A reflexive democracy must be both self-focused and other-focused to enable citizens and their representatives, along with investors, producers and consumers, to critically reflect on the consequences that their decisions and practices have on others, in space and time' (Eckersley, 2017, p. 13).

11.4.2 A Convivial Citizenship Rooted in the Vitality of the Biosphere and Combatting Hybris

The Anthropocene brings with it a set of major types of knowledge: those on planetary boundaries, those relating to the need for a change in lifestyles (following a certain type of reading of the Anthropocene), and those relating to the characteristics of modernity which have progressively cut us off from the Earth to which we owe our existence. Thinking about education in the Anthropocene leads to profound changes in priorities (allowing the human adventure to continue, rather than just learning to read, write and do arithmetic),[45] in pedagogical methods (should education in schools take place in concrete classrooms or in the fields and forests?), in the social foundations (is it to allow the development of individual freedom or the sharing of wealth and goods?), in its proximity or distance to the market (is professional integration the goal at all costs, or is it possible to hope for the advent of another world – one that is sustainable, equitable and based on other foundations of justice?).

Thinking about education in the Anthropocene, based on data from the Earth System Sciences and philosophers, showing the error and unsustainability of the anthropology of modernity's withdrawal from nature, implies thinking about a radically new education. It is a question of allowing an anthropological shift. Educating is linked to the existential anthropological roots worked on by Hartmut Rosa, David Abram, Maurice Bellet, Christian Arnsperger and Andreas Weber. 'The image of reality learned in school says nothing about the feeling of being alive, about why relationships are important and how they can be built – the relationship with

[45] Learning to write, read and calculate is absolutely fundamental, but here it is a means to an end: the perpetuation of the adventure of vitality as a whole and of the human adventure.

oneself, with others, and with all those who are not human, such as animals, plants, fungi, bacteria, stones and rivers, the air, the sea. The image[46] that we teach our children is of a world without love. Their mechanics are solely focused on success: what is there has prevailed, putting everything else aside. We are guided by an ideology of death' (Weber, 2017, p. 19).[47] Another quote from Weber here shows the importance of sensuality[48]: 'Individuation is the expression of the vital force that fills each being. This life force, in order to become itself completely, wants to come into contact with others. It must be protected, assured, nourished; it requires contact, tenderness, connection in order to become involved, to give itself and to grow' (Weber, 2017, p. 21).[49] This sensitive, sensory, if not sensual dimension is one of the components of an education in the Anthropocene. Educating in the Anthropocene in fact implies apprehending nature with a form of poetry. How can we learn to be grateful for the wonder of nature without becoming esoteric and irrational? Yet we must recognise that the sunlight never ceases to shine 'freely' and that the diversity of the biosphere is based on the gratuity and superabundance of this gift. Education in the Anthropocene presupposes openings to rationales other than the western Cartesian scientific rationality that is largely responsible for the entry into this new geological epoch, unfavourable to life.[50]

The notion of existential citizenship appears very interesting, but its limitation is that it does not refer to the inclusion of the human adventure within this living world as a whole. This is why we propose the adjective 'convivial' to describe the type of citizenship which needs to be exercised in the Anthropocene. It also well reflects the importance of educating against the nascent *hybris*.[51]

---

[46] Image is *Bild* from which comes *Bildung* – education/training; in this German sentence, there is an implicit reference to education.

[47] 'Das Bild der Wirklichkeit, welches in der Schule gelernt wird, sagt nichts darüber, wie es sich anfühlt, am Leben zu sein, warum Beziehungen wichtig sind und wie man si aufbauen kann – Beziehungen zu mir selbst, zu anderen Menschen, und zu allen, die keine Menschen sind, zu Tieren, Pflanzen, Pilzen, Bakterien, Steinen und Flüssen, der Luft, dem Meer. Das Bild, dass wir unseren Kindern beibringen, ist das einer lieblosen Welt. Deren Mechanik kommt es allein auf Erfolgt an: Das, was da ist, hat sich durchgesetzt, indem es andere aus dem Weg räumte. Wir lassen uns von einer Ideologie des Toten leiten'(Weber, 2017, p. 19)

[48] This is also clearly highlighted by David Abram in *The Spell of the Sensuous* (2013).

[49] 'Individuation ist der Ausdruck der Lebenskraft, die jedes Wesen erfüllt, und diese Lebenskraft will, um ganz zu sich selbst zu kommen, in Kontakt mit anderen treten, muss von diesen geschützt, geborgen, genährt werden, verlangt Berührung, Zärtlichkeit, Gehaltensein, will sich einbringen, sich verschenken, Größer werden lassen' (Weber, 2017, p. 21).

[50] On this point, Weber is particularly close to David Abram. Andreas Weber regularly references the impact of the elements (sun, sea, wind, etc.) on his senses. It is on the basis of this experience of immersion in nature that he wro *Sein und Teilen*.

[51] We can identify two main poles in the way of thinking politically about education: between an educational action with a formative aim (or, to put it more soberly, professional training), and an educational action with a subversive aim (or, to put it more soberly, training of the critical mind). In the contemporary postmodern period, we could position these two poles respectively on a conception of a capitalist society or a post-capitalist society. In education, these two conceptions of society are based on the choice of individualism, on the one hand, and convivialism, on the other.

The conception of convivial citizenship developed in this research (against the background of Arnsperger's 'existential citizenship') assumes the integration of constraints and limits: the limits of the biosphere, the finitude of individual existence, the finitude of our Promethean civilisational model, and the finitude of the human adventure. The integration of these constraints stands in opposition to the limitless emancipation promised by modernity, starting with the emancipation of the human adventure from the constraints of its environment. Thus, this work proposes educational goals other than emancipation, which are in part opposed to it. The proposed learning of convivial citizenship does not aim at the emancipation of the learner, but at responsibility towards a collective and hospitality towards the other (human or otherwise). What is important with the learning of convivial citizenship is entry into the political vocation, through participation in the emergence of a concerted action marked by a post-Promethean 'between-us'. The learning of convivial citizenship thus constitutes a paradigmatic break with the usual conceptions of learning in education. This is a point of divergence with Curnier's thinking about 'emancipatory citizenship education' (2017, p. 234) – even if, as noted, emancipation is not thought of at the level of the individual.[52]

11.4.3 Learning Together

Extending this Arendtian idea and the proposal for an anthropological model of the human adventure (Chap. 10), there are two structural components in convivial citizenship: the integration of plurality and the capacity to be responsible for it; and the capacity to focus on the other and to be hospitable towards them. Responsibility and hospitality are the political ends of learning convivial citizenship. These two dimensions also generate habitability of spaces.

The notion of convivial citizenship is a conceptual tool for thinking about some of the educational conditions for the consolidation of a *homo collectivus* against the background of a *homo religatus* that allows us to distance ourselves from the hegemonic desire of *homo oeconomicus*. Convivial citizenship is learned 'together'. Thus, the other can be apprehended as an object, a means and an end of learning. Firstly, the other as an object of learning means learning about the other through listening and hospitality. Secondly, the other can also be a means of learning: 'learning together', thus assuming that the subject of learning can be a collective, and that it is possible not to consider only learning on an individual basis. Such a perspective breaks with the place usually reserved for the other in individual learning; the other is generally understood as a facilitator or participant in the context necessary for

[52] Moreover, Curnier regularly uses the adjective 'humanistic' to describe education for sustainable development based on strong sustainability. For our part, we prefer not to use this equivocal adjective, insofar as humanism can be modernist, Promethean and anthropocentric. We prefer to outline an Arendtian conception of politics as concerted action at the heart of plurality – human and non-human.

individual learning. This presupposes an understanding of the learning dynamic of the 'between-us' (which is as much the arena of *homo collectivus* as of *homo religatus*).[53] From this perspective, the other is not a facilitator of the individual's learning, but rather participates, in his or her connection with the individual, in the subject of learning. Thirdly, the other can also become the goal of learning. This implies thinking about education from the point of view of the other (Lamarre, 2006, 2012) – i.e. education as a cosmopolitics.

Before considering citizenship as an education, we propose to consider it as something that is *learned together*. Convivial citizenship is viewed in this work as a learning process that takes place together, rather than the education of a few by a few. This means that this learning is achieved through the involvement of older people in the act of learning within a dissymmetrical relationship.

Learning together for a convivial citizenship constitutes learning of the political, which the study of the Anthropocene shows to be in profound crisis in the contemporary period. This period is understood here as favourable to the investment of a political function against the background of the sharing of existential finitude, by differentiation from the economic paradigm structuring the social space. If the contemporary period is to become that of convivial citizenship,[54] it is because its timescale is based on the long-term preparation of the future against the background of the world's responsibility towards future generations, different from the instantaneous and emergency timescales that dominate the postmodern period.

Learning to Die, Learning the World and Learning the Other

The function of learning convivial citizenship is to contribute to the emergence of a post-Promethean society. The 'between-us' is the real object of convivial citizenship in the dual dimension of the world and the sharing of existence. This 'between-us' is learned through contact with people who assume responsibility for the world and share the finitude of their existence with others.[55] The 'between-us' is learned directly in the world as well as 'outside the world' – i.e. at school (Foray, 2004). Based on the definition of the human adventure at the crossroads between *homo oeconomicus*, *homo collectivus*, and *homo religatus*, it is possible to propose three

---

[53] We perceive the consolidation of a *homo religatus* partly in opposition to the *homo oeconomicus*, the usual subject of learning. However, we are also aware of the way in which the consideration of the 'between-us' in education puts the political at work and consolidates *homo collectivus* as a subject of learning, also partly opposed to *homo oeconomicus*. These two paradigms of learning about others and sharing existence with them, on the one hand, and learning about the world and preparing for the future, on the other, can find a place in the pedagogy of politics. They can be at the heart of what we can *learn together*.

[54] Alain Caillé, for example, believes that the contemporary period is conducive to the development of citizenship (2015).

[55] Existential finitude is associated with that of the biosphere. The finitude of the biosphere is understood here as 'the macroscopic reflection of the microscopic finitude of human existence' (Arnsperger, 2013, p. 95). The Earth plays host to human finitude: the Earth is the place of our birth and our roots as well as our death (Payan, 2011, p. 115).

categories of learning for convivial citizenship in the Anthropocene: learning to die, learning the world, and learning the other.

Learning to die – homo oeconomicus. Convivial citizenship in the Anthropocene presupposes an integration of our anthropological finitude, 'our insurmountable state', which 'does not diminish us [but] is our being in the world and the condition of our creativity' (Bourg and Papaux, 2015b, p. 44). One of the first things to learn in existential citizenship lies in the acceptance of limits and resistance to the hegemonic component of *homo oeconomicus*. From this point of view, it is not so much a question of learning to live as learning to die (Plato, Phaedo; Montaigne, *The Essays*, Book I-20) – learning that is linked to learning about self-limitation. More than ever, the acceptance and integration of our finitude appear necessary and salutary.[56] If we are to learn to live collectively, we must begin by accepting individual death (Scranton, 2013, 2015).[57] Convivial citizenship implies learning an ethical and political act of discernment between *sapiens* and *hybris*. It is a matter of not descending into *hybris* but of learning to identify and satisfy our desire for eternity, to be distinguished from a demiurgic desire for omnipotence.

Learning the world – homo collectivus. The anthropocenic rupture calls for learning about the plurality of the world. This leads to a distance from contemporary conceptions of education – for example, centred on self-emancipation. Learning about the world includes learning about the space that is home to humanity in all its variety, learning about responsibility towards the collective and the adventure of vitality, learning about the preparation for the future, and learning about the courage to participate. It is a question of learning to consider one's own existence in relation to the long timescale of the world rather than only in the brief temporal window of one's own life, as well as in the whole fabric of solidarity that is life, as opposed to only within human societies.

Learning the other – homo religatus. Since the Enlightenment, has education in the west not been conceived against the background of an individualistic ethic, with self-emancipation and self-formation as its goal? What and whom should we learn? Is it to become oneself? To learn to live one's life? On the contrary, is it to learn about the other? The Anthropocene calls for us to think of education from the point of view of the other (from the point of view of future generations, who do not yet exist; from the point of view of the animals with whom we share sentience; from the point of view of the trees, which can sometimes live for thousands of years). Learning about the other presupposes the agreement of a primacy of listening over

---

56 Here, 'learning to die' carries proximities to learning to live in the awareness of one's own finitude, or learning to live with the knowledge that one is mortal (Chartrin, 2015). There is nothing in this idea of 'learning to die' that is 'mortifying'. This dynamic bears some resemblance to the illustration by Jean-Philippe Pierron, in the article '*Pour le meilleur et pour le pire*?' and his staging of guidance of life until death (2017).

57 For Scranton, the author of *Learning to Die in the Anthropocene*, 'learning to die is difficult' and 'it takes practice' (2015, p. 91)! This is all the more true in the contemporary period marked by the transhumanist perspective opened up by the techno-scientific and digital revolution, where the future of the human adventure is dependent on the acceptance of individual death.

speaking. Learning about the other presupposes hospitality, consisting of welcoming the other, and especially the three radical figures of the other: the foreigner (the strange, the other), the generations to come (those who do not yet exist but will one day) and the non-human. From the perspective of learning convivial citizenship together, in the relationship with the other, it is a question of learning to share vulnerability. Vulnerability is understood here as the necessary means for the advent of a relational fabric at the foundation of a post-Promethean society.

The next three chapters will explore three functions necessary for Anthropocene education to learn this convivial citizenship: a function of resistance, a function of critique and a function of utopia.

Bibliographical References

Abram, D. (2013). *Comment la Terre s'est tue – Pour une écologie des sens*. Paris La Découverte. (original edition 1996), French translation.

Afeissa, H.-S. (2014). *La fin du monde et de l'humanité, Essai de généalogie du discours écologique*. PUF.

Arendt, H. (1972). Qu'est-ce que la liberté ?. In H. Arendt, *La crise de la culture*. Gallimard (pp. 186–222) (ed. or. 1961d), tr. fr.

Arendt, H. (1983). *Condition de l'homme moderne*. Calmann-Lévy. (original edition 1958, French translation 1961), French translation.

Arnsperger, C. (2001). Entre impartialité, horizons de sens et précarité existentielle: les fondements de l'éthique économique et sociale. In C. Arnsperger, C. Larrère, & J. Ladrière (Eds.), (pp. 19–67). *Trois essais sur l'éthique économique et sociale*, Editions Quae.

Arnsperger, C. (2002). Justice et économie. Latitudes d'égalisation et obstacles existentiels. *Revue de métaphysique et de morale, 33*, 7–26.

Arnsperger, C. (2003). Peut-il exister une critique sociale rawlsienne? *Mouvements, 27–28*, 165–170.

Arnsperger, C. (2005a). Voie large ou voie étroite de l'économie politique? Une interrogation toujours pertinente. *Reflets et perspectives de la vie économique, 44*, 49–55.

Arnsperger, C. (2005b). La "moralisation du capitalisme": une approche du problème. L'expérimentalisme démocratique et la coévolution des normes. *Revue internationale des sciences sociales, 185*, 477–488.

Arnsperger, C. (2006a). Les sciences sociales face aux problèmes sociaux. In C. Arnsperger (Ed.), *L'économie c'est nous, pour un savoir citoyen* (pp. 33–42). Eres.

Arnsperger, C. (2006b). Qu'est-ce qu'être libre dans un système? In R. de Borchgrave (Ed.), *Le philosophe et le manager* (pp. 157–170). De Boeck.

Arnsperger, C. (2008). L'angoisse dans la financiarisation: les causes existentielles de la crise. *Finance & Bien Commun, 31–32*, 94–102.

Arnsperger, C. (2010). Monnaie, dette et croissance sans prospérité: portée et limites du 'tournant' jacksonnien. *Etopia, 8*, 109–116.

Arnsperger, C. (2011a). *L'homme économique et le sens de la vie*. Textuel.

Arnsperger, C. (2011b). Dépasser le capitalisme, mais par étapes. *Revue Projet, 324–325*, 73–81.

Arnsperger, C. (2011c, October 3). *Economie existentielle, critique sociale et plasticité anthropologique*. Texte accompagnant le séminaire "Anthropologie et Justice Sociale" au Centre Sèvre.

Arnsperger, C. (2013). Fonder l'économie écologique. *Revue d'éthique et de théologie morale, 276*, 93–120.

Arnsperger, C. (2016). Progrès et conscience, Eléments pour une anthropologie économique non réductionniste de la durabilité. In G. Hess & D. Bourg (Eds.), *Science, conscience et environnement, Penser le monde complexe* (pp. 179–206). PUF.

Aubert, N. (Ed.). (2005). *L'individu hypermoderne*. Eres.

Avanzini, G. (1996). *L'éducation des adultes*. Anthropos.

Badré, M., Bourg, D., Brunetière, J.-R., Gadrey, J., Granjean, A., Perret, A., & Villalba, B. (2011). Les droits de l'avenir. *Esprit, 3*, 205–235.

Balibar, E. (2001). *Nous, citoyens d'Europe, Les frontières, l'Etat, le peuple*. La Découverte.

Balibar, E. (2009). Insurrection et constitution: la citoyenneté ambiguë. In *Revue Mouvements, Pensée critiques* (pp. 9–28). La Découverte.

Balibar, E. (2012). L'introuvable humanité du sujet moderne. L'universalité "civique-bourgeoise" et la question des différences anthropologiques. *L'Homme, 203–204*, 19–50.

Balibar, E., & Gebauer, G. (2012). L'anthropologie philosophique et l'anthropologie historique en débat. *Rue Descartes, 75*, 81–101.

Barnosky, A. D., Haldy, E. A., Bascompte, J., Berlow, E. L., Brown, J. H., Fortelius, M., Getz, W. M., Harte, J., Hastings, A., Marquet, P. A., Martinez, N. D., Mooers, A., Roopnarine, P., Vermeij, G., Williams, J. W., Gillespie, R., Kitzes, J., Marshall, C., Matzke, N., Mindell, D. P., Revilla, E., & Smith, A. B. (2012). Approaching a state shift in Earth's biosphere. *Nature, 486*, 52–58.

Bergier, B. (1992). *Compagnons d'Emmaüs – Sociologie du quotidien communautaire*. Editions Ouvrières.

Bourg, D. (2013). Dommages transcendantaux. In D. Bourg, P.-B. Joly, & A. Kaufmann (Eds.), *Du risque à la menace, Penser la catastrophe* (pp. 109–126). PUF.

Bourg, D., & Papaux, A. (Eds.). (2015a). *Dictionnaire de la pensée écologique*. Paris.

Bourg, D., & Papaux, A. (2015b). Anthropologie de la finitude. In D. Bourg & A. Papaux (Eds.), *Dictionnaire de la pensée écologique* (pp. 44–46). PUF.

Boutinet, J.-P. (1998). *L'immaturité de la vie adulte*. PUF.

Boutinet, J.-P. (2004). *Vers une société des agendas*. PUF.

Boutinet, J.-P. (2006). Interview avec Jean-Pierre Boutinet. *Communication et organisation, 28*, 194–196.

Boutinet, J.-P., & Dominicé, P. (Eds.). (2009). *Où sont passés les adultes?* Téraèdre.

Brière, L., Sauvé, L., & Jickling, B. (2010). Vivre ensemble sur Terre: un projet éducatif à dimension politique. *Education relative à l'environnement, 9*, 251–263.

Brougère, G., & Ulmann, A.-L. (Eds.). (2009). *Apprendre de la vie quotidienne*. PUF.

Cabanes, V. (2017). *Homo natura – En harmonie avec le vivant*. Buchet Chastel.

Caillé, A. (2015). *Le convivialisme en dix questions. Un nouvel imaginaire politique*. Le Bord de l'Eau.

Chartrin, T. (2015). *Apprendre à vivre, c'est se savoir mortel: récits "épiphaniques » d'autoformation existentielle*. Doctoral thesis in Educational Science, University of Nantes, defended in December 2015.

Circulaire no. 77-300 du 29 August 1977, Programmation et Coordination: bureau DGPC 9, Instructions pédagogiques, Article 525-0.

Colin, L., & Le Grand, J.-L. (Eds.). (2008). *L'éducation tout au long de la vie*. Anthropos.

Cornu, L. (2004). Espace du monde, lieux de citoyenneté, liens d'hospitalité. In H. Vincent (Ed.), *Citoyens du monde: Enjeux, responsabilités, concepts* (pp. 284–300). L'Harmattan.

Cornu, L. (2007). Confiance, étrangeté et hospitalité. *Diogène, 220*, 15–29.

Curnier, D. (2017) *Quel rôle pour l'école dans la transition écologique? Esquisse d'une sociologie politique, environnementale et prospective du curriculum prescrit*. Doctoral thesis in Environmental Sciences from the University of Lausanne, supervised by Dominique Bourg and Philippe Hertig.

Deléage, J.-P. (2010). En quoi consiste l'écologie politique? *Ecologie & Politique, 40*, 21–30.

Delory-Momberger, C. (2003). *Biographie et éducation*. Anthropos.

Delory-Momberger, C. (2004). *Les histoires de vie*. Anthropos.

Delory-Momberger, C. (2005). *Histoire de vie et recherche biographique en éducation*. Anthropos.

Dominicé, P. (2007). *La formation biographique*. L'Harmattan.

Dominicé, P. (2016). *Au risque de se dire*. Téraèdre.

Dubet, F. (2002). *Le déclin de l'institution*. Seuil.
Dumont, L. (1991). *L'idéologie allemande*. Galimard.
Eckersley, R. (2017). La démocratie à l'ère de l'Anthropocène. *lapenseeecologique.com, 1*(1), 1–19. French translation.
Ehrwein Nihan, C. (2013). Penser la crise ou plaidoyer pour une réflexion critique sur la crise de la crise à partir de l'œuvre de Hannah Arendt. *Revue d'éthique et de théologie morale, 276*, 43–60.
Foessel, M. (2004). Kant: du droit cosmopolitique à l'habitation du monde. In H. Vincent (Ed.), *Citoyens du monde: Enjeux, responsabilités, concepts* (pp. 19–31). L'Harmattan.
Foray, P. (2004). L'école et la citoyenneté du monde. In H. Vincent (Ed.), *Citoyens du monde, Enjeux, responsabilités, concepts* (pp. 179–193). L'harmattan.
Gemenne, F. (2015). L'Anthropocène et ses victimes: une réflexion terminologique. In F. Gemenne (Ed.), *L'enjeu mondial* (pp. 233–240). Presses de Sciences Po, "Annuels".
Génard, J.-L. (1999). *La grammaire de la responsabilité*. Cerf.
Giddens, G. (1994). *Les conséquences de la modernité*. L'Harmattan. French translation.
Gotman, A. (1997). La question de l'hospitalité aujourd'hui. *Communications, 65*, 5–19.
Holston, J. (2009). *Insurgent citizenship, disjunctions of democracy and modernity in Brazil*. Princeton University Press.
Jackson, T. (2010). *Prospérité sans croissance: la transition vers une économie durable*. De Boeck-Etopia. French translation.
Jonas, H. (1990). *Le principe responsabilité*. Cerf. French translation 1990 (or. 1979).
Kant, I. (1991). *Qu'est-ce que les Lumières?* Flammarion. (original 1784), French translation.
Kant, I. (2006). *Vers la paix perpétuelle*. Flammarion. (original 1795), French translation.
Lamarre, J.-M. (2002a). La personne. *Le Télémaque, 21*, 19–28.
Lamarre, J.-M. (2002b). Fichte et l'éducation: devenir homme parmi les hommes. *Le Télémaque, 21*, 65–80.
Lamarre, J.-M. (2006). Seule l'altérité enseigne. *Le Télémaque, 29*, 69–78.
Lamarre, J.-M. (2012). L'éducation cosmopolitique: apprendre le propre, apprendre l'étranger. *Le Télémaque, 41*, 31–46.
Lamarre, J.-M. (2020). « Citoyenneté mondiale et éducation cosmopolitique ». In N. Wallenhorst & E. Mutabazi (dir.), *D'une citoyenneté empêchée à une éducation citoyenne*. Le Bord de l'eau.
Lapassade, G. (1997). *L'entrée dans la vie, Essai sur l'inachèvement de l'homme*. Anthropos. (original edition 1963).
Lévinas, E. (1972). *L'Humanisme de l'autre homme*. Fata Morgana.
Lyotard, J.-F. (1979). *La condition postmoderne*. Editions de minuit.
Maigné, C. (2002). J. F. Herbart: Pédagogie humaniste et critique du sujet. *Le Télémaque, 21*, 51–64.
Manifet, C. (2015). L'éducation des adultes dans les universités en France. Une mission hétérodoxe? *Cahiers de la recherche sur l'éducation et les savoirs, 14*, 267–294.
Martuccelli, D. (2002). *Grammaires de l'individu*. Gallimard.
Marx, K. (2014). *Thèses sur Feuerbach*. Syllepse. (or. 1888), French translation.
Masschelein, J. (2002). Education et humanisme? *Le Télémaque, 21*, 37–50.
Monjo, R. (2013). L'école peut-elle être, à la fois, libératrice et obligatoire? *Le Télémaque, 43*, 87–102.
Morin, E. (1990). *Introduction à la pensée complexe*. Seuil.
Morin, E. (2014). *Enseigner à vivre*. Actes Sud/Play Bac.
Murard, N., & Tassin, E. (2006). La citoyenneté entre les frontières. *L'Homme et la société, 160–161*, 17–35.
Neveu, C. (2004). Les enjeux d'une approche anthropologique de la citoyenneté. *Revue européenne des migrations internationales, 20*(3), 89–101.
Pal, J. S., & Eltahir, E. A. B. (2015, October 26). Future temperature in southwest Asia projected to exceed a threshold for human adaptability. *Nature Climate Change, 6*, 1–4.

Payan, S. (2011). De l'autochtonie à la citoyenneté: vers une construction psychique de la citoyenneté. *Topique, 114*, 115–123.
Pélabay, J. (2011). Former le "bon citoyen" libéral. *Raisons politiques, 44*, 117–138.
Pelluchon, C. (2017). L'éthique des vertus: une condition pour opérer la transition environnementale. *lapenseeecologique.com, PUF, 1*(1), 1–18.
Pelluchon, C. (2018). *Ethique de la considération*. Seuil.
Pierron, J.-P. (2017). Pour le meilleur et pour le pire? *Jusqu'à la mort accompagner la vie, 129*, 93–103.
Pineau, G. (dir). (1992). *De l'air. Essai sur l'éco-formation*. Païdéia/Sciences et culture.
Pineau, G. (2003). Vers une anthropo-formation en deux temps trois mouvements. *Spirale, 31*, 35–46.
Ricken, N. (2002). La philosophie de l'éducation de langue allemande. Rupture avec l'un: différence – pluralité – socialité. *Le Télémaque, 21*, 121–142.
Ricœur, P. (1961). Préface. In H. Arendt (Ed.), *Condition de l'homme moderne* (pp. I–XXVIII). Calmann-Lévy.
Rosa, H. (2013). *Accélération. Une critique sociale du temps*. La découverte. (original edition 2010), French translation.
Rosa, H., & Wallenhorst, N. (interview). (2017). Apprendre ensemble dans la société de l'accélération. *Bildungsforschung, 2017*(1), 1–7.
Sanchez-Mazas, M., & Gély, R. (2005). Des appartenances aux identités: vers une citoyenneté politique européenne. *Connexions, 84*, 73–86.
Sauvé, L. (2009). Le rapport entre éthique et politique: un enjeu pour l'éducation relative à l'environnement. *Education relative à l'environnement, 8*, 147–162.
Sauvé, L. (2015). Education à l'environnement. In D. Bourg & A. Papaux (Eds.), *Dictionnaire de la pensée écologique* (pp. 376–379). PUF.
Schérer, R. (1997). Cosmopolitisme et hospitalité. *Communications, 65*, 59–68.
Scranton, R. (2013, November 10). Learning how to die in the Anthropocene. *The New York Times*.
Scranton, R. (2015). *Learning to die in the Anthropocene*. City Light Publisher.
Semal, L., & Villalba, B. (2013). Obsolescence de la durée. La politique peut-elle continuer à disqualifier le délai? In F.-D. Vivien, J. Lepart, & P. Marty (Eds.), *L'évaluation de la durabilité* (pp. 81–100). Editions Quæ.
Sigaut, O. (2010). L'éducation à l'environnement, entre politique et politiques publiques. *Education relative à l'environnement, 9*, 59–75.
Sinaï, A. (2013). Le destin des sociétés industrielles. In A. Sinaï (Ed.), *Penser la décroissance* (pp. 23–48). Presses de Science Po.
Taleb, M. (2016). Ecoformation. In A. Choné, I. Hajek, & P. Hamman (Eds.), *Guide des Humanités environnementales* (pp. 83–91). Presses universitaires du Septentrion.
Tassin, E. (2004). Le cosmopolitique à l'épreuve de l'étranger. In H. Vincent (Ed.), *Citoyens du monde: Enjeux, responsabilités, concepts* (pp. 269–282). L'Harmattan.
Touraine, A. (1997). *Pourrons-nous vivre ensemble? Egaux et différents*. Fayard.
Touraine, A. (2010). *Après la crise*. Seuil.
Touraine, A. (2015). *Nous, sujets humains*. Seuil.
Vincent, H. (2004). Citoyen du monde: éléments de problématiques. In H. Vincent (under supervision), *Citoyens du monde: Enjeux, responsabilités, concepts* (pp. 7–16). L'Harmattan.
Wallenhorst, N., Poché, F., Robin, J.-Y., Heslon, C., Bergier, B., & Boutinet, J.-P. (2017). Il est grand temps de rallumer les étoiles. revue-*esprit.fr. Online*. http://www.esprit.presse.fr/actualites/nathanael-wallenhorst-bertrand-bergier-jean-pierre-boutinet-christian-heslon-fred-poche-jean-yves-robin/il-est-grand-temps-de-rallumer-les-etoiles-543
Weber, A. (2016). *Enlivenment. Eine Kultur des Lebens. Versuch einer Poetik für das Anthropozän*. Matthes und Seitz.
Weber, A. (2017). *Sein und Teilen – Eine Praxis schöpferischer Existenz*. Transcript Verlag.
Wilhelm Graf, F. (2002). Le politique dans la sphère intime, Protestantisme et culture en Allemagne au XIXème siècle. *Annales. Histoire, Sciences Sociales, 3.*, 57ème année, 773–787.

Chapter 12
Resilient Education: Dealing with Nascent *Hybris*

Abstract This chapter is divided into three sections. The first section examines the aims of education in the Anthropocene: ow they can be positioned between the two poles of learning to live and to become oneself, or conversely, to learn about the exteriority of the world? While it may be relatively easy to set the goal of education in the Anthropocene as learning about the world, and therefore about politics, the second section goes into more detail on the need to learn to live, and to bring life to others by supporting them with limits (Hétier, R. (2019). Apprendre à faire vivre en Anthropocène. In N. Wallenhorst & J.-P. Pierron (Eds.), *Eduquer en Anthropocène*. Le Bord de l'eau) starting with the structuring limits on human actions that are, in fact, planetary boundaries. The third section then focuses on the shift we need to make in the way we learn about the world.

Keywords Hybris · Anthropocene · Resilient education · Political education · Convivialism

12.1 *Hybris* and the World

The geoscientific teachings of the Anthropocene put the emphasis on human *hybris*. They illustrate how a burgeoning and unbridled desire for more, when combined with this same desire in others, contributes to potentially global levels of destruction. Thus, it becomes an imperative for education to rein such *hybris* in. The *Second Convivialist Manifesto* identifies the imperative of *hybris* control in Chap. 2. It states that 'The first condition for rivalry to serve the common good is that it be devoid of desire for omnipotence, excess, *hybris* (and a fortiori *pleonexia*, the desire to possess ever more).' The convivialists continue, 'On this condition, it becomes rivalry to cooperate better. This principle of controlling *hybris* is in fact a metaprinciple, the principle of principles. It permeates all the others and is intended to serve as a regulator and safeguard for them. For each principle, pushed to its extreme and not tempered by others, risks being reversed into its opposite: the love of nature or

© The Author(s), under exclusive license to Springer Nature Switzerland AG 2023

N. Wallenhorst, *A Critical Theory for the Anthropocene*, Anthropocene – Humanities and Social Sciences, https://doi.org/10.1007/978-3-031-37738-9_12

that of abstract humanity in hatred of concrete men; the common sociality in corporatism, clientelism, nationalism, or racism; individuation in individualism indifferent to others; the creative opposition in the struggle of egos, in the narcissism of the small difference in destructive conflicts.'

The first section examines the aims of education in the Anthropocene: ow they can be positioned between the two poles of learning to live and to become oneself, or conversely, to learn about the exteriority of the world? Though education is not primarily aimed at the well-being of the learner, this does not mean that we should retain archaic methods built around knowledge, imparted from on high..

While it may be relatively easy to set the goal of education in the Anthropocene as learning about the world, and therefore about politics, the second section goes into more detail on the need to learn to live, and to bring life to others by supporting them with limits (Hétier, 2019) starting with the structuring limits on human actions that are, in fact, planetary boundaries. The current and future systemic ruptures, when examined through the scientific articles on the Anthropocene, are profoundly significant. We must learn to live again, by supporting each other within these limits brought about by these systemic ruptures. These limits structure our thinking about education. This second section offers a selection of some of the insight from the Anthropocene to be passed down through education, allowing *hybris* to be openly acknowledged and discussed (rather than hidden, as is usually the case). Passing on the knowledge about the Anthropocene is a real challenge. Indeed, there has been a huge gulf between political and scientific discourse on the environmental situation for several decades. Whilst the Anthropocene is a new period born of ruptures, it can also be heralded as a new period marked by solidarity and conviviality. This is a thesis developed by Pablo Servigne, Raphaël Stevens and Gauthier Chapelle in *Une autre fin du monde est possible* (2018) (*Another End of the World is possible*, Polity, 2021). Though learning about the Anthropocene may lead to feelings of desolation, sadness, resentment, anger or disgust, these emotions can be overcome through sharing, which generates conviviality through shared activity. This is one potential approach as to how to manage *hybris*.

The third section then focuses on the shift we need to make in the way we learn about the world. We need to move from learning about the accelerating world (i.e. learning about the capitalist system, and falling into it as it is perpetuated) to learning about resonance. The contemporary period is marked by threefold acceleration (technological, social changes and pace of life), highlighted by the German sociologist Hartmut Rosa (2013, 2014), in the wake of the analyses carried out in the late nineteenth and early twenteith centuries on the accelerating pace of life, by his compatriot, the sociologist Georg Simmel. This acceleration is even perceptible in the sediment records, and it is this acceleration that has propelled Earth into the Anthropocene, with the uncertain future it holds. Now, at the heart of this acceleration with so many geological implications, what do we need to learn to help us curb this problematic acceleration? What must we learn to bring about remedial action, when we are at the heart of the Great Acceleration that is the Anthropocene? Hartmut Rosa, through his intellectual journey and the evolution of his thinking, proposes to move from this acceleration towards resonating with the world – and this resonance appears to be fundamentally post-Promethean and convivialist. This shift from acceleration to resonance is the second proposal for how to manage *hybris*.

12.2 The Purpose of Convivialist Education: Learning About the World or Learning to Live?

In order for the knowledge we pass on to 'make sense', it must be integrated into the learners' reading. Edgar Morin, in his work on education and in his recent book, *Enseigner à vivre* (2014), clearly captures the pedagogical challenge of integrating an existential component into education. Without actually naming it as such, he updates the ideal of *Bildung* presented in Chap. 11. However, the introduction of this existential component could be ambiguous if it is not linked to an examination of the purpose of education. Is that purpose to learn about the world or to learn how to live?

For Morin, a teachers' role is to pass on 'a culture that allows us to understand our condition and to help us live with it' (Morin, 1999, p. 11). Education must 'teach us to put up with the prosaic aspect and live the poetic aspect of our lives' (Morin, 1999, p. 11); it must 'teach us how to live' (Morin, 1999, p. 51). While teaching us how to live is one possible purpose of education, teaching about the world and guiding the learner as they learn about the world is another. Following in the footsteps of Part II of this work in an Arendtian reading of the political, learning about the world means: Firstly, becoming aware of the cultural heritage preceding our existence. Secondly, discovering plurality – human and non-human. Thirdly, learning from what arises in relation. Indeed, we learn about the world through relationships and sharing – the world referring less to geopolitical space and more to the 'between-us'. Learning to live together in order to act as one is a fundamental component of this way of learning about the world.

12.2.1 Interiority and Exteriority in Education

Defining teaching how to live as the potential goal of education harks back to our notion of *Bildung*, which the Franco-German educationalist Christine Delory-Momberger defines as 'the self-taught movement by which every human as a unique being ascribes to his or her own traits and thus participates in constructing 'humanity' (as a universal value)'(2009, p. 152). The notion of *Bildung* (which we have already said is difficult to translate into English because it refers to a conceptual universe so different from our own) is frequently translated as self-training, self-educating or self-culturing. We explained previously that what sets this notion (which is closer to an ideal than a conceptual tool) apart is the existential, internal component within the teaching.[1] This idea of Protestant origin is far removed from French ideas on education – historically rooted in Catholicism (Wallenhorst, 2013) – where the institution is central to the learner's future success in life. In truth, it would be more accurate to translate this concept as training *of the* self, education *of the* self, or culture *of the* self. This individualistic component of *Bildung* is often overlooked,[2] though it is at its core. This notion emerged in Protestant circles against a backdrop of a move towards personal salvation: ideas referred to in the dogma on salvation were replaced by more transformative teaching whereby each person must work personally on his or her own learning (Wilhelm Graf, 2002, p. 784). As identified in Chap. 11, the self is the goal of *Bildung*. The other and the world are called upon but as part of the training of the self. They are not the end goal of *Bildung*. *Bildung* is within, while the 'other' refers to the world, the outside.

An important moment in the development of this 'value-idea' (Dumont, 1991, p. 108) is to be found in the *Bildungsromane* from the late eighteenth to late nineteenth century. In it, journeys of self-teaching are made through discovering the world, the most famous being *Wilhelm Meister's Journeyman Years* by Goethe (1795–1796). Learning about the world stirs up existential elements within the subject: geographical mobility is accompanied by self-discovery. Exploring the external world brings with it an internal exploration of the self. The subject learns about their own existence by setting out to discover the world. They respond to the call of the world to broaden their horizons and grow into who they are. The world is what enables them to enter adult life and find their place in existence; it is, in a way, to serve the development of the self. For Humboldt, it is the internal self that needs the external world (Von Bonsdorff & Grenaudier-Klijn, 2011, p. 185), rather than the other way around. He does not set out to contribute to the world and assume responsibility for safeguarding it for future generations.

12.2.2 Experiencing Existence to Open Up the World

In the spirit of *Bildung*, 'man's work is himself' (Hess & Weigand, 2008, p. 14). However, is man's greatest work really himself? From a convivialist perspective, would the greatest work of man not be the relationship to the other, the one over which we have no control and which eludes us? The German traditions of Educational Sciences (*Erziehungswissenschaften*) and Spiritual Sciences (*Geisteswissenschaften*) can again be called upon, from a critical perspective, in thinking about this interweaving of the pedagogical challenge of introducing an existential component into

[1] This is the reason why it, rightly, intrigues French people interested in education and teaching.

[2] There is little criticism of the concept of *Bildung* as a whole. However, the UK-based educationalist Gert J. J. Biesta (2006, p. 2–8) criticises the intrinsic idealisation of the concept of *Bildung*, and the American philosopher Richard Rorty substitutes the notion of edification for *Bildung* in order to mitigate the excessive attention given to the self, to the detriment of community and solidarity (1990, p. 396).

education with the purpose being underpinned not by an individualistic ethic that has largely been shown to be limited, but on the contrary, by a convivialist ethic. In German, there are two concepts which refer to the notion of experience: *Erlebnis* and *Erfahrung*. In German thinking, *Erfahrung* is *Bildung*. *Erfahrung* is an experience that has made its mark on existence, but it is also the experience of existence or existence as an experience of life. It is in this sense an 'authentic experience' (Lavelle, 2007, p. 111), lived through subjectivity. *Bildung is* conceived on the basis of *Erfahrung*. In German educational concepts, where the notion of *Bildung* is central, the subject as a whole is involved. Education in the German conceptual universe is based on the density of experience (*Erfahrung*) and the subject's experience (*Erlebnis*), without ignoring the aesthetic or poetic component of existence: 'The experiential is the overall experience, taken in its personal totality. [...] It is an experience built and grasped in the lucidity of a consciousness with self-possession, and in the generosity of a love that gives of itself. It is a fully personal experience, in the strict sense of the word. It is a lifelong education. Experience is acquired, over and over again. In this sense, all authentic intellectual experience is experiential' (Hess & Weigand, 2008, p. 12).

Experience defined in this way (the *Erfahrung,* which is *Bildung*) calls into question the subject's relation to the world: 'we can define experience as an act, or set of acts, by which man perceives himself in relation to the world, space and time' (Hess & Weigand, p. 13). *Erfahrung* is the interface between the self and the world. Conversely, the limitation of *Bildung* is that it does not open the subject up to the importance of his or her role in the world from the point of view of the other. A convivialist re-thinking of education here would imply thinking of it against the backdrop of the ethic of the other, while upholding the importance of *Erfahrung* (which is *Bildung*) teaching – a real challenge. It would imply learning about the world (the life space of the other) from an inner experience – which is what the knowledge of the Anthropocene produces – thus generating intense implication. It would be to switch from the logic of 'constructing one's self from the existing world' (individualist ethic) to 'constructing the world that will be there for the generations to come' (convivialist ethic). The experience of the world, which precedes and succeeds the learner, enables them to experience their own human experience as an entry into something beyond themselves.[3]

[3] Another possible way of thinking about the purpose of education and the tensions within which they may be embedded may be to use the categories of the familiar and the foreign. Using these categories, we can narrow down the political figure of the citizen by reflecting on the cosmopolitan component of the purpose of education, that of the aim of 'sharing a common world by bringing cultures, political communities and citizens together' (Lamarre, 2012, p. 32) Education is cosmopolitan only if it 'experiences and tests the foreign' (Lamarre, p. 41), while also being pluralistic and allowing for learning about the familiar. Cosmopolitan education 'is a form of hospitality – educational hospitality.' The foreign is not to be assimilated, but welcomed (Lamarre, p. 44). Conviviality as an educational style implies a voyage through the foreign, but not only for the purpose of a return to the familiar Lamarre draws heavily on the German poet Friedrich Hölderlin, who, at the beginning of the nineteenth century, distances himself from German idealism by giving equal importance to the familiar and the foreign in the notion of *Bildung*.

12.2.3 Taking on Responsibility for the World

The world is so important to Arendt because it is where birth takes place, which represents the power to continue the human adventure and the power to renew the world. The world, though ancient and permanent, is simultaneously renewing itself through birth, which gives it its youth (1972, p. 251). Birth here does not refer primarily to a biological, but rather an ontological understanding – what is foremost is the individuality of each human being. Birth symbolises the possibility of the human being acting, beginnings and the power of initiative. While having to learn about the world and its permanence, learners, by their birth, are an 'irruption from the future' in a present so marked by the past. They have the capacity to act on the world. Learning about the world enables learners to put into practice the power of initiative with which they are born.. This is what will enable them to live out their political vocation to renew the world.[4]

To Arendt's followers, education is pre-political – it precedes and prepares for politics. The aim of education is to learn about the world and, 'to learn what the general anthropological condition of political activity is within the world' (Foray, 2001, p. 93). For Arendt, the teacher's mission is to teach about the world by basing their teaching on their own ability to take responsibility for the world for children so that they can learn about the world and go on to take responsibility for it. She speaks of the entirely pre-political role of education. 'The Crisis in Education', a text that is nearly 70 years old, is interesting when thinking about contemporary pedagogical issues. In it, Arendt criticizes 'pedagogical' approaches to education, which is why this text has been wrongly labelled conservative. In fact, the flaw that any good teacher risks falling into is the confusing the ends with the means.

To be responsible for the world means to try, alongside others, to make it hospitable, to take care of that 'between-us' which is manifested, in particular, by the word 'shared', allowing a viable world to emerge among humans – one which is also hospitable to the non-human. The aim of education is not for the learner to learn to live, but to learn about the world. Can the self be the goal of education? Would the purpose not be something else: to learn the world in order to assume responsibility for it, in the face of the other – whose three radical figures, as mentioned previously, are the stranger (the different other from elsewhere), future generations (those who are not yet here but whom we know will be one day), and the

[4] An underlying issue in widespread contemporary existential *krisis* (Wallenhorst, 2015) is that of vocation (Prouteau, 2004). The idea of vocation, in its secular sense, is based on that of orientation but incorporating the idea of an existence that has meaning through its social utility in a field of competence. In education, to speak of vocation rather than of orientation involves a notional proximity with Arendt's natality It means giving primacy to the singularity of each person and to his or her capacity for creativity. It also implies a relational conception of existence, based on a call (the etymological origin of vocation is *vocare*: to call): each person has their place within the plurality of humanity. In this conception of humanity, there can be no question of anyone's existence being pointless. To think in terms of vocation is to situate oneself in a paradigmatic space where no-one is condemned not to find a place, value, and usefulness.

whole of non-human life? It seems to us that only the world can constitute this finality.

The need to experience existence is not the same as the end; the purpose is not to enable the learner to learn to live (learning underpinned by an individualist ethic) but to learn the world, in order to assume responsibility for the world (learning underpinned by a convivialist ethic). The teacher's goal is to ensure that the learner's existence is mobilised and altered by the educational dynamic, without the formation of the self becoming the aim and without a form of personal development consisting of ego development (Schmitt, 2014). This aim must be set without ambiguity so as not to risk reifying and sacralising the experience of existence; the other (human and non-human) living in the world, matters just as much as oneself – if not more.

12.3 Learning to Live (and to Bring Life) by Existing with Limits

12.3.1 Living Within the Limits of the Biosphere

The direst threat of the contemporary period is humanity's potential to destroy itself (Les Convivialistes, 2013). Thus, one of the challenges of education is to teach us how to control our power, and also keep our *hybris* under control. How can humans, wielding demiurgic power, learn moderation? Hence, the function of education is to learn to identify and control the nascent *hybris*. The Promethean *hybris* to which we refer here is that of the individualism of the modern *homo oeconomicus*. The hypothesis in this work is that a safeguard against *hybris* is the sharing of existence with others, within plurality (in other words, that convivialism is a bulwark against *hybris*). We must shift from Promethean mastery to mastery of our Prometheism, starting with mastery of our *hybris*.[5] Moreover, it is absolutely necessary, in the sharing of existence, to support ourselves within limits (Hétier, 2019), starting with the planetary boundaries, of which we must be aware.

The simple and clear diagram of planetary boundaries can serve as an important indicator to guide human behaviour. Indeed, it has particularly powerful pedagogical potential, and clearly illustrates the need to cease the ongoing violation of

[5] Moreover, on a more individual scale, it is necessary to learn to identify and satisfy one's desire for eternity, which must be distinguished from a mortifying demiurgic desire for immortality and omnipotence. In Arendt's view, the greatest desire of *homo faber* is the desire for permanence and immortality. This 'cannot be fulfilled in his undertakings, but only when he understands that the beautiful and the eternal cannot be manufactured' (1983, p. 342). For this purpose, it is necessary to enter into a new understanding of mastery, no longer thought of as domination but as 'adjustment, adaptation, adequacy' – that is, 'modelled on the gardener rather than the engineer' (Flahault, 2005, p. 47). This refers to the 'choice of non-power' proposed by the French intellectual Jacques Ellul as a critical attitude, and renunciation of all possibilities in favour of prudence.

planetary boundaries. As an extension of the pedagogical work of the planetary boundaries diagram, another device for combatting nascent *hybris* through education is to look at humanity's geological footprint in comparison to the planet's carrying capacity. The proposed planetary boundary transgressions alter the major regulatory mechanisms of the biosphere and we see the depletion of the resources necessary for our economic functioning as well as the 'saturation of the carrying capacities of the biosphere system that sustains us' (Bourg, 2013, p. 63). The concept of 'carrying capacity' primarily applies locally for a species, and reflects the environmental constraints that limit its growth. When this concept is used on a planetary scale for the human species, matters become considerably more complex, and it is then necessary to integrate it with the issue of lifestyles (Leandri, 2015).[6]

In a 1992 article, the American environmentalist William Rees proposed the concept of ecological *footprints*, based on the idea of natural capital available for economic life, corresponding to what nature can generate in one year. In line with this concept, each year, an American NGO, the *Global Footprint Network*, calculates the '*Earth Overshoot Day*', corresponding to the day of the year when human beings have consumed more resources than the Earth can produce or regenerate in one year. When Earth Overshoot Day has passed, it means that human beings are consuming non-renewable earth resources; the objective is that this day should not arrive until 31 December. According to the calculation methods defined in 2017, Earth Overshoot Day was defined as 24 December in 1971, and 2 August for 2017, which corresponds to the 'consumption' of the resources of 1.7 planets in one year. From an educational perspective, it is important to regularly calculate this geological footprint (the book by Babeth Porcelijn, *The Hidden Impact*, is interesting for this purpose): individual, family, social, afference to the different political choices in progress, etc. and to systematically examine it in relation to the planet's carrying capacity.

The main environmental issue we face is the management of material and energy flows on the scale of the entire biosphere – in other words, the sustainability of our production methods (Faber et al., 2002; Bourg, 2009; Bourg & Papaux, 2010, p. 108; Arnsperger & Bourg, 2017).[7] The management of these material flows is indicative of a problem for humanity in the relationship to the limits and finitude of its

[6] With current lifestyles, the saturation of the biosphere's capacities could overload our ecosystems: 'We are hacking away at the branch on which we sit. […] We are heading towards a situation worse than that experienced by western Europe in the fourteenth century with the advent of the Little Ice Age and its consequences for food and health, the Hundred-Years War (1337) and the Black Death (1347–1351), which killed half of all city dwellers' (Bourg et al., 2017, p. 17).

[7] Human beings, using tools and technology, are able to modify their environment on a large scale, which other living species are not capable of doing to such an extent. Nitrogen and CO_2 are not pollutants, *per se* – it is the rate of their emission that is the problem (Bourg, 2012). How we uncouple non-renewable resource consumption and growth is a problem (Perret, 2014, p. 72). We are dealing with an unreasonable level of growth in material and energy flows. Mined mineral deposits may be exhausted within 10–50 years, with half of the water tables depleted (Bourg, 2013, p. 64).

environment.[8] Arnsperger and Bourg, in *Ecologie intégrale – Pour une société permacirculaire* (2017), mention the absolute necessity of reducing our ecological footprint in order to return to within the limits of the biosphere. To this end, they propose the development of a permacircular economy whose objective is 'the preservation of the biosphere in order to maintain its viability, first and foremost for the human species' (p. 48), and which involves net reductions in material flows by 'seeking to loop human activities back on themselves in order to spare as much as possible the biosphere that hosts them' (p. 57).

The world population is often mentioned as a determining factor when calculating environmental impact. In the year 1000, the planet had a population of 300 million human beings, rising to 500 million in 1500 and to about 790 million by 1750. The world population then continued to grow at a dizzying rate for over 200 years, rising from one billion in 1800 to six billion in 2000, and sits at more than eight billion today, with the UN estimating that the world's population will be around 9.5 billion in 2050 (United Nations, 2017), allowing for a decline in the birth rate. In the space of a century, the world's human population has almost quadrupled. However, singling out growth in world population when calculating ecological footprints often masks other, much more significant, growth: that of living standards and consumption.[9] It is essential that these points be the subject of discussion and educational debate. Indeed, while the population has increased sixfold, in the same period energy consumption has increased 40-fold and economic production 50-fold. While world population doubled in the 50 years from 1950 to 2000, aluminium production (to take one example) increased 118-fold between 1930 and 1990 (Bourg, 2001, p. 94). The land area used intensively for human activity has increased from 10% to 25–30% (Steffen et al., 2007, p. 616, 2011, p. 848). It can therefore be argued that 'the limit to the planet's capacity to accommodate, feed, and support human populations has much less to do with nature and much more to do with our social organisation. […] The planet's carrying capacity is more a social than a natural phenomenon' (Paillard, 2017, pp. 3–5).[10]

[8] Hans Jonas was able to gauge the seriousness of the environmental situation early on. He was a real pioneer in the development of his political ethics, with the 1979 publication of *Imperative of Responsibility* (*Das Prinzip Verantwortung*), emphasising the need for a principle of prudence in our relationship with the environment.

[9] Over the past 50 years, changes in nutrition and food security have 'almost halved undernutrition while doubling the proportion of overweight people'. However, these changes 'have also reduced the resilience of the biosphere', transgressing four of the six planetary boundaries to food production (Gordon et al., p. 1). We need to shift the emphasis from quantity to quality of food production and make significant changes in our food-production systems. The challenge for the coming decades is to ensure that food production can grow, while reducing the ecological footprint. Foley et al. (2011) demonstrate, in an article in *Nature*, that this goal is achievable by reducing agricultural expansion, increasing crop efficiency, changing diets and reducing waste.

[10] For example, researchers have identified scenarios for feeding 9–10 billion people in 2050 without any additional deforestation (Erb et al., 2015).

12.3.2 Recognising the Earth as Our Master

Hannah Arendt's intuition in 1958 as to the extent of the environmental destruction in progress has proven to be perspicacious. 'Should the emancipation and secularization of the modern age, which began with a turning-away, not necessarily from God, but from a god who was the Father of men in heaven, end with an even more fateful repudiation of an Earth who was the Mother of all living creatures under the sky?' (1983, p. 8). The exercise of developing a self-sufficient 'between-us' is complex, since it is part of the cultural foundations of our civilisation. Indeed: 'Today, it is often said […] that humanity is part of the planet. Easy to say, but difficult to conceive of! We are ill-prepared to mean what we say – i.e. to put into practice all that this environmental formula implies' (Flahault, 2008, p. 60). For Flahault, entering a post-Promethean era means giving up 'an overarching and distant position from our environment' (Flahault, 2005, p. 378). The question of the 'between-us; reflects what is between human beings, what connects them and sets them apart, not neglecting the material, bio-geochemical space that is between humans. Nature and the environment are therefore integral parts of the 'between-us'. Ecological thinking attests to a decline in Prometheism and is an essential foothold for reflection on these post-Promethean spaces: the planet's limitations are an obstacle to the expansion of the Promethean project, as we are forcefully reminded (Flahault, 2008, p. 17).

Since the 'between-us' is formed by 'nature', the political also emerges in the relationship with 'nature', in which we can 'grow together' (Berque, 2016, p. 170).[11] Humanity is politically active in a world, which is made possible by bio-geochemical matter-flow. It is the act of joining this flow of matter between the Earth and humanity that makes political activity possible. Our political activity is dependent on our link to matter, and therefore to the Earth. Another way of saying much the same thing is that 'The Earth is our master' (Westbroek, 2015, p. 962). Assigning a central role to the environment in educational thinking gives primacy to the notion of hospitality – that enjoyed by humanity, welcomed on Earth as an extension of the welcoming of the other. This is one of the virtues of ecological thinking, which stems from the idea of human and non-human hospitality – of Earth towards humans, of humans towards each other, and of humans towards non-humans who inhabit the earth with them. The recognition of humanity's role in this flow of biological, geological and chemical matter is necessary in understanding the concept of a post-Promethean 'between-us', but it is also important in understanding the extent to which co-existence – constituted by social and cultural life – comes before individual existence.

[11] As Valerie Cabanes writes: 'We must find a way of cohabiting harmoniously with trees, plants and animals, but also with each other' (Cabanes, 2017, 4th page).

12.3.3 Understanding Civilisations as the Result of Climate Stability

Indeed, the Earth must become our master. For this reason, it is important, when teaching future generations, to gauge climate impact on cultures. As mentioned in the introduction, the great civilisations emerged through having gained control of ecosystems – something made possible by climate stability. What will become of our civilisation as a result of the current major climate upheaval?

The question of global warming as an unprecedented event in human history must be analysed in depth in education.[12] We need to quash burgeoning climate-scepticism as it is so destructive. It is important to recognise that the average increase in degrees Celsius compared to the twentieth century is a good indicator of global warming for scientists, but it is a very poor indicator for public opinion. Increase in temperature peaks would be a better indicator, as would loss of agricultural areas.[13]

Indeed, in a prospective study published in 2017, Margot Bador et al. identify what the heat peaks could be in 2100 were there to be a heatwave comparable to the one in 2003, with a global surface temperature increase of 3.7 °C; according to the IPCC, this is a probable scenario. The authors use temperature simulation models to identify temperature peaks in eastern France that are 12 °C higher than those experienced in 2003, which would generate heat peaks of 55 °C. The current world record is 58 °C, recorded in a desert in Libya (El Fadli et al., 2013).[14]

---

[12] For example, Steffen et al. (2011) highlight the climate variations on Earth's surface over the last 70 million years and identify the importance of the more than 10,000 years of climate stability in the Holocene. In the diagram they provide in their article, we see a long period of cooling over the last 70 million years. This coincides with the decrease in atmospheric CO_2 levels. Over the last three million years, temperature has repeatedly oscillated over 100,000-year cycles, alternating long cold periods with short warm periods. These oscillations are due to changes in the earth's orbit. We then see the transition to the Holocene and the onset of stability. It is interesting to view the last 11,700 years of Earth's history and the long interglacial period that is the Holocene.

[13] Recent studies on the extent of heat waves at Earth's surface have led several researchers to conduct further research that measures the impacts of these heat waves on crop yields. Stefan Siebert and Frank Ewert (2014) have shown that future agricultural production may be threatened by extreme heat waves (especially due to a decrease in photosynthesis, pollen production and the number and weight of grains). None of the effects of climate change are particularly easy to predict. Forecasting climate change is hampered by a series of uncertainties with interannual, decadal and multidecadal fluctuations in addition to anthropogenic changes. In addition, the science of predicting changes that we will encounter in years to come is at the crossroads between two forecasts – climate change *per se*, and the responses from the various species, readjustments to their niches and interactions between species (Jackson et al., 2009). To date, there has been some work done on forecasting the effects on species, but little work on the ways in which interactions between species are altered. This is an important prediction for agricultural food production, for example. In addition to modifying agricultural yields, climate change will lead to changes for pollinators and rodents that we should try to anticipate (Gilman et al., 2010).

[14] Contrary to the illusion that humans could gradually adapt to a warming climate, Sherwood and Huber (2010) demonstrate that humans cannot withstand excessive heat stress (temperature and humidity, or wet-bulb temperature). Like all living species, humans cannot survive above a certain

According to Sherwood and Huber (2010), the threshold beyond which human life is no longer possible is 35 °C TW (wet-bulb temperature). The current climate very rarely exceeds 31 °C TW at the Earth's surface. In their study of the Arabian Gulf, a key area for oil exports, they show that the suitability of Earth for human habitation is likely to be significantly affected and that, were there to be a temperature increase of 3.7 °C by the end of the century, this area would run the risk of no longer being habitable. In this region, several places would risk exceeding 35 °C TW and becoming uninhabitable for humans.[15]

Currently, there are three regions where the wet-bulb temperature regularly exceeds 28 °C TW. These are Southwest Asia (the Persian Gulf and the Red Sea), South Asia (the Indus and Ganges valleys), both of which are covered in the article by Im, Pal and Eltahir (2015), and eastern China. The predicted heatwaves are in places with very high population density and low GDP per capita. Though only 4% of the current populations of South Asia in the geographical areas these authors examined would be exposed to +35 °C wet-bulb temperatures in 2100, 75% would be exposed to wet-bulb temperatures over 31 °C (in the range considered dangerous for humans). It is important to note that these deadly heatwaves are not only part of the prospective scenarios but are already being encountered.[16] Working on this 'knowledge of the Anthropocene' is crucial to education in the Anthropocene (which includes education about the Anthropocene).

temperature. The human body has the ability to adapt to outside temperatures through transpiration. However, when heat and humidity thresholds are exceeded, the body can no longer adapt, the perspiration system becomes overwhelmed and metabolic heat can no longer be dissipated, with death from hyperthermia occurring within hours. Camilo Mora et al. (2017) conducted worldwide analysis of lethal heat events between 1980 and 2014. They estimate that they have identified 783 occurrences of excess human mortality associated with excessive heat. On that basis, the authors estimate that 30% of the world's population is currently exposed to potentially fatal heatwaves 20 days a year. By 2100, this figure will have risen to between 48% and 74% of the world's population, depending on whether greenhouse gases are reduced.

[15] In addition, the areas where the temperature would exceed 60 °C dry heat would also be significant. In dry heat conditions, the human body runs a high risk of not surviving temperatures even well below 60 °C. It should also be noted that these heat peaks would lead to a series of malfunctions in our technical infrastructures. Above 40 °C DBT, many machines like trains or planes cannot function. In their 2017 article, Eun-Soon Im, Jeremy Pal and Elfatih Eltahir note that in the summer of 2015, the wet-bulb temperature in Bandar-e Mahshahr (in the Persian Gulf, in Iran) reached 35 °C TW. and mention that the critical threshold of 35 °C TW by the end of the century according to the business-as-usual assumption may even be reached much earlier. Im, Pal and Eltahir also consider that peak wet-bulb temperatures will be approaching this critical threshold in south-east Asia by the end of the century. The regions most likely to be exposed to these risks are heavily populated agricultural areas in the Ganges and Indus basins. This causes the authors to claim that 'Climate change, without mitigation, presents a serious and unique risk in South Asia, a region inhabited by about one-fifth of the global human population, due to an unprecedented combination of severe natural hazard and acute vulnerability.' (p. 1).

[16] Thousands of humans deaths and livestock loss have been reported in eastern India on different occasions as a result of heat waves: in 1998 in eastern India, in 2003 in western India, and in 2010.

12.4 From Acceleration to Resonating with the World

12.4.1 Confronting the 'Great Acceleration'

One of the challenges of the Anthropocene is the relationship to time. Based on their analysis of the length of time in the past, Earth System scientists can identify long-term scenarios to be used to prepare, in the here and now. However, the Anthropocene, which is also known as the Great Acceleration, in part linked to social, technological and scientific acceleration, raises the question of our relationship to speed. Part I of this work on the Anthropocene highlighted a Promethean problem, which is discussed in Part II. Acceleration is the clearest embodiment of *hybris* in our time. It is the very characteristic, in the life of each individual, of the grip of capitalism, which an education in the Anthropocene ultimately seeks to overcome. Capitalism generates an imperative for dynamic stabilisation within our societies. Thus, everything accelerates: technological evolution, social changes and rhythms of life. What effect does this have on our learning? How do we learn and internalise this dynamic stabilisation of capitalism? Studying the acceleration that entering the Anthropocene exposed continues the questioning of the way in which environmental limitations are revealing of anthropological limitations. Is it possible to envisage civilisational evolution that does not have these harmful effects on the Earth system? The concept of resonance developed by Hartmut Rosa is the counterpart to acceleration, and its function is to address the problem of *hybris*.

Part I of this chapter consists of a critical reading of the impact of contemporary acceleration on learning. We will first work on the question 'What do we learn in a society of acceleration?' In his work on acceleration, Rosa addresses the topic of learning only on rare occasions, sometimes summed up in one paragraph.[17] After looking at how the society of acceleration promotes learning, and what kinds of learning in particular, we shall work on the concept of resonance, which comes down to the question, 'What kinds of learning might it be important to promote, given the Great Acceleration?' Rosa has invested in the notion of an 'oasis of deceleration' as the proponent of deceleration. It could be said that Rosa proposed deceleration as a solution to the alienation generated by contemporary acceleration. This is an erroneous interpretation that Rosa has repeatedly denied, notably in his book, *Resonance* (2018), where he is particularly clear about his position. He states that it is not deceleration but resonance that would serve as a form of antidote to, or way out of, the alienation generated by acceleration.[18] The type of learning Rosa puts

[17] His book *Resonance* has more for readers from an education background, with a chapter on schools.

[18] It would seem that the term 'oasis of deceleration' in his works was equivocal – though he used it critically, showing that they were only temporary oases for individuals (yoga, weekends in the countryside, etc.) to better re-enter the race in a competitive way. In *Resonance*, Rosa no longer uses this term, modifying it to 'oases of resonance' which are distinct from 'deserts of alienation' (p. 130).

forward is this learning of resonance with the world, in the critical tradition of the Frankfurt School and never far from the ideas of post-Promethean 'between-us' or convivialism. In Rosa's idea of resonance with the world, the world is just two of the three dimensions of the human adventure we propose in this book, having defined both the world and coexistence. In the following paragraphs, when Rosa refers to the world, he means both what we have termed the 'world' and what we have called 'coexistence' – i.e. this political component opposed to the economic component.

12.4.2 *What Do We Learn in the Contemporary Period of Acceleration?*

12.4.2.1 The Current Acceleration Provides Many Opportunities to Learn

In line with what has already been mentioned, the concept of acceleration is brought to bear here following Rosa's definition. There are three component parts – technical acceleration, acceleration of social change and the accelerating pace of life. Furthermore, the concept of learning is used to mean that which refers to the acquisition (and to the processes of acquisition), by the learning subjects, of knowledge, know-how, techniques, practices or skills, as well as the acquisition of attitudes (and their processes) and social and cultural codes.[19]

The first thing that can be said about the impact of contemporary acceleration on learning is that acceleration forces us to learn. It is an opportunity for learning. Firstly, from the point of view of technological acceleration, we are dealing with new objects we must learn how to operate. There may even be new social practices related to technological developments (e.g. social networks) that require an understanding of the logic, and this must be learned. Paradoxically, we learn by having the impression of not knowing: 'things become more complicated, [and] I become more ignorant about them' (Rosa, 2014, p. 119). Secondly, social change is accelerating. In early modernity, changes in family and work realms affected the type of 'intergenerational pace […… [and then we transitioned] to a generational pace in 'classical' modernity, then to an intragenerational pace in late modernity' (Rosa, 2014, p. 23). The consequence for vocational learning is that an individual has to learn several new trades during their lifetime and may also be confronted several times over with a reconfiguration of family life, meaning they must learn all over again. They can certainly redeploy some of the learning they have already done, but they must still learn again and reorganise their learning. Thirdly, the pace of life is accelerating. We can see a 'multiplication of episodes of action and experience per

[19] For Arendt , modern acceleration is the direct consequence of the contemporary domination of a regime of consumption: 'In modern conditions, it is not destruction that causes ruin, but conservation, for the durability of the objects we keep is in itself the greatest obstacle to the process of replacement, whose constant acceleration is all that remains constant once it has established its domination' (1983, p. 284).

unit of time' (Rosa, 2014, p. 25). The acceleration of life rhythms is 'the consequence of the desire or perceived need to do more things in less time' (Rosa, 2014, p. 25). This means that success in new learning belies the very structure of contemporary social life. We are compelled to learn throughout our lives. Furthermore, from a neurological point of view, it would seem that we learn differently than our predecessors and that our neurological capacities have developed. 'Young people today are developing multitasking abilities that the brains of older generations were not capable of' (Rosa, 2014, p. 95).

We are increasingly confronted with unforeseen situations that we have to get to grips with quickly, putting us in a situation of having to learn. People 'no longer attempt to follow a life plan, but begin to "ride the waves": as soon as a new and attractive opportunity presents itself, we must be ready to grab it on the hop' (Rosa, 2014, ca. p. 95). Contemporary acceleration confronts us with new situations from which we must learn. However, contemporary acceleration is also accompanied by what Hermann Lübbe calls a compression of the present (2009). The present is becoming denser and more saturated. This contemporary change in our relationship to time has a direct impact on learning because learning takes place in the present – and even occupies a large part of the present. The present is increasingly being replaced by the instant, with the instantaneous timescales of the urgent and ephemeral (Boutinet, 2004). We have entered a society of reflexivity (Giddens, 2005) and at the same time the space of time in which this reflexivity is exercised is becoming increasingly restricted. There is less room for looking back at the recent past or projecting into the medium- or long-term future. Yet the exercise of reflexivity is a fundamental component for successful learning.

12.4.2.2 Characteristics of Acceleration-Enhanced Learning

We are learning more, and the rate of learning is increasing, but it is mostly to keep us afloat. For example, the everyday practical knowledge individuals accumulate quickly becomes obsolete and needs to be updated regularly: 'Friends' addresses and phone numbers, office and shop opening hours, insurance premiums and telephone tariffs, television stars, parties' and politicians' levels of popularity, the jobs people do and the relationships in which they are involved' (Rosa, 2014, p. 22), not to mention the multiple times-daily monitoring of news high on information and low on analysis. What surrounds us is changing and it is important to update our knowledge to keep up with our environment. The key feature of this effort to stay afloat is that it is individual. On the one hand, no-one can keep us in our place; on the other, keeping afloat is only looking after ourselves. The energy put into keeping afloat does not generate any collective, social, or political progress (this refers to the image used by Rosa of the hamster in its wheel that goes very fast but goes nowhere, or a swimmer who is 'treading water').

An acceleration of social change also leads to more frequent socio-professional transitions for individuals, which are a source of fragility, and which require a series of learning experiences. During these transitions, people are particularly focused on themselves and reconfiguring their lives. 'This prodigious speed of events and transformations seems to be a superficial phenomenon, barely hiding the deep cultural and structural inertia of our time' (Rosa, 2014, p. 52). One consequence of contemporary acceleration is that 'It is very rare that we meet someone who witnesses all of our biographical life' (Rosa, 2014, p. 60). Individuals have to recreate spaces of relational and emotional security several times over the course of their lives. Yet, periods of stability are particularly important for more in-depth learning. Contemporary society, marked by acceleration, is moving towards precariousness of the social positions and individuals must put significantly more energy into maintaining their position. Rosa effectively states that we have to 'run as fast as we can to stand still' (Rosa, 2014, p. 37).[20] People must learn to give an appraised performance in a competitive space. However, this is for a solely individual purpose, to preserve one's own position. Some authors identify the limits of this acceleration society by speaking of 'hyper-accelerated immobilisation'[21] (Rosa, 2014, p. 51). 'The "position"' that an individual occupies in modern society is not predetermined by birth, nor is it stable over the course of an (adult) lifetime, only for the length of an ongoing competitive negotiation' (Rosa, 2014, p. 36). Everyone moves forward in their position, but we do not move forward collectively. The ability to stay afloat comes down to relative learning of survival – with all the individual and competitive components of the logic of survival.

We are now in a secular society where we expect more from life before death than after. 'According to this view of life, a *good life* is a *fulfilled life* – i.e. a life rich in experience and developed skills' (Rosa, 2014, p. 38). For this reason, contemporary individuals aspire to 'Taste life in all its flavours' (Rosa, 2014, p. 39). Consequently, the diversity of experiences sought by individuals are learning spaces. Yet they are still only learning about their own lives. The individual learns about life, about his own life and how to live. This need for accomplishment in life places too much value on certain types of learning.

12.4.3 Learning to Resonate with the World

12.4.3.1 Presentation of the Concept of Resonance

The initial question: 'What are we learning in the society of acceleration?' leads us to a second question: 'What would we need to learn?' What would we need to learn in order to break free from this great social and techno-economic acceleration that has such disastrous impacts on the Earth system? What can we do with this present time that is being compressed, from which we can learn? In the preceding pages, we have seen that two of the characteristics of the learning favoured by contemporary

[20] 'The 'position' that an individual occupies in modern society is not predetermined by birth, nor is it stable over the course of an (adult) lifetime, only for the length of an ongoing competitive negotiation.' (Rosa, 2014, p. 36).

[21] Examples include Paul Virilio, Jean Baudrillard, Fredric Jameson and Francis Fukuyama.

acceleration are that it is learning to stay afloat, and learning that is solely related to one's own life or to something akin to survival. These types of learning are underpinned by an individualistic ethic. We identify that some are less favoured by contemporary acceleration: namely political lessons that could be categorised as inculcating plurality and coexistence. In a way, below, we propose theoretical remedies to the 'social pathology' identified by Rosa.

Thus, because acceleration brings only individual learning, it is important to allow individuals to develop world learning and existential learning – or coexistence. Working on these two categories of learning seems necessary to Rosa, who views our relationship to the world and our being in the world as being profoundly altered by contemporary acceleration. 'Acceleration serves […] as a strategy to erase the difference between the timescale on which the world operates and the timescale on which our lives run. *The eudaemonic promise of modern acceleration therefore lies in the (unspoken) idea that the acceleration of the "pace of life" is our (i.e. modernity's) answer to the problems of finitude and death.* This is one of the tragedies of modern man: while he feels trapped in an endless race like a hamster in its wheel, his hunger for life and the world is not satisfied, but increasingly frustrated' (Rosa, 2014, p. 40). Political learning, through the integration of finitude, makes it possible to reposition existence in the timescale of the world, and no longer only in the narrow time window of human life.

One of the key elements in Rosa's thinking is the dynamic stabilisation of contemporary societies, which only stabilise themselves in the quest for constant acceleration in the continuation of an incessant search for growth. For Rosa, the heart of the contemporary period lies in this ever more rapid pace of relations to others, to oneself, to things or to the material world. This process of dynamisation is characteristic of the social acceleration of modernity, which Rosa highlighted in *Acceleration* (2010). As mentioned, this dynamic stabilisation is directly related to the globalisation of capitalism.[22] Societies can only become stable in the movement of growth and accelerating innovation. Competition is central to the interactions between individuals and organisations, but also to the very core of psychological functioning. It is to this intrinsic functioning of modernity that Rosa proposes a form of alternative. The dynamic stabilisation of contemporary societies threatens to render our relationships to the world 'mute'. Resonance offers a way out of the contemporary *hybris* which characterises our ways of life.

As mentioned above, in order to delve more deeply into Rosa's idea and articulate it with a type of learning that could be interesting to promote in order to come through the anthropological *krisis* into which the Anthropocene is forcing us, we went to meet the German sociologist for a discussion. In the following paragraphs, we describe this concept of resonance, based on the interview with Rosa and our reading of *Resonance*. For Rosa, it is not so much a matter of decelerating as of

[22] 'Acceleration, growth, and densification of innovation refer respectively to a temporal, material, and social dimension of a single process of dynamisation which, in turn, is defined as a quantitative increase per unit time' (Rosa, 2018, p. 466). In a way: '*we must run faster and faster in order to maintain our place in the world*' (2018, p. 478).

entering into a new relationship with the world marked by a relationship of resonance with it that allows us to hear the world.[23] The resources of the German notional universe within which the concept of *Resonanz* emerges are marked by two characteristics that are unusual in the French sociological tradition: the use of the existential component[24] of the human condition, and the formulation of a prospective thought against a background of social criticism.

We need to broaden the meaning of what it is to be in this world, which is all too often restricted to the subject itself. The process of resonance runs counter to *homo oeconomicus*' capitalisation. The relationship between the subject and the world is key, and the process of resonance is defined as the meeting of one with the other. There is a deep and meaningful relationship with the world, and the effects of this relationship transform the subject (and constitute an *Erfahrung*).[25] Rosa differentiates the appropriation of knowledge from assimilation: 'I can acquire knowledge and learn to use machines or programmes. When it comes to music, I can recognise variations, patterns and nuances. This is all appropriating knowledge or acquiring skills. When I appropriate things, I can control them. Assimilation, though, is a completely different kind of process that leads to transformation (*Erfahrung*):

> The difference between appropriation and assimilation is very clear for music or poetry. I can appropriate interpretations of poems or pieces of music. Equally, I can quickly tell what literary movement a poem belongs to, what era it is from, how the rhymes or lines are composed, pick out stylistic devices, and so on.. However, if I am affected, this poem means something to me, it does something to me. It means that it causes me to have a new experience, it opens up new horizons or a relationship with the world that I did not have before. In other words, after encountering this poem or this piece of music, I am differently present in the world. I have allowed myself to be transformed and I am, in part, someone else. I believe that truly formative processes – such as assimilation – are more like this. When

[23] Etymologically, 'resonance' comes from *re-sonare* meaning to resound, to echo (p. 189). It is an acoustic phenomenon in which the vibration of one body causes another body to vibrate. Resonance is a musical metaphor. Music is understood by Rosa as that which allows one to intensively feel the relationship to the world and its transformative power.

[24] The notion of existence has links with that of resonance, though they do not overlap completely (the existential component is drawn on particularly heavily in the notion of resonance). The idea of existence cannot be thought of without the notion of experience, which comes from the Latin *experiri*, which means to experience; *periri* having also given *periculum*, danger, trial. The root *per* refers to the idea of crossing (Lacoue-Labarthe, 1997, p. 30). The experience of the world represents a risk and is necessarily an ordeal which we must survive. In German, two conceptual tools refer to the notion of experience. *Erlebnis* means, in a certain way, lived experience; and *Erfahrung*, experience in its broadest and longest-lasting sense, having an impact on the course of a person's life or their identity (the experience of fatherhood, the professional experience, etc.). *Erfahrung* is the experience that has marked existence, but it is also the experience of existence, or existence as life experience, which is *Bildung*. From more of a sociological perspective, the concept of experience refers to Dubet's concept of social experience (Dubet, 1994), which weaves together the three logics of action: integration, strategy and subjectivation (Dubet, 1991; Dubet et al., 1991; Dubet & Martuccelli, 1996, 1998).

[25] Taking up Walter Benjamin's thinking, Rosa suggests that we may be entering a period high in *Erlebnissen* and low in *Erfahrungen*. *Erlebnisse* have a more superficial quality. They are part of a series of things people do. *Erfahrungen,* however, leave a deeper impression.

> assimilating, I come into contact with the world. It is as if I am talking to it. It has an impact on me, affects me and transforms me (Rosa & Wallenhorst, 2017b, p. 3).

The concept of resonance, like Arendt's concept of action, is not controlled. It simply emerges, does not last beyond what is happening when it occurs, and we do not know how long it will last:

> The assimilation that leads to transformation is a process that always includes something unavailable and unmanageable. One does not know exactly when it begins or what the outcome will be. Say a teacher is leading a class on literature, and a transformation happens. It is not clear what the students will do with it, what it will mean to them, how long it will last or what the final outcome will be. It is identifiable as a process of resonance. First of all, resonance takes time. Secondly, the outcome is unknown. Thirdly, you have to be willing to make yourself vulnerable – fragile, even – because it is a process that opens you up to something in which you are invested. I believe that in most cases, the education we received does not bring about any encounters with the world (Rosa & Wallenhorst, 2017b, p. 3).

The central focus of this notion of resonance is neither the world nor the subject. It is the relationship of one with the other. '[S]uccessful resonant interaction is when we are willing to listen to the voice of the other and make our own more perceptible, so that there is a horizontal resonance' (Rosa & Wallenhorst, 2017b, p. 3). This ties in with the idea of the relational void of the post-Promethean 'between-us' (discussed in Chap. 9). Resonance is decidedly existential, to judge by Rosa's tone and the terms he uses. When he gives examples of resonance, these are existential and emotion-based, not unlike the 'essential' the Maurice Bellet describes:

> The moment I fall in love, whether it's with another person, or an idea, or with music or a landscape, both become alive. I feel that the landscape is beautiful. All of a sudden, I hear the birds singing again. I notice the sun and the flowers. On the inside, everything in me seems to be singing and chattering again. Thus, loss of self and loss of the world are intimately connected, but it depends on the quality of the relationship with the world. I speak of vibrant resonance or a vibrating string. It seems to me that everyone can understand this image. I sometimes feel so connected to the world that it touches me. It comes to me and moves me. But I don't want to be totally pessimistic, because I think humans have always been good at resonance. All of us have experienced this and know what the world is like, which brings answers as well as true encounters. We can all be comfortable, healthy and wealthy, with a lot of knowledge and happy relationships, and still feel like something is missing. In this age of acceleration, we need the other (Rosa & Wallenhorst, 2017b, p. 4).

In concrete terms, we experience resonance through different types of activities in our daily lives:

> Politics and democracy are forms of resonance, as is work. Resonance has to do with what forms the basis for our existence. I believe that human beings need to be sure of their relationship with the world. However, we lack the resources to answer this question. Albert Camus (and, to a lesser extent, Friedrich Nietzsche) wrote that at the source of our existence there is a silence that we cannot understand but cannot ignore. In James, as well as in Martin Huber, there is some very interesting thinking about the conviction or hope that there might be answers at the source of existence. Religion, nature, art and history would be forms of that (Rosa & Wallenhorst, 2017b, p. 3).

A particularly important place to experience resonance with the world is in our connection to nature. For Rosa, we need nature:

> not only as a resource or refuge from the world, but also as a sphere for resonance. I think environmental movements are linked to idea of potentially losing nature as a sphere of resonance – plus, of course, as a space for environmental action in the hope of turning things around. […] In many ways, we commodify nature (without always perceiving how much we need it for our own understanding), as something that has an organic relationship of resonance with us. […] We used to see nature as a resource. We should see it, instead, as something we fundamentally need. If we rediscover our organic connection with it, we will no longer destroy it ruthlessly as we are doing now. This is something worth thinking about (Rosa & Wallenhorst, 2017b, p. 5).

In terms of what we can or should do to make the future a better place for future generations, Rosa says that:

> it's hard to find the right answer. We're all a bit confused when we look at political and planning issues. I don't think that the revolution will start by thinking cognitively [...]. What should we do? It seems to me that it is actually the attitude underlying this question that we have to learn to master. That is why I think (even though I am well aware that it is not a perfect answer) that we first have to listen. Then we can try out responses that acknowledge the presence of the other in order to find new ways of coexisting. What do we need to learn in the years to come? To listen and to reconnect with others and the world. What we first need to clarify is how we want to relate to others and the world (Rosa & Wallenhorst, 2017b, p. 5).

With the concept of resonance, Rosa goes so far as to question whether some of the contemporary ills actually do stem from a Promethean need for control. To a certain extent, he distances himself from the paradigm of the project.

The concept of resonance is a paradigm shift from the Promethean. It is a matter of learning to listen to the world rather than seizing and subjugating it. For Rosa, 'a better world is possible, and it will be characterised by the transition from a relationship with the world that strives for the power to dispose of and control things to an attitude towards the world focused on listening' (Rosa & Wallenhorst, 2017a, p. 28).[26]

12.4.3.2 A Political Sociology of the 'Good Life'

The 'Good Life'

The starting point of Rosa's book was 'a rethinking of the concept of the good life as different for everybody' (p. 11).[27] One of the characteristics of the contemporary iteration of life as an individual is that it is 'increasingly guided by the care a person takes in maintaining and improving the state of their resources, and especially in

---

[26] As Arendt notes, *homo faber* is a craftsman whose goal is to build a permanent home for humans by waging war on nature (1983, p. 342). This violence has had such an impact on the Earth that it has been changed forever. After the emergence of *homo faber*, many authors are calling for *vita contemplative*, including Rosa with his concept of resonance, which is paradigmatically opposed both to manufacture and, to a certain extent, action.

[27] This is an observation Hannah Arendt had already made *The Human Condition* (1983).

broadening their horizons' (p. 11). It shows that the causal link between maximising individual resources and enjoying the good life is erroneous. Rosa, in his endeavour to define what a good or successful life might look like, goes further than a simple definition. The very first pages reveal the thesis for the tome, 'everything in life depends on the quality of our *relationship with the world*, i.e. the way in which we, as subjects, experience the world and position ourselves in relation to it – in short, on the quality of our *appropriation of the world*' (p. 12). Rosa identifies certain contemporary functions that he believes contribute to a good life, in order to analyse and describe them in detail. His aim is to bring about a type of anthropological shift.

The empirical question he is trying to answer is 'what is the difference between successful and unsuccessful relationships with the world?' However, the answer here diverges radically from that which positive psychology has been telling us since the 2000s, fundamentally in the service of the market (Cabanas and Illouz 2018).[28] It does not espouse the neoliberal paradigm of perpetual happiness. Rosa's approach, on the contrary, repoliticises our existence without sacrificing the individual to the causes of justice, knowledge, the future, the environment, etc. Rosa believes it is not a question of allowing individuals to have a good life based on the accepted idea of what a successful life is. His approach is political rather than emotion-based,[29] and his conception of the political is in line with Arendt's.[30] Rosa accepts common hegemonic individual freedoms, but rejects the meeting and sharing of freedoms. His approach aligns strongly with the convivialists in this area.[31] Rosa encourages us to reflect on the possibility of a good life without it being necessarily a question of emancipation or a neoliberal individualistic quest for happiness. Resonance is a criterion for a successful life in Rosa's social theory. It counters the idea that it is up to each individual to decide for themselves what a good life means – something that has even become a maxim of educational institutions (Rosa, 2016, p. 11). Indeed, he shows that the good

---

[28] Edgar Cabanas and Eva Illouz's *Happycracy* (2018) does a good job of highlighting how the capitalist happiness industry is trying to 'take control' of our lives.

[29] Resonance, which is learning a form of relationship to the world, has politics at its core. Like citizenship, resonance is as much a political as an educational concept. This is why it is particularly interesting for our political anthropology of education.

[30] Rosa regularly references Arendt in defining the political component of resonance, as is apparent in these excerpts: 'such views on democratic political action are perhaps most saliently expressed in the writings of Hannah Arendt' (Rosa, 2018, p. 248); 'The idea that the very nature of the capacity for human action is to be able to *start anew* and *find* ourselves anew, regardless of our past or predefined logic about action, tallies with Hannah Arendt's view that only this *forgiving* form of (resonant) action is world-creating (and thus world-relating). However, Arendt believes taking action is, first and foremost, a matter of taking *political* action and politics. Therefore, it is the central sphere of resonance, or the axis of the modern relationship to the world *par excellence*' (Rosa, 2018, p. 244).

[31] The concept of resonance centres around a 'We' and converges with the convivialist perspective, as Hartmut Rosa himself points out. 'The demand for resonance may well be, it seems to me, the hidden spring of communitarian movements, and maybe of recent "convivialist" movements too. The subtitle *Declaration of Interdependence* to the *Convivialist Manifesto* is a clue: interdependence is brandished as a weapon against separatist independence and indifference' (Rosa, 2018, p. 244).

life is not the result of resources we accumulate over the course of our existence, and that capitalism is an illusion: '(capitalist) distributive relationships seem only to legitimise themselves in a society that remains blind and deaf to the question of the good life, and believes that unlimited growth and accumulating private resources are sufficient to define well-being' (p. 14).[32]

In a contemporary period marked by the decline of the institution (Dubet, 2002) where individuals have to build themselves from within (Martuccelli, 2002), it is important to find ways of protecting and supporting people to live their lives. This is a characteristic usually attributed to the institutions that are in decline. 'The modern subject [...] has pumped himself up. It is up to *him* to grow and innovate; his motivational and creative energy alone can sustain the game of growth. The result is a constant need for self-invention and the flexibility as to one's own position in the world – something that has already been analysed extensively. If he is not prepared to change his professional and family, religious and political views, title and aesthetic position, an individual risks losing his place in the running order and a large chunk of his resources (and thus of his access to the world) as a result' (Rosa, 2018, p. 478). Intensifying competition in the labour market, together with growing liberalisation and flexibility in employment, is constantly making individuals more fragile. Work cannot provide this protection or reduce the tension inherent in the contemporary social experience. The concept of resonance is interesting as, although the good life provides moments of happiness, it is far from being as reductive as a 'happy' life. Indeed, it is through a relationship with the world within that people feel welcomed, heard and protected. Resonance includes a component of structural solidarity.

Rosa's Relationship to Sociology

Rosa bases his approach, which he describes as 'A sociology of our Relationship with the World' (this is the book's subtitle), on the phenomenology he inherited from Maurice Merleau-Ponty,[33] thinkers at the Frankfurt School (Herbert Marcuse, Theodor Adorno, Walter Benjamin and Alex Honneth) and a revival of Albert Bandura's concept of self-efficacy. Rosa diverges from the more classical sociological approach to individual resources: the starting point of *Resonance* is that it is not the quality of resources that makes it possible to appreciate life, but the type of relationship we have with the world. In the sociological tradition, it is the distribution of resources which determines the ideal of justice, with the presupposition that

[32] Here, he is in line with Christian Arnsperger's approach, which demonstrates that capitalism is based on the fear of lack and, ultimately, of death.

[33] The first chapter focuses on the importance of our upright position (walking with our feet on the ground) and on the predominance of instrumental relationships where hands produce technical artefacts over resonant relationships where the voice prevails. Rosa questions the type of relationship to the world induced by the omnipresence of screens. Are they not blocking bodily resonance? What will be the long-term effects of the 'culture of the lowered gaze' (p. 58, based on Plessner), that we have created with our constant use of smartphones in public?

it is better to have more resources than fewer. This is how twentieth-century sociology has addressed the question of what a successful life is. Rosa's additional piece of the puzzle marks a paradigm shift in sociology and renews the sociological attention paid to a good or successful life. Rosa's sociology of the relationship with the world is built on two fundamental concepts that have a dialectical relationship with each other – resonance and alienation. When a subject is invested in and affected by a situation, how do we distinguish a state of resonance from a state of non-resonance? Rosa uses the examples of a stomach-ache, a storm or a dispute over inheritance (p. 195). This is a fundamental question. Resonance is characterised as a form of 'unison with the self and with the world' (p. 195). Here, resonance is differentiated from discordance. When a subject's affectation takes the form of discordant accord, Rosa calls this a repulsive relationship to the world, which is a form of alienation. The subject has a mute relationship with the world.

Throughout his book, Rosa is updating, developing, criticising and refining the Critical Theory of the Frankfurt School for the present day. From its beginnings, Critical Theory has sought a better relationship to the world in the wake of Marx's reflection on alienation in the 1944 *Manuscripts*. György Lukács developed the idea of successful appropriation of the world in opposition to a relationship of reification. Herbert Marcuse was inspired by Orpheus and Narcissus to think of a good life as being the opposite of Prometheus'. Theodor Adorno wrote of a mimetic relationship to the world characterised by warmth (he was also a composer) that would oppose instrumental reason. Benjamin continues to inject some messianic hope in spite of disasters. For Erich Fromm, what differentiates successful relationships with the world from failed ones is mainly the intensity of our connections to others. Axel Honneth distinguishes between relationships to the world marked by recognition and those marked by contempt (Rosa, 2018, p. 36). Throughout the book, Hartmut Rosa persists in positioning his thinking within the field of sociology as opposed to social philosophy. Sociology, with its tradition of grounding theory in empirical data, is less subject to interpretation than is philosophy. It would seem that there is a form of militancy here on the author's part to assert the normative component of sociology. The aim of sociology, as he sees it, is to transform society.[34]

12.4.3.3 Resonance as a Convivialist Anthropological Shift – Reconfiguring the Concept of Learning

One of the interesting features of the concept of resonance is that it takes intersubjectivity (and thus relationship) at its anthropological foundation.[35] The anthropology of resonance is relational; coexistence is primary. This underpins the proposed re-politicisation of society with the call for a post-growth society. Rosa draws upon

[34] The sociology of the relationship to the world developed by Rosa is imbued with existentialism, as it is not rooted in the interests of individuals but in their existential sensibilities.

[35] Otherness is intrinsic to the concept of resonance, since it is dissimilar people who respond to each other.

Randall Collins' work in developing the idea that human beings are first shaped by solidary interactions, then by competitive and violent interactions. These ideas are similar to those of convivialists, such as François Flahault, Fred Poché, Corine Pelluchon, Alain Caillé, Pablo Servigne, etc. without denying the importance of opposition and conflict (but in a form of opposition that can be tolerated without massacring one another).[36]

In the different illustrations of how to position oneself in the world, Rosa highlights two typical ideals: extending access to the world, through domination; and the quest for creative interactions that foster connections, characterised by assimilation (p. 18). His research works through the polarisation of the Promethean and post-Promethean. In this three-dimensional classification we are putting forward, *homo oeconomicus*' profit-maximising logic represents domination (giving a green light to pre-emption) and reflects a Promethean characteristic; *homo religatus*' responsibility-driven logic represents participation in the world (and actively taking part in building the world); and *homo religatus*' logic of hospitality represents assimilation. Resonance stands out as being non-Promethean and in opposition to the Prometheism of domination. It is a characteristic of both *homo collectivus* and *homo religatus*.

The concept of learning is central to education. Education is based on the idea that learning is possible and desirable, and that learning is focused on mastery. When an individual has learned, he or she has acquired something that he or she owns; education is fundamentally developmental. It develops the self while having little focus on the other, and rarely involves the point of view of the other. When the other is invoked in education, it is to facilitate learning (whether it is the educator, the teacher, the peer or the group). This 'other' that is invoked is the otherness that enables learning to take place, rather than the neighbour (the one we get close to so we can help, support, listen to them, etc.) – except in the case of the reflection on care that is currently taking shape. Here, there are the twin issues of decentration to work on: a decentration in space towards the other and in time towards the neighbour (the one who is to come, who may represent future generations). Paul Ricœur has described this tension between the *socius* and the neighbour, and highlights the importance of a relational style both for the one who is there and the one who does not yet exist.

[36] Like Jürgen Habermas before him, Rosa identifies four elementary cultural forms of relationship to the world: (1) a relationship when adapting to the world, characterised by an active relationship with and affirmation of the world (this is the case for Confucianism); (2) a relationship of world domination, characterised by an active relationship with and negation of the world (e.g. Western Protestantism); (3) a relationship of observation or contemplation of the world characterised by a passive relationship to and an affirmation of the world (e.g. Athenian Greece); (4) a relationship of escaping from the world, characterised by a passive relationship to and a negation of the world (e.g. Eastern spiritualities such as some types of Hinduism). (Rosa, 2018, p. 112–115). Contemporary Western societies are characterised by world domination relationships that reify and silence the world. The search for harmony with nature and with others then becomes subversive, because it interrupts the extractivist and maximising nature of this relationship of domination.

To think in a post-Promethean manner in education means not to attach absolute primacy to self-emancipation, but to combine it with inculcating responsibility for the other. 'In the modern West, to be unbound has come to mean a form of absolute existence (etymologically, "absolute" means "unbound" or "untied")', whereas in antiquity, 'unbound was a term for the slave as opposed to the free man. The latter was free precisely because he had a place in society and could count on the binds that tied him to others' (Flahault, 2008, p. 64). In education, responsibility – and the connection to others this creates – is a counterpart to the logic of emancipation, which is hegemonic in educational paradigms. As such, the real target of education in the Anthropocene is not the subject himself/herself, but the other. How can we take care of the other and give them the attention they need in order to exist? In addition to learning responsibility as a necessary means to ushering in post-Promethean relationships, we also note the role of acceptance and sharing of human vulnerability.[37]

The novelty of the concept of resonance – something of a paradigm shift in educational thinking – is its opposition to alienation. It differentiates opposition to emancipation (of the self) from alienation. Traditional education views affirmation of the self as separate from its liberation from alienating societal constraints, in terms of reification. It has been difficult to counter this mainstay in educational thinking, which is driven by a concern for social justice. The limitation on the educationalists' contrasting self-emancipation[38] and alienation is the perpetuation of this logic of reification. In self-emancipation, the self can sometimes be absolutised and commodified. It may become the purpose of the act of educating, which is then depoliticised and refers solely to a strict individualist economic logic. However, the aim of education is to combat the alienation of individuals in the here and now. This alienation can take different forms (adaptation to market constraints, unequal access to social and financial resources, poor prospects of integrating into society, etc.). The concept of resonance offers this alternative and a counterargument to alienation without reification of the self or excessive focus on the individual. The relationship

[37] Corine Pelluchon has developed a reflection on the ethics of consideration, promoting environmental transition, as already mentioned in this research (2018). She states that the ethics of consideration aim 'to direct the impulses [of beings], so that they have pleasure in loving what is lovable or that their desires are no longer only those of *homo oeconomicus*' (Pelluchon, 2017, p. 14). Pelluchon contrasts the virtue of consideration (involving the acceptance and sharing of our vulnerability) with that of domination. The Promethean individual, cut off from what connects him to others, 'feeds a system that inevitably involves overproduction, inbuilt obsolescence and the ceaseless exploitation of nature, other living things, other humans, and even nations by other nations' (Pelluchon, 2017, p. 15). This distances the individual from 'the magnanimous man who knows joy and is kind to others whose weaknesses also remind him of his own' (p. 15). Recognising and sharing vulnerability and accepting coexistence has a directly political impact, in Pelluchon's eyes. For her, it is 'the condition for living better with and for others by committing to passing on a habitable world' (2017, p. 10).

[38] If not the emancipation of the self, as is the case in Germany with the notion of *Selbstbildung*, which further accentuates the importance of the self, already central to the notion of *Bildung*.

is the real pull in the concept of resonance.[39] Thinking (in education) with the aim of learning resonance implies starting not with the subject of learning but with the relationship between the subject and the other, with things and with the world. This shift in focus is unusual in education. It is a matter of thinking of learning as something other than education, inculcating knowledge. The relational focus of the concept of resonance makes it possible to re-politicise and 'de-economise' the act of educating. We can now position the educational act or educational thought in a post-Promethean domain where mastery and domination of the world no longer matter. The concept of resonance provides creative and stimulating dissonance for reflections on education in the Anthropocene.

From Resistance to Critique
Having explored the resistant function of an education in the Anthropocene, the next chapter brings to bear the necessary critical function. This will be based on a further appropriation of the concept of resonance, articulated through interpretations of work by Andreas Weber and David Abram.

Bibliographical References

Arendt, H. (1972). Qu'est-ce que la liberté ?. In H. Arendt, *La crise de la culture*. Gallimard (pp. 186–222) (ed. or. 1961d), tr. fr.

Arendt, H. (1983). *Condition de l'homme moderne*. Calmann-Lévy. (original edition 1958, French translation 1961), French translation.

Arnsperger, C., & Bourg, D. (2017). *Ecologie intégrale, Pour une société permacirculaire*. PUF.

Berque, A. (2016). Perception de l'espace ou milieu perceptif? *L'Espace géographique, 45*, 168–181.

Biesta, G. J. J. (2006). *Beyond learning. Democratic education for a human future*. Paradigm Publishers.

Bourg, D. (2001). Le nouvel âge de l'écologie. *Le Débat, 113*, 92–105.

Bourg, D. (2009, December). L'impératif écologique. *Esprit*, 59–71.

Bourg, D. (2012). Transition écologique, plutôt que développement durable. *Vraiment durable, 1*, 77–96.

Bourg, D. (2013). Peut-on encore parler de crise écologique? *Revue d'éthique et de théologie morale, 276*(HS), 61–71.

Bourg, D., & Papaux, A. (2010). Ecologie, 1980–2010: de l'exception française à la normalisation. *Le Débat, 160*, 94–114.

Bourg, D., et al. (2017). *Inventer la démocratie du XXIe siècle*. Paris, Les Liens qui Libèrent.

Boutinet, J.-P. (2004). *Vers une société des agendas*. PUF.

Cabanas, E., & Illouz, E. (2018). *Happycratie*. Premier Parallèle.

Cabanes, V. (2017). *Homo natura – En harmonie avec le vivant*. Buchet Chastel.

Delory-Momberger, C. (2009). Postface, Etre adulte ou la condition impossible. In J.-P. Boutinet & P. Dominicé (Eds.), *Où sont passés les adultes?* Téraèdre.

[39] In the documentary directed by a former trader-turned-teacher Gilles Vernet, *Tout s'accélère* (April 2016), inspired in particular by Rosa's work, we clearly see this tension between acceleration and resonance in education, and the political importance of this choice about education. We perceive how students develop their resonance with the world together. They learn to listen to the world (to listen to and understand their parents' behaviour and attitudes, how a globalised society works, both in terms of its institutions, but also their peers and their emotions, etc.).

Dubet, F. (1991). *Les lycéens*. Seuil.
Dubet, F. (1994). *Sociologie de l'expérience*. Seuil.
Dubet, F. (2002). *Le déclin de l'institution*. Seuil.
Dubet, F., & Martuccelli, D. (1996). *A l'école ! Sociologie de l'expérience scolaire*. Seuil.
Dubet, F., & Martuccelli, D. (1998). Sociologie de l'expérience scolaire. *L'orientation scolaire et professionnelle, 27*(2), 169–187.
Dubet, F., Cousin, O., & Guillemet, J.-P. (1991). Sociologie de l'expérience lycéenne. *Revue française de pédagogie, 94*, 5–12.
Dumont, L. (1991). *L'idéologie allemande*. Galimard.
El Fadli, K. I., Cerveny, R. S., Burt, C. C., Eden, P., Parker, D., Brunet, M., Peterson, T. C., Mordacchini, G., Pelino, V., Bessemoulin, P., Stella, J. L., Driouech, F., Wahab, M. M. A., & Pac, M. B. (2013). World Meteorological Organization assessment of the purported world record 58°C temperature extreme at El Azizia, Libya (13 September 1922). *Bulletin of the American Meteorological Society, 94*, 199–204.
Erb, K.-H., Lauk, C., Kastner, T., Mayer, A., Theurl, M. C., & Haberl, H. (2015). Exploring the biophysical option space for feeding the world without deforestation. *Nature Communications, 7*, 1–9.
Faber, M., Petersen, T., & Schiller, J. (2002). Homo oeconomicus and homo politicus in Ecological Economics. *Ecological Economics, 40*, 323–333.
Flahault, F. (2005). Vers une nouvelle pensée sociale. *Revue du MAUSS, 26*, 377–382.
Flahault, F. (2008). *Le crépuscule de Prométhée – Contribution à une histoire de la démesure humaine*. Mille et une nuit.
Foley, J. A., Ramankutty, N., Brauman, K. A., Cassidy, E. S., Gerber, J. S., Johnston, M., Mueller, N. D., O'Connell, C., Ray, D. K., West, P. C., Balzer, C., Bennett, E. M., Carpenter, S. R., Hill, J., Monfreda, C., Polasky, S., Rockström, J., Sheehan, J., Siebert, S., Tilman, D., & Zaks, D. P. M. (2011). Solutions for a cultivated planet. *Nature, 478*, 337–342.
Foray, P. (2001). Hannah Arendt, l'éducation et la question du monde. *Le Télémaque, 19*, 79–101.
Giddens, A. (2005). *La constitution de la société*. PUF. French translation.
Gilman, S. E., Urban, M. C., Tewksbury, J., Gilchrist, G. W., & Holt, R. D. (2010). A framework for community interactions under climate change. *Trends in Ecology & Evolution, 25*(6), 325–331.
Hess, R., & Weigand, G. (2008). L'éducation tout au long de la vie: une théorie de l'expérience comme processus d'apprentissage et de connaissance. In L. Colin & J.-L. le Grand (Eds.), *L'éducation tout au long de la vie* (pp. 5–22). Anthropos.
Hétier, R. (2019). Apprendre à faire vivre en Anthropocène. In N. Wallenhorst & J.-P. Pierron (Eds.), *Eduquer en Anthropocène*. Le Bord de l'eau.
Jackson, S. T., Betancourt, J. L., Booth, R. K., & Gray, S. T. (2009). Ecology and the ratchet of events: Climate variability, niche dimensions, and species distributions. *Proceedings of the National Academy of Sciences, USA, 106*(suppl. 2), 19685–19692.
Lacoue-Labarthe, P. (1997). *La poésie comme expérience*. Christian Bourgeois.
Lamarre, J.-M. (2012). L'éducation cosmopolitique: apprendre le propre, apprendre l'étranger. *Le Télémaque, 41*, 31–46.
Lavelle, P. (2007). Marino Pulliero, *Le désir d'authenticité Walter Benjamin et l'héritage de la* Bildung *allemande*. *Labyrinthe, 26*, 107–112.
Leandri, M. (2015). Capacité de charge. In D. Bourg & A. Papaux (Eds.), *Dictionnaire de la pensée écologique* (pp. 121–123). PUF.
Les convivialistes. (2013). *Manifeste convivialiste*. Le bord de l'eau.
Lübbe, H. (2009). The contraction of the present. In H. Rosa & W. Scheuerman (Eds.), *High-speed society. Social acceleration, power and modernity* (pp. 159–178). Pennsylvania State University.
Martuccelli, D. (2002). *Grammaires de l'individu*. Gallimard.
Mora, C., Dousset, B., Caldwell, I. R., Powell, F. E., Geronimo, R. C., Bielecki, C. R., Counsell, C. W., Dietrich, B. S., Johnston, E. T., Louis, L. V., Lucas, M. P., McKenzie, M. M., Shea, A. G., Tseng, H., Giambelluca, T. W., Leon, L. R., Hawkins, E., & Trauernicht, C. (2017). Global risk of deadly heat. *Nature Climate Change, 7*, 501–506.
Morin, E. (1999). *La tête bien faite*. Seuil.

Morin, E. (2014). *Enseigner à vivre*. Actes Sud/Play Bac.
Paillard, S. (2017). Anthropocène: la planète va-t-elle craquer? *Revue Projet, 359*, 1–6.
Pal, J. S., & Eltahir, E. A. B. (2015, October 26). Future temperature in southwest Asia projected to exceed a threshold for human adaptability. *Nature Climate Change, 6*, 1–4.
Pelluchon, C. (2017). L'éthique des vertus: une condition pour opérer la transition environnementale. *lapenseeecologique.com, PUF, 1*(1), 1–18.
Pelluchon, C. (2018). *Ethique de la considération*. Seuil.
Perret, B. (2014). De la difficulté de penser un avenir sous contrainte écologique. *Transversalités, 130*, 69–81.
Prouteau, F. (2004). *Anthropologie et pédagogie de la vocation*. Doctoral thesis in Educational Sciences, University of Lyon 2, ANRT, Lille.
Rorty, R. (1990). *L'homme spéculaire*. Seuil. (original edition 1979), French translation.
Rosa, H. (2013). *Accélération. Une critique sociale du temps*. La découverte (ed. or. 2010), tr. fr.
Rosa, H. (2014). *Aliénation et accélération – Vers une théorie critique de la modernité tardive*. La Découverte (or. 2012), tr. fr.
Rosa, H. (2016). *Resonanz, Eine Soziologie der Weltbeziehung*. Suhrkamp.
Rosa, H. (2018). *Résonance – Une sociologie de la relation au monde*. La découverte. (original edition 2016), French translation.
Rosa, H., & Wallenhorst, N. (interview). (2017a). Apprendre à écouter le monde. *Chemins de formation, 21*, 19–31.
Rosa, H., & Wallenhorst, N. (interview). (2017b). Apprendre ensemble dans la société de l'accélération. *Bildungsforschung, 2017*(1), 1–7.
Schmitt, L. (2014). *La valse des égo*. Odile Jacob.
Sherwood, S. C., & Huber, M. (2010). An adaptability limit to climate change due to heat stress. *Proceedings of the National Academy of Sciences, USA, 107*(21), 9552–9555.
Steffen, W., Crutzen, P. J., & McNeil, J. R. (2007). The Anthropocene: Are humans now overwhelming the great forces of nature? *Ambio, Royal Swedish Academy of Sciences, 36*(8), 614–621.
Steffen, W., Grinevald, J., Crutzen, P., & McNeill, J. (2011). The Anthropocene: Conceptual and historical perspectives. *Philosophical Transactions of the Royal Society, 369*, 842–867.
Steffen, W., Persson, A., Deutsch, L., Zalasiewicz, J., Williams, M., Richardson, K., Crumley, C., Crutzen, P., Folke, C., Gordon, L., Molina, M., Ramanathan, V., Rockström, J., Scheffer, M., Schellnhuber, H. J., & Svedin, U. (2011). The Anthropocene: From global change to planetary stewardship. *Ambio, 40*(7), 739–761.
United Nations, Department of Economic and Social Affairs, Population Division. (2017). *World population prospects*. https://esa.un.org/unpd/wpp/Download/Standard/Population/
Von Bonsdorff, P., & Grenaudier-Klijn, F. (2011). Esthétique et *Bildung*. *Diogène, 233–234*, 178–195.
Wallenhorst, N. (2013). *L'école en France et en Allemagne, Comparer des expériences scolaires*. Peter Lang.
Wallenhorst, N. (2015). L'étudiant face à la désorientation. In M.-H. Jacques (Ed.), *Les transitions en contexte scolaire* (pp. 281–293). Rennes.
Westbroek, P. (2015). Système terre. In D. Bourg & A. Papaux (Eds.), *Dictionnaire de la pensée écologique* (pp. 957–962). PUF.
Wilhelm Graf, F. (2002). Le politique dans la sphère intime, Protestantisme et culture en Allemagne au XIXème siècle. *Annales. Histoire, Sciences Sociales, 3.*, 57ème année, 773–787.

Chapter 13
A Critical Education: We Are Not Separate from the Earth – We *are* the Earth

Abstract There are three reasons for continuing to consider the concept of resonance when thinking about education in the Anthropocene. Firstly, as we have just seen, resonance, thought of as the counterpart to acceleration, tells us something about what we need to learn in this geological epoch of the 'Great Acceleration' (the other name by which the Anthropocene is known). Secondly, in Rosa's thinking, nature is an important sphere of resonance. The subject is not thought of without its ineluctable link to nature, and Rosa explicitly evokes the entry into the Anthropocene. Thirdly, resonance is understood as a relationship that develops and is learned. Thus, it is linked to an educational dynamic.

Keywords Anthropocene · Critical education · Political education · Political anthropology

13.1 Resistance, Critique and Utopia: Three Functions of Education in the Anthropocene

The concept of resonance, which has been described as a musical metaphor, is about the 'string' that connects us to the world and that sometimes begins to vibrate. How is it that, from time to time, we find ourselves in tune with the world, with others, with causes, with aesthetic or spiritual elements? Resonance is not an emotional state but a way of relating to others and the world, which includes the emotional component.[1]

[1] Rosa stresses this point regularly in his book so that there is no ambiguity. For example, having already mentioned it on page 188, he states on page 193 that: '*resonance is not an emotional state, but a relational mode*'. Drawing on the theories of mirror neurons and Canadian psychologist Albert Bandura's theory of self-efficacy, resonance is defined as 'a cognitive, affective, and bodily relationship to the world in which the subject, on the one hand, is touched – and sometimes 'shaken' to their neural foundations – by a fragment of the world, and in which, reciprocally, he or she 'responds' to the world by acting concretely upon it, thus experiencing its efficacy' (2018, p. 187).

© The Author(s), under exclusive license to Springer Nature Switzerland AG 2023

N. Wallenhorst, *A Critical Theory for the Anthropocene*, Anthropocene – Humanities and Social Sciences, https://doi.org/10.1007/978-3-031-37738-9_13

Resonance is a form of relationship to the world in which the subject and the world are both affected and transformed. What is at stake in this relationship is a response of the world to the subject and *vice versa*. Resonance is distinguished from an emotional state or the type of emotion experienced in this relationship (Rosa, 2018, p. 200). Resonance is distinguished from harmony: it is not the beauty that characterises resonance, but the response. This concept refers to those moments of living relationship between the world and the subject. Resonance also has the advantage of being non-essentialist, and thus differs from the concepts of identity or authenticity which are regularly drawn upon as resources in education.

As we began to identify in Chap. 12, the concept of resonance is descriptive of a part of social reality: some subjects feel more than others that they are living a 'good' or 'worthwhile' life, regardless of the amount of material resources available to them. It is also a normative concept, though, where a life marked by resonance with the world is preferable to a life characterised by silence from the world. Finally, resonance is a prospective concept: the author seeks ways out of the dynamic stabilisation of contemporary societies (which results in relentless and endless acceleration), leading to the reification of subjects and alienation, due to the imperative of growth of resources. It is a post-growth society that the author calls for, based on another relational habitus than the maximisation of individual interests characteristic of *homo oeconomicus*: '*another type of being-in-the-world is possible*, but it can only result from a political, economic and cultural revolution that is both simultaneous and concerted' (Rosa, 2016, p. 38).

The appropriation of the concept of resonance proposed in Chaps. 12, 13 and 14 of this work is organised around three educational functions which we deem important to reactivate and develop, in light of the entry into the Anthropocene. The first is a function of resistance, characterised by its anchorage in reality and the pursuit of struggles in the here and now. The second is a critical function, referring to the need to understand and rectify some of the errors that have been committed in modernity. The third and final function is utopian, whereby it is important to give ourselves the means to continue to hope, and to believe in a possible future. It is important to stand in opposition (resistance) to what is identified as problematic (critique) so that the desired future can come about (utopia). The functions of critique and utopia run the risk of being sterile if they are not thought of in combination with resistance; those of resistance and critique run the risk of violence if not articulated with utopia; and finally those of utopia and resistance run the risk of error and illusion if not triangulated with critique.

13.2 Hartmut Rosa, Andreas Weber and David Abram

13.2.1 Anthropological Convergences

Rosa's concept of resonance is rooted in the idea of nature, but it is only loosely articulated and anchored in the concept of the Anthropocene. In *Resonance*, Rosa seems to have a certain awareness of contemporary environmental issues, and in particular, of the extent of the anthropogenic alteration of the Earth system. Indeed, it quickly becomes apparent that he does not have an in-depth knowledge of the work in Earth System Sciences on the Anthropocene. This is why the educational reception of the concept of resonance proposed in this chapter is linked with the reading of the 2017 book *Sein und Teilen* (Being and Sharing) by another German 50-something, familiar with Merleau-Ponty: Andreas Weber, whose work we have already come across in this book. Weber's ideas provide necessary complementary articulations with Rosa's thinking, and reinforce his anthropology of immersion in nature, through his biological understanding of the living and his reading of the geoscientific work of the Anthropocene.[2] The thinking of a third author, the philosopher David Abram, completes this educational reception of the concept of resonance. Almost every chapter of *Sein und Teilen* begins with a relationship to the natural elements (the air, the sea, etc.). The style here is similar to that of David Abram. Moreover, it may be noted that David Abram wrote the foreword to Andreas Weber's *The Biology of Wonder – Aliveness, Feeling, and the Metamorphosis of Science* (2016a, b, New Society Publishers). Merleau-Ponty's phenomenology, examined by these three authors, is a helpful pillar when thinking about the organisation of life in the Anthropocene. Andreas Weber draws on the French philosopher's concept of the 'flesh of the world' developed in *The Visible and the Invisible*, where he thinks of the human body in terms of its belonging to the world. These three authors develop particularly radical thoughts,[3] which have in common the fact that they break away from two paradigms of modernity: firstly, the way in which contemporary societies can only stabilise themselves in a dynamic way through acceleration (which is particularly visible in the constant race for growth); and secondly, the separation between a silent non-human world and a human world which, alone, is capable of speaking.

In Weber's texts, we find some of the characteristics of Bellet's writings, combining existential depth, a poetic style of writing, and an opening to transcendence. For Weber, it is matter that opens the human adventure to something greater than itself, while being what allows it to exist. Weber describes the experience of resonance with the world and nature. There are times when moments are 'given' as 'gifts', when our small existence appears to be connected to the cosmos, and we experience the vastness of the world around us, not as something outside of ourselves, but something within: 'When we ourselves are the world, we feel good'[4] (Weber, 2017, p. 96). In his writing, he does not hesitate to bring up examples from his life experience.[5] In his argument, he draws on his own experience, which he describes in a

[2] *Sein und Teilen* is a much briefer and less academic philosophical essay than Rosa's proposed theoretical formula, which has already been particularly well received in Germany.

[3] Weber does not quote Rosa in *Sein und Teilen*, as he was writing concurrently with the publication of *Resonanz* (released in April 2016, when Weber was finalizing *Sein und Teilen*), but he does occasionally mention Rosa's work in lectures.

[4] 'Wenn wir selbst Welt sind, fühlen wir uns richtig' (Weber, 2017, p. 96).

[5] The writing is particularly well crafted to be accessible to a wide audience (it is evocative and poetic, and in this sense, bears some resemblance to that of Maurice Bellet).

poetic way not devoid of a form of spirituality[6] in his relationship with nature. For example, he writes of a childhood experience of an encounter with a butterfly: 'But from the encounter with [this butterfly], I retain a sense of living life in the fullness of the world'"[7] (Weber, 2017, p. 44).

Weber's book *Sein und Teilen* is constructed in the same way as Hartmut Rosa's book *Resonance*: the further we read, the clearer the political component of the author's agenda, with the proposal of an alternative to capitalism, becomes. In their opposition to capitalism, these two authors rely in particular on a poetic component or a metaphysics of the link to nature.[8] Moreover, in their work, Andreas Weber and Hartmut Rosa manage to take account of the 'self' while breaking with the individualism of *homo oeconomicus* and the predominance given to individual emancipation.[9] In this sense, their work is fruitful in the development of an educational theory for the Anthropocene. The characteristic shared by Weber, Rosa and Abram is that they open up the philosophy of intersubjectivity to non-humans by highlighting the fact that sharing conviviality with non-humans helps to make us more human.

13.2.2 The Playing Down of the Anthropocene in Rosa's Work

Rosa identifies the concept of resonance as being in line with Critical Theory. Extending that idea, we see that an education in resonance is critical and one of its functions is to generate critique. Among all critical possibilities, one of them appears to be particularly fundamental: the critique of modernity's ideology of withdrawal from nature, instead favouring an anthropology of immersion in nature. This critique is of primary importance because it is one of the most powerful revelations of the Anthropocene.

Rosa's anthropology is marked by an immersion of subjects in the world, who do not stand 'in front of the world, but *in the world*' (p. 43). This anthropology incorporates a cosmic component. Rosa is concerned with identifying what is at the root of the problem in our relationship to nature. Another relationship to the world is possible, other than that of modernity, based on rationalised distancing, exteriority from nature and the appropriation of its resources in order to develop our own. In addition to generating alienation, this type of relationship to the world compromises the future survival of the human adventure, because of the error on which it is based: the distinction which it draws between the subject and the world. For Rosa, it is important to develop responsive and reciprocal relationships between the subject and the world.

[6] From time to time, Andreas Weber quotes theologians.

[7] 'Aber von der Begegnung mit ihr bleibt das Gefühl, ein Leben in der Fülle der Welt zu führen' (Weber, 2017, p. 44).

[8] Weber, like Rosa, makes use of poetic, artistic and musical metaphors.

[9] Weber thinks of the self in sharing with the cosmos, and Rosa thinks primarily not of the individual but of the relationship of the individual with the world.

Rosa identifies several 'axes of resonance'. One of these is our relationship with things. He points out that western modernity, characterised by rational thinking, cannot recognise the fact that it might be possible to establish a relationship with an inanimate object. What Rosa, in line with Tobias Röhl, calls 'the *reification of things* through their objectification' (p. 258), is an attitude that is taught in school. This separation between the world of humans and that of objects is indicative of the distinction drawn between nature and culture, and impoverishes our relationship to the world. It is one of the reasons for what Rosa calls the current ecological 'crisis'. This radical reification of the non-human has a twofold consequence: the destruction of nature, which is seen merely as an insignificant thing, and the exhaustion of humanism as an end in itself. What is at stake in establishing resonant relationships with things is not a vague spiritualism or sentimentalism. It is directly related to our capacity to continue becoming human within the environment that plays host to the human adventure.

Rosa's environmental interpretation is interesting but lacks data from the Earth System Sciences. While he does identify some of the anthropological causes of a dysfunctional relationship with nature, he downplays the issue of transgressing the boundaries of the Earth system: 'What lies at the heart of the ecological crisis is not our unreasonable treatment of natural resources, but the fact that, in treating nature as a mere resource, we deny it its character as a sphere of resonance' (Rosa, 2018, p. 53). This minimisation is problematic. It is indicative of Rosa's apprehension of the concept of resonance as a concept with a totalising tendency. Everything is seen as solely a problem of resonance, and the adoption of resonant relationships would allow for living together in the Anthropocene. This is only partially true and does not consider the systemic risks involved in passing the tipping points. Rosa does not seem to identify that what is at stake is not only the establishment of a new type of relationship with nature, but also the reorganisation of human societies to allow us to continue to live within a reorganised Earth system. For example, he writes: 'It is here, it seems to me, that the great ecological anxiety of late modernity finds its origin: what lies at the heart of the deep environmental anxieties of our present is not so much the loss of nature as a *resource* as the threat of seeing this nature understood as a *sphere of resonance*, as an autonomous counterpart capable of responding to and orienting us, and of being silenced' (Rosa, 2016, p. 314). Whilst Rosa's analysis 'resonates' with us, we nevertheless believe that he overlooks the profound concern inherent in the fact that it would be impossible for a large sector of humanity to live on a planet whose climate is 5 or 6 °C warmer than at the beginning of the twentieth century. Thus, after noting how the media has staged a discourse holding that recent environmental disasters have been manmade while acknowledging that 'a considerable amount of climate change is manmade' (2018, p. 315), Rosa goes on to say, 'My point is only to show that the interpretations regularly given to these natural events and which portray them, far from any empirical explanation, as *nature's revenge* and *retaliation*, or a *cry of abused nature*, emanate from a cultural conception with powerful effects and from a dimension of experience where nature *can speak with its own voice* and *has something to tell us*. Testifying to a relationship to nature 'counterfactually' conceived as resonant, or *desired to be so*, they result from an unfortunate absence of positive resonant experiences and the fear of losing nature as a sphere of resonance' (Rosa, 2018, p. 315). Again, it is apparent here that Rosa has not read the Earth System Sciences articles highlighting the extent of the anthropogenic nature of the Earth system change. When he writes that 'a considerable part of climate change is human-induced' (2018, p. 315) or that 'The causal link between the emission of carbon dioxide, or other industrial discharges, and the occurrence of such natural disasters is not scientifically demonstrated' (2018, p. 315) he is playing down the magnitude of the anthropogenic component in the current crisis in the functioning of the Earth system (though his position is not actually one of climate-scepticism).

13.2.3 Weber's Overcoming of the Limits of Rosa's Work

In the face of the limitations of Rosa's thinking on the scope of the concept of resonance as a critical theory for the Anthropocene, Andreas Weber's thinking is particularly interesting. Weber is in agreement with Rosa when he mentions how fundamentally important it is to find or regain a relationship with what is non-human: 'We must preserve nature because we ourselves are nature; and we must preserve nature because it is everything we are not' (Weber, 2007, p. 294).[10] Weber, while maintaining a singularity and uniqueness of the subject, shifts the boundary between the subject and the world by showing that the subject is also part of the world. He bases his theoretical framework on the relationships that he describes as sharing – nothing short of absolutely vital – between the world and the subject, based on breathing and eating. Breathing, for example, is understood as sharing with the biosphere – ecosystems rely solely on exchanges. Thus, becoming oneself is a bio-geo-physical process of interaction with one's environment: 'The umbrella pines and the blue-green algae in the aquarium water also participate in the vast alchemy of existence (*Dasein*) that permeates us as our chests rise and fall. They absorb what they are, their outer space, their environment, and transform that exterior into something that they themselves are. It is a process that takes place on a completely natural and physical level' (Weber, 2017, p. 26).[11]

In his philosophical stance, Andreas Weber means that the Anthropocene can represent an opportunity for the human adventure. He is informed by the data of the Earth System Sciences on the entry into the Anthropocene and develops an

[10] 'Wir müssen Natur bewahren, weil wir sie selbst sind, und wir müssen Natur bewahren, weil sie alles ist, was wir nicht sind'.

[11] 'Auch die Schirmpinien und die Blaualgen im Aquarium Wasser beteiligen sich an der umfassenden Alchemie des Daseins, die uns durchherrscht, während unsere Brust auf und ab geht. Sie nehmen das, worin sie sich aufhalten, ihren Außenraum, ihre Umwelt, in sich hinein und verwandeln dieses Äußere in etwas, was sie selbst sind. Das ist ein Prozess, der auf einer ganz natürlichen und körperlichen Ebene stattfindet'.

anthropology for the Anthropocene that is characterised by a poetic and confident component, without ever lapsing into a technomodernist paradigm. It is the opposite of a *Great Anthropocene*. The opportunity it represents is to become more fully integrated into the fabric of life and in solidarity with it. The Anthropocene can be an opportunity to foster greater solidarity with humans and non-humans.[12] It is a question of becoming more alive, cooperative, feeling, poetic, creative: Andreas Weber outlines the anthropological shift which we are compelled to make by the entry into the Anthropocene us. Solidarity with all beings is fundamental: 'To become human, we must become animal, stone, water, yes: world'[13] (Weber, 2016a, b, p. 137). The Anthropocene can provide an opportunity to break with the sterile dualism between nature and culture. In particular, at the end of his book *Enlivenment*, Weber proposes to understand culture as the imagination of our naturalness. How can we learn to modify our representations and strengthen our imaginary sense of belonging to nature, which is one of the aspects which will allow us to navigate in the Anthropocene? Here, death is an inescapable passage and an anchor point of our naturalness.

The creativity and vitality of the living world is the foundation of Andreas Weber's anthropology. Thus, he attaches great strength and importance to the paradigm of life, understood on the basis of its interaction with all living things. Life is only possible through and in sharing. Weber's approach is philosophical, based on his biological knowledge, as can be seen in the following quote, 'The sea, the ocean with its smallest as well as its most gigantic beings, which regulates the climate and supplies the continents with water, is the very embodiment of a being that is only realised in sharing' (Weber, 2017, p. 16).[14] Bodies are understood not in isolation and 'in themselves', but on the basis of their interaction with the environment or milieu.

Andreas Weber's anthropology contains the concept of the omnipresent whole. Individuals are understood as members of a whole: 'Whether we like it or not: through our metabolism, which requires us to feed on other living beings and to incorporate and transform the atmosphere within us, we participate in the totality of the biosphere'[15] (Weber, 2017, p. 27). Not only do we incorporate the elements that surround us and transform them into ourselves, but, based on this same very simple observation of bio-physical exchange, Weber goes further: we transform ourselves back into what surrounds us: 'But the fabric we compose at a time "t" will have

[12] In this sense, his thought is particularly valuable and interesting for thinking about a political anthropology for education – hope is a necessary condition for thinking in education.

[13] 'Um Mensch zu sein, müssen wir Tier werden, Stein, Wasser, ja, Welt' (Weber, 2016a, b, p. 137).

[14] 'Das Meer, der Ozean mit seinen kleinsten und gigantischen Wesen, der das Klima reguliert und die Kontinente mit Wasser versorgt, ist der Inbegriff eines Seins, das sich nur im Teilen realisiert' (Weber, 2017, p. 16).

[15] 'Ob es uns gefällt oder nicht: durch unseren Stoffwechsel, der verlangt, dass wir uns von anderen Lebewesen ernährend und die Atmosphäre in uns hineinziehen und in uns verwandeln, haben wir an der Totalität der Biosphäre teil'.

become air again the next moment, then it will take the form of a plant or a shell, and 1 day it will become limestone sediments, rocks or sand. From the point of view of matter, this world is a great body, whose individuals represent momentary outgrowths'[16] (Weber, 2017, p. 27). This leads Weber to say that 'our singular capacities are but a very insignificant variation of the whole'[17] (Weber, 2017, p. 27). We are both much less and much more than the *homo oeconomicus* of modernity, which is an individual who has the capacity to magnify through pre-emption. Andreas Weber's anthropological conceptions are opposed to *homo oeconomicus*, not in the setting aside of his feelings and sensations that lead him to turn inward and seek to maximise his own interests, but in showing that, fundamentally, we are not the isolated actors we think we are. Sharing is what constitutes us: it does not diminish us but enhances us. Beyond exchange with our environment, there is no life possible.[18] 'We are both, the world and individuals'[19] (Weber, 2017, p. 29). Also, in Weber's thinking, while we are the earth, it remains an exterior entity with which it is possible to enter into a relationship, and which it is possible to hear speak.

Andreas Weber distinguishes sharing (*Teilen*) from separation (*Trennung*). Sharing has to do with love, it organises human identity, as the very fact of being. Based on biological conceptions, Weber only thinks philosophically about the notion of identity in solidarity. We are dealing with a form of biophenomenology whose political intention is to propose an alternative to capitalism, whose tenet is destruction (not sharing).[20]

---

[16] 'Aber der Stoff, aus dem wir in diesem Moment bestehen, wird im nächsten wieder Luft sein, dann Körper einer Pflanze oder Schale einer Muschel, und eines Tages Kalksediment, Felsen, Sand. Stofflich ist diese Welt ein Großer Körper, von dem die Einzelnen momentane Auswüchse darstellen'.

[17] 'Unsere besonderen Fähigkeiten sind nur eine sehr unbedeutende Variation des Ganzen'.

[18] The approach of thinking of the social body from another reading of the living and of inscribing the political body in this interaction with the biosphere has some similarities with recent works, which have been cited several times herein. We find this in the work of French biologists Pablo Servigne and Gauthier Chapelle in *Mutual Aid – The Other Law of the Jungle* (2021), or in *Le vivant comme modèle: la voie du biomimétisme* (by Chapelle & Decoust, 2015). This is the same approach taken by the French anthropologist François Flahault in 2018 in *L'homme, une espèce déboussolée – Anthropologie générale à l'écologie*. It is this same dynamic that was worked on, philosophically and without referring to biology as a science, by Corine Pelluchon in *Les nourritures* (2015).

[19] 'Wir sind beides, Welt und Einzelner'.

[20] At first, the cosmic anthropology of Andreas Weber appears relatively depoliticised. Where are the injustices to be denounced and the battles to be fought? Gradually, his anthropology becomes truly political and represents an alternative proposal to capitalism, in the same vein as the proposal made by Hartmut Rosa. The political scope of Andreas Weber's *Sein und Teilen* is significant. It underpins a break with individualism by showing the primacy of coexistence and giving it biological foundations.

13.3 A Political Anthropology of Education Rooted in the Biosphere

13.3.1 Enlivenment as a Practice of the Commons

Andreas Weber regularly uses the term '*commons*', which he understands as the network of relationships that are woven into vitality (*Lebendigkeit*). Thus, his political and economic conception of the commons is biologically based; it is founded on metabolic transactions between the human body and the environment: 'The budget of the commons is not an economy comparable to neoliberalism. It is not based on extraction, but on the nourishment of a metabolism within which one's own body is connected to the surrounding Earth and through which one's own experience of being becomes possible'[21] (Weber, 2017, p. 78). The commons are, in Weber's view, the natural or normal form of human societies for millennia, as opposed to the market as it is now. It is on the basis of the commons that he considers it necessary to think of an organisation of contemporary societies, starting with landscapes or forests.[22] Everything that has to do with vitality is of the order of the commons, starting with ecosystems. One of the benefits of his approach is that it associates matter with

[21] 'Der Haushalt der Allmende ist kein Wirtschaften wie der Neoliberalismus. Er heißt nicht, zu extrahieren, sondern einen Metabolismus zu nähren, in dem der eigene Körper mit dem umgebenden Land zusammenhängt, und durch den die eigene Seinserfahrung erst ermöglicht wird' (Weber, 2017, p. 78).

[22] Weber's reflection on the commons is opposed to the conventional thinking about the market as the basis of the organisation of contemporary human societies, which is an error still conveyed in the introductory teachings on the functioning of our societies in school: 'This reflection [on the commons] does not consist of bartering but of an exchange of vital necessities with the environment. Emotional experience is also part of this exchange, with the feeling of being accepted, of being nourished, of having an identity. This myth that people are traders and merchants, which children learn in primary school, is false. What matters most to people is not to trade in order to overcome 'scarcity'. The first cultural act was not: I have mussels, give me coconuts. The market, as we know it today, has not always existed. The concept was invented 300 years ago, particularly by the Anglo-Saxon economist Adam Smith, around the idea that free trade between supply and demand is the most efficient way to regulate the distribution of goods. Since then, this thinking has sunk into our ideas so deeply that we cannot question the belief that every aborigine is a trader. In school, this myth is still often part of the introduction of thinking about the social' (Weber, 2017, pp. 79–80) ('(« Diese Arbeit besteht nicht in Tauschgeschäften, sondern in einem Austausch des Lebensnotwendigen mit der Umgebung. Zu diesem Austausch gehört explizit auch die emotionale Erfahrung: das Gefühl, angenommen zu sein, ernährt zu sein, eine Identität zu haben. Der Mythos von Menschen als Händler und Schacherer, den Kindern heute schon in der Grundschule lernen, ist falsch. Menschen fällt es nicht als erstes einzutauschen, um "Knappheit"zu überwinden. Die früheste Kulturhandlung war nicht: Ich habe Muscheln, gib du mir Kokosnüsse. Der Markt, wie wir ihn heute kennen, existierte in historischer Zeit nicht. Sein Konzept wurde vor 300 Jahren insbesondere von angelsächsischen Nationalökonomen Adam Smith erfunden, mit der Idee, dass die freie Balance von Angebot und Nachfrage am effizientesten die Verteilung der Güter regeln können. Inzwischen ist uns dieses Denken so in Fleisch und Blut übergegangen, dass wir den Glauben, dass jede Ureinwohner Krämer sei, kaum hinterfragen. In der Schule bildet dieser Mythos häufig immer noch die Einführung in das Fach Sozialkunde' (Weber, 2017, pp. 79–80)).

meaning. From the standpoint of *Enlivenment*, nature as such is a commons. Nature is not only made up of matter, but also of subjects who interact, and it also covers a set of meanings or sensations.[23]

It describes the situation of the Singhs, a people living in India, who feel that they need to give something to the forest in order for it to remain healthy and develop. The forest is not a resource, but a partner, if not a part of themselves. Through his work, Andreas Weber attempts to attenuate or erase this separation between nature and society, in the continuation of what has been masterfully achieved by Philippe Descola (2005). He tries to make the world speak again, by giving life and vitality to the earth.[24] Here, it is our senses that are at stake. It is a question of learning to hear again the vitality of the living, with which we have such a close relationship that we are also part of this living fabric, apprehended as being outside of us.

With the demonstration that contemporary liberalism is rooted in a certain Darwinian conception of nature, Weber then proposes a natural conception of anti-capitalism, this time based on other biological postulates (notably marked by cooperation or interdependence). To the concept of identity, the economy of the commons opposes that of relation, to empiricism or abstraction, it opposes poetic objectivity, competition, mutual gifts, etc. Thus, Weber develops an economy of the commons that has been the daily life of 99% of the adventure of *Homo sapiens*. It is a thought and a management of the commons that allow us to think and to make a post-capitalist and post-neoliberal world emerge. Here love is understood as a commons good. 'To love means to abandon oneself entirely to the cycle of giving. Conversely, not to love means to exclude one's neighbour from the commons'[25] (Weber, 2017, p. 97).

From the perspective of *Enlivenment* and the economy of the commons which it opens up, Andreas Weber revisits a set of postulates or concepts. Thus, health does not mean harmony or integrity, but the ability to establish creative relationships with other organisms in the environment. The question of ownership is also reviewed: 'Nothing in nature can be claimed or controlled exclusively as property. Nothing

[23] Material resources are understood as also being invested with immaterial meanings to be taken into consideration. In this sense, it is in line with the concept of milieu as proposed by Augustin Berque.

[24] Weber takes up the analyses of Marx's ecological dimension that are currently being discussed increasingly widely. Continuing in this vein, he understands Marx's 'communism' ('*Kommunismus*') as a 'communalism' ('*Commonismus*'). For this, he relies on a reading of the *Economic and Philosophic Manuscripts of 1844* where 'The worker can do nothing without nature, without the sensual/sensory outer world' (Marx, *1844 Manuscripts*, cited in Weber, 2017, p. 84). The Marx who wrote the *1844 Manuscripts* is viewed as a humanist, considering man on the basis of his belonging to nature: 'man is a part of nature' (Marx, *1844 Manuscripts*, p. 47, cited by Weber, 2017, p. 98), and he retains the relationship of reciprocal transformation between nature and humans, at a distance from the dualism nature/culture or nature/society.

[25] 'Lieben heißt, sich den Kreislauf der Gabe ganz zu überlassen. Umgekehrt heißt, nicht zu lieben, sein Gegenüber aus der Allmende auszuschließen' (Weber, 2017, p. 97).

can fall under monopoly, everything in it is "open source"'[26] (Weber, 2016a, b, p. 78).

The break with anthropocentrism developed by Andreas Weber is particularly clear. The human being can only be understood in a relational fabric with other living organisms and with the environment. He does not exist in his own right, but only through his relationships. Andreas Weber's intellectual effort is to rethink nature as a commons asset in order to develop other anthropological conceptions that would then allow us to imagine another world. Weber's biological reading is directly political. He stresses the fact that biological evolution is a creative process. The dynamics of this process allow a whole to grow as much as each of its individual members. There is thus no opposition between the individual and the whole to which he belongs. Individual freedom is not opposed to the freedom of other individuals; it grows with them. Thus, Weber proposes to rethink the notion of freedom, starting from that of need. He moves from a concept of freedom vis-à-vis necessities (liberalism) to a freedom through necessities (biological freedom based on relationships).[27] The anchorage of Andreas Weber's thought in the living fabric gives rise to a political stance that is neither that of liberalism and the exacerbation of individual freedoms, nor a Marxist orientation of constraint of freedoms with a view to an end that would surpass them, but that of togetherness and a possible sharing of individual freedoms. They can be woven together, just as living beings are woven from a multitude of organisms. In this sense, Andreas Weber's political position is that of the convivialists.[28] Inter-reliance or inter-dependence is central: the self is thus as much a function or part of the whole as the whole is a function or part of the whole.

13.3.2 Sharing as a Way of Being Oneself

Weber places importance on being oneself. The approach is existential and does not ignore basic human aspirations such as to love and be loved, or to be saved from despair: 'The despair aroused by love, which has always been "long lost", is in truth the sadness that we cannot really be, that we have lost reality. We live in exile, but we do not know it because we are looking for something that cannot save us' (Weber,

[26] 'Nichts in der Natur kann exklusiv als Besitz in Anspruch genommen oder kontrolliert werden. Nichts in ihr ist Monopol, alles ist "Open Source"' (Weber, 2016a, b, p. 78).

[27] His approach is similar to the Ecuadorian approach of *Buen Vivir*, which is based on harmonious relations between humans and nature with a distribution of wealth within communities of solidarity. Coexistence is at the heart of *Buen Vivir*: between humans, but also with humans and animals, and between societies and the *Pachamama*, Mother Earth.

[28] Weber has a closeness to convivialism around this togetherness and the possible sharing of individual freedoms: 'When a community is organised as a common, this means an increase in individual freedom through the development of communal freedom' (Weber, 2016a, b, p. 84) ('Wenn eine Gemeinschaft als Allmende organisiert wird, bedeutet dies eine Zunahme individueller Freiheit durch das Anwachsen gemeinschaftlicher Freiheit').

2017, p. 18).[29] An initial reading might not identify the depth of Andreas Weber's political thinking, where we may sometimes feel that his object of study is identity or being. Becoming oneself is a bio-geo-physical process of interaction with one's environment: 'We are able to experience interiority because our interior (what constitutes us) is shared with others. We can only grasp ourselves because we do not belong to ourselves. As long as we do not belong to ourselves, we can be ourselves. Identity resides only in the making available'[30] (Weber, 2017, p. 46). By showing that one becomes oneself by being a part of a whole and by being physically shared, he lays the foundation for a society based on the sharing of goods.

This sharing at the foundation of what makes it possible to be oneself is also to be understood as the sharing of the earth with a set of other living beings, and as the sharing of the power of the very vitality that makes it possible to be us – to be human. It is necessary, in the very pursuit of the process of humanisation and for the perpetuation of the resilience of the adventure of life, that humans enter into a different type of relationship with animals, plants, and also with bacteria and fungi, whose importance in vitality we are constantly realising.

Moving away from the imperialist posture of destroyers, we must learn to adopt that of participants in this power of vitality that is far greater than us. Let us take a few moments to gauge the extent of the destruction in progress; we could receive the unpleasant and false impression that it is the inevitable mark of the human adventure. Between 50,000 and 3000 years ago, two-thirds of the mammals weighing more than 44 kg, and half of the species weighing more than 44 kg, died out. This episode was the Quaternary megafauna extinction. Between 50,000 and 12,000 years ago, Barnosky (2008) considers that the growth of human biomass roughly corresponds to the loss of megafaunal biomass (excluding humans).[31] The number of species that have lived on the Earth's surface is estimated to be 3.5 billion, and 99% of them are believed to have become extinct (Barnosky et al., 2011, p. 51). Thus, the extinction of species is a phenomenon that has occurred frequently since life first

[29] 'Die Verzweiflung über die Liebe, die immer schon "längst verloren" ist, ist in Wahrheit die Trauer darüber, dass wir nicht wirklich sein können, das wir die Wirklichkeit verloren haben. Wir leben im Exil, aber wir wissen es nicht, weil wir dahin nach dem suchen, was uns nicht retten kann' (Weber, 2017, p. 18).

[30] 'Wir sind zur Erfahrung von Innerlichkeit in der Lage, weil unser Inneres mit anderen geteilt ist. Wir können uns nur erfassen, weil wir uns nicht gehören. Sobald wir uns nicht selbst gehören, können wir selbst sein. Allein in der Veräußerung liegt die Identität' (Weber, 2017, p. 46)

[31] Humans are considered to belong to the megafauna, with an average weight of 67 kg for *Homo sapiens*. Barnosky directly links the extinction of the megafauna biomass with the growth of the human population (and therefore of human biomass). Today, 'the global ecosystem has gradually reorganised itself into a new state with the megafaunal biomass concentrated around one species, humans, rather than being distributed among many species' (Barnosky, 2008, p. 11543). The continuation of this dynamic is highly problematic for Barnosky: 'with the continued growth of human biomass and unprecedented global warming, only extraordinary and intensified conservation efforts will prevent a new round of extinctions'. Furthermore, he believes that this collapse in animal biomass would have a significant impact on humans and domestic animals. For Barnosky, the Quaternary megafauna extinction is an example of a threshold crossing and a radical break in the ecosystem.

appeared on Earth, but it is one that is offset by the creativity of life and the emergence of new species. On-land spaces have experienced five mass extinctions, since their colonisation by animals 541 million years ago (Barnosky et al., 2011). Barnosky et al. (2011) show that current species extinction rates are higher than the 'baseline' or 'reference rate',[32] requiring species conservation action with particular urgency. Given the current rates of species extinction, these authors estimate that we could be facing a sixth mass extinction event[33] within a few hundred years, which is particularly short on a geological time scale.[34]

A recent study shows that, in addition to the rate of species extinction due to human activity, the rate of population loss among terrestrial vertebrates is very high, even among species not under threat of extinction. Thus, out of 177 mammal species surveyed, 40% have experienced a significant decline in their overall population (Ceballos, Ehrlich and Dirzo, 2017). These authors show that the Earth is

---

[32] The 'background rate' is what could be called the normal rate of species extinction, before humans drove a number of species to extinction (and after the last mass extinction event).

[33] While the mass extinction that occurred 65 million years ago with the end of the Cretaceous took place in less than 100 years, following an asteroid strike on Earth, it may have taken two million years for the other transitions to occur (Barnosky et al., 2012, p. 53; Barnosky et al., 2011). As mentioned in the introduction, a mass extinction is defined by palaeontologists as the disappearance of 75% of living species in a relatively short geological timeframe (within two million years). Between the 1500s and 1700s, we saw the extinction of 50 species (both animal and plant). During the nineteenth century, 125 species disappeared. Then it was 500 species during the twentieth century. Today the rate of species extinction is 3–12 times higher than a baseline rate; and some even estimate it to be 1000 times higher (Pimm et al., 2014). According to the IUCN (*International Union for the Conservation of Nature*), 41% of amphibians, 33% of coral reefs, 34% of conifers, 25% of mammals and 13% of birds are at risk of extinction (IUCN, 2017). In the oceans, a number of animal species are disappearing and the total number of marine animals is also decreasing significantly due to overfishing, with almost half as many marine animals in 2012 compared to 1970 (Federau, 2017, p. 36). Furthermore, in the last 27 years, in Germany there has been a decline of over 75% in insect biomass (the proportions of decline are likely to be the same as in Europe as a whole) (Hallmann et al., 2017). This decline should be considered in estimates of biodiversity declines because of the importance of insects in the food chain. The current rate of biodiversity loss has rarely been paralleled in the history of the biosphere, and in any case, the situation of biodiversity loss due to the dominance of one species (*Homo sapiens*) has never been encountered.

[34] If all currently threatened species became extinct in the next 100 years and this rate remained constant, we would be looking at a mass extinction within 240–540 years (241.7 years for amphibians, 536.6 years for birds, 334.4 years for mammals) (Barnosky et al., 2011, p. 55). For Barnosky et al. (2012), 'First, the recent loss of species is dramatic and serious but does not yet constitute a mass extinction in the palaeontological sense of the *Big Five*. Over the course of historical times, we have lost only a few percent of the species assessed (although we have no way of knowing how many species we have lost that have never been described). […] The second point is particularly important. Even allowing for the difficulties of comparing the fossil and modern records, and applying conservative comparative methods that favour minimising the differences between fossil and modern extinction measures, it is clear that the loss of critically endangered species is propelling the world to a state of mass extinction that has been observed only five times in the last 540 million years. Further losses of species in the 'endangered' and 'at-risk' categories could bring about the sixth mass extinction in just a few centuries. Of particular concern is that this extinction trajectory is occurring under conditions that coincide perfectly with past mass extinctions: multiple atypical high-intensity ecological stressors, including rapid and unusual climate change'.

encountering a major episode of population and species extinction that will have consequences for the functioning of ecosystems and for all the vital services necessary for the sustainability of our civilisation. This is how they describe a 'biological annihilation'. This term highlights how significant this sixth round of species extinction that the Earth is experiencing is (Ceballos et al., 2017, p. 6089).

In this great transition towards a conception of sharing (rather than pre-emption and destruction) as what allows one to be oneself, Corine Pelluchon, in the *Animalist Manifesto* (2017a, b), stresses the importance of education and the adoption of new cultural practices. On this point, we can only stress how fundamental it seems that the school should position itself in relation to the means of enduring in the Anthropocene. To date, educational thinking has been characterised by a form of axiological neutrality, which ultimately amounts to relaying the market logic of pre-emption of vitality. This is a problematic situation.

13.3.3 The Inclusion of Life in the Carbon Cycle

The grounding of Weber's anthropology in the limits of the biosphere underlies his existential philosophy. We are but an insignificant part of a whole that is completely beyond us, and at the same time, in bio-physical terms, we are that whole: 'Here, breathing is something that is both completely trivial and absolutely non-trivial. Its secret is that each of us breathes not simply through the displacement of a gas, but through the transformation of the surrounding world. By feeling the way we breathe, we can gain an understanding of the fact that we are this world'[35] (Weber, 2017, p. 278). Weber emphasises this idea that we are the world.[36] This is an important support point for thinking about a critical theory for the Anthropocene and developing education in the Anthropocene: 'We can understand [*erfassen*: to grasp, to experience] that we do not merely walk upon the Earth, nor even in it, within its body, but that *we are this Earth*. In particular, we can see that we are not fundamentally separate from other living beings, for we feed on them and thereby transform their bodies into our own'[37] (Weber, 2017, p. 29). Weber continues, 'We

[35] 'Der Atem ist darum etwas ganz und gar nicht Triviales. Sein Geheimnis besteht darin, dass jeder von uns atmend nicht bloß ein Gas hin- und her bewegt, sondern sich in die umgebende Welt verwandelt. Indem wir spüren, wie wir atmen, können wir erfahren, dass wir diese Welt sind' (Weber, 2017, p. 27).

[36] Breathing represents the primordial experience of belonging to the world – more so, even, than eating: 'In breathing, we do not feel ourselves to be merely an individual subject through the pulsations of the body, but we experience ourselves to be the world itself through its rhythm' (Weber, 2017, p. 30) ('Im Atmen spüren wir uns nicht nur als einzelnes Subjekt in seinem pulsierenden Körper, sondern wir erfahren uns als die Welt selbst in ihrem rhythmischen Auf und Ab' (Weber, 2017, p. 30)).

[37] 'Wir können erfassen, dass wir nicht auf der Erde herumlaufen, und auch nicht nur in ihr, innerhalb ihres Körpers, sondern *dass wir diese Erde sind.* Wir können insbesondere sehen, dass wir nicht grundsätzlich von den anderen Lebewesen getrennt sind, denn wir ernähren uns ja von ihnen und verwandeln darum ihre Körper in unseren eigenen' (Weber, 2017, p. 29).

are both, the world and individuals'[38] (Weber, 2017, p. 29). The point here is to be careful to base our reasoning on accurate biological and physical data, so as not to descend into esoteric approaches such as that generated by the image of the Earth as *Gaia*.[39]

A profound understanding of the carbon cycle is important here for two reasons: firstly, because carbon is one of the essential elements of life and is present everywhere in the Earth system; and secondly, because disruption of the carbon cycle is one of the main causes of global warming (although not the only one). From an educational perspective, the aim is to enable learners to understand, but also to feel, that their existence is intimately linked to the carbon cycle and to these carbon 'reservoirs' in the atmosphere, in the lithosphere, in the oceans and, of course, in the living world. Today, the main problem is that the rate of anthropogenic carbon emissions exceeds the absorption capacity of continental or oceanic carbon 'sinks'.[40] We need to become fully aware of our integration, as humans, into this carbon cycle (half of the CO_2 produced by humanity is absorbed by terrestrial (a quarter) and marine ecosystems (also a quarter); the other half remains in the atmosphere – Rockström & Klum, 2015).[41] The absorption of carbon by the oceans generates acidification, causing changes in marine environments. Alongside this acidification, the oceans are experiencing another change: sea level rise due to water expansion, melting continental ice and changes in the water cycle (Federau, 2017, p. 58).[42]

Understanding the footprint of human activity mediated by technology (industry, transport, clothing, construction, etc.) is obviously a hugely important goal. However, it is also important to understand and experience the extent to which we are one with these oceans, and we the atmosphere that we are saturating with

[38] 'Wir sind beides, Welt und Einzelner' (Weber, 2017, p. 29).

[39] In the above quotations, we perceive the importance of working in education on the conceptions of the human body as an interrelated 'transforming body'. This has an impact on school curricula, especially in biology.

[40] The pre-industrial level of CO_2 in the atmosphere sat between 270 and 275 ppm, rising to 310 ppm in 1950, 379 ppm in 2005 and now 400 ppm. Human activity has been responsible for returning 555 billion metric tons of carbon to the atmosphere since 1750 (a metric ton is equal to 1000 kg – it is called a metric ton to differentiate it from other types of tons used in some countries). This level has not previously been reached for at least 800,000 years (and perhaps many more) (Lewis & Maslin, 2015, p. 172).

[41] As noted above, the release of 555 billion metric tons of carbon into the atmosphere since 1750 has resulted in levels of ocean acidification not seen in 300 million years (Lewis & Maslin, 2015, p. 172). On a scale of thousands of years, this acidity will gradually be absorbed, but currently, the increase in CO_2 is too rapid not to cause a change in pH. This is a process that will last about 10,000 years, while continuing to leave 25% of CO_2 in the atmosphere that will gradually be absorbed by the lithosphere over the next 100,000 to 200,000 years (Federau, 2017, p. 63).

[42] Since the last ice age on Earth, sea levels have risen by about 120 metres. During the twentieth century, the sea level has increased globally by 3.4 millimetres per year (this increase is roughly proportional to global warming). By 2100, according to IPCC scenarios, sea-level rise is estimated to be between 0.5 and 1.4 metres above the 1990 level (Rahmstorf, 2007, p. 368). Given the significant inertia, the German oceanographer Stefan Rahmstorf (2007) estimates that, in the long term, sea levels will stabilise at +10 to +30 m per one degree increase in global temperature.

carbon. The very function of breathing is important when thinking about a political anthropology of education in the Anthropocene. We are the world, as our breath reveals to us. From one moment to the next, we need to transform molecules from the atmosphere and integrate them into ourselves; this is what keeps us alive. The matter of our body and that of the world pass from one to the other in a never-ending series of exchanges. We have a vital need for this transformation. We are all part of the same living matter. Biological knowledge allows us to make a change to a different conception of the living world: no longer monadic but rhizomatic. This lays the foundations for alternative political conceptions for living in the Anthropocene. These biological conceptions allow us to shift our thinking away from the monadic and functionalist conceptions of transhumanism.[43]

We are made up through constant exchange. Without sharing with our environment, there could be no life: 'This is the secret of biological life: its form is not its fabric or substance but an activity'[44] (Weber, 2017, p. 32). The biological interpretation which Weber puts forward allows us to lay the foundations for a certain conception of social relations that can be categorised as 'convivialist': 'Everything that evolutionary researchers have described over the past 150 years exclusively as competition is more: it actually refers to a relationship of unconditional reciprocity. This is not to say that there is no competition in the biosphere, that there is only harmonious cooperation. But this simplistic idea of competition still does not account for the significance of what is taking place'[45] (Weber, 2017, p. 35).

[43] Indeed, we can implicitly see the extent to which the proponents of transhumanism are primarily computer scientists before they are biologists or neurobiologists, as the neurobiologist Catherine Vidal shows plainly in *Nos cerveaux resteront-ils humains?* (2019). Our bodily existence is the result of countless symbiotic relationships with different lifeforms. It is not possible to think of ourselves as not involved in this complex fabric of the living that transhumanist theories and research over-reduce: 'Our body is not only composed of cells of a single species, Man, *Homo sapiens*, but of a huge number of foreign species. We carry bacteria in our intestines, bacteria on our skin and mucous membranes, and amoebae in our mouths. The number of these creatures exceeds the number of our cells tenfold. The genes of the microorganisms that make us up are a hundred times more numerous than our own' (Weber, 2017, p. 36) ('Unser Körper besteht nicht nur aus Zellen einer einzigen Art, dem Menschen, Homo sapiens, sondern aus Dutzenden fremden Spezies. Wir tragen Bakterien im Darm, Bakterien auf der Haut und den Schleimhäuten, und Amöben im Mund. Die Zahl dieser Wesen übertrifft die Menge unserer Zellen um das Zehnfache. Die Gene der Mikroorganismen, die uns gemeinsam hervorbringen, sind sogar 100 Mal so zahlreich wie unsere eigenen').

[44] 'Das ist das Geheimnis biologischen Lebens: Seine Form ist nicht sein Stoff, sondern eine Tätigkeit' (Weber, 2017, p. 32).

[45] 'All das, was Evolutionforscher in den letzten 150 Jahren ausschließlich als Wettkampf beschreiben, ist mehr als das: ein Verhältnis von bedingungsloser Gegenseitigkeit. Das soll nicht Heißen, dass in der Biosphäre keine Konkurrenz herrsche, sondern nur die Harmonie hilfreicher Kooperation. Aber die bloße Idee des Wettkampfes löst den Sinn des Geschehens noch nicht ein' (Weber, 2017, p. 35).

13.3.4 *The Importance of Feeling in the Anthropocene*

For Weber, sensations, and the fact of experiencing feelings and being existentially affected by the experience of the world, are what allows us to connect with all other living beings. Feeling is understood as that which is absolutely primary: through feeling, we share existence under the same sun. It is through feeling that we can experience ourselves as being in a relationship with the world. Feeling is an experience shared with the living: 'Organisms are matter which are concerned with a specific goal – namely to continue to exist in a certain form and with a certain individuality. Feeling is therefore the fundamental phenomenon of vitality. Everything that lives feels: every being strives to maintain itself and develop'[46] (Weber, 2017, p. 44). Thus, learning to feel and sense is an important feature of education in the Anthropocene.

The experience of feeling allows us to become aware of our belonging to the world where the political emerges. Weber gives primacy to the political in his anthropological conceptions: 'We – like all beings – feel because the fabrics from which we create our individuality are part of the world'[47] (Weber, 2017, p. 45). However, he does not subsume identity in a collectivist whole in order to offer an alternative path to neoliberalism and its individualism. Instead, he shows that the self is dependent on others. Founding a philosophical anthropology on the sharing of the organic tissues that constitute us reveals a political relationship to the foreign as that which makes us what we are: 'It is what we are not that makes us develop an identity around something. Because we share our tissues with everyone, yes, because we are in fact the whole world, at every moment within these flows, our self is intensely fired up for its own survival. We – and I mean *all* life, *all* cells without exception – are worried about ourselves, not because enemy invaders might destroy our firmly established identity, but because, from the beginning, we do not belong to ourselves. This is what identity is: a process of self-concern'[48] (Weber, 2017, p. 47).

Weber links the consideration given to a child to its ability to care for the world: 'It is therefore of the utmost importance to take a child seriously, to give it the right to be, itself, a piece of the world with its own colour, so that the child can later take

---

[46] 'Organismen sind Materie, der es um etwas geht, nämlich darum, in einer bestimmten Form und Individualität weiter zu existieren. Fühlen ist somit das Grundphänomen der Lebendigkeit. Alles, was lebt, fühlt: Jedes Wesen strebt danach, sich zu erhalten und zu entfalten' (Weber, 2017, p. 44).

[47] 'Wir – wie alle Wesen – fühlen, weil der Stoff, aus dem wir unsere Individualität erschaffen, der Welt gehört' (Weber, 2017, p. 45).

[48] 'Es ist das, was wir nicht sind, was uns dazu bringt, eine Identität auszubilden, der es um etwas geht. Weil wir unseren Stoff mit allem teilen, ja, weil wir eigentlich die ganze Welt sind, im jeweiligen momentanen Ausschnitt ihres Durchflusses, flammt unser Selbst im intensiven Interesse an seinem eigenen Fortbestand auf. Wir – und damit meine ich ausnahmslos *alles* Leben, *alle* Zellen – sind um uns besorgt, nicht weil feindliche Angreifer unsere festgefügte Identität zerstören könnten, sondern weil wir uns von Anbeginn nicht gehören. Identität ist das: ein Prozess fühlender Sorge um sich' (Weber, 2017, p. 47).

care of this piece of the world it has in its own hands'[49] (Weber, 2017, p. 62). On this point, things seem to us rather more complex than Weber's explanation. The educator's attention to the child is indeed geared towards an end that goes beyond himself: to allow the child to take care of the world in turn. On the other hand, this link is not systematic, as Arendt showed in her critique of 'pedagogical approaches' to teaching, positioning the child too centrally in the educational act. Moreover, in his conception of sharing, Weber sometimes rushes the argument somewhat, mentioning logical links that are not necessarily self-evident. For example, he writes: 'In order to be able to share, I must want to be able to share. If I can be, I want to share'[50] (Weber, 2017, p. 101). Here, it is open to question whether the ability to be is necessarily accompanied by sharing.[51] It happens, and this is currently one of the norms, that being is accompanied by the desire to be more (and more, and so on), without the figure of the other intervening and opening up this process that turns out to be marked by domination.

13.3.5 A Proposal to Go Beyond the Enlightenment

Faced with the way in which the Anthropocene has become synonymous with eco-modernism or transhumanism, Andreas Weber develops an alternative: the poetics[52] of vitality. It is to this end that he dedicates the book entitled *Enlivenment* (2016a, b) – a term that he translates into German as '*Verlebendigung*', while retaining the English term *Enlivenment* in his text. Through the term *Enlivenment*, Andreas Weber signifies that the category of life or vitality is at the foundation of the understanding of our world. This short book has an ambitious aim: to replace the biolib-eral principles that guide our scientific, political and educational decisions with the dynamics and principles of *Enlivenment*. Thus, Weber explicitly evokes the possibility of developing a political thought, and an educational thought that opposes

[49] 'Es ist also von zentraler Wichtigkeit, dass ein Kind wahrgenommen wird, dass ihm zugestanden wird, selbst ein Stück Welt mit seinem spezifischen Kolorit zu sein, damit das Kind später die Sorge für dieses Stück Welt in die eigenen Hände nehmen kann' (Weber, 2017, p. 62).

[50] 'Um teilen zu können, muss ich sein wollen. Wenn ich sein kann, will ich teilen' (Weber, 2017, p. 101).

[51] It should be noted, though, that the above sentence is to be related to his conception of being as sharing.

[52] Weber proposes a definition of the poetic that he regularly evokes: 'The poetic dimension is the dimension of our organic existence that we deny. It is the world of our emotions, of our social ties and of everything we experience as significant. The poetic is thus inseparable from everyday social communication, from exchanges and interactions, from laughter and dismay or even from our own flesh' ('Die poetische Dimension ist die Dimension unserer organischen Existenz, die wir verleugnen. Es ist die Welt unserer Gefühle, unserer sozialen Bindungen und von allem, was wir als bedeutsam und sinnvoll erleben. Das Poetische ist deshalb untrennbar mit der alltäglichen sozialen Kommunikation, mit Austausch und Interaktionen verbunden, mit Lachen und Betroffenheit, mit unserem Fleisch.') (Weber, 2016a, b, p. 65).

bioliberal principles on the basis of *Enlivenment*. Being aware of one's vitality, of the fact that we are alive, is the basis for establishing a link with nature and other living organisms. What is at stake here is educational. The function of the term *Enlivenment* is to relate to and be articulated with the term *Enlightenment*. It is by no means a question of replacing rational thought and empirical observation with poetry, but of enabling different types of rationality to be articulated with each other. It is a matter of allowing science, politics and our societies to regain interest in the sensitive component of the existences of humans and other beings.[53]

In his introduction, Andreas Weber refers to the American poet and ecophilosopher Gary Snyder and his understanding of the wild as a process beyond human control. It is not control of the earth that will make it better, but participation. This is an extremely important shift. The human adventure has a role to play, having the possibility of participating more fully in the web of solidarity between all beings that is life. Participation is possible, but it should not be thought of as a takeover. Natural existence is poetic, says Weber, and this component of existence is too often neglected. In the dark times of the Anthropocene, Weber proposes to rely on this radical component to get us through.

There is something wrong with the way the world is run in the contemporary period. There is something missing in the way we conduct our lives, and we do not want to know what it is. For Andreas Weber, the crisis of meaning or significance in the contemporary period is really a crisis of vitality resulting from a difficulty, as an individual, in exercising our freedom to exist in relation to other beings: 'We have forgotten what it means to be alive'[54] (Weber, 2016a, b, p. 24). That is, everything around us is considered dead – as inert matter. In a way, this ideology of death suits us in our economic functioning. It is then a question of rediscovering what it means to be alive, as much in the conduct of our lives as in the scientific paradigms, and in the organisation of human societies. *Enlivenment* constitutes a form of second Enlightenment ('*Aufklärung 2.0*') (Weber, 2016a, b, p. 25). Andreas Weber, in his critique of the Enlightenment and its ideology of death, argues in the same vein as the critique of Max Horkheimer and Theodor W. Adorno in their *Dialectic of Enlightenment*. In this work, Horkheimer and Adorno point out how the ideology of the Enlightenment did not only bring freedom, but contributed to totalitarian catastrophes. This interesting critique of the Enlightenment, however, was not accompanied by an alternative concept. This is the exercise undertaken by Andreas Weber with the concept of *Enlivenment*.[55] In this respect, he is directly in line with

[53] Weber does not wish to establish a new utopia, but simply to generate a form of tenderness ('*Zärtlichkeit*') with what is real – what lives.

[54] 'Wir haben vergessen, was es heißt, am Leben zu sein' (Weber, 2016a, b, p. 24).

[55] The concept of *Enlivenment* is, as such, a critique of transhumanist theories and ideologies, against which Andreas Weber sometimes takes a stance (such as, for example, on page 27, 2016). Thus, it is a matter of learning to understand the human being differently than algorithms do, as a machine devoid of feelings. Andreas Weber reframed the principles of life by showing that they do not obey algorithms, but rather a principle of creativity. This is driven by intentional behaviours marked by freedom and autonomy. 'Subjectivity is not an illusion that helps organisms maximise

Hartmut Rosa's approach, who also follows the Critical Theory of the Frankfurt School and who, with the concept of resonance, develops an alternative concept – a form of positive political proposal.[56] *Enlivenment* is a form of 'corrective' concept.[57] While education in the Anthropocene presupposes education for resonance, it also presupposes an education for *Enlivenment* (both experience-based, allowing us to feel our immersion in the living world, and cognitive, with a lesson in how to overcome the dominant Cartesian rationality).

While the Enlightenment emphasised emancipation and self-determination,[58] Weber focuses on the contrary, with the concept of *Enlivenment*, on the right of each being to relate to others and its environment, to anchor its own vitality in the fabric of life – *Enlivenment* is based on what we have in common and share between beings: the very fact of living and feeling.

The *Enlivenment* perspective does not consider nature as a mechanical phenomenon characterised by its exteriority; we are immersed in it and involved in it.[59] The concept of *Enlivenment* differs from that of sustainable development ('*Nachhaltigkeit*') in its radicalness. *Enlivenment* is a critique of the neo-Darwinian perspective with the idea of biological optimisation and a critique of neo-liberalism and its quest for economic efficiency (Weber, 2016a, b p. 45). Indeed, these two paradigms, which may sometimes seem sufficient to identify our contemporary knowledge of how the world works, overlook any common and shared dimension between humans, and between humans and living things (and disregard cooperative dynamics). In *Enlivenment*, Andreas Weber shows that the anthropological conceptions based on competition and individual competition at the foundation of our biological (Darwinism) and economic (liberalism) functioning are linked to a historical way of viewing reality. His work consists proposes another reading of reality.

their evolutionary success, but it is the power that makes biological existence possible' ('Subjektivität ist keine Illusion, die Organismen in der Evolution zur Erfolgsmaximierung verhilft, sondern sie ist die Kraft, die biologische Existenz überhaupt ermöglicht') (Weber, 2016a, b, p.62). Andreas Weber's poetics constitute a critique of transhumanism. Among the biological principles he states, the last one appears to be a stone laid against the development of transhumanist madness: 'Death is real' ('Der Tod ist wirklich') (Weber, 2016a, b, p. 65).

[56] Indeed, the concepts of *Enlivenment* and resonance are particularly close. Andreas Weber uses the term '*Verbindung*' very frequently; the word can be translated as relationship, link or connection. The concept of *Enlivenment* gives an additional biological basis to the concept of resonance (in addition to, for example, mirror neurons).

[57] 'Das Enlivenment wird sich als Korrektiv verstanden wissen' (Weber, 2016a, b, p. 28).

[58] Let us remember here how diverse they were, which Weber seems to overlook somewhat in order to support his argument.

[59] The Anthropocene represents the end of Enlightenment dualism: 'nature does not lie outside of us. It is within us – and we are within it' ('Nature liegt nicht außerhalb von uns. Sie liegt in uns – und wir sind in ihr') (Weber, 2016a, b, p. 37). Weber sometimes uses Maurice Merleau-Ponty's expression 'flesh of the world'.

The world is not a war of one against all. Weber proposes to overhaul the foundations of political thought from another biological reading than Darwinism.

Thus, he takes the opposite view of a set of ideas about nature from which political positions are developed. Firstly, nature is not efficient. On the contrary, it is constantly wasteful: fish, amphibians and insects have to lay millions of eggs in order for just a few offspring to reach maturity. Another example of this inefficiency is the fact that warm-blooded animals use 90% of their energy to maintain their metabolism. Secondly, the biosphere is not growing. The biomass of the biosphere is at equilibrium (with very slight variations). Thus nature is characterised, rather, by a steady state. Thirdly, no new species has ever emerged as a result of competition for resources. On the contrary, it is new cooperations and new symbioses (or simply chance) that allow for the emergence of new lifeforms. Fourthly, nature provides sufficient resources for all, starting with solar energy, which is sufficiently abundant for all living things. From there, a set of symbioses and cooperations are possible to allow all species to live. Fifthly, property is not a characteristic of the biosphere, starting with the body, which is not the property of the living being, since it is so much in interaction with its environment and characterised by exchanges of matter (Weber, 2016a, b, pp. 55–57). On this basis, Andreas Weber develops a form of biopoetics as a model for living relationships.

13.3.5.1 From Critique to Utopia

The foundations of *Comment la terre s'est tue*, *Resonance*, and *Sein und Teilen* are the same: sensory experience of the world. Merleau-Ponty and the importance given to the sensitive experience represent the common point between them. Is it possible to take seriously the fact that these three authors mention that the Earth speaks and that it is possible to hear and heed it?

Bibliographical References

Barnosky, A. D. (2008). Megafauna biomass tradeoff as a driver of Quaternary and future extinctions. *Proceedings of the National Academy of Sciences, USA, 105*(Suppl 1), 11543–11548.

Barnosky, A. D., Matzke, N., Tomiya, S., Wogan, G. O. U., Swartz, B., Quental, T. B., Marshall, C., McGuire, J. L., Lindsey, E. L., Maguire, K. C., Mersey, B., & Ferrer, E. A. (2011). Has the Earth's sixth mass extinction already arrived? *Nature, 471*, 51–57.

Barnosky, A. D., Haldy, E. A., Bascompte, J., Berlow, E. L., Brown, J. H., Fortelius, M., Getz, W. M., Harte, J., Hastings, A., Marquet, P. A., Martinez, N. D., Mooers, A., Roopnarine, P., Vermeij, G., Williams, J. W., Gillespie, R., Kitzes, J., Marshall, C., Matzke, N., Mindell, D. P., Revilla, E., & Smith, A. B. (2012). Approaching a state shift in Earth's biosphere. *Nature, 486*, 52–58.

Ceballos, G., Ehrlich, P. R., & Dirzo, R. (2017). Biological annihilation via the ongoing sixth mass extinction signaled by vertebrate population losses and declines. *Proceedings of the National Academy of Sciences, USA, 114*(30), 6089–6096.

Chapelle, G., & Decoust, M. (2015). *Le vivant comme modèle: la voie du biomimétisme*. Albin Michel.

Descola, P. (2005). *Par-delà nature et culture*. Gallimard.

Federau, A. (2017). *Pour une philosophie de l'Anthropocène*. PUF.

Hallmann, C. A., Sorg, M., Jongejans, E., Siepel, H., Hofland, N., Schwan, H., Stenmans, W., Müller, A., Sumser, H., Hörren, T., Goulson, D., & de Kroon, H. (2017). More than 75 percent decline over 27 years in total flying insect biomass in protected areas. *PLOS One, 12*, 1–21.

IUCN. *The IUCN red list of threatened species, Version 2017-2*. http://www.iucnredlist.org

Lewis, S. L., & Maslin, M. A. (2015). Defining the Anthropocene. *Nature, 519*, 171–180.

Pelluchon, C. (2017a). L'éthique des vertus: une condition pour opérer la transition environnementale. *lapenseeecologique.com, PUF, 1*(1), 1–18.

Pelluchon, C. (2017b). *Manifeste animaliste – Politiser la cause animale*. Alma.

Pimm, S. L., Jenkins, C. N., Abell, R., Brooks, T. M., Gittleman, J. L., Joppa, L. N., Raven, P. H., Roberts, C. M., & Sexton, J. O. (2014). The biodiversity of species and their rates of extinction, distribution, and protection. *Science, 344*, 987–998.

Rahmstorf, S. (2007). A semi-empirical approach to projecting future sea-level rise. *Science, 315*, 368–370.

Rockström, J., & Klum, M. (2015). *Big world, small planet, abundance within planetary boundaries*. Yale University Press.

Rosa, H. (2016). *Resonanz, Eine Soziologie der Weltbeziehung*. Suhrkamp.

Rosa, H. (2018). *Résonance – Une sociologie de la relation au monde*. La découverte. (original edition 2016), French translation.

Weber, A. (2007). *Alles fühlt: Mensch, Natur und die Revolution der Lebenswissenschaften*. Berlin Verlag.

Weber, A. (2016a). *Enlivenment. Eine Kultur des Lebens. Versuch einer Poetik für das Anthropozän*. Matthes und Seitz.

Weber, A. (2016b). *The biology of Wonder – Aliveness, Feeling, and the Metamorphoses of Science*. New Society Publishers.

Weber, A. (2017). *Sein und Teilen – Eine Praxis schöpferischer Existenz*. Transcript Verlag.

Chapter 14
Utopian Education: Earth and the World Speak

Abstract This last chapter is composed of three parts. The first is relatively simple for contemporary Cartesian rationality: the Earth preserves the memory of human activities. It is particularly striking to note, here, that we find traces, in the sediments, of the major milestones in the human adventure. The earth speaks about us. If it is possible to read what the earth says about us, it is important to go further and learn to listen to the Earth as it speaks to us. This is the second part of this chapter: the Earth speaks to us – if we are capable of listening. Finally, this chapter concludes with a pedagogical reflection on the importance of listening in education and its primacy over speaking.

Keywords Anthropocene · Utopian education · Political education · Education in the Anthropocene

14.1 Making Earth and the World Sing in the Anthropocene

Along with resistance and critique, the third function to be integrated into the framework of an Anthropocene education is utopia. An Anthropocene education is utopian and has the function of bringing utopia into being. To make Earth and the world sing, in the Anthropocene, would be one of the ambitious goals of education in this disenchanted postmodernity. The idea that the teacher is there to enable students to hear the music of the world, rather than merely to understand and grasp it, is present in the history of educational thinking – notably in the work of Wilhelm von Humboldt, Johann Gottfried von Herder and Friedrich von Schiller. This means that in opposition to the technicist utopia, conceived of as the source of humanity's salvation, we place the utopia of dialogue with the non-human world, based on listening to it. Earth and the world may have things to tell us if we are able to break away from the paradigms of techno-scientific mastery of the world and the pre-emption of nature as a resource: 'The subjects of late modernity lose the ability to hear the world speak and respond as they extend their instrumental access to it.

© The Author(s), under exclusive license to Springer Nature Switzerland AG 2023

N. Wallenhorst, *A Critical Theory for the Anthropocene*, Anthropocene – Humanities and Social Sciences, https://doi.org/10.1007/978-3-031-37738-9_14

Their personal efficacy is not experienced as a means of accessing the world in the mode of resonant sensibility, but as reifying domination' (Rosa, 2016, 493).

The first part of this chapter is relatively simple for contemporary Cartesian rationality: the Earth preserves the memory of human activities. It is particularly striking to note, here, that we find traces, in the sediments, of the major milestones in the human adventure. The earth speaks about us. It is geologists who read the earth in this way, using their methodological and epistemological arsenal. In fact, they organise Earth's history into different temporal categories based on changes in the global state of the Earth, which it is possible to perceive in the sediment records. Stratigraphic signals allow for the definition of chronostratigraphic units that provide geologists with a form of common language to advance our knowledge of Earth's history. In the debate over dating the entry into the Anthropocene,[1] what drives Earth System researchers is the identification of currently discernible features that we know (or can assume with a high degree of probability) will still be discernible in several hundred thousand or several million years. 'What features of the Earth system have been modified by humans and will continue to be visible over time?'[2] is the question that stratigraphers ask.[3] The stratigraphic debate on the dating of the entry into the Anthropocene is generally less involved in political science than the debate on the systemic alteration of Earth's functioning, with the work on the crossing of planetary boundaries. However, it seems important to understand this stratigraphic debate for several reasons: firstly, stratigraphy adds a further institution to the usual scientific institutionalization of editorial spaces, with the production of the geological time scale by the International Union of Geological Sciences. Secondly, this debate is particularly important from an educational perspective because its outcome will determine the inclusion of the notion of the Anthropocene in the school curriculum. It is not primarily a political decision on the part of Ministries of Education that will cause the Anthropocene to be included on the

---

[1] Within the international scientific community of stratigraphers, we can identify two main criticisms of the definition of a date of entry into the Anthropocene. First is the fact that this geological time unit is based more on the observation of human history than on geological events. Is the Anthropocene a unit of time in the history of the Earth or merely a unit of time in human history? Secondly, the reasons for defining an entry into this new period would have more to do with politics than with geology (Autin & Holbrook, 2012; Finney & Edwards, 2016; Edwards, 2015; Visconti, 2014): 'the drive to officially recognize the Anthropocene may, in fact, be political rather than scientific' (Finney & Edwards, 2016: 4). While it is undeniable that the concept is used by social and political groups, its stratigraphic underpinnings appear to be formally demonstrable and distinct from its social and political use, and the changes currently underway within the Earth system appear partly irreversible in the eyes of the geologists in the Anthropocene Working Group (Zalasiewicz et al., 2017).

[2] The logic driving the search for stratigraphic markers for the Anthropocene is quite different from that for other geological time units. Geologists are looking for stratigraphic evidence of events in human history that we know about independently of geology (Federau, 2016, p. 49).

[3] Entering the stratigraphic debate on dating the entry into the Anthropocene, it is important to note that the dates of the beginning of the Anthropocene refer not to the starting point of human influence on the Earth but to the perceptibility of this influence in the sediments (Zalasiewicz et al., 2017, p. 57).

curriculum, but an institutional decision (with profoundly political implications) of the International Union of Geological Sciences. When the Anthropocene is recognised as an official geological epoch, it will appear on the geological time scale and will, *de facto*, be addressed early on in secondary school, in Life and Earth Sciences (where the geological time scale is included in the school curriculum). Here, it is important not to reduce the dating of the entry into the Anthropocene to the official date alone, but to teach pupils the complexities of the debate on the dating of the Anthropocene, enabling them to take the measure of the geological traces of human activities (for example, if the dating of the first nuclear bomb detonations is adopted as the official marker, it is important not to forget the impact of the use of fossil fuels from 1800 onwards, or the dawn of a consumer society in the mid-twentieth century).

If it is possible to read what the earth says about us, it is important to go further and learn to listen to the Earth as it speaks to us. This is the second part of this chapter: the Earth speaks to us – if we are capable of listening. Educational and political rationalities may appear to be shaken here, but the magnitude of the issue of the Anthropocene necessitates changes in the very field of rationality. Here, our reasoning is in line with the double filiation of Maurice Merleau-Ponty's phenomenology and the Critical Theory of the Frankfurt School – the same intersection where the work of Hartmut Rosa, Andreas Weber and David Abram[4] lies.

Finally, this chapter concludes with a pedagogical reflection on the importance of listening in education and its primacy over speaking. What is important to listen to here is not first of all the learned word of the teacher (which would be tantamount to granting primacy to speech over listening), but to listen to the unheard, so that this 'as-yet-unheard' speech can reach the learners. What is at stake here is that we may receive ourselves anew as humans, in a profound relationship with the whole of the non-human living world and the earthly environment which hosts us.

14.2 The Earth Speaks for Itself: The Different Candidate Dates for the Entry Into the Anthropocene

While we do not yet know precisely what the Anthropocene will be like in a few hundred or a few thousand years, and the major anthropogenic changes in the Earth system are yet to come, it is important for a body of Earth system scientists to declare the entry into a new geological epoch on the basis of a sufficiently strong stratigraphic record. One of the difficulties in defining a GSSP is that the effects of anthropogenic events on the Earth system as a whole are not immediate, but instead are sometimes delayed by decades or even centuries (Lewis & Maslin, 2015,

[4] David Abram does not fit directly into the framework of the Frankfurt School, but his work follows in the footsteps of Maurice Merleau-Ponty's thinking, interwoven with animist ideas.

p. 173).[5] For Waters et al. (2016) and Steffen et al. (2016, p. 11), the stratigraphic approach is defined by the simple question, 'Have humans changed the Earth system to such an extent that geological deposits forming now and in the recent past include a signature distinct from that of the Holocene and earlier epochs which will remain in the geological record? If so, when did the stratigraphic signal (not necessarily the first detectable anthropogenic change) become recognizable worldwide?'.[6]

Waters et al. (2014, b) claim to provide evidence for a stratigraphic basis to the Anthropocene from the identification of a set of anthropogenic sediments. For these authors, the matter of the stratigraphic perception of a different geological epoch marking the boundary between the Holocene and Anthropocene is clear-cut: 'Humans are altering the planet at an increasing rate, including long-term, global geologic processes' (p. 137). Candidate stratigraphic indicators differentiating the Holocene from the Anthropocene have recently been observed in sediments from a lake in western Greenland,[7] including plastics, radionuclides, fly ash, pesticides, reactive nitrogen and metals (Waters et al., 2016, p. 137). In these authors' view, the anthropogenic signatures attesting to the entry into the Anthropocene and its distinction from the Holocene hinge on the combination of three force multipliers: 'accelerated technological development, rapid growth of human population, and increased consumption of resources' (Waters et al., 2016, p. 139).

[5] A GSSP marker consists of matching a date with a physical marker correlated with other secondary markers. It must be positioned at a specific location on the globe while having a proven correlation with a new global context, and benefit from continuous and clearly discernible sedimentation (Lewis & Maslin, 2015, p. 172). Stratigraphy is also an interdisciplinary science, drawing upon data from sedimentology, palaeontology, geochemistry, geochronology, archaeology, pedo-stratigraphy, palaeomagnetism, and even palaeoclimatology (Steffen et al., 2016, p. 3).

[6] A range of indicators can be used as a basis for dating the onset of the Anthropocene, such as the abundance of anthropogenic deposits, biotic changes, geochemical changes, oceanic changes (such as ocean geochemistry, ocean biodiversity, sea-level changes), and catastrophic events – both 'natural' and anthropogenic. Isotopes can also help in this debate on the date of our entry into the Anthropocene. Changes in isotopes can be used to reconstruct data such as climate or atmospheric chemistry. For example, the time of the Ancient Greeks, marked by innovations and the use of heavy metals, is reflected in the lead isotopes. Thus, it is not only the concentration of gases in the atmosphere that has been changed by humans, but also the isotopic composition. While there is evidence of discernible anthropogenic isotope changes in the last few millennia, isotopes alone do not clearly show where to position the boundary between the Holocene and the Anthropocene (Dean et al., 2014, p. 284).

[7] Earth System researchers are also investigating Arctic and Alpine lakes, far from urban environments or direct human influence, to see if changes in sediment composition can be observed. The human impact on biogeochemical cycles (especially the nitrogen cycle) is well documented in lake sediments (Wolfe et al., 2013; Waters et al., 2016).

14.2.1 The Stone Age

The earliest proposed dates of entry into the Anthropocene date back to the Stone Age[8] (Doughty, 2013). The frequent use of fire during the Late Pleistocene geological epoch, about 400,000 years ago in Africa, is certainly the first landmark event whereby humans impacted their environment (Roebroeks & Villa, 2011). However, because fire use is always local, it does not produce a global GSSP and therefore cannot be used as a global marker (Lewis & Maslin, 2015, p. 173). On the other hand, another very significant event that the biosphere experienced was the megafauna extinction, with the disappearance of half of the species of large mammals. However, this event, which took place between 50,000 years and 12,000 years before the present, cannot give rise to a GSSP either, because the extinction was very uneven on the different continents and took place at different times (Lewis & Maslin, 2015, p. 174).

14.2.2 Agricultural Development

Some authors, such as the American science journalist Michael Balter (2013), the American-Belgian ecologist Jed O. Kaplan et al. (2011), and the American palaeoclimatologist William F. Ruddiman (2003, 2013; Ruddiman et al., 2014), believe that the Anthropocene began with the development of agriculture and the subsequent change in the chemical composition of the atmosphere. As already discussed, the climate stabilisation marking the entry into the Holocene allowed for the development of agriculture in different parts of the globe. Agriculture began simultaneously on three continents 10,000 years ago: in Southeast Asia, South America, and northern China (Lewis & Maslin, 2015, p. 174).[9]

The early Anthropocene hypothesis relies on a body of evidence showing terrestrial habitat modification and terrestrial biotic changes, marine microbiotic changes and changes in the constitution of the atmosphere as a result of agriculture (Zalasiewicz et al., 2014b). It is not impossible that the development of agriculture in different parts of the globe was responsible for the maintenance of stable Holocene temperatures and a non-return to an ice age due to the increase in CO_2 8000 years

[8]The Stone Age began with the creation of stone tools by the first hominids (2.4–3.2 million years ago).

[9]Agriculture has progressively led to the replacement of natural vegetation by another type of vegetation, and the disappearance of certain species in favour of the development of others domesticated by humans. Agriculture has also changed biogeochemical flows. (The ice, which offers a true record of the Earth's recent history, has retained air bubbles from the atmosphere, allowing us to determine the exact chemical make-up of the atmosphere over the last few hundred thousand years).

ago and the increase in methane 5000 years ago (Mac Farling Meure et al., 2006).[10] Two authors (Certini & Scalenghe, 2011) propose an original dating of the Anthropocene some 2000 years ago. This does not correspond exactly to the development of agriculture, but to a stage in the development of the great civilisations, of which traces can be found in the soils.[11]

14.2.3 *The Meeting of the Old and the New World*

Another event that could help define the entry into the Anthropocene is what Lewis and Maslin (2015) call the collision between the Old World and the New. So far, though, they are the only ones to propose this GSSP. This event marked the beginning of a global organisation of humanity across the planet, with common food products. This led to the reorganisation of animal and plant life. Above all, the arrival of the Europeans in America in 1492 resulted in a significant decline in the world population. The population in America is estimated to have been between 54 and 61 million in 1492, and hit a low of six million 158 years later, in 1650. This population decline in America was due to wars, slavery, diseases brought by the Europeans and famines. This decrease in population led to a decrease in agricultural land and an increase in the area of forests, estimated at 50 million hectares. This resulted in a decrease in the level of CO_2 in the atmosphere by 7 to 10 ppm,[12] which can be seen in the sediments of Antarctic ice between 1570 and 1620. Both authors consider 1610 to be an appropriate GSSP marker.[13] In addition to being perceptible in the sediment record and being a good candidate for a GSSP, the meeting of the old and new world has the interest of being a major event in human history,

[10] However, this hypothesis is not yet fully confirmed. The unusual length of the current interglacial period could also be due to non-anthropogenic changes in solar radiation and changes in Earth's orbit, altering methane emissions from tropical wetlands (Steffen et al., 2011, p. 847; Lewis & Maslin, 2015, p. 174). Earth's rotational axis, like its orbit, is not perfectly stable and experiences periodic changes that have impacts on the surface climate. These changes are known to have been largely responsible for the alternating glacial and interglacial phases.

[11] They consider the pedosphere (non-ice-covered soils) to be 'the best indicator of the rise to dominance of human impacts on the total environment because it reflects strongly the increasing impact of early civilizations on much of the Earth's surface' (2011, p. 1269). Finally, they believe that soils are a better stratigraphic indicator than changing atmospheric chemical makeup, when deciding where to position a golden spike. Soils have undergone numerous anthropogenic modifications in order to increase crop fertility. On the other hand, the limitation of their proposal is that they do not pursue the Anthropocene logic of systemic modification of the Earth through to its conclusion, but rather apprehend it as the anthropisation of the planet.

[12] Two different Antarctic ice cores show this 7–10 ppm drop in CO_2 in the atmosphere, exceeding the possible margins of error (of the order of 1–2 ppm).

[13] Secondary stratotypes are associated with this date, such as the impact of the meeting of biotopes between the Old and New Worlds, with the fossil record showing the arrival of maize in Europe and the Huaynaputina eruption, which can be seen in the sediments of both poles and the tropics. It is also possible to perceive a decrease in atmospheric methane and an increase in Arctic ice.

representing the starting point of globalisation, which is a real characteristic of the contemporary period.[14]

Through the different dating proposals presented so far, we can see how the Earth carries the traces of the stages of human history. The GSSP proposal of 1610 is particularly striking and deserves special attention in education because of the way in which the Earth reflects this characteristic of domination (here perceptible through imperialism and colonialism).

14.2.4 The Industrial Revolution

In his early articles, Crutzen (Crutzen & Stoermer, 2000; Crutzen, 2002) identified the entry into the Anthropocene at the Industrial Revolution and, if he had to date it precisely, he would choose 1769, marking the invention of the steam engine.[15] Indeed, for a number of authors, including Crutzen, industrial technologies are responsible for the entry into the Anthropocene (Steffen et al. 2011; Waters et al., 2014; Zalasiewicz et al., 2011), and the Industrial Revolution has supplanted agriculture in its overall effects on the environment.[16]

In recent years, less attention has been paid to the Industrial Revolution as the most serious candidate for the entry into the Anthropocene, as it was considered when the concept was first being fleshed out. In 2007, Steffen, Crutzen and McNeil proposed that the Industrial Age be considered the first stage of the Anthropocene. This began in Great Britain with the use of the steam engine, and helped transform the country around 1850 before converting the rest of the world.[17] Steffen, Grinevald, Crutzen and McNeil, in a 2011 article, propose to formally define the Anthropocene as a new era with a starting point around 1800, with the beginning of the Industrial Revolution. Their position would later evolve, as all four of them signed an article

[14] The benefit of Lewis and Maslin's (2015) approach is that in proposing a GSSP dating the entry into the Anthropocene, they analyse geological data rather than human history data. This is why the proposed GSSP is 1610, with the inflection of CO_2 in the atmosphere, as opposed to 1492. They reject the other possible dates of entry into the Anthropocene because they are not derived from a global marker. In their view, this date is the most serious candidate for a GSSP because of the radical and global change which occurred from that point. Since the meeting of the Old and New Worlds also facilitated the Industrial Revolution, it is a date compatible with Crutzen and Stoermer's (2000) initial hypothesis.

[15] 1769 was when James Watt filed the patent.

[16] For Zalasiewicz et al. in 2008, the eruption of Mount Tambora in April 1815, which led to a year of darkening in the Northern Hemisphere, could serve as a 'natural' stratigraphic marker of the entry into the Anthropocene, corresponding to the advent of the Industrial Revolution.

[17] The issue with the Industrial Revolution as a proposed date of entry into the Anthropocene is that it took place in the early eighteenth century in the United Kingdom, then in the years 1820–1880 on the East Coast of the United States and in Western Europe, and then at the very end of the eighteenth century, in the rest of Europe and the United States. The rest of the world was not affected by the Industrial Revolution until the twentieth century.

in 2014 in which they propose 1945 as the entry into the Anthropocene from a stratigraphic point of view, with the positioning of a golden spike (Zalasiewicz et al., 2014).[18]

14.2.5 The Great Acceleration

Among the social changes that have occurred since the mid-twentieth century, one of the most striking is the massive increase in world population, and in particular in the urban population, rising from 730 million to 3.7 billion people living in cities between 1945 and 2014 (Zalasiewicz et al., 2014, b, p. 3).[19] The Great Acceleration refers to an acceleration of human-induced sediment production. A range of authors also believe it can serve as a relevant stratigraphic marker (Waters et al., 2014, b; Holtgrieve et al., 2011; Wolfe et al., 2013; Zalasiewicz et al., 2014).[20]

When they updated the Great Acceleration graphs, including extending the curves to 2010, Steffen et al. (2015) revisited the debate on dating the entry into the Anthropocene, and argue that the onset of the Great Acceleration in the 1950s is the most convincing candidate from an Earth System Science perspective for dating the entry into the Anthropocene, due to an overshoot of Holocene variability and the evidence that these Earth System changes are driven by human activity.[21]

---

[18] For Steffen et al. (2011), the concentration of CO_2 in the atmosphere is a good indicator of the entry into the Anthropocene, which may be disputed by some stratigraphers who are particularly attentive to the data which are perceptible in rock sediments. The concentration of atmospheric CO_2 was 277 ppm in 1750; it then rose to 279 ppm in 1775; then 283 ppm in 1800; and 284 ppm in 1825. These rates remain within the Holocene range of variability of 260 to 285 ppm. The natural variability limit of the Holocene was reached in 1850, when the atmospheric concentration hit 285 ppm, and then it was exceeded in 1900 with 296 ppm (Steffen et al., 2011, p. 848; Etheridge et al., 1998; Kleinen et al., 2010). Metal pollution associated with the Industrial Revolution cannot be used as a GSSP, in the view of Lewis and Maslin, because of its local and diachronic nature (it is possible to perceive traces of this metal exploitation between 8000 BP and the Industrial Revolution) (2015, p. 175). While a set of markers are perceptible from 1800 onwards from North America and Northern Europe (including the increase in the concentration of atmospheric CO_2 from the nineteenth century onwards), these authors feel that none of these presents a clear GSSP.

[19] The Great Acceleration curves were presented by Steffen et al. in 2004, though the term 'Great Acceleration' was not used at that time. They were republished with the name, in 2007, to illustrate the evolution of human activity on Earth.

[20] Zalasiewicz et al. (2014, p. 4) identify a set of possible stratigraphic markers afferent to the Great Acceleration: the disruption of the nitrogen cycle following the development of the Haber-Bosch process; technofossils or plastic or aluminium waste spread across the globe; the explosion of human-induced deposits on land; the explosion of pollutants related to the development of industrial activity; the passing of a significant point in the biotic changes of anthropic origin, with a series of living species becoming extinct; the acceleration of hydrocarbon consumption, leading to an increase in the level of CO_2 in the atmosphere; the fossil traces linked to oil drilling; oil spills on the coasts; or the large dams which cause the disappearance of a series of river deltas.

[21] From a strictly stratigraphic point of view, Zalasiewicz and Williams (2013) also consider the Great Acceleration, beginning in the 1950s, as the point of entry into the Anthropocene.

14.2.6 Nuclear Detonations

Another hypothesis explored as a possible date of entry into the Anthropocene is the first detonation of a nuclear bomb on 16 July 1945 in Alamogordo, New Mexico[22] or the radioactivity peak that occurred some 20 years later. Following this first detonation, other bombs exploded at an average of one every 9.6 days, subsequently changing the chemical make-up of the atmosphere (Zalasiewicz et al., 2014, b, p. 1). The radioactivity peak can be seen in 1964, just after the Partial Test Ban Treaty signed on 5 August 1963 in Moscow was put into practice. For Lewis and Maslin (2015), and Masco (2010), atomic detonations define a good entry into the Anthropocene and the radiocarbon peak (Δ^{14} C) in the atmosphere represents an unambiguous GSSP.[23] For this reason, they propose the 1964 peak as a possible GSSP for the entry into the Anthropocene.[24]

While the atomic detonations are discernible in the stratigraphic record, they did not radically change the Earth system – although the power that they hold endows them with that possibility. The proposed date of entry is roughly the same as that of the Great Acceleration, but we wish to distinguish it – which not all authors do – because it refers not to the acceleration of the consumption of industrial production, but to the crossing of a techno-scientific point directly linked to the capacity for self-destruction.[25] The great stratigraphic benefit of nuclear bomb fallout is that it offers a clear, abrupt, and widespread signature, consistent with the kinds of signals of which stratigraphers are fond (Waters et al., 2015).[26]

---

[22] It was a test conducted by the US Army; its success led to the bombing of the cities of Hiroshima and Nagasaki on 6 and 9 July 1945, only weeks later.

[23] In addition to Δ^{14} C, plutonium 239 (^{239}Pu) could be an interesting marker, as it is particularly rare in the natural state and present in large proportions in the fallout from atomic detonations (Waters et al., 2015).

[24] The GSSP proposed by Lewis and Maslin (2015) is a pine tree at King Castle, 25 km east of Kraków in Poland, within whose rings this radiocarbon peak is found. The appeal of Lewis and Maslin's approach to the 1964 GSSP proposal is the same as for the 1610 proposal. They focus on stratigraphic data, rather than human history – even if the former are a direct reflection of the latter, delayed by a few years or decades. The point of choosing 1964 as the date of entry into the Anthropocene is that it refers to a set of markers of human impact on the Earth system that were discernible during the Great Acceleration.

[25] Crutzen has gradually revised his proposal for dating the entry into the Anthropocene. Initially, he proposed the invention of the steam engine and the approximate date of 1800, with the development of the Industrial Revolution in Britain. He later joined other researchers in proposing the first nuclear bomb detonation as the GSSP because of stronger stratigraphic evidence.

[26] Zalasiewicz et al. (2014, p. 6) in a paper entitled 'When did the Anthropocene begin? A mid-twentieth century boundary is stratigraphically optimal', state that the significance of the Anthropocene lies not so much in seeing within it the "first traces of our species", but in the scale, significance and longevity of change to the Earth system'. This is why they propose a GSSA as opposed to a GSSP, with the boundary marking the entry into the Anthropocene defined in 1945 with the first nuclear test at Alamogordo, which has the advantage of opening up a whole range of multidisciplinary research on the Anthropocene, rather than confining the concept to the sole field

14.2.7 Somewhere in the Future

For British climatologist Eric Wolff (2014), there is no doubt that human activity has altered elements of the Earth system, which he sees unambiguously in the ice cores taken from the ice caps, the polar glaciers that are particularly important for climate stability.[27] On the other hand, the current anthropogenic impact on the ice caps does not yet justify a new geological boundary between the Holocene and the Anthropocene, as the current change may be reversible within 10,000 to 100,000 years, although he believes that we can say that human activity is likely to delay the next ice formation by at least 100,000 years.[28] For Wolff, it is clear that humans have taken control of large parts of the Earth system. However, since it is not yet clear what the Anthropocene will be like in a few hundred or a few thousand years, it seems geologically more prudent to him that future generations should decide when its entry should be placed, when time has given them a clearer perspective.

14.2.8 The Systemic Approach: The anthropocene Rather Than the Anthropocene

All the GSSPs or candidate dates for entry into the Anthropocene have the characteristic of referring to key moments in human history. However, the matter of identifying an optimal stratigraphic point becomes meaningless in the view of some researchers, such as Ruddiman et al. (2015), or Gibbard and Walker (2013), who question whether it is a good thing to define the epoch in which we live in. Indeed, identifying that the Anthropocene Working Group was looking at 1945 as the date for entry into the Anthropocene could mean overlooking the thousands of years of human alteration of the environment – and even tens of thousands of years, if we include the extinction of large mammals. The Anthropocene Working Group chose 1945 or 1950 (which is more likely) because of its clarity as a stratigraphic marker in ice cores, ocean and lake sediments, and soils. On the other hand, it seems strange to define a date thousands of years after human beings effected major agricultural transformations that caused a change in the chemical constitution of the atmosphere, and two centuries after the Industrial Revolution. Thus, the 'stratigraphically optimal' choice of 1945 as the

of geology. Their argument appears interesting, and is unusual for geologists and Earth system scientists, who represent the majority of the 26 signatories of this article.

[27] Ice caps are large glaciers located at the poles. They are sometimes several kilometres thick and extend over tens of kilometres.[2] These ice caps are an important part of the Earth system and are visible from space. They have a high albedo and thus contribute to global cooling.

[28] Wolff considers that the nuclear explosions have left significant global traces of an anthropogenic nature. While Wolff remains very cautious, he does believe that a GSSP could be placed in the ice of Greenland or Antarctica – although he is aware that this is not ideal from a stratigraphic point of view because of the impermanence of the ice caps.

beginning of the Anthropocene cannot be considered 'environmentally optimal' (Ruddiman et al., 2015, p. 39). In view of this situation, these researchers formulate a proposal: to use the term '*a*nthropocene' with a lowercase a, and in an informal way, to refer to all anthropogenic environmental transformations, without restricting it.[29] This is one of the reasons why it appears necessary to work on the elements of this debate in the framework of an education for the Anthropocene – whose purpose, indeed, is to be 'environmentally optimal' – in order to take the measure of the extent of humanity's systematic alteration of the Earth.

In most debates, we see tension between two conceptions of the Anthropocene as anthropisation or as systemic alteration. While anthropisation began with the first human beings and intensified with sedentarisation, the alteration of the Earth system as a whole is more recent. Steffen et al. (2016, p. 13) propose to understand the Anthropocene from the perspective of Earth System Sciences (rather than from a stratigraphic perspective).

This systemic approach is important for Anthropocene education because it allows for the integration of all impacts of human activities on the Earth system over the past decades, unlike the stratigraphic approach which only looks at a major element (the location of the golden spike), with which secondary stratigraphic markers are then associated. The systemic approach provides a more exhaustive anthropological indicator, allowing us to propose possible forms of remediation. It can be a decisive support in the framework of an education in the Anthropocene, whose purpose is to identify the extent tó which the Earth speaks about us. These things are fundamentally important to know. However, what also appears to be important is to learn to listen to the Earth as it speaks to us in the present.

14.3 The Earth Speaks to Us

One of the aims of this critical theory for the Anthropocene is to think about ways of bringing about a hospitable and responsive world. This aim involves an effort to leave behind the contemporary *hybris* that is so characteristic of our way of life. To edify the theory of resonance, Rosa goes so far as to say that the world speaks to human beings.[30] On this point, it is important to bear in mind that *Resonance* can be read primarily as a book of political theory and that it is in no way a work of personal development. The utopian component of a pedagogy of resonance does not aim to

[29] Some geologists would like to see the Anthropocene recognised as a unit of time in human history that is not formally stratigraphically defined. They do not question the anthropogenic record in the sediments, but feel that we are not dealing with a global and distinct signature of the Holocene that is sufficient on the three registers of change that they believe are necessary to justify a change of epoch (biotic, sedimentary and geochemical) (Gibbard & Walker, 2013).

[30] In the September 2018 issue of *Philosophy Magazine*, where he was interviewed, Rosa responds to the reporter who questions this claim that the world speaks, 'Oh, right… doesn't the world speak to you?'

conquer the world, but to make it audible. In this context, it is not a matter of expanding our access to the world, subjugating it to control or conquering it, but, on the contrary, about making the world available: 'A better world is possible, a world where it is no longer primarily a matter of exploiting others but of hearing and responding to them' (Rosa, 2016, p. 527).

The two authors mentioned thus far – Hartmut Rosa with the concept of resonance, and Andreas Weber with his conception of existence as sharing with the biosphere – have in common in their intellectual goal to allow the Earth and the world to speak again. They are particularly close to the American philosopher David Abram and his analysis of writing as the basis of the world's silence in Western societies, developed in *The Spell of the Sensuous*. In these three ideas, sensory experience is central, and their arguments build upon the phenomenology of Maurice Merleau-Ponty. Is it possible to take these three authors' idea that the Earth speaks and that it is possible to listen to it? Hartmut Rosa explicitly mentions that it is a matter of allowing the world to sing again. Andreas Weber aims to make the silent world speak once more, to give vitality back to the Earth. In *The Spell of the Sensuous*, David Abram begins by recounting his experience of a relationship with nature and living in Asia, which he compares with his experience of returning to the United States (where he mentions having lost this relationship with nature due to the omnipresence of technology in Western civilisations, which he identifies as being rooted in writing).

14.3.1 Cutting Ourselves off from the Earth Gradually Leads Us Into Madness

Do the Earth and the world speak? What could it mean to learn to listen to the Earth and the world? This is indeed the profound change that Hartmut Rosa proposes in our interview with him. When asked about the actions needed in today's world to prepare for the future, he mentioned that it seemed fundamental to him to listen to the world before trying to act upon it. David Abram, in his book *The Spell of the Sensuous*, shows that the Earth speaks. How can we enter into a sensory, sensitive, experiential relationship with the world, rather than only techno-scientific rationality marked by distance – which Rosa explains has led to a 'catastrophe of resonance'. What does the sensory offer us? What do the Earth and the world give us? Are we so numbed in our relationship to the world and to the Earth that we do not realise that they nourish us, clothe us, and bind us together? Do our thoughts really only come from ourselves? Is our brain activity not energetically nourished by the food that the Earth gives us?

More fundamentally, how do we become human? For David Abram, this happens only by being in a relationship and 'conviviality with that which is not human' (2013, p. 16). For him, this means that we have to enter into a relationship with the sensual world – the very world that produces all of our technical artefacts. Indeed:

'Without the oxygen and breath of the forests, without the embrace of gravity, without the tumultuous magic of the rivers, we have no distance from our technologies, no way to assess their limits, no way to avoid their grip' (Abram, 2013, p. 16). For David Abram, Western societies cut off from the Earth have descended into a form of madness.

> Sadly, our culture's relation to the earthly biosphere can in no way be considered a reciprocal or balanced one: with thousands of acres of nonregenerating forest disappearing every hour, and hundreds of our fellow species becoming extinct each month as a result of our civilization's excesses, we can hardly be surprised by the amount of epidemic illness in our culture, from increasingly severe immune dysfunctions and cancers, to widespread psychological distress, depression, and ever more frequent suicides, to the accelerating number of household killings and mass murders committed for no apparent reason by otherwise coherent individuals' (Abram, 2013, p. 45). Abram then continues: 'From an animistic perspective, the clearest source of all this distress, both physical and psychological, lies in the aforementioned violence needlessly perpetrated by our civilization on the ecology of the planet; only by alleviating the latter will we be able to heal the former. While this may sound at first like a simple statement of faith, it makes eminent and obvious sense as soon as we acknowledge our thorough dependence upon the countless other organisms with whom we have evolved (Abram, 2013, p. 45).

The Spell of the Sensuous is not a strictly academic work. The author shares his sensory experience, which is, in a way, an experience of openness to transcendence based on the relationship with the living world. Here, we are dealing with the same characteristic as the works of Andreas Weber (especially *Sein und Teilen*). After his experience in Asia, where David Abram tells of his experience of a deep relationship with the earth and living things, he explains his return to the United States went and how this experience gradually faded away: 'Indeed, the more I spoke *about* animals, the less possible it became to speak *to* them. I gradually came to discern that there was no common ground between the unlimited human intellect and the limited sentience of other animals, no medium through which and they might communicate and reciprocate one another' (Abram, 2013, p. 49). He elaborates further: 'As the expressive and sentient landscape slowly faded behind my more exclusively human concerns, threatening to become little more than an illusion or fantasy, I began to feel – particularly in my chest and abdomen – as though I were being cut off from vital sources of nourishment. I was indeed reacclimating to my own culture, becoming more attuned to its styles of discourse and interaction, yet my bodily senses seemed to be losing their acuteness, becoming less awake to subtle changes and patterns' (Abram, 2013, p. 49). David Abram's account of this experience of comparing two cultural spaces (one of which promotes an experience of relationship with nature and the living and the other of which only offers an experience of relationship with the human world) is particularly interesting. Abram points out some very simple factors that this comparative experiment highlights: the fact that we deny any form of intelligence to non-human life, that the constant noise of engines prevents us from listening to birdsong, that the artificial lights of cities prevent us from seeing the immensity of starry skies, and that, finally, nature is only encountered through the medium of our technologies.

14.3.2 Writing, and Then the Complexity of Our Technical Artefacts, Have Distanced Us from the Biospheric Web

Abram builds the argument of his book around the fact that language is not exclusive to humans, and shows how writing has gradually silenced the Earth. The author shows that we are in an environment that speaks:

> We regularly talk of howling winds, and of chattering brooks. Yet these are more than mere metaphors. Our own languages are continually nourished by these other voices – by the roar of waterfalls and the thrumming of crickets. It is not by chance that, when hiking in the mountains, the English terms we spontaneously use to describe the surging waters of the nearby river are words like "rush," "splash," "gush," "wash." For the sound that unites all these words is that which the water itself chants as it flows between the banks. If language is not a purely mental phenomenon but a sensuous, bodily activity born of carnal reciprocity and participation, then our discourse has surely been influenced by many gestures, sounds, and rhythms besides those of our single species. Indeed, if human language arises from the perceptual interplay between the body and the world, then this language "belongs" to the animate landscape as much as it "belongs" to ourselves (Abram, 2013, p. 113).

Language is the result of this speaking Earth, of which it is a form of invisible extension.

The study of nature since the mid-nineteenth century shows how the environment can only be understood as a complex network of interrelated organisms that cannot be analysed separately from one another, being intrinsically related. Nature is a biospheric web – a web of relationships, a network of interrelated organisms of which our bodies are a part. The structure of language must be sought within the whole sensuous world, within which we are immersed, along with other living things: 'It is no more true that *we* speak than that the things, and the animate world itself, *speak within us*' (2013, p. 117). From this observation, it is possible to associate the complexity of our language with the complexity of the web of the biosphere rather than the superiority of our species (2013, p. 117). Thus, when we wreck biodiversity with bulldozer, wrecking ball and reinforced concrete, we are affecting our language. When the birds stop singing and the springs become silent, our language loses its evocative capacity, becomes cut off from its living source, the living and speaking diversity that is the Earth. Merleau-Ponty's contribution, which is most important in David Abram's thinking, is the way in which language is not apprehended as being produced by humans and outside of them, but rather the fact that humans are enveloped inside language.

The question that *The Spell of the Sensuous* then attempts to answer is how nature has been lost to our senses. How have we managed to blinker our eyes and muffle our ears to such a degree?, 2013, p. 125). Abram questions the way in which European civilisation has neglected and silenced the natural world. He shows how the written word has contributed to this distance from the world around us. Thus, it was long before Descartes or the modern period that we silenced the Earth.

> By asking the speaker to explain himself or to repeat his statement in different terms, Socrates forced his interlocutors to separate themselves, for the first time, from their own words – to separate themselves, that is, from the phrases and formulas that had become habitual through the constant repetition of traditional teaching stories. Prior to this moment, spoken discourse was inseparable from the endlessly repeated stories, legends, and myths that provided many of the spoken phrases one needed in one's daily actions and interactions. To speak was to live within a storied universe, and thus to feel one's closeness to those protagonists and ancestral heroes whose words often seemed to speak through one's own mouth. Such, as we have said, is the way culture preserves itself in the absence of written records. But Socrates interrupted all this. By continually asking his interlocutors to repeat and explain what they had said in other words, by getting them thus to listen to and ponder their own speaking, Socrates stunned his listeners out of the mnemonic trance demanded by orality, and hence out of the sensuous, storied realm to which they were accustomed. (Abram, 2013, pp. 149–150).

Abram shows that the cultures that have developed from the alphabet have the characteristic of being particularly anthropocentric. In a way, in reading *The Spell of the Sensuous*, one understands how deep some of the roots of the Anthropocene are: 'Once the stories are written down, however, the visible text becomes the primary mnemonic activator of the spoken stories – the inked traces left by the pen as it traverses the page replacing the earthly traces left by the animals, and by one's ancestors, in their interactions with the local land' (Abram, 2013, p. 238). Writing has had a defining influence on the sensory experience of the Earth and non-human life: '*wherever the alphabet advanced, it proceeded by dispelling the air of ghosts and invisible influences – by stripping the air of its anima, its psychic depth*' (Abram, 2013, p. 323). Consequently, we have forgotten a set of fundamental elements, starting with the air that reminds us of the invisible richness of the present. If language is a bodily and sensory phenomenon, rooted in the environment and profoundly animistic, writing, for its part, makes a decisive break through which civilisations have closed in on themselves.[31]

The question that challenges David Abram, and to which he tries to provide an answer is: how is it that our civilisation is so insensitive to non-human nature? How is it even possible that we humans can accept such extermination of other living species? Human beings do not seem to be aware of their debt to nature which provides them with shelter and nourishment.

[31] Indigenous cultures that do not have a written language have a different experience of language than our western cultures. They do not see language as a human construct, but as a feature of the environment and of life.

14.4 Learning to Listen (to the Earth, to the World and to Each Other)[32]

14.4.1 Overcoming the Reification of Things and Nature Through Their Objectification

Rosa identifies several 'axes of resonance'. The first concerns our relationships with things. He points out that Western modernity, characterised by rational thinking, cannot recognise the possibility to establish a relationship with an inanimate object (Rosa, 2018, p. 257). The reification of things through their objectification is an attitude that is taught in school. This separation between the world of humans and that of objects is indicative of the distinction between nature and culture, and impoverishes our relationship to the world. It is one of the reasons for our entry into the Anthropocene. This radical reification of the non-human has a twofold consequence: the destruction of nature, which is seen merely as a 'thing' of little importance, and the exhaustion of humanism as an end in itself. What is at stake in establishing living relationships with things is not vague spiritualism or sentimentality. It is directly related to our capacity to continue to become human, within the environment that hosts the human adventure. This ties in with Bruno Latour's idea of the parliament of things – which is based on Michel Serres' natural contract. What does it mean, in concrete terms, to establish resonant relationships with things? It means establishing a lasting relationship with them and transforming our mental representations of the object, which is no longer just a consumer product that satisfies transient desire, but, as it is of the world, becomes a vector of relationship with it.[33] How can we think about this relationship to things in the contemporary period, characterised as it is by the technosphere? Things mark and condition us much more than we are willing to admit: they 'produce us in the literal sense' (Rosa, 2018, p. 264). Here, it is a matter of learning the lessons of *homo religatus* that oppose *homo oeconomicus*' instrumental relationship to things.[34]

The purpose of a critical theory for the Anthropocene is to contribute to the conception of modes of relating to the world and to the Earth that are not only non-reified, but vitalising, which help lend vitality to both the world and the Earth. It is the Earth itself that needs to be brought to life. In David Abram's narrative where he explains the context in which the Earth and the living animal and plant world spoke to him, we identify how silence was an important feature, and also how important it was that western technology was not greatly present. Here, we can see that an education in the Anthropocene requires silence (which is not easy to achieve

[32] Here, 'Earth' refers to the earth's expanse and all of its biodiversity; 'the world' refers to the plurality of human political communities; and 'others' refers to human subjects.

[33] A ball, for example, can speak to us of excelling ourselves, of building bonds with our teammates, of dreams of victory or of the widespread desire to make a mark on human history.

[34] The technologisation of things contributes to the fact that they leave less and less of an impression on us; *high-tech* objects are characterised by the fact that we never fully master them and cannot assimilate them.

in education, where classrooms witness an uninterrupted torrent of words). For the Earth and the world to speak, and for resonance with them to become possible, it is necessary to listen to them, and we begin to do that by, ourselves, being silent. However, it often happens that the world, which is central to the educational act, does not speak – it is silent. Thus, the child may be presented with long stretches of boredom and meaninglessness at school. How can we conceive of an education that allows students to experience a warm, responsive relationship with the world? This is one of the questions we can ask in the Anthropocene. How can we allow the world to speak, in contemporary societies marked by hyper-technicisation, rationalisation of thought and reification of everything in the world – starting with human subjects?

One of the problems of schooling in the author's native France is precisely that it is built on silent relations with the world. Yet, an educational relationship is one of resonance. The teacher does not primarily transmit information, but responds to a request, an expectation on the part of the pupils.[35] This is an important hinging point in the development of political education in the Anthropocene. The learning of resonance is neither in the centring on the self (intrinsic to the notion of *Bildung*, for example), nor in the integration of cold and external knowledge of the world. The learning of resonance takes place in this living and responsive relationship with the world.[36]

14.4.2 Listening to the Earth's Living and Speaking Biodiversity

David Abram draws on the animist tradition discovered during his time in Asia, and on the phenomenology of Edmond Husserl and Maurice Merleau-Ponty, who highlighted the different modes of relating to the world between modern western societies and indigenous societies. In the west, scientific culture has imposed a perception of the world that disregards the sensory experience of immersion in the world, in favour of distancing which can be particularly problematic and destructive – and of which the Anthropocene is an indicator. One of Abram's hypotheses is that rediscovering the sensual aspect of our relationship to the world will allow us to make space for the Earth, which will then be able to speak again. By renewing the perceptive dimension of our sensibility and a form of carnal empathy through Anthropocene education, we shall be able to develop responsibility for non-human nature and attentiveness to life. How can we learn to speak *to* the world if we do not speak *about* the world in education? It starts with listening to it. Going further in this argument, we can say that thinking about education in the Anthropocene necessitates a radically alternative education. It is this radicality which we must have the courage

[35] Every teacher knows it well: in front of an attentive class, they are inexhaustible; on the other hand, the minutes seem like hours, the hours go so slowly, with a class that has no expectations or questions related to teaching.

[36] We can see here that the school can be a space for resonance as well as alienation.

to adopt, and which is far from a reality –so much so that education is part of a system of reifying rationalities, which reduce the Earth and the world to silence.

In establishing a lively and dynamic relationship to the world, listening is fundamental. Listening allows for the emergence of speech, and in this case, a speech from the world to the subject, which allows for emergence from a state of mute alienation in the subject's relationship to the world. The world will not be able to speak amid the incessant clamour of contemporary societies. For our part, we differentiate between silence and mutism. Silence is the fact of remaining silent so that unheard speech (that is, which we are not yet capable of hearing) can be discerned, whereas mutism is the absence of speech, even when listening. Mutism is the inability to speak. Silence is listening – the first element in allowing speech to emerge.[37]

The muteness of the world in contemporary societies is directly related to the logic of maximising individual interests, typical of *homo oeconomicus*: 'While the continual increase in the possibilities of choice may be effective from the point of view of maximising individual profit, the latter also fosters a mute relationship to the world, in which singular action leaves no trace and receives no response' (1996, p. 185). In order to allow the world to speak again, it is necessary to learn lessons that are opposed to the characteristics of *homo oeconomicus* – above all, those of hospitality and responsibility of *homo collectivus* and *homo religatus*.[38]

For Rosa, 'The resonant relationship to nature is *not* established through cognitive processes of learning or rational knowledge: it results from active and emotionally meaningful practical experiences' (2018, p. 313). While we fully agree with the second part of this sentence, it turns out that the cognitive transmission of elements related to the Anthropocene can be a means, and a particularly important one if not a fundamental one, to enable us to experience nature and enter a process of resonance.[39]

---

[37] From an educational perspective, we can say that listening, in this context, is different from attentive silence as a teacher speaks. Listening is an active process – it is a question of showing hospitality towards the other or the world, of welcoming it in and then trying to understand the mystery of its otherness and strangeness, so that the other and the world then begin to 'speak', to borrow a term from Rosa, Weber and Abram.

[38] The following excerpt from Rosa confirms the importance of these lessons learned in responsibility and hospitality: 'These reflections were corroborated from another perspective by the findings of a long-term study carried out at my research laboratory at the University of Jena, on the reasons why we engage in voluntary activities, often expending considerable energy, time and resources, without any social or economic benefit. Regardless of the fields studied, all the people interviewed confirmed, each in their own way, that their motivation was a need for resonance: in working with teenagers, disadvantaged people, refugees, within cultural structures or local associations, volunteers experience their personal effectiveness through "all that they receive in return"; in the fact that they feel able to make things happen – or that they manage to "leave a trace of their passage" and "make a difference"' (2018, pp. 185–186). The development of this learning in opposition to individual interest maximisation go so far as to enable school dropouts to return to education (Wallenhorst, 2016).

[39] On this point, we have regularly seen how the transmission of Anthropocene knowledge, based on the 'great articles' of the Anthropocene, has allowed the establishment or reinforcement of resonance with nature.

As we enter the Anthropocene, a double-edged catastrophe is looming on the horizon. The first is the development of uninhabitable zones in different parts of the globe. The second is the widespread collapse of resonance with nature. As mentioned in Chap. 13, in describing this possible collapse, Hartmut Rosa downplays the magnitude of the first, bio-geo-physical catastrophe. In his description of the collapse of resonance, Rosa extends the analysis of David Abram (whom he does not quote) as to the mutism of the world (Rosa, 2018, pp. 314–315).

Education in the Anthropocene must aim to move away from the strict logic of action on the environment – whether it be the pre-emptive action of *homo oeconomicus* or the environmental organisation action of *homo collectivus*. It is a matter of enabling *homo collectivus* to enter into a relationship with the environment – which is not to be understood as a resource. Establishing a responsive relationship with nature and hearing it speak does not mean the individual enjoyment of the beauty of nature during a walk in the forest or the mountains. It is not a matter of confining nature to non-daily moments of life, on weekends or during holidays. The challenge of education in the Anthropocene is to allow the world to speak by leaving behind the reifying approach to the world inherited from modernity, in which a subject studies an object, looks at it, and ultimately dominates, to exploit what is a mere object. As Hans Blumenberg points out in *The Readability of the World*, modernity attempts to make the world as a whole readable. As scientific discoveries progress, the world actually seems to be slipping away from our control and becoming more and more enigmatic or mysterious (Rosa, 2018, pp. 484–485). Yet the metaphor of the readability of the world is also an expectation or need for an intimate connection with the world (and in this sense, is reminiscent of the theory of resonance). Thus, it becomes necessary, in Hans Blumenberg's view, to renounce our domination over nature in order to regain its trust. One of the challenges of education in the Anthropocene is, quite simply, to reduce our instrumental access to the world and to recover a form of indigeneity (Arnsperger, 2019).

14.4.3 Learning to Listen to the Other

The French political scientist and educationalist Pierre Statius, examining certain features of contemporary political anthropology, questions the way in which the figure of the expressive individual dominates the debate in the field of education: 'the question of self-expression is even a prerequisite for any discussion on democratic education. The idea is as follows: it is a question of manifesting our authentic interiority, of being attentive to our true nature without ever giving in to the mirages of artifice. From that point on, the modern individual, who is immediately all that he or she should be, undermines the very conditions of any democratic education, which is necessarily artificialist and requires a long time, invested in exploring culture, step by step. In other words, the evolution of the contemporary individual undermines the conditions of democratic education, even though there can be no democracy without education!' (Statius, 2017, pp. 10–11).

Thus, learning to listen is particularly important in education. Here, debates can be a highly useful tool[40] – in addition to other equally essential pedagogical methods (such as sitting in the natural world and listening to what is going on). Debates are not viewed as a quest for victory over one's counterpart through argumentation where only the 'loudest' are heard.[41] The idea is that, as far as possible, everyone should be able to express themselves and everyone should show openness. What is important is not so much the art of rhetoric and learning to speak, but first and foremost, learning to listen.[42] One of the objectives of debate systems based on listening to the other with a view to expressing disagreement is that pupils learn to value listening to the other's speech – i.e. listening to their peers as they express themselves. Listening comes first: the teacher or educator (the one who is in the know, or is supposed to be) is careful to remain silent so that the learner (the one who does not know, or does not know that he/she knows) learns to know, and is listened to by the teacher and his/her peers. Listening must allow the other to become visible (Honneth, 2004). The idea is to achieve three types of learning: firstly, to learn to recognise the speaker as an alter ego. Even if I do not agree with him, he has the capacity for thought and subjectivation (Touraine & Khosrokhavar, 2000). Secondly, to learn hospitality by trying to be receptive to what is said and to open up one's own representations to make room for the other's ideas, without rejecting them *a priori* and without feeling threatened by this alien thinking. The pedagogical objective of interactions is not to learn to overcome, but to exhibit courteous hospitality so that the classroom becomes a place where differences are welcomed. The goal is for everyone to be able to express themselves and for everyone to learn to show

---

[40] Allowing democratic debates to continue to exist, and thus learning to debate whilst listening to others and accepting their positions, is a real challenge for democracy in the Anthropocene (Villalba, 2015, p. 60). 'The public arena, the space in the world which men need in order to emerge, is thus "man's work" more specifically than is the work of his hands or the labour of his body' (Arendt, 1983/1958, p. 234). Indeed, this is the very crux of the matter: debate is an arena in which individuals can be shaped and from which they can emerge as citizens. The *polis* is not a physical location in the city. It is there when men begin to speak or act together. Wherever men go, they will be a *polis*: 'action and speech create, amongst the participants, a space whose proper location may be found almost any time and anywhere' (Arendt, 1983/1958, p. 223).

[41] We do not think of debates as a contest to be won over others by the superiority of one's arguments (though this is what debates often look like): 'I can emerge "victorious" from a heated discussion with family members or colleagues, or from a political debate, because I have succeeded in imposing my objectives: everything went as I wanted. From the point of view of *rational choice theory*, I should be 'happy' because I have maximised my profit. However, there may be a bitter aftertaste if I know that the other parties, although of a different opinion, have only kept quiet or backed down. Quite different is the situation when we come out of a discussion with the feeling that we *really had something to say to each other*, that we *moved* in the course of the conversation, that *new perspectives opened up* and that we really *touched* one another' (Rosa, 2018, p. 402).

[42] One of the elements of institutionalist political ecology that brings together such as Dominique Bourg, the American politician Kerry Whiteside, the Australian politician Robyn Eckersley and many others, is the capacity for 'inclusive and egalitarian dialogue among all citizens' (Bourg & Whiteside, 2017, p. 12). These are elements that can be learned.

hospitality.[43] Thirdly, to learn to relate in a confrontational way, whilst be respectful of mutual differences.

Bibliographical References

Abram, D. (2013). *Comment la Terre s'est tue – Pour une écologie des sens.* La Découverte (ed. or. 1996), tr. fr.

Arendt, H. (1983). *Condition de l'homme moderne.* Calmann-Lévy (original edition 1958, French translation 1961), French translation.

Arnsperger, C. (2019). Serons-nous enfin, un jour, indigènes? Permaculture et éducation des profondeurs. In N. Wallenhorst & J.-P. Pierron (Eds.), *Eduquer en Anthropocène*. Le Bord de l'eau.

Autin, W. J., & Holbrook, J. M. (2012). Is the Anthropocene an issue of stratigraphy or pop culture? *GSA Today, 22*(7), 60–61.

Balter, M. (2013). Archaeologists say the 'Anthropocene' is here – But it began long ago. *Science, 340*, 261–262.

Bourg, D., & Whiteside, K. (2017). Ecologies politiques: essai de typologie. *lapenseeecologique.com, PUF, 1*(1), 1–26.

Certini, G., & Scalenghe, R. (2011). Anthropogenic soils are the golden spikes for the Anthropocene. *The Holocene, 21*(8), 1269–1274.

Crutzen, P. J. (2002). Geology of mankind: "The Anthropocene". *Nature, 415*, 23.

Crutzen, P. J., & Stoermer, E. F. (2000). The Anthropocene. *Global Change, Newsletter, 41*, 17–18.

Dean, J. R., Leng, M. J., & Mackay, A. W. (2014). Is there an isotopic signature of the Anthropocene? *The Anthropocene Review, 1*, 276–287.

Doughty, C. E. (2013). Preindustrial human impacts on global and regional environment. *Annual Review of Environment and Resources, 38*, 503–527.

Edwards, L. E. (2015). What is the Anthropocene? *Eos*. https://eos.org/opinions/what-is-the-anthropocene, consulted online on 26 January 2018.

Etheridge, D. M., Steele, L. P., Langenfelds, R. L., Francey, R. J., Barnola, J.-M., & Morgan, V. I. (1998). Historical CO_2 records from the Law Dome DE08, DE08-2, and DSS ice cores. In *Trends: A compendium of data on global change*. Oak Ridge National Laboratory, U.S. Department of Energy.

Federau, A. (2016). *Philosophie de l'Anthropocène – Interprétation et épistémologie*. Doctoral thesis supervised jointly by the University of Lausanne and the University of Bourgogne, by Professors Dominique Bourg and Jean-Claude Gens.

Finney, S., & Edwards, L. E. (2016). The 'Anthropocene' epoch: Scientific decision or political statement? *GSA Today, 26*(3-4), 4–10.

Gibbard, P. L., & Walker, M. J. C. (2013). The term "Anthropocene" in the context of formal geological classification. *Geological Society (London, Special publications), 395*, 1–9.

Holtgrieve, G. W., Schindler, D. E., Hobbs, W. O., Leavitt, P. R., Ward, E. J., Bunting, L., Chen, G., Finney, B. P., Gregory-Eaves, I., Holmgren, S., Lisac, M. J., Lisi, P. J., Nydick, K., Rogers, L. A., Saros, J. E., Selbie, D. T., Shapley, M. D., Walsh, P. B., & Wolfe, A. P. (2011). A coherent signature of anthropogenic nitrogen deposition to remote watersheds of the Northern Hemisphere. *Science, 334*, 1545–1548.

Honneth, A. (2004). La Théorie de la reconnaissance: une esquisse. *Revue du MAUSS, 23*, 133–136. French translation.

[43] Indeed, learning to listen to oneself is fundamentally important, and this is the 'challenge facing postmodern societies' (Robin, 2015, p. 48). Trying to understand the other is a challenge for education in the contemporary period (Morin, 2000, p. 103).

Kaplan, J. O., Krumhardt, K. M., Ellis, E. C., Ruddiman, W. F., Lemmen, C., & Goldewijk, K. (2011). Holocene carbon emissions as a result of anthropogenic land-cover change. *The Holocene, 21*(5), 775–791.

Kleinen, T., Brovkin, V., von Bloh, W., Archer, D., & Munhoven, G. (2010). Holocene carbon cycle dynamics. *Geophysical Research Letters, 37*, 1–5.

Lewis, S. L., & Maslin, M. A. (2015). Defining the Anthropocene. *Nature, 519*, 171–180.

Mac Farling Meure, C., Etheridge, D., Trudinger, C., Steele, P., Langenfelds, R., van Ommen, T., Smith, A., & Elkins, J. (2006). Law Dome CO_2, CH4 and N2O ice core records extended to 2000 years BP. *Geophysical Research Letters, 33*, 1–4.

Masco, J. (2010). Bad weather: On planetary crisis. *Social Studies of Science, 40*(1), 7–40.

Morin, E. (2000). *Les sept savoirs nécessaires à l'éducation du futur*. Seuil.

Robin, J.-Y. (2015). Adulte certes, mais postmoderne ! Le recours biographique. In A. Slowik & O. Czerniawska (Eds.), *Trajets de formation et approche biographique* (pp. 29–50). L'Harmattan.

Roebroeks, W., & Villa, P. (2011). On the earliest evidence for habitual use of fire in Europe. *Proceedings of the National Academy of Science, USA, 108*(13), 5209–5214.

Rosa, H. (2016). *Resonanz, Eine Soziologie der Weltbeziehung*. Suhrkamp.

Rosa, H. (2018). *Résonance – Une sociologie de la relation au monde*. La découverte. (original edition 2016), French translation.

Ruddiman, W. F. (2003). The anthropogenic greenhouse era began thousands of years ago. *Climatic Change, 61*, 261–293.

Ruddiman, W. F. (2013). The Anthropocene. *The Annual Review of Earth and Planetary Sciences, 41*, 45–68.

Ruddiman, W. F., Vavrus, S., Kutzbach, J., & Feng, H. (2014). Does pre-industrial warming double the anthropogenic total? *The Anthropocene Review, 1*, 1–7.

Ruddiman, W. F., Ellis, E. C., Kaplan, J. O., & Fuller, D. Q. (2015). Defining the epoch we live in. *Science, 348*, 38–39.

Statius, P. (2017). "La transition écologique et la démocratie: Quelques remarques philosophiques, politiques, et anthropologiques", *lapenseeecologique.com* (PUF), *1*(1), pp. 1–15.

Steffen, W., Sanderson, R. A., Tyson, P. D., Jäger, J., Matson, P. A., Moore, B., III, Oldfield, F., Richardson, K., Schellnhuber, H.-J., Turner, B. L., & Wasson, R. J. (2004). *Global change and the earth system. A planet under pressure* (The IGBP Book Series). Springer.

Steffen, W., Grinevald, J., Crutzen, P., & McNeill, J. (2011). The Anthropocene: Conceptual and historical perspectives. *Philosophical Transactions of the Royal Society, 369*, 842–867.

Steffen, W., Richardson, K., Rockström, J., Cornell, S. E., Fetzer, I., Bennett, E. M., Biggs, R., Carpenter, S. R., de Vries, W., de Witt, C. A., Folke, C., Gerten, D., Heinke, J., Mace, G. M., Persson, L. M., Ramanahan, V., Reyers, B., & Sörlin, S. (2015). Planetary boundaries: Guiding human development on a changing planet. *Science, 347*, 736–746.

Steffen, W., Leinfelder, R., Zalasiewicz, J., Waters, C. N., Williams, M., Summerhayes, C., Barnosky, A. D., Cearreta, A., Crutzen, P., Edgeworth, M., Ellis, E. C., Fairchild, I. J., Galuszka, A., Grinevald, J., Haywood, A., Ivar do Sul, J., Jeandel, C., McNeill, J. R., Odada, E., Oreskes, N., Revkin, A., Richter, D. d B., Syvitski, J., Vidas, D., Wagreich, M., Wing, S. L., Wolfe, A. P., & Schellnhuber, H. J. (2016). Stratigraphic and Earth system approaches to defining the Anthropocene. *Earth's Future, 4*, 1–22.

Touraine, A., & Khosrokhavar, F. (2000). *La recherche de soi: dialogue sur le sujet*. Fayard.

Villalba, B. (2015). Au fondement matériel de la démocratie. *Revue Projet, 344*, 56–63.

Visconti, G. (2014). Anthropocene: Another academic invention? *Rendiconti Lincei: Science Fisiche e Naturali, 25*(3), 381–392.

Wallenhorst, N. (2016). Citoyenneté existentielle et reconfiguration du politique – Les pratiques écologiques de deux jeunes professionnels. *Revue des Sciences Sociales, 55*, 116–123.

Waters, C. N., Zalasiewicz, J., Williams, M., Price, S. J., Ford, J. R., & Cooper, A. H. (2014). Evidence for a stratigraphic basis for the Anthropocene. In R. Rocha, J. Pais, J. Kullberg, & S. Finney (Eds.), *STRATI 2013* (pp. 989–993). Springer Geology, Springer.

Waters, C. N., Zalasiewicz, J. A., Williams, M., Ellis, M. A., & Snelling, A. M. (2014). A stratigraphical basis for the Anthropocene? *Geological Society (London, Special Publications), 395*, 1–21.

Waters, C. N., Syvitski, J. P. M., Gałuszka, A., Hancock, G. J., Zalasiewicz, J., Cearreta, A., Grinevald, J., Jeandel, C., McNeill, J. R., Summerhayes, C., & Barnosky, A. (2015). Can nuclear weapons fallout mark the beginning of the Anthropocene Epoch? *Bulletin of the Atomic Scientists, 71*(3), 46–57.

Waters, C. N., Zalasiewicz, J., Summerhayes, C., Barnosky, A. D., Poirier, C., Galuszka, A., Cearreta, A., Edgeworth, M., Ellis, E. C., Ellis, M., Jeandel, C., Leinfelder, R., McNeill, J. R., Richter, D. d B., Steffen, W., Syvitski, J., Vidas, D., Wagreich, M., Williams, M., Zhisheng, A., Grinevald, J., Odada, E., Oreskes, N., & Wolfe, A. P. (2016). The Anthropocene is functionally and stratigraphically distinct from the Holocene. *Science, 351*, 137–147.

Wolfe, A. P., Hobbs, W. O., Birks, H. H., Briner, J. P., Holmgren, S. U., Ingólfsson, O., Kaushal, S. S., Miller, G. H., Pagani, M., Saros, J. E., & Vinebrooke, R. D. (2013). Stratigraphic expressions of the Holocene–Anthropocene transition revealed in sediments from remote lakes. *Earth-Science Reviews, 116*, 17–34.

Wolff, E. W. (2014). Ice sheets and the Anthropocene. In C. N. Waters, J. Zalasiewicz, M. Williams, M. A. Elis, & A. Snelling (Eds.), *A stratigraphical basis for the Anthropocene* (pp. 255–263). Geological Society, Special Publications 395.

Zalasiewicz, J., & Williams, M. (2013, December). The Anthropocene: A comparison with the Ordovician-Silurian boundary. In *Rendiconti Lincei: Scienze Fisiche e Naturali* (pp. 3–11). Springer.

Zalasiewicz, J., Williams, M., Smith, A., Barry, T. L., Coe, A. L., Brown, P. R., Brenchley, P., Cantrill, D., Gale, A., Gibbard, P., Gregory, F. J., Hounslow, M. W., Kerr, A. C., Pearson, P., Knox, R., Powell, J., Waters, C., Marshall, J., Oates, M., Rawson, P., & Stone, P. (2008). Are we now living in the Anthropocene? *GSA Today, 18*(2), 4–8.

Zalasiewicz, J., Williams, M., Haywood, A., & Elis, M. (2011). The Anthropocene: A new epoch of geological time? *Philosophical Transactions of the Royal Society, 369*, 835–841.

Zalasiewicz, J., Waters, C. N., Williams, M., Barnosky, A. D., Cearreta, A., Crutzen, P., Ellis, E., Ellis, M. A., Fairchild, I. J., Grinevald, J., Haff, P. K., Hajdas, I., Leinfelder, R., McNeill, J., Odada, E. O., Poirier, C., Richter, D., Steffen, W., Summerhayes, C., Syvitski, J. P. M., Vidas, D., Wagreich, M., Wing, S. L., Wolfe, A. P., Zhisheng, A., & Oreskes, N. (2014b). When did the Anthropocene begin? A mid-twentieth century boundary level is stratigraphically optimal. *Quaternary International, 30*, 1–8.

Zalasiewicz, J., Williams, M., & Waters, C. N. (2014). Can an Anthropocene series be defined and recognized? *Geological Society (London, Special Publications), 395*, 39–53.

Zalasiewicz, J., Waters, C. N., Summerhayes, C. P., Wolfe, A. P., Barnosky, A. D., Cearreta, A., Crutzen, P., Ellis, E., Fairchild, I. J., Gałuszka, A., Haff, P., Hajdas, I., Head, M. J., Ivar do Sul, J. A., Jeandel, C., Leinfelder, R., McNeill, J. R., Neal, C., Odada, E., Oreskes, N., Steffen, W., Syvitski, J., Vidas, D., Wagreich, M., & Williams, M. (2017). The Working Group on the Anthropocene: Summary of evidence and interim recommendations. *Anthropocene, 19*, 55–60.

Zalasiewicz, J., Waters, C. N., Wolfe, A. P., Barnosky, A. D., Cearreta, A., Edgeworth, M., Ellis, E. C., Fairchild, I. J., Gradstein, F. M., Grinevald, J., Haff, P., Head, M. J., Ivar do Sul, J. A., Jeandel, C., Leinfelder, R., McNeill, J. R., Oreskes, N., Poirier, C., Revkin, A., Richter, D., Steffen, W., Summerhayes, C., Syviski, J. P. M., Vidas, D., Wagreich, M., Wing, S., & Williams, M. (2017). Making the case for a formal Anthropocene Epoch: An analysis of ongoing critiques. *Newsletters on Stratigraphy, 50*, 205–226.

Chapter 15
Educating to Change the World in the Anthropocene

Abstract 'Can we change the world or not?' That is the question that underlies this work. The answer is yes. How can this be done? By changing humans: No more, no less. It is the very essence of the human being that needs to be transformed. Now, the most profound essence of what makes us who and what we are actually lies outside of ourselves. It is in that 'in-between' which makes us human: in the relationships between us, in the resonant relationship with the Earth, in the convivial relationship with the trees, poppies, earthworms and finches. To transform humanity is to bring forth a common and shared world. Nurturing this hope that a transformation of the world is possible is the challenge of the twenty-first century, and indeed of centuries to come. Nurturing this hope is the fundamental challenge we face in the Anthropocene.

Keywords Anthropocene · Education in the Anthropocene · Political theory · Changing the world

Utopia. Indeed, the advent of the Anthropocene leads humans into a black night. However, the darkness of the night does not necessarily plunge us into darkness. The hope of transforming the world sustains us and brings light. In the midst of the darkening of the future caused by the Anthropocene, this work proposes to invest education as a glimmer of hope accompanying a regenerative 'between-us'. The glimmer may seem faint and ephemeral, but it is there. Its function is to allow some hope to shine through to illuminate a path – that of the transformation of the world and of human beings. Perhaps the emergence of concerted action, against a background of shared speech, facilitating a profound transformation of humans, is a miracle; at least, that is what Hannah Arendt regularly suggests. However, miraculous though it may seem, it has happened many times in human history – such is the case with the American War of Independence, the fall of the Berlin Wall, the movements in the subsoil of Maghrebi societies and that we have seen since the beginning of the 'Arab Spring' in the winter of 2010, and so it is with the various marches or strikes to spotlight climate issues. The hope that keeps us going is the fact of

© The Author(s), under exclusive license to Springer Nature Switzerland AG 2023

N. Wallenhorst, *A Critical Theory for the Anthropocene*, Anthropocene – Humanities and Social Sciences, https://doi.org/10.1007/978-3-031-37738-9_15

waiting for a miracle: a historical miracle, made of human hands cooperating with the uncontrollable power of life, which, five times already in the long history of terrestrial life, has been able to rise from the ashes of mass extinctions.

Utopia. Yes, in the face of the 'mother of all threats' (Les convivialistes, 2013, p. 12), the capacity we have to destroy ourselves, it is 'high time to relight the stars' (Apollinaire). Guillaume Apollinaire wrote this line in 1917 for the play *Les mamelles de Tirésias* (The Breasts of Tiresias), when he was seriously wounded and the First World War was taking its toll. Here again, night is differentiated from darkness by the presence of lights – even faint ones. Today, we fundamentally need the stars for our lives to be human. These stars represent the world outside of the human adventure, with its cosmic component, allowing humanity to be part of a whole that goes infinitely beyond it, and as a result, defining it as human in its relationship to the infinite. However, the stars that are to be relit also represent those collective horizons of hope necessary for humanity to continue its adventure, enabling it to work on preparing for the future. It is this 'essential' aspect that this book has dealt with, the first stage of which is coming to an end.

Critique. Let there be no misunderstanding here: such hope is worth nothing if it is not anchored in a critical analysis of that which so easily takes hold of each of us – that inordinate mania of unlimited possession. We all lack criticality. How many of us – citizens, researchers in various disciplines or elected officials in a position of responsibility – have read these 'great articles of the Anthropocene' which, often, are still owned by private groups? With these scientific articles of the Anthropocene, we have a treasure trove of information concerning humanity that should be made public and included in UNESCO's world heritage. Making public the knowledge of the Anthropocene is vital. What matters more? The historic centre of Venice, the Great Wall of China, the banks of the Seine or even Mont Saint-Michel, listed as World Heritage sites? or the articles by Barnosky et al. (2011) on the sixth mass extinction, Rockström et al. (2009a, b) and Steffen et al. (2015a, b) on the importance of respecting planetary boundaries, Barnosky et al. (2012) on how our biosphere is approaching a systemic tipping point, or those of Im, Pal and Eltahir (2015), Mora et al. (2017), and Bador et al. (2017) on the frequency and intensity of future heat waves?

Critique. The starter's pistol for this research work was the entry into the Anthropocene. While *homo sapiens* has inhabited the Earth for nearly 200,000 years, we had to wait 190,000 years before agriculture and animal husbandry were made possible by a stabler and milder climate, enabling us to settle down.[1] The mastery of ecosystems allowed us to generate agricultural surpluses whose management gave rise to the development of our civilizations: because we did not all need to eke out our existence (subsistence), some were able to build pyramids; others invented vaccines; still others devoted themselves to the education of children. Because the foundations of the way in which we currently live together on Earth are changing, it

[1] This rises to 340,000 years, if *Homo sapiens* is indeed 350,000 years old – which seems increasingly to be accepted (Hubelin et al., 2017).

is necessary to put elements of political anthropology to work again. This is the task that this book set out to accomplish, with the risks that the creativity of interdisciplinarity required by the exercise entails.

Critique. With the category of action to conceive and bring about a non-totalitarian space, Arendt's anthropology can be put to work in a prospective way. Through the three fundamental dimensions of the human condition, Arendt is not content with describing our relationship to the world: she proposes to combat economic domination and to think about the conditions for the development of a public space marked by plurality – a bulwark against totalitarianism. This intellectual posture of Arendt's, like that of the Critical Theoreticians of the Frankfurt School, has marked the one adopted in this research. What the first part of this work brought to light is none other than the potential ending of the human adventure. Therefore, the aim of this work is to contribute to re-enchantment with the world and the Earth, with the idea of concerted action and the use of modes of collective mobilisation based on a 'between-us' mentality open to sharing conviviality with non-human entities.

Critique. Finding common ground between Arendt's theory of politics as *inter homines esse* which creates humans as *zoa politika*, Bellet's focus on the 'between-us' as the original place of humanity, Abram's exhortation to listen to the Earth, the convivialists' sharing of food and the very fact of living, Rosa's concept of resonance, or Weber's conception of being as sharing, we can treat intersubjectivity as the anthropological foundation of politics. The anthropological works of Arendt, Bellet, Abram, the Convivialists, Rosa and Weber are relational: coexistence is of prime importance. This element is the basis for the repoliticisation of society which these authors propose. This requires the advent of post-Promethean, post-capitalist or post-growth societies. Once again, the question arises: 'How can this be done?' For our authors, this remains an open question. That does not mean, however, that this transition is not possible: indeed, we have, in the past, moved on from the Middle Ages to the Modern Era. We need to move from modernity to a reorganisation of societies in the Anthropocene, based on a renewed political anthropology.

Resistance. The transformation of the world will be equal to our hope and our critical analysis, but it is conditional on our resistance. In entering the Anthropocene, the planet is challenging us to a fight – and this is something we could certainly have worked on more in this research. Brazilian President Jair Bolsonaro's response to the Amazon fires of summer 2019 can be interpreted as a declaration of war on the world, in a period characterised by the rise of a totalitarian-leaning madness (Jair Bolsonaro, Vladimir Putin, Xi Jinping, Donald Trump, etc.) in an unstable biosphere. Faced with this increasing instability of the biosphere and the geopolitical context, we should go to war and choose our weapons. Our choice is made: education, characterised by convivial and democratic radicalism. We must learn to coexist and to love the world and the Earth that accommodates us. Yes, but this must be done against the backdrop of a war – there is surely no more appropriate term than this – to be fought and won. We must fight against this madness of exploitation of every corner of the earth, of every human being able to work to produce ever more profit. This madness is murderous. Capitalism – which easily possesses many of us – has blood on its hands. This war will not be fought without weeping and gnashing of teeth, without sweat and tears. What will happen when teachers talk to children about how every car journey contributes to the realisation of a certain future, about the reality of ongoing global warming (which is taught almost nowhere), about the seriousness of deforestation for the human adventure, about how we exploit animals in factory farms, or about the scandalous power of the oil lobbies? Children will look to their elders and ancestors and scream, 'What have you done? You have consumed the world. There are only a few crumbs left. Scandal!' When children realise how bloody their parents' hands are, the resistance will be immense. A war is being waged, though, and educators cannot refuse to fight.

15.1 Educating in the Anthropocene So That We Can Disagree Without Killing Each Other

Utopia, critique. Throughout *Resonance*, we perceive the care Rosa takes in distancing himself from Axel Honneth's thinking. Hartmut Rosa did his doctorate under the supervision of Axel Honneth, who influenced him in anchoring his thinking in the Critical Theory of the Frankfurt School. In *Resonance*, however, he steps away very significantly from the work of his mentor. He writes: 'It seems to me reductive to assume that the goal of all our efforts, all our actions and all our struggles is ultimately to be recognised and to avoid being despised. Such an assumption is insufficient, to say the least, to explain what drives us toward religion, aesthetics, or nature' (2018, p. 224). In Honneth's idea of recognition, what matters is to be seen, and this quest for recognition is the struggle of existence. Resonance is an attempt to theoretically integrate all the dynamics of subjects marked by desire and hope. For Rosa, it is not so much about being seen as it is about hearing what the world has to say – which can be assumed to be a message of understanding. Where Rosa disagrees with Honneth is on the centrality of struggle in the experience of everyday existence. Resonance is less combative than recognition, which can only be understood in the context of a struggle (in which there is inevitably a winner and a loser). Rosa also proposes to rediscover a peaceful relationship with the world, which the struggle for recognition hampers. The theory of resonance, compared to that of recognition, breaks with individualism by its focus on the 'we' where resonance occurs. Rosa's stance is one of rupture with Honneth (but also with Habermas), and he returns to the origins of critical theory, grounded in concern, in identifying modernity as the generalisation of a catastrophe of reification. Thus, it becomes imperative to get out of modernity, which carries no potential for increased resonance. In spite of the innumerable promises of resonance made to subjects, reality is found in the alienation from a world that we have ceased to hear.

Utopia. Feeling and experiencing the world are important abilities that must be learned in order for us to oppose each other without killing each other. Their learning is not favoured by the contemporary capitalist context, in which calculation is the central hinge around which the world rotates. This way of being exercises a form of control over identity and means individuals cannot be themselves, because the very fact of being and feeling must feed the added value. Andreas Weber supports the search for poetic objectivity that allows us to enter into a different relationship with reality than the dominant materialist, productivist and extractivist one. Our existences, like all our experiences, are corporeal, and therefore tangible. We are a body, with which we enter into a relationship with our environment. We are the experiences we have. What is interesting about Weber and Abram's proposals is that the development of the emotional component of subjectivity is integrated with the proposal of an alternative form of politics to capitalism. While the combination of the emotional and affective component with the market is particularly problematic (with the obligation to be happy highlighted by Edgar Cabanas and Eva Illouz in 2018 in *Happycracy*), Rosa and Weber manage to integrate it without it being on the side of the problem and merely whetting *homo oeconomicus*' voracious appetite.

Utopia, critique, resistance. Resonance proposes to go beyond the deadly component of the competition of capitalist modernity – competition whose traces can still be seen in this struggle for recognition. Resonance has the merit of being infinite. It proposes that individuals do something non-destructive with their need for boundlessness. The objective is not to conquer the world, but to allow us to hear it. From a political perspective, there is a form of novelty here where the aim of solidarity and social justice (the real objective of the theory of resonance aspiring to the advent of a post-capitalist world) could come to pass without a fight. Resonance can only be understood with a certain degree of poetry – therein lies the richness and also the limitation of this concept. On the other hand, one of the major limits of resonance is the way in which conflicts appear to be excluded from the process of construction and advent of a renewed world. One of the limitations of our work is that we have not sufficiently explored the importance of conflict and opposition (although these are central to the idea of convivialism, which hinges on the possibility of disagreeing without killing each other). Plainly, we have devoted more attention to the importance of learning hospitality than to learning conflictuality. Yet learning conflictuality – learning how to be in conflict – is one of the conditions that allow concerted action to emerge in a plural space. Conflictuality is, however, at the heart of this post-Promethean 'between-us', which is by no means a harmonious 'between-us'. It is the arena in which dialogue, critique and confrontation all take place.[2]

Resistance. This is all very well: massacre must be avoided at all costs. However, it must be recognised that we are no longer living in the peaceful geological epoch that is the Holocene. This is not the time for appeasement: it is time for combat. Learning to resonate will not be enough. We are at war: against human *hybris*, of

[2] Education in the Anthropocene is not fundamentally non-competitive. Indeed, is it not necessary to continue to surpass ourselves by making progress in medicine, justice, education, thought, interculturality, etc.? Competition can be a means of choice, if sufficiently thought out and regulated, to channel the nascent *hybris* in a non-destructive way.

course, but also, concretely, against the oil lobbies; against leaders who, for the sake of provocation, are destroying decades of cooperation between humans (for example, the United States' withdrawal from the Paris agreements during Trump's term – which Biden changed as soon as he took office) or millennia of cooperation between species (allowing the Amazon rainforest to burn by rejecting all offers of international aid); against the financialisation of capitalism; against intensive fishing; against globalised food exports leading to famine in many regions of the world; against the criminal ignorance of climate scepticism which sows confusion and helps blind people to the reality of the situation; against fake news produced for the sake of buzz (and therefore of gain); etc. Yes, we are at war, and we must learn to fight. Identifying the war to be waged, the weapons to be wielded and those to be avoided no matter what, is a matter for learning at school. We must learn to proscribe any physical violence in the conduct of this struggle and use the weapons of law, knowledge and democracy.

15.2 Educating in the Anthropocene for a Love of the World and the Earth

Critique. It sometimes happens that a few voices are raised in education to point out the love that is woven into the educational relationship, which is as if embroidered throughout the fabric of education with a barely perceptible golden thread. The educational professions are often considered vocations to which the educator is committed wholeheartedly. What Part III of this work illustrated is not so much the love of the pupil for the educator or of the educator for the pupil, but, through the political learning of hospitality and responsibility, the learning of this necessary love for the world so that a genuinely human life remains possible, even in the Anthropocene.

Critique. Perhaps some individuals are satisfied, with the development of psychology, with how they have learned to love their children or their students more. But, frankly, can we be satisfied with loving our children or our students? Is there not dramatic blindness in this? The illusion that a life led by the love of one's loved ones is enough? The illusion that the act of educating is a guarantee of using one's existence right – in the arena where we cannot be reproached. Educating with love is neither sufficient nor satisfying. For the love of humans alone does not hinder the destructive fire that we leave on Earth. Loving one's children or pupils is not the issue of education in the contemporary period characterised by the Anthropocene. On the contrary, love is one of the fundamental issues of the Anthropocene: love for the world and for Earth.

Critique, utopia, resistance. The concept of the world in Arendt's work is indeed invested with a decision: the decision to love it. As the French philosopher Véronique Albanel (2010) states, Arendt has been grappling with the question of love of the world since her doctoral research in the 1920s on Augustine's concept of love. She did not formulate the concept as such until March 1955 – after having gone through the ordeal of the Gurs internment camp and then written *The Origins of Totalitarianism* – in her Journal of Thought: '*Amor mundi* – Why is it so difficult to love the world?' (2002, p. 714). She used the same expression 5 months later in a letter to Karl Jaspers, on 6 August: 'I started so late, only a few years ago, to really love the world. […] Out of gratitude, I would like to call my book of political theory *Amor mundi*'[3] (p. 264). The idea proposed in this work is marked by a shift away from the political to the choice of loving the world. This decision actualises the power to begin as the power to forgive. The love of the world preserves the place of genesis of humanity, and in this way, allows us to survive moments of destruction of the political. Because this love of the world helps to preserve us from the temptation to desert the world, it is absolutely fundamental for the dark times of the Anthropocene that lie ahead of us.

Critique, utopia. Education in the Anthropocene appears to be the repository of a treasure. Not the treasure of emancipation made possible by a 'successful' integration into the world of work in accordance with the criteria of the market economy, nor the treasure of world knowledge, but the treasure of the world itself, which it is possible to love, to showcase emerge, to share with one's contemporaries and to offer to future generations as what will allow them to live – no more, no less. At the heart of the human adventure is a dynamic of anthropological and political transformation. Having now come to the end of this work, education in the Anthropocene appears to wield the first great power: that of liberating the power to make a start, to foster relationships between individuals in a certain climate conducive to the emergence of action. From an educational perspective, action is the power that humans have to actualise their freedom: 'Men *are* free […] when they act – neither before nor after; indeed, *being* free and acting are one and the same' (Arendt, 1972, p. 198). Here, freedom is tied to others and to the world. This political understanding of freedom is particularly important in education. It stands in opposition to a common understanding of freedom that typically evokes individual freedoms, a form of inner feeling of escape from constraint. A political conception of freedom is opposed to this inner freedom that is sometimes reified in education. This freedom allows for the emergence of the second great power that education in the Anthropocene must aim to instil: the power to forgive and to promise. The power to forgive, which is a remedy for irreversible action, and the power to promise, which gives humans some certainty and provides a stabilising framework for unpredictable action. Such education establishes relationships among humans and opens up a space in which it is possible to speak and act, allowing the power of politics to be exercised. For this to happen, education as an institution must give up the power of domination, and sometimes even the power of knowledge (understood as decontextualised disciplinary knowledge of the Anthropocene).

Critique, utopia. Arendt begins the prologue of *Between Past and Future* with this quotation from her contemporary, the French poet René Char, who was also a

[3] Hannah Arendt refers here to the book that would be published as *The Human Condition*.

Resistance fighter during the Second World War: 'Our heritage was left to us without a testament'.[4] This is a particularly significant phrase in Arendt's work, manifesting the contemporary crisis, between the past and the future. The present generations are dealing with a break in the usual thread of generational transmissions. This analysis of Arendt's has a strong contemporality. The political function of education is precisely to accompany, with testament, the appropriation of this heritage that is the world, while not being afraid to think about the advent of another type of world. A better world is possible, in which no-one owns another, where freedoms can be opposed without fear, where it is possible to meet and share in a convivial way. A better world is possible, where *hybris* makes more room for responsibility and hospitality. To educate in the Anthropocene is to help bring such a world about.

Critique, resistance, utopia. Having now come to the end of this work, we can formulate the hypothesis that in what could be identified as an anthropological root, the feeling of insufficiency is present. This is a feeling to work with in education. How can we allow humans to have the feeling of 'being enough' – of not needing to become more? Where do we find that which will allow humans to simply be, so that they can in turn work to bring life? As an extension of this idea, as we have mentioned previously, education in the Anthropocene cannot ignore the question of love – one of the central elements of our existences – with the difficulty of dissociating this form of existential love from the feeling upon which contemporary emotional capitalism (in this association of happiness, self-love and the market) is founded. The question is: how can we love? It seems to us that love can be positioned on the side of vitality: it is a matter of shaping the link between oneself and others in such a way that the vitality of each is strengthened. In our apprehension of love, we have to be careful because the deepest of our needs can become the worst of threats. Indeed, love can turn into consumption, which means that the other person becomes the means for the satisfaction of one's own needs. On the other hand, we could certainly consider the experience of love for the world – in spite of the injustices and violence it harbours – as the goal of education. This love of the world is not an abstract love – it is the love of the other which is manifested in the three-faceted figure of the stranger, of those one who have not been born yet but whose possibilities of existence depend on the how responsibly we prepare the world today, and of the whole of the non-human world. Love is the desire for life in the world. Here, love is not only a feeling but an action. It is a desire in action that embeds subjectivity and instils it in the world so that it becomes more alive.

Critique, utopia. There are a raft of possible readings of modernity, ranging from a story of social and economic progress to a story of alienation of the human adventure from the rest of the world, in pursuit of technico-economic ideologies that destroy nature. We can consider growth (particularly perceptible in the relentless pursuit of economic growth) to be the main characteristic of modernity. Such growth

[4] She used this quotation again when writing *On Revolution* (published in 1963), which she put at the beginning of the Chap. 6, 'The Revolutionary Tradition And Its Lost Treasure', and again when she wrote the introduction to *The Life of the Mind in the* 1970s.

has caused the world to fall silent, which is catastrophic from the point of view of resonance. Modernity has been permeated by a fundamental fear: that our relationships to the material world, to others, to nature, to ourselves or to our own bodies are characterised by a relational void. This fear is that the world will become mute and that we will end up being merely atoms, entirely isolated from others, from the world and from the Earth.[5] This muteness disturbs our relationship to the world, and it is this that education must seek to overcome as we head into the Anthropocene. Indeed, this 're-enchantment' of our relationship to the world is the political task required of education in the Anthropocene. Max Weber's idea *Entzauberung* (translated 'disenchantment', literally meaning that the world has stopped singing) originally meant in German that the world has lost the magic of its expression. Enabling the world and the Earth to sing again are the ambitious but vital political goals of education in the Anthropocene, for which love and care for both the world and the Earth are necessary.

Critique, resistance. In learning about this love of the world, it is important here not to focus on individualism, but instead to think of 'learning together' so as not to reduce the love of the world to individual subjective capacity. The education in the Anthropocene proposed in this book assumes the possibility of a mode of existence that is not based on maximisation of individual interests and the extension of access to the world driven by calculating and Promethean rationality. In the advent of this other mode of existence, characterised by coexistence, nature is not only what we have to take care of, but also the means of disarming Prometheus – not least because it has no other purpose than to *be there*, with its trees, oceans and sky. Getting out of the rut of hegemony of instrumental reason is an eminently complex issue. Although publications have, for many decades, been highlighting this problematic rationality (especially publications by the Critical Theoreticians of the Frankfurt School), we cannot say that the situation has changed much since the condemnation of this organising rationality of the Holocaust and the nuclear bomb, which put the world and nature at the disposal of humans as a result of our scientific, technical, economic and political means. Indeed, the situation is actually getting worse, with the spread of algorithms at the heart of our lives and social spaces. What makes it all the more difficult to leave this instrumental rationality behind is that any other rationality appears to be madness or unreasonable. The Anthropocene marks the start of a real emergency for the human adventure: we urgently need to end this economic hegemony. Some view our path to that goal in rediscovering our indigeneity (Arnsperger, 2019), or in anchoring ourselves in the 'wild part of the world' (Maris, 2018).

[5] It would be simplistic to view the history of modernity solely as a catastrophe of resonance. This is why, Rosa, in Chap. 10, proposes a counter-history, in which modernity is seen as a story of heightened resonant sensitivity: 'Modernity is not simply marked by historically unprecedented reification and alienation: it is also characterised by unprecedented resonant sensitivity' (Rosa, 2018, p. 412). The idea that the world speaks is one of the fundamental elements of Romanticism.

15.3 Educating in the Anthropocene to Bring About a Post-Promethean Society

Resistance, critique. Education in the Anthropocene has the function of developing social resistance. It is a way of repoliticising our lives without sacrificing the individual to the causes of justice, knowledge, the future, or the environment. It is not about enabling individuals to have a good life by developing a sense of successful living. The approach is not emotional but political. To counter the hegemonic individual freedoms, the education in the Anthropocene developed in this book does not present the commons as an alternative, but togetherness and the sharing of freedoms. Thus, it is possible envisage the possibility of a good life without it being marked either by the drive for emancipation or by a neoliberal individualist project of a quest for happiness. At the end of this work, it seems possible to say that the structural crisis of modernity caused by the world's mutism can be overcome by the advent of post-growth societies, which is the goal of education in the Anthropocene.

Critique, resistance. An education in the Anthropocene hinges not on the individual or the subject, but on a We: the We formed with others, the world, objects and nature. Indeed, the centre of gravity of education in the Anthropocene, which, among other things, aims to instil a form of resonant relationship to the world, is the political. It is political action in concert that makes the world emerge, thus rescuing us from the silent wilderness in which we currently live, to which we are not irredeemably condemned – despite the Anthropocene. Among the political lessons proposed by such education in the Anthropocene is that of listening and responding, in a process of reciprocal transformation between the subject and the world. This presupposes hospitality towards others, and not conformity. The courage to participate and the capacity to engage in a collective struggle are also fundamental political teachings of this education in the Anthropocene.

Critique, resistance. The proposed education in the Anthropocene takes account of the emotional component of subjectivity and sensitivity, while breaking with a driving logic of action with emancipation as its goal, and permeated by the individualism of *homo oeconomicus*. We propose not to think primarily of the subject, but of its relation to the world, and not to think of the subject in isolation from its sharing with the cosmos. It is a question of philosophically considering the notion of identity from the point of view of solidarity and sharing. Here the critique of neoliberal capitalism is radical. Our critique focuses not on the monopolisation or consumption of resources, but on the annihilation of life itself. It is what prevents being, by enslaving the vital and creative energy of subjects, whose bodies appear to be consumed. What has been identified as capitalist exploitation is this integration of the vitality of human subjects. Capitalism is largely responsible for the entry into the Anthropocene. It is, in fact, what annihilates our very capacity to exist in the world and maintain a healthy relationship with it. When we reduce the Earth to silence, a set of objects, produced by human hands, begin to speak to our senses in turmoil. When the Earth is silent, it is mass consumption, with planned obsolescence, that begins to speak to us with the voice of desire – none other than the desire to be and to have more. Basically, the many objects that adorn the storefronts of international chains tell us that we are 'not enough'. What is the Earth telling us, though? What is the world telling us? Is theirs a different message?

Resistance, critique, utopia. The dynamic stabilisation of modernity, the never-ending race for growth and 'ever more', has an unprecedented destructive capacity. This is one of the realities of the Anthropocene that the Great Acceleration curves show us. Today the desire to extend access to the world and maximise our own interests is hegemonic. Even when we consider a reform, it is viewed through the lens of the extension of this access to the world. The proposed education in the Anthropocene aims to move away from maximising the incorporation of fragments of the world to improving the quality of our relationship to the world, by listening to it, and to the Earth, which accommodates the political space.

Critique, resistance, utopia is a leitmotif for education in the Anthropocene. These three functions need to be woven together in order to change the world by transforming ourselves – through each other, against each other, and for each other. The complex challenge of education in the Anthropocene is to reconstitute the utopian, critical and resistant energies that develop visions of the future, not based on technophile fantasy, but on political overhaul. Thinking about education in the Anthropocene necessarily involves imagining radical alternatives. While these are sometimes relatively easy to think about, they are much more complex to implement. And yet, and yet…! A political and educational hope remains possible. It is possible, necessary and intellectually honest to maintain a political and educational hope that is rooted in the possibility of a refounding of the world and of an anthropological shift that allows us to become more human in sharing conviviality and 'between-us' with the non-human world.

Bibliographical References

Albanel, V. (2010). *Amour du monde – Christianisme et politique chez Hannah Arendt*. Cerf.

Arendt, H. (1972). « Qu'est-ce que la liberté ? ». In H. Arendt, *La crise de la culture*. Gallimard (pp. 186–222) (ed. or. 1961d), tr. fr.

Arendt, H. (2002). *Les origines du totalitarisme III. Le système totalitaire*. Seuil. (original edition 1948), French translation.

Arnsperger, C. (2019). Serons-nous enfin, un jour, indigènes? Permaculture et éducation des profondeurs. In N. Wallenhorst & J.-P. Pierron (Eds.), *Eduquer en Anthropocène*. Le Bord de l'eau.

Bador, M., Terray, L., Boé, J., Somot, S., Alias, A., Gibelin, A.-L., & Dubuisson, B. (2017). Future summer mega-heatwave and record-breaking temperatures in a warmer France climate. *Environmental Research Letters, 12*, 1–12.

Barnosky, A. D., Matzke, N. Tomiya, S., Wogan, G. O. U., Swartz, B., Quentall, T. B., Marshall, C., McGuire, J. L., Lindsey, E. L., Maguire, K. C., Mersey, B., & Ferrer, E. A. (2011). « Has the Earth's sixth mass extinction already arrived? ». *Nature, 471*, 51–57.

Barnosky, A. D., Haldy, E. A., Bascompte, J., Berlow, E. L., Brown, J. H., Fortelius, M., Getz, W. M., Harte, J., Hastings, A., Marquet, P. A., Martinez, N. D., Mooers, A., Roopnarine, P., Vermeij, G., Williams, J. W., Gillespie, R., Kitzes, J., Marshall, C., Matzke, N., Mindell, D. P., Revilla, E., & Smith, A. B. (2012). Approaching a state shift in Earth's biosphere. *Nature, 486*, 52–58.

Hubelin, J.-J., Ben-Ncer, A., Bailey, S. E., Freidline, S. E., Neubauer, S., Skinner, M. M., Bergmann, I., Le Cabec, A., Benazzi, S., Harvati, K., & Gunz, P. (2017). New fossils from Jebel Irhoud, Morocco and the pan-African origin of Homo sapiens. *Nature, 546*, 289–292.

Les convivialistes. (2013). *Manifeste convivialiste*. Le bord de l'eau.

Maris, V. (2018). *La part sauvage du monde*. Seuil.

Mora, C., Dousset, B., Caldwell, I. R., Powell, F. E., Geronimo, R. C., Bielecki, C. R., Counsell, C. W., Dietrich, B. S., Johnston, E. T., Louis, L. V., Lucas, M. P., McKenzie, M. M., Shea, A. G., Tseng, H., Giambelluca, T. W., Leon, L. R., Hawkins, E., & Trauernicht, C. (2017). Global risk of deadly heat. *Nature Climate Change, 7*, 501–506.

Pal, J. S., & Eltahir, E. A. B. (2015, October 26). Future temperature in southwest Asia projected to exceed a threshold for human adaptability. *Nature Climate Change, 6*, 1–4.

Rockström, J., Steffen, W., Noone, K., Persson, Å., Chapin, F. S., III, Lambin, E., Lenton, T. M., Scheffer, M., Folke, C., Schellnhuber, H., Nykvist, B., De Wit, C. A., Hughes, T., van der Leeuw, S., Rodhe, H., Sörlin, S., Snyder, P. K., Costanza, R., Svedin, U., Falkenmark, M., Karlberg, L., Corell, R. W., Fabry, V. J., Hansen, J., Walker, B., Liverman, D., Richardson, K., Crutzen, P., & Foley, J. (2009a). Planetary boundaries: Exploring the safe operating space for humanity. *Ecology and Society, 14*(2), 1–33.

Rockström, J., Steffen, W., Noone, K., Persson, Å., Chapin, F. S., III, Lambin, E., Lenton, T. M., Scheffer, M., Folke, C., Schellnhuber, H., Nykvist, B., De Wit, C. A., Hughes, T., van der Leeuw, S., Rodhe, H., Sörlin, S., Snyder, P. K., Costanza, R., Svedin, U., Falkenmark, M., Karlberg, L., Corell, R. W., Fabry, V. J., Hansen, J., Walker, B., Liverman, D., Richardson, K., Crutzen, P., & Foley, J. (2009b). A safe operating space for humanity. *Nature, 461*, 472–475.

Rosa, H. (2018). *Résonance – Une sociologie de la relation au monde*. La découverte. (original edition 2016), French translation.

Steffen, W., Broadgate, W., Deutsch, L., Gaffney, O., & Ludwig, C. (2015a). The trajectory of the Anthropocene: The great acceleration. *The Anthropocene Review, 2*(1), 81–98.

Steffen, W., Richardson, K., Rockström, J., Cornell, S. E., Fetzer, I., Bennett, E. M., Biggs, R., Carpenter, S. R., de Vries, W., de Witt, C. A., Folke, C., Gerten, D., Heinke, J., Mace, G. M., Persson, L. M., Ramanahan, V., Reyers, B., & Sörlin, S. (2015b). Planetary boundaries: Guiding human development on a changing planet. *Science, 347*, 736–746.

Bibliographical References

A. Anthropocene

I. Concept of the Anthropocene

1. Definition of the Anthropocene

Collective Work. (2020). *Dictionnaire critique de l'anthropocène*. CNRS Editions.
Crutzen, P. J. (2002). Geology of mankind: "The Anthropocene". *Nature, 415*, 23.
Crutzen, P. J. (2007). La géologie de l'humanité: l'Anthropocène. *Ecologie & Politique, 34*, 143–145.
Crutzen, P. J., & Schwägerl, C. (2011). Living in the Anthropocene: Toward a new global ethos. *Yale Environment 360*.http://e360.yale.edu/features/living_in_the_anthropocene_toward_a_new_global_ethos, 2011, consulted online on 25 October 2017.
Crutzen, P. J., & Stoermer, E. F. (2000). The Anthropocene. *Global Change, Newsletter, 41*, 17–18.
DellaSalla, D., & Goldstein, M. (Eds.). (2017). *Encyclopaedia of the Anthropocene*. Elsevier.
Gemenne, F., Rankovic, A., Ansart, T., Martin, B., Mitrano, P., & Rio, A. (2019). *Atlas de l'Anthropocène*. Presses de Sciences Po.
Gibbard, P. L., & Walker, M. J. C. (2013). The term "Anthropocene" in the context of formal geological classification. *Geological Society (London, Special publications), 395*, 1–9.
Grinevald, J. (2007). *La Biosphère de l'Anthropocène: climat et pétrole, la double menace*, Repères transdisciplinaires (1824–2007). Georg/Editions Médecine et Hygiène.
Guillaume, B. (2015a). Anthropocène (point de vue 1). In D. Bourg & A. Papaux (Eds.), *Dictionnaire de la pensée écologique* (pp. 32–35). PUF.
Haraway, D. (2015). Anthropocene, capitalocene, plantationocene, chthulucene: Making kin. *Environmental Humanities, 6*, 159–165.
Keats, J. (2011). Anthropocene. In J. Keats (Ed.), *Virtual words* (pp. 18–22). University Press of Oxford.
Lewis, S. L., & Maslin, M. A. (2015). Defining the Anthropocene. *Nature, 519*, 171–180.
Lorius, C., & Carpentier, L. (2010). *Voyage dans l'Anthropocène*. Actes Sud.
Magny, M. (2019). *Aux racines de l'Anthropocène – Une crise écologique reflet d'une crise de l'homme*. Le bord de l'eau.

© The Author(s), under exclusive license to Springer Nature Switzerland AG 2023
N. Wallenhorst, *A Critical Theory for the Anthropocene*, Anthropocene – Humanities and Social Sciences, https://doi.org/10.1007/978-3-031-37738-9

Masco, J. (2010). Bad weather: On planetary crisis. *Social Studies of Science, 40*(1), 7–40.
Nature. (2011). Editorial. The human epoch. Official recognition for the Anthropocene would focus minds on the challenges to come. *Nature, 473*, 254.
Oldfield, F., Barnosky, A. D., Dearing, J., Fischer-Kowalski, M., McNeill, J., Steffen, W., & Zalasiewicz, J. (2014). The Anthropocene review: Its significance, implications and the rationale for a new transdisciplinary journal. *The Anthropocene Review, 1*, 3–7.
Robin, L. (2013). Histories for changing times: Entering the Anthropocene? *Australian Historical Studies, 44*(3), 329–340.
Robin, L., Avango, D., Keogh, L., Möllers, N., Scherer, B., & Trischler, H. (2014). Three galleries of the Anthropocene. *The Anthropocene Review, 1*(3), 207–224.
Ruddiman, W. F. (2013). The Anthropocene. *The Annual Review of Earth and Planetary Sciences, 41*, 45–68.
Ruddiman, W. F., Ellis, E. C., Kaplan, J. O., & Fuller, D. Q. (2015). Defining the epoch we live in. *Science, 348*, 38–39.
Steffen, W., Crutzen, P. J., & McNeil, J. R. (2007). The Anthropocene: Are humans now overwhelming the great forces of nature? *Ambio, Royal Swedish Academy of Sciences, 36*(8), 614–621.
Steffen, W., Grinevald, J., Crutzen, P., & McNeill, J. (2011a). The Anthropocene: Conceptual and historical perspectives. *Philosophical Transactions of the Royal Society, 369*, 842–867.
Steffen, W., Sanderson, R. A., Tyson, P. D., Jäger, J., Matson, P. A., Moore, B., III, Oldfield, F., Richardson, K., Schellnhuber, H.-J., Turner, B. L., & Wasson, R. J. (2004). *Global change and the earth system. A planet under pressure* (The IGBP Book Series). Springer.
Testot, L., & Wallenhorst, N. (2023). *Vortex – Faire face à l'Anthropocène*. Payot.
Wallenhorst, N. (2019). *L'Anthropocène décodé pour les humains*. Le Pommier.
Wallenhorst, N. (2020a). *La vérité de l'Anthropocène*. Le Pommier.
Wallenhorst, N., & Wulf, C. (Eds.). (2023). *Handbook of the Anthropocene*. Springer.
Westbroek, P. (2015). Système terre. In D. Bourg & A. Papaux (Eds.), *Dictionnaire de la pensée écologique* (pp. 957–962). PUF.
Wilkinson, B. H. (2005). Humans as geologic agents: A deep-time perspective. *Geology, 33*, 161–164.
Williams, M., Zalasiewicz, J., Haff, P. K., Schwägerl, C., Barnosky, A., & Ellis, E. C. (2015). The Anthropocene biosphere. *The Anthropocene Review*, 1–24.
Zalasiewicz, J., Waters, C. N., Summerhayes, C. P., Wolfe, A. P., Barnosky, A. D., Cearreta, A., Crutzen, P., Ellis, E., Fairchild, I. J., Gałuszka, A., Haff, P., Hajdas, I., Head, M. J., Ivar do Sul, J. A., Jeandel, C., Leinfelder, R., McNeill, J. R., Neal, C., Odada, E., Oreskes, N., Steffen, W., Syvitski, J., Vidas, D., Wagreich, M., & Williams, M. (2017a). The Working Group on the Anthropocene: Summary of evidence and interim recommendations. *Anthropocene, 19*, 55–60.
Zalasiewicz, J., Waters, C. N., Williams, M., Barnosky, A. D., Cearreta, A., Crutzen, P., Ellis, E., Ellis, M. A., Fairchild, I. J., Grinevald, J., Haff, P. K., Hajdas, I., Leinfelder, R., McNeill, J., Odada, E. O., Poirier, C., Richter, D., Steffen, W., Summerhayes, C., Syvitski, J. P. M., Vidas, D., Wagreich, M., Wing, S. L., Wolfe, A. P., Zhisheng, A., & Oreskes, N. (2014a). When did the Anthropocene begin? A mid-twentieth century boundary level is stratigraphically optimal. *Quaternary International, 30*, 1–8.
Zalasiewicz, J., Williams, M., Smith, A., Barry, T. L., Coe, A. L., Brown, P. R., Brenchley, P., Cantrill, D., Gale, A., Gibbard, P., Gregory, F. J., Hounslow, M. W., Kerr, A. C., Pearson, P., Knox, R., Powell, J., Waters, C., Marshall, J., Oates, M., Rawson, P., & Stone, P. (2008). Are we now living in the Anthropocene? *GSA Today, 18*(2), 4–8.
Zalasiewicz, J., Williams, M., Steffen, W., & Crutzen, P. (2010). The new world of the Anthropocene. *Environmental Science & Technology, 44*, 2228–2231.

2. History of the Idea of Humanity as a Geological Force

Adams, W. H. D. (1856). *The history, topography, and antiquities of the isle of wight*. Smith Elder.
Bergson, H. (1996). *L'évolution créatrice*. PUF. (original edition 1907).
Buffon, G. L. L. (1998). *Les époques de la nature*. Diderot éditions. (original edition 1778).
Dana, J. D. (1863). *Manual of geology: Treating of the principles of the science with special reference to American Geological history, of the use of colleges, academies, and schools of science*. Theodore Bliss & Co./Trübner & Co..
de Tribolet, M. (1886, January–April). Les animaux disparus depuis l'apparition de l'homme. *Revue savoisienne*, 45–114.
Fischer, E. (1915). Der Mensch al geologischer Faktor. *Zeitschrift der Deutschen Geologischen Gesellschaft, 67*(2), 106–148.
Hamilton, C., & Grinevald, J. (2015). Was the Anthropocene anticipated? *The Anthropocene Review, 2*(1), 59–72.
Huzar, E. (2008). *La fin du monde par la science*. Editions Ere. (original edition 1855).
Jenkyn, T. W. (1854). *Lessons in geology, XLIX*. Chapter V. On the classification of rocks section. *Popular Educator, 4*, 312–316.
Le Conte, J. (1884). *Religion and Science: A series of Sunday lectures*. D. Appleton.
Lyell, C. (1990). *Principles of geology* (Vol. I, II, III). University of Chicago Press. (original edition 1830-1833).
Marsh, G. P. (1970). *The earth as modified by human action: A new edition of "Man and Nature"*. Arno Press. (original edition 1874).
Reclus, E. (1881). *La Terre: Description des phénomènes de la vie du globe (1868–1869), in Les continents, t2*. Hachette.
Schellnhuber, H. J. (1999). 'Earth system' analysis and the second Copernican revolution. *Nature, 402*, C19–C23.
Shantser, E. V. (1973). The Anthropogenic system (period). In *The Great Soviet Encyclopedia* (Vol. 2, pp. 139–144). Macmillan.
Sherlock, R. L. (1922). *Man as a geological agent*. H. F. & G. Witherby.
Stoppani, A. (1873). *Corso di geologia, vol. 2, Geologia Stratigrafica*. Bernardoni Brigola Editori.
Suess, E. (1875). *Die Entstehung der Alpen*. W. Braunmüller.
Suess, E. (1924). *La face de la Terre*. Armand Colin. French translation.
Teilhard de Chardin, P. (2007). *Le phénomène humain*. Seuil. (original edition 1955).
Thomas, W. L. (Ed.). (1962). *Man's role in changing the face of the earth*. University of Chicago Press. (original edition 1956).
Vernadsky, V. I. (1925). L'autotrophie de l'humanité. *Revue générale des sciences pures et appliquées, 36*(17–18), 495–502. French translation.
Vernadsky, V. I. (2014). Some words about the Noösphere. In J. Ross (Ed.), *150 Years of Vernadsky* (Vol. 2). 21st Century Science Associates.
Vernadsky, V. I. (1945). Biosphere and noosphere. *American Scientist, 33*, 1–12.
Vernadsky, V. (1929). *La biosphère*. Librairie Félix Alcan. (original edition 1926, réed. Paris, Seuil, 2002), French translation.
Vernadsky, V. (1924). *La géochimie*. Librairie Félix Alcan. French translation.
Whewell, W. (1853). *Of the plurality of worlds*. Library of Alexandria.
Woeikof, A. I. (1901). De l'influence de l'homme sur la Terre. *Annales de Géographie, t1*(50), 193–215.

3. Critique of the Anthropocene

Autin, W. J., & Holbrook, J. M. (2012). Is the Anthropocene an issue of stratigraphy or pop culture? *GSA Today, 22*(7), 60–61.
Baskin, J. (2015). Paradigm dressed as Epoch: The ideology of the Anthropocene. *Environmental Values, 24*, 9–29.

Baskin, J. (2014). The ideology of the Anthropocene? *MSSI Research Paper (Melbourne Sustainable Society Institute), 3*, 3–19.

Brondizio, E. S., O'Brien, K., Bai, X., Biermann, F., Steffen, W., Berkhout, F., Cudennec, C., Lemos, M. C., Wolfe, A., Palma-Oliveira, J., & Arthur Chen, C.-T. (2016). Re-conceptualizing the Anthropocene: A call for collaboration. *Global Environmental Change, 39*, 318–327.

Ellis, E. C., Maslin, M., Boivin, N., & Bauer, A. (2016). Involve social scientists in defining the Anthropocene. *Nature, 540*, 192–193.

Finney, S., & Edwards, L. E. (2016). The 'Anthropocene' epoch: Scientific decision or political statement? *GSA Today, 26*(3-4), 4–10.

Fressoz, J.-B. (remarks collected by T. Henneton and J. Confavreux). (2013). L'apocalypse et l'Anthropocène, *Vacarme, 65*, 202–233.

Fressoz, J.-B. (2016). Colón también descubrió el cambio climático. *El Pais*. https://elpais.com/elpais/2016/06/09/ciencia/1465489189_275680.html, consulted online on 10 January 2018.

Gemenne, F. (2015). L'Anthropocène et ses victimes: une réflexion terminologique. In F. Gemenne (Ed.), *L'enjeu mondial* (pp. 233–240). Presses de Sciences Po, "Annuels".

LeCain, T. J. (2015). Against the Anthropocene. A neo-materialist perspective. *International Journal for History, Culture and Modernity, 3*(1), 1–28.

Lövbrand, E., Beck, S., Chilvers, J., Forsyth, T., Hedrén, J., Hulme, M., Lidskog, R., & Vasileiadou, E. (2014, September 6). Taking the human (sciences) seriously: Realizing the critical potential of the Anthropocene. In *ECPR General Conference*, Glasgow.

Malm, A. (2013). The origins of fossil capital: From water to steam in the British Cotton Industry. *Historical Materialism, 21*(1), 15–68.

Malm, A., & Hornborg, A. (2014). The geology of mankind? A critique of the Anthropocene narrative. *The Anthropocene Review, 1*, 62–69.

Malm, A. (2017). *L'anthropocène contre l'histoire*. La Fabrique éditions. French translation.

Moore, J. W. (Ed.). (2016). *Anthropocene or capitalocene? Nature, history and the crisis of capitalism*. PM Press.

Palsson, G., Szerszynski, B., Sörlin, S., Marks, J., Avril, B., Crumley, C., Hackmann, H., Holm, P., Ingram, J., Kirman, A., Pardo Buendia, M., & Weehuizen, R. (2013). Reconceptualizing the 'Anthropos' in the Anthropocene: Integrating the social sciences and humanities in global environmental change research. *Environmental Science and Policy, 28*, 3–13.

Visconti, G. (2014). Anthropocene: Another academic invention? *Rendiconti Lincei: Science Fisiche e Naturali, 25*(3), 381–392.

Zalasiewicz, J., Caerreta, A., Crutzen, P., Ellis, E., Ellis, M., Grinevald, J., McNeill, J., Poirier, C., Price, S., Richter, D., Scholes, M., Steffen, W., Vidas, D., Waters, C., Williams, M., & Wolfe, A. P. (2012). Response to Autin and Holbrook on 'Is the Anthropocene an issue of stratigraphy or pop culture?'. *GSA Today, 22*(7), e21–e22.

Zalasiewicz, J., Waters, C. N., Wolfe, A. P., Barnosky, A. D., Cearreta, A., Edgeworth, M., Ellis, E. C., Fairchild, I. J., Gradstein, F. M., Grinevald, J., Haff, P., Head, M. J., Ivar do Sul, J. A., Jeandel, C., Leinfelder, R., McNeill, J. R., Oreskes, N., Poirier, C., Revkin, A., Richter, D., Steffen, W., Summerhayes, C., Syviski, J. P. M., Vidas, D., Wagreich, M., Wing, S., & Williams, M. (2017b). Making the case for a formal Anthropocene Epoch: An analysis of ongoing critiques. *Newsletters on Stratigraphy, 50*, 205–226.

4. Philosophy of the Anthropocene

Chalier, J., & Schmid, L. (2015, December). Comment penser l'Anthropocène? *Esprit*, 5–7.

Corbin, A., Courtine, J.-J., & Vigarello, G. (Eds.). (2017). *Histoire des émotions, tome 3, De la fin du 19^{ème}siècle à nos jours*. Seuil.

Dubois, J. (2016). Comprendre l'Anthropocène, entre données quantitatives et choix sociaux. *Espaces et sociétés, 164–165*, 227–233.

Federau, A. (2016). *Philosophie de l'Anthropocène – Interprétation et épistémologie*. Doctoral thesis supervised jointly by the University of Lausanne and the University of Bourgogne, by Professors Dominique Bourg and Jean-Claude Gens.

Federau, A. (2017). *Pour une philosophie de l'Anthropocène*. PUF.

II. Other Geological Works

1. Geology and Palaeontology

Dalrymple, G. B. (2001). The age of the earth in the twentieth century: a problem (mostly) solved. *Journal of the Geological Society of London (Special Publication), 190*, 205–221.

Gibbard, P. L., & Lewin, J. (2016). Partitioning the quaternary. *Quaternary Science Reviews, 151*, 127–139.

Haughton, S. (1865). *Manual of geology, F.R.S., & c*. Longmans, Green, and Co.

McDougall, I., Brown, F. H., & Fleagle, J. G. (2005). Stratigraphic placement and age of modern humans from Kibish, Ethiopia. *Nature, 433*, 733–736.

McPherron, S., Alemseged, Z., Marean, C. W., Wynn, J. G., Reed, D., Geraads, D., Bobe, R., & Béarat, H. A. (2010). Evidence for stone-tool-assisted consumption of animal tissues before 3.39 million years ago at Dikika, Ethiopia. *Nature, 466*, 857–860.

Renne, P. R., Sprain, C. J., Richards, M. A., Self, S., Vanderkluysen, L., & Pande, K. (2015). State shift in Deccan volcanism at the Cretaceous-Paleogene boundary, possibly induced by impact. *Science, 350*(6256), 76–78.

Roebroeks, W., & Villa, P. (2011). On the earliest evidence for habitual use of fire in Europe. *Proceedings of the National Academy of Science, USA, 108*(13), 5209–5214.

Schulte, P., Alegret, L., Arenillas, I., Arz, J. A., Barton, P. J., Bown, P. R., Bralower, T. J., Christeson, G. L., Claeys, P., Cockell, C. S., Collins, G. S., Deutsch, A., Goldin, T. J., Goto, K., Grajales-Nishimura, J. M., Richard, A. F., Grieve, R. A. F., Gulick, S. P. S., Johnson, K. R., Kiessling, W., Koeberl, C., Kring, D. A., MacLeod, K. G., Matsui, T., Melosh, J., Montanari, A., Morgan, J. V., Neal, C. R., Nichols, D. J., Norris, R. D., Pierazzo, E., Ravizza, G., Rebolledo-Vieyra, M., Reimold, W. U., Robin, E., Salge, T., Speijer, R. P., Sweet, A. R., Urrutia-Fucugauchi, J., Vajda, V., Whalen, M. T., & Willumsen, P. S. (2010). The chicxulub asteroid impact and mass extinction at the cretaceous-paleogene boundary. *Science, 327*(5970), 1214–1218.

Semaw, S., Rogers, M. J., Quade, J., Renne, P. R., Butler, R. F., Dominguez-Rodrigo, M., Stout, D., Hart, W. S., Pickering, T., & Simpson, S. W. (2003). 2.6-Million-year-old stone tools and associated bones from OGS-6 and OGS-7, Gona, Afar, Ethiopia. *Journal of Human Evolution, 45*, 169–177.

2. Stratigraphy and Dating of the Entry into the Anthropocene

Balter, M. (2013). Archaeologists say the 'Anthropocene' is here – But it began long ago. *Science, 340*, 261–262.

Barnosky, A. D. (2013). Palaeontological evidence for defining the Anthropocene. In C. N. Waters, J. Zalasiewicz, M. Williams, M. A. Ellis, & A. Snelling (Eds.), *A stratigraphical basis for the Anthropocene, geological society* (Vol. 395, pp. 1–17). Special Publications.

Certini, G., & Scalenghe, R. (2011). Anthropogenic soils are the golden spikes for the Anthropocene. *The Holocene, 21*(8), 1269–1274.

Dean, J. R., Leng, M. J., & Mackay, A. W. (2014). Is there an isotopic signature of the Anthropocene? *The Anthropocene Review, 1*, 276–287.

Doughty, C. E. (2013). Preindustrial human impacts on global and regional environment. *Annual Review of Environment and Resources, 38*, 503–527.

Holtgrieve, G. W., Schindler, D. E., Hobbs, W. O., Leavitt, P. R., Ward, E. J., Bunting, L., Chen, G., Finney, B. P., Gregory-Eaves, I., Holmgren, S., Lisac, M. J., Lisi, P. J., Nydick, K., Rogers, L. A., Saros, J. E., Selbie, D. T., Shapley, M. D., Walsh, P. B., & Wolfe, A. P. (2011). A coherent signature of anthropogenic nitrogen deposition to remote watersheds of the Northern Hemisphere. *Science, 334*, 1545–1548.

Ruddiman, W. F. (2003). The anthropogenic greenhouse era began thousands of years ago. *Climatic Change, 61*, 261–293.

Steffen, W., Broadgate, W., Deutsch, L., Gaffney, O., & Ludwig, C. (2015a). The trajectory of the Anthropocene: The great acceleration. *The Anthropocene Review, 2*(1), 81–98.

Steffen, W., Leinfelder, R., Zalasiewicz, J., Waters, C. N., Williams, M., Summerhayes, C., Barnosky, A. D., Cearreta, A., Crutzen, P., Edgeworth, M., Ellis, E. C., Fairchild, I. J., Galuszka, A., Grinevald, J., Haywood, A., Ivar do Sul, J., Jeandel, C., McNeill, J. R., Odada, E., Oreskes, N., Revkin, A., Richter, D. d B., Syvitski, J., Vidas, D., Wagreich, M., Wing, S. L., Wolfe, A. P., & Schellnhuber, H. J. (2016). Stratigraphic and Earth system approaches to defining the Anthropocene. *Earth's Future, 4*, 1–22.

Walker, M., Johnsen, S., Rasmussen, S. O., Popp, T., Steffensen, J.-P., Gibbard, P., Hoek, W., Lowe, J., Andrews, J., Björck, S., Cwynar, L. C., Hughen, K., Kershaw, P., Kromer, B., Litt, T., Lowe, D. J., Nakagawa, T., Newnham, R., & Schwander, J. (2009). Formal definition and dating of the GSSP (Global Stratotype Section and Point) for the base of the Holocene using the Greenland NGRIP ice core, and selected auxiliary records. *Journal of Quaternary Science, 24*, 3–17.

Waters, C. N., Zalasiewicz, J., Summerhayes, C., Barnosky, A. D., Poirier, C., Galuszka, A., Cearreta, A., Edgeworth, M., Ellis, E. C., Ellis, M., Jeandel, C., Leinfelder, R., McNeill, J. R., Richter, D. d B., Steffen, W., Syvitski, J., Vidas, D., Wagreich, M., Williams, M., Zhisheng, A., Grinevald, J., Odada, E., Oreskes, N., & Wolfe, A. P. (2016). The Anthropocene is functionally and stratigraphically distinct from the Holocene. *Science, 351*, 137–147.

Waters, C. N., Zalasiewicz, J., Williams, M., Price, S. J., Ford, J. R., & Cooper, A. H. (2014a). Evidence for a stratigraphic basis for the Anthropocene. In R. Rocha, J. Pais, J. Kullberg, & S. Finney (Eds.), *STRATI 2013* (pp. 989–993). Springer Geology, Springer.

Waters, C. N., Zalasiewicz, J. A., Williams, M., Ellis, M. A., & Snelling, A. M. (2014b). A stratigraphical basis for the Anthropocene? *Geological Society (London, Special Publications), 395*, 1–21.

Waters, C. N., Syvitski, J. P. M., Gałuszka, A., Hancock, G. J., Zalasiewicz, J., Cearreta, A., Grinevald, J., Jeandel, C., McNeill, J. R., Summerhayes, C., & Barnosky, A. (2015). Can nuclear weapons fallout mark the beginning of the Anthropocene Epoch? *Bulletin of the Atomic Scientists, 71*(3), 46–57.

Williams, M., Zalasiewicz, J., Waters, C. N., Edgeworth, M., Bennett, C., Barnosky, A. D., Ellis, E. C., Ellis, M. A., Cearreta, A., Haff, P. K., Ivar do Sul, J. A., Leinfelder, R., McNeill, J. R., Odada, E., Oreskes, N., Revkin, A., Richter, D. d B., Steffen, W., Summerhayes, C., Syvitski, J. P., Vidas, D., Wagreich, M., Wing, S. L., Wolfe, A. P., & Zhisheng, A. (2016). The Anthropocene: A conspicuous stratigraphical signal of anthropogenic changes in production and consumption across the biosphere. *Earth's Future, 4*, 1–20.

Wolfe, A. P., Hobbs, W. O., Birks, H. H., Briner, J. P., Holmgren, S. U., Ingólfsson, O., Kaushal, S. S., Miller, G. H., Pagani, M., Saros, J. E., & Vinebrooke, R. D. (2013). Stratigraphic expressions of the Holocene–Anthropocene transition revealed in sediments from remote lakes. *Earth-Science Reviews, 116*, 17–34.

Wolff, E. W. (2014). Ice sheets and the Anthropocene. In C. N. Waters, J. Zalasiewicz, M. Williams, M. A. Elis, & A. Snelling (Eds.), *A stratigraphical basis for the Anthropocene* (pp. 255–263). Geological Society, Special Publications 395.

Zalasiewicz, J., Waters, C., & Head, M. J. (2017c). Anthropocene: Its stratigraphic basis. *Nature, 541*, 289–289.

Zalasiewicz, J., & Williams, M. (2013, December). The Anthropocene: A comparison with the Ordovician-Silurian boundary. In *Rendiconti Lincei: Scienze Fisiche e Naturali* (pp. 3–11). Springer.

Zalasiewicz, J., Williams, M., Fortey, R., Smith, A., Barry, T. L., Coe, A. L., Bown, P. R., Rawson, P. F., Gale, A., Gibbard, P., Gregory, F. J., Hounslow, M. W., Kerr, A. C., Pearson, P., Knox, R., Powell, J., Waters, C., Marshall, J., Oates, M., & Stone, P. (2011a). Stratigraphy of the Anthropocene. *Philosophical Transactions of the Royal Society, 369*, 1036–1055.

Zalasiewicz, J., Williams, M., Haywood, A., & Elis, M. (2011b). The Anthropocene: A new epoch of geological time? *Philosophical Transactions of the Royal Society, 369*, 835–841.

Zalasiewicz, J., Williams, M., & Waters, C. N. (2014b). Can an Anthropocene series be defined and recognized? *Geological Society (London, Special Publications), 395*, 39–53.

III. Limits of the Biosphere

1. Global Warming and Atmospheric CO_2Levels

Arrhenius, S. (1896). On the influence of carbonic acid in the air upon the temperature of the ground. *Philosophical Magazine and Journal of Science, Series 5, 41*, 237–276.

Bador, M., Terray, L., Boé, J., Somot, S., Alias, A., Gibelin, A.-L., & Dubuisson, B. (2017). Future summer mega-heatwave and record-breaking temperatures in a warmer France climate. *Environmental Research Letters, 12*, 1–12.

Caldeira, K., & Wickett, M. E. (2003). Anthropogenic carbon and ocean pH. *Nature, 425*, 365.

Calendar, G. S. (1938). The artificial production of carbon dioxide and its influence on temperature. *Quarterly Journal of the Royal Meteorological Society, 64*, 223–240.

Edwards, L. E. (2015). What is the Anthropocene? *Eos*. https://eos.org/opinions/what-is-the-anthropocene, consulted online on 26 January 2018.

El Fadli, K. I., Cerveny, R. S., Burt, C. C., Eden, P., Parker, D., Brunet, M., Peterson, T. C., Mordacchini, G., Pelino, V., Bessemoulin, P., Stella, J. L., Driouech, F., Wahab, M. M. A., & Pac, M. B. (2013). World Meteorological Organization assessment of the purported world record 58°C temperature extreme at El Azizia, Libya (13 September 1922). *Bulletin of the American Meteorological Society, 94*, 199–204.

Etheridge, D. M., Steele, L. P., Langenfelds, R. L., Francey, R. J., Barnola, J.-M., & Morgan, V. I. (1998). Historical CO_2 records from the Law Dome DE08, DE08-2, and DSS ice cores. In *Trends: A compendium of data on global change*. Oak Ridge National Laboratory, U.S. Department of Energy.

Im, E.-S., Pal, J. S., & Eltahir, E. A. B. (2017). Deadly heat waves projected in the densely populated agricultural regions of South Asia. *Science Advances, 3*(8), 1–7.

IPCC. (2013). *Changements climatiques 2013 – Les éléments scientifiques, Résumé à l'intention des décideurs*. Contribution du Groupe de travail I au cinquième rapport d'évaluation du Groupe d'experts intergouvernemental sur l'évolution du climat.

IPCC. (2021). *Climate change 2021. The physical science basis. Summary for Policymakers*. Working Group I contribution to the Sixth Assessment Report of the Intergovernmental Panel on Climate Change.

IPCC. (2022a). *Climate change 2022. Impacts, adaptation and vulnerability*. Working Group II contribution to the Sixth Assessment Report of the Intergovernmental Panel on Climate Change.

IPCC. (2022b). *Climate change 2022. Mitigation of climate change*. Working Group III contribution to the Sixth Assessment Report of the Intergovernmental Panel on Climate Change.

Kaplan, J. O., Krumhardt, K. M., Ellis, E. C., Ruddiman, W. F., Lemmen, C., & Goldewijk, K. (2011). Holocene carbon emissions as a result of anthropogenic land-cover change. *The Holocene, 21*(5), 775–791.

Kleinen, T., Brovkin, V., von Bloh, W., Archer, D., & Munhoven, G. (2010). Holocene carbon cycle dynamics. *Geophysical Research Letters, 37*, 1–5.

Lenton, T. M., Livina, V. N., Dakos, V., van Nes, E. H., & Scheffer, M. (2012). Early warning of climate tipping points from critical slowing down: Comparing methods to improve robustness. *Philosophical Transactions of the Royal Society, London, 370*, 1185–1204.

Le Quéré, C., Raupach, M. R., Canadell, J. G., Marland, G., Bopp, L., Ciais, P., Conway, T. J., Doney, S. C., Feely, R. A., Foster, P., Friedlingstein, P., Gurney, K., Houghton, R. A., House, J. I., Huntingford, C., Levy, P. E., Lomas, M. R., Majkut, J., Metz, N., Ometto, J. P., Peters, G. P., Prentice, I. C., Randerson, J. T., Running, S. W., Sarmiento, R. L., Schuster, U., Sitch, S., Takahashi, T., Viovy, N., van der Werf, G. R., & Woodward, F. I. (2009). Trends in the sources and sinks of carbon dioxide. *Nature Geoscience, 2*, 831–836.

Mac Farling Meure, C., Etheridge, D., Trudinger, C., Steele, P., Langenfelds, R., van Ommen, T., Smith, A., & Elkins, J. (2006). Law Dome CO_2, CH4 and N2O ice core records extended to 2000 years BP. *Geophysical Research Letters, 33*, 1–4.

Mora, C., Dousset, B., Caldwell, I. R., Powell, F. E., Geronimo, R. C., Bielecki, C. R., Counsell, C. W., Dietrich, B. S., Johnston, E. T., Louis, L. V., Lucas, M. P., McKenzie, M. M., Shea, A. G., Tseng, H., Giambelluca, T. W., Leon, L. R., Hawkins, E., & Trauernicht, C. (2017). Global risk of deadly heat. *Nature Climate Change, 7*, 501–506.

Pal, J. S., & Eltahir, E. A. B. (2015, October 26). Future temperature in southwest Asia projected to exceed a threshold for human adaptability. *Nature Climate Change, 6*, 1–4.

Petit, J. R., Jouzel, J., Raynaud, D., Barkov, N. I., Barnola, J.-M., Basile, I., Bender, M., Chappellaz, J., Davisk, M., Delaygue, G., Delmotte, M., Kotlyakov, V. M., Legrand, M., Lipenkov, V. Y., Lorius, C., Pépin, L., Ritz, C., Saltzmank, E., & Stievenard, M. (1999). Climate and atmospheric history of the past 420 000 years from the Vostok ice core, Antartica. *Nature, 399*, 429–436.

Raftery, A. E., Zimmer, A., Frierson, D. M. W., Startz, R., & Liu, P. (2017). Less than 2 °C warming by 2100 unlikely. *Nature Climate Change, 7*, 637–641.

Rahmstorf, S. (2007). A semi-empirical approach to projecting future sea-level rise. *Science, 315*, 368–370.

Revkin, A. C. (1992). *Global warming: Understanding the forecast*. American Museum of Natural History, Environmental Defense Fund, Abbeville Press.

Ruddiman, W. F., Vavrus, S., Kutzbach, J., & Feng, H. (2014). Does pre-industrial warming double the anthropogenic total? *The Anthropocene Review, 1*, 1–7.

Satterthwaite, D. (2009). The implications of population growth and urbanization for climate change. *Environment & Urbanization, 21*(2), 545–567.

Scheffer, M., Bascompte, J., Brock, W. A., Brovkin, V., Carpenter, S. R., Dakos, V., Held, H., van Nes, E. H., Rietkerk, M., & Sugihara, G. (2009). Early-warming signals for critical transitions. *Nature, 461*, 53–59.

Sherwood, S. C., & Huber, M. (2010). An adaptability limit to climate change due to heat stress. *Proceedings of the National Academy of Sciences, USA, 107*(21), 9552–9555.

Siebert, S., & Ewert, F. (2014). Future crop production threatened by extreme heat. *Environmental Research Letters, 9*, 1–4.

Williams, J. W., Jackson, S. T., & Kutzbach, J. E. (2007). Projected distributions of novel and disappearing climates by 2100 AD. *Proceedings of the National Academy of Sciences, USA, 104*, 5738–5742.

2. Extinction of Biodiversity

Barnosky, A. D. (2008). Megafauna biomass tradeoff as a driver of Quaternary and future extinctions. *Proceedings of the National Academy of Sciences, USA, 105*(Suppl 1), 11543–11548.

Barnosky, A. D., Matzke, N., Tomiya, S., Wogan, G. O. U., Swartz, B., Quental, T. B., Marshall, C., McGuire, J. L., Lindsey, E. L., Maguire, K. C., Mersey, B., & Ferrer, E. A. (2011). Has the Earth's sixth mass extinction already arrived? *Nature, 471*, 51–57.

Ceballos, G., Ehrlich, P. R., & Dirzo, R. (2017). Biological annihilation via the ongoing sixth mass extinction signaled by vertebrate population losses and declines. *Proceedings of the National Academy of Sciences, USA, 114*(30), 6089–6096.

Hallmann, C. A., Sorg, M., Jongejans, E., Siepel, H., Hofland, N., Schwan, H., Stenmans, W., Müller, A., Sumser, H., Hörren, T., Goulson, D., & de Kroon, H. (2017). More than 75 percent decline over 27 years in total flying insect biomass in protected areas. *PLOS One, 12*, 1–21.

IPBES. (2019). *Summary for policymakers of the global assessment report on biodiversity and ecosystem services of the Intergovernmental Science-Policy Platform on Biodiversity and Ecosystem Services.*

IPBES-IPCC. (2021, June). *Biodiversity and climate change workshop report.* Co-sponsored workshop IPBES and GIEC.

IUCN. *The IUCN red list of threatened species, Version 2017-2.* http://www.iucnredlist.org

Kolbert, E. (2015). *La 6ème Extinction. Comment l'homme détruit la vie.* La librairie Vuibert.

Pimm, S. L., Jenkins, C. N., Abell, R., Brooks, T. M., Gittleman, J. L., Joppa, L. N., Raven, P. H., Roberts, C. M., & Sexton, J. O. (2014). The biodiversity of species and their rates of extinction, distribution, and protection. *Science, 344*, 987–998.

3. Tipping Point of the Earth system

Barnosky, A. D., Haldy, E. A., Bascompte, J., Berlow, E. L., Brown, J. H., Fortelius, M., Getz, W. M., Harte, J., Hastings, A., Marquet, P. A., Martinez, N. D., Mooers, A., Roopnarine, P., Vermeij, G., Williams, J. W., Gillespie, R., Kitzes, J., Marshall, C., Matzke, N., Mindell, D. P., Revilla, E., & Smith, A. B. (2012). Approaching a state shift in Earth's biosphere. *Nature, 486*, 52–58.

Leandri, M. (2015). Capacité de charge. In D. Bourg & A. Papaux (Eds.), *Dictionnaire de la pensée écologique* (pp. 121–123). PUF.

Persson, L., Carney Almroth, B. M., Collins, C. D., Cornell, S., de Wit, C. A., Diamond, M. L., Fantke, P., Hassellöv, M., MacLeod, M., Ryberg, M. W., Jørgensen, P. S., Villarrubia-Gómez, P., Wang, Z., & Zwicky Hauschild, M. (2022). Outside the safe operating space of the planetary boundary for novel entities. *Environmental Science and Technology, 56*(3), 1510–1521.

Rockström, J. (2015, April). Bounding the planetary future: Why we need a great transition. *Great Transition Initiative*, 1–14.

Rockström, J., & Klum, M. (2015). *Big world, small planet, abundance within planetary boundaries*. Yale University Press.

Rockström, J., Steffen, W., Noone, K., Persson, Å., Chapin, F. S., III, Lambin, E., Lenton, T. M., Scheffer, M., Folke, C., Schellnhuber, H., Nykvist, B., De Wit, C. A., Hughes, T., van der Leeuw, S., Rodhe, H., Sörlin, S., Snyder, P. K., Costanza, R., Svedin, U., Falkenmark, M., Karlberg, L., Corell, R. W., Fabry, V. J., Hansen, J., Walker, B., Liverman, D., Richardson, K., Crutzen, P., & Foley, J. (2009a). Planetary boundaries: Exploring the safe operating space for humanity. *Ecology and Society, 14*(2), 1–33.

Rockström, J., Steffen, W., Noone, K., Persson, Å., Chapin, F. S., III, Lambin, E., Lenton, T. M., Scheffer, M., Folke, C., Schellnhuber, H., Nykvist, B., De Wit, C. A., Hughes, T., van der Leeuw, S., Rodhe, H., Sörlin, S., Snyder, P. K., Costanza, R., Svedin, U., Falkenmark, M., Karlberg, L., Corell, R. W., Fabry, V. J., Hansen, J., Walker, B., Liverman, D., Richardson, K., Crutzen, P., & Foley, J. (2009b). A safe operating space for humanity. *Nature, 461*, 472–475.

Steffen, W., Richardson, K., Rockström, J., Cornell, S. E., Fetzer, I., Bennett, E. M., Biggs, R., Carpenter, S. R., de Vries, W., de Witt, C. A., Folke, C., Gerten, D., Heinke, J., Mace, G. M., Persson, L. M., Ramanahan, V., Reyers, B., & Sörlin, S. (2015b). Planetary boundaries: Guiding human development on a changing planet. *Science, 347*, 736–746.

Steffen, W., Rockström, J., Richardson, K., Lenton, T. M., Folke, C., Liverman, D., Summerhayes, C. P., Barnosky, A. D., Cornell, S. E., Crucifix, M., Donges, J. F., Fetzer, I., Lade, S. J., Scheffer,

M., Winkelmann, R., & Schellnhuber, H. J. (2018). Trajectories of the Earth system in the Anthropocene. *Proceedings of the National Academy of Sciences, USA, 115*(33), 8252–8259.

Syvitski, J., Waters, C. N., Day, J., Milliman, J. D., Summerhayes, C., Steffen, W., Zalasiewicz, J., Cearreta, A., Gałuszka, A., Hajdas, I., Head, M. J., Leinfelder, R., McNeill, J. R., Poirier, C., Rose, N. L., Shotyk, W., Wagreich, M., & Williams, M. (2020). Extraordinary human energy consumption and resultant geological impacts beginning around 1950 CE initiated the proposed Anthropocene Epoch. *Nature. Communications Earth & Environment, 1*(32), 1–13.

Wang-Erlandsson, L., Tobian, A., van der Ent, R. J., Fetzer, I., te Wierik, S., Porkka, M., Staal, A., Jaramillo, F., Dahlmann, H., Singh, C., Greve, P., Gerten, D., Keys, P. W., Gleeson, T., Cornell, S. E., Steffen, W., Bai, X., & Rockström, J. (2022, April). A planetary boundary for green water. *Nature Reviews Earth and Environment, 26*, 1–13.

4. Nitrogen Cycle

Canfield, D. E., Glazer, A. N., & Falkowski, P. G. (2010). The evolution and future of Earth's nitrogen cycle. *Science, 330*, 192–196.

Cornell, J. (2010). Fertilizers. *Encyclopedia of the Earth*. https://editors.eol.org/eoearth/wiki/Fertilizers

Gruber, N., & Galloway, J. N. (2008). An Earth-system perspective of the global nitrogen cycle. *Nature, 451*, 293–296.

5. Inhabiting the Earth

Ellis, E. C. (2011b). Anthropogenic transformation of the terrestrial biosphere. *Philosophical Transactions of The Royal Society, 369*, 1010–1035.

Ellis, E. C. (2013, September 13). Overpopulation is not the problem. *New York Times*. http://www.nytimes.com/2013/09/14/opinion/overpopulation-is-not-the-problem.html

Ellis, E. C. (2015). Ecology in an anthropogenic biosphere. *Ecological Monographs, 85*(3), 287–331.

Ellis, E. C., Goldewijk, K. K., Siebert, S., Lightman, D., & Ramankutty, N. (2010). Anthropogenic transformation of the biomes, 1700 to 2000. *Global Ecology and Biogeography, 19*, 589–606.

Ellis, E. C. (2017). Physical geography in the Anthropocene. *Progress in Physical Geography, 41*(5), 525–532.

Erb, K.-H., Lauk, C., Kastner, T., Mayer, A., Theurl, M. C., & Haberl, H. (2015). Exploring the biophysical option space for feeding the world without deforestation. *Nature Communications, 7*, 1–9.

Fischer-Kowalski, M., Krausmann, F., & Pallua, I. (2014). A sociometabolic reading of the Anthropocene: modes of subsistence, population size and human impact on Earth. *The Anthropocene Review, 1*, 8–33.

Foley, J. A., Ramankutty, N., Brauman, K. A., Cassidy, E. S., Gerber, J. S., Johnston, M., Mueller, N. D., O'Connell, C., Ray, D. K., West, P. C., Balzer, C., Bennett, E. M., Carpenter, S. R., Hill, J., Monfreda, C., Polasky, S., Rockström, J., Sheehan, J., Siebert, S., Tilman, D., & Zaks, D. P. M. (2011). Solutions for a cultivated planet. *Nature, 478*, 337–342.

Fredrik, A. J. (2015). Anthropocene blues: Abundance, energy, limits. In F. Felcht & K. Ritson (Eds.), *The imagination of limits: Exploring scarcity and abundance* (Vol. 2, pp. 55–63). RCC Perspectives.

Gilman, S. E., Urban, M. C., Tewksbury, J., Gilchrist, G. W., & Holt, R. D. (2010). A framework for community interactions under climate change. *Trends in Ecology & Evolution, 25*(6), 325–331.

Gordon, L. J., Bignet, V., Crona, B., Henriksson, P. J. G., Van Holt, T., Jonell, M., Lindahl, T., Troell, M., Barthel, S., Deutsch, L., Folke, C., Haider, L. J., Rockström, J., & Queiroz, C. (2017).

Rewiring food systems to enhance human health and biosphere stewardship. *Environmental Research Letters, 12*, 1–12.
Haberl, H., Erb, K. H., Krausmann, F., Gaube, V., Bondeau, A., Plutzar, C., Gingrich, S., Lucht, W., & Fischer-Kowalski, M. (2007). Quantifying and mapping the human appropriation of net primary production in Earth's terrestrial ecosystems. *Proceedings of the National Academy of Sciences, USA, 104*, 12942–12947.
Hardin, G. (1968). The tragedy of the commons. *Science, 162*, 1243–1248.
Hibbard, K. A., Crutzen, P. J., Lambin, E. F., Liverman, D. M., Mantua, N. J., McNeill, J. R., Messerli, B., & Steffen, W. (2006). Decadal-scale interactions of humans and the environment. In R. Costanza, L. Graumlich, & W. Steffen (Eds.), *Sustainability or collapse? An integrated history and future of people on Earth*. MIT Press.
Hubelin, J.-J., Ben-Ncer, A., Bailey, S. E., Freidline, S. E., Neubauer, S., Skinner, M. M., Bergmann, I., Le Cabec, A., Benazzi, S., Harvati, K., & Gunz, P. (2017). New fossils from Jebel Irhoud, Morocco and the pan-African origin of Homo sapiens. *Nature, 546*, 289–292.
Jackson, S. T., Betancourt, J. L., Booth, R. K., & Gray, S. T. (2009). Ecology and the ratchet of events: Climate variability, niche dimensions, and species distributions. *Proceedings of the National Academy of Sciences, USA, 106*(suppl. 2), 19685–19692.
Jambeck, J. R., Geyer, R., Wilcox, C., Siegler, T. R., Perryman, M., Andrady, A., Narayan, R., & Lavender Law, K. (2015). Plastic waste inputs from land into the ocean. *Science, 347*, 768–771.
Lussault, M. (2010). *L'avènement du Monde – Essai sur l'habitation humaine de la Terre*. Seuil.
Lynas, M. (2011). *The god species: How the planet can survive the age of humans*. Harper Collins.
Maris, V. (2018). *La part sauvage du monde*. Seuil.
Millenium Ecosystem Assessment. (2005). *Ecosystems and human well-being – Synthesis*. Island Press.
Paillard, S. (2017). Anthropocène: la planète va-t-elle craquer? *Revue Projet, 359*, 1–6.
Thompson, R. C., Olsen, Y., Mitchell, R. P., Davis, A., Rowland, S. J., John, A. W. G., McGonigle, D., & Russell, A. E. (2004). Lost at sea: Where is all the plastic? *Science, 304*, 838.
United Nations, Department of Economic and Social Affairs, Population Division. (2017). *World population prospects*. https://esa.un.org/unpd/wpp/Download/Standard/Population/
Vitousek, P. M., Mooney, H. A., Lubchenco, J., & Melillo, J. M. (1997). Human domination of Earth's ecosystems. *Science, 277*, 494–499.
Wohlleben, P. (2017). *La vie secrète des arbres – Ce qu'ils ressentent, comment ils communiquent – Un monde inconnu s'ouvre à nous*. Les arènes. (original edition 2015), French translation.

IV. Politics

1. Politics in the Anthropocene

Beau, R., & Larrère, C. (Eds.). (2018). *Penser l'Anthropocène*. Les Presses de Sciences Po.
Blanc, G., Demeulenaere, E., & Feuerhahn, W. (2017a). Difficile interdisciplinarité. In G. Blanc, E. Demeulenaere, & W. Feuerhahn (Eds.), *Humanités environnementales* (pp. 271–275). Publications de la Sorbonne.
Bonneuil, C. (2015). Anthropocène (point de vue 2). In D. Bourg & A. Papaux (Eds.), *Dictionnaire de la pensée écologique* (pp. 35–40). PUF.
Bonneuil, C. (2014). L'Anthropocène et ses lectures politiques. *Les Possibles, 3*, 1–7.
Bonneuil, C., & Fressoz, J.-B. (2013). *L'événement Anthropocène*. Seuil.
Chartier, D., & Deléage, J.-P. (2010). Mise à jour des écologies politiques pour une politique de l'Anthropocène. *Ecologie & Politique, 40*, 15–20.
Cochet, Y. (2013). Trois modèles du monde. In A. Sinaï (Ed.), *Penser la décroissance* (pp. 49–72). Presses de Science Po.

Dalby, S. (2016). Framing the Anthropocene: The good, the bad and the ugly. *The Anthropocene Review, 3*(1), 33–51.
Eckersley, R. (2017). La démocratie à l'ère de l'Anthropocène. *lapenseeecologique.com, 1*(1), 1–19. French translation.
Ecomodernism. (2015). *An ecomodernist manifesto*. www.ecomodernism.org
Foster, J. B. (2000). *Marx's ecology – Materialism and nature*. Monthly Review Press.
Karlsson, R. (2013). Ambivalence, irony, and democracy in the Anthropocene. *Futures, 46*, 1–21.
Lesourt, E. (2018). *Survivre à l'Anthropocène*. PUF.
Pena-Ruiz, H. (2018). *Karl Marx, penseur de l'écologie*. Seuil.
Sinaï, A., & Mathilde, S. (Eds.). (2017). *Gouverner la décroissance – Politiques de l'Anthropocène III*. Les Presses de Sciences Po.
Wallenhorst, N. (2021). *Mutation. L'aventure humaine ne fait que commencer*. Le Pommier.
Wallenhorst, N. (2022). *Qui sauvera la planète?* Arles.

2. Political Ecology

Deléage, J.-P. (2010). En quoi consiste l'écologie politique? *Ecologie & Politique, 40*, 21–30.
Meadows, D. (2013). Il est trop tard pour le développement durable. In A. Sinaï (Ed.), *Penser la décroissance* (pp. 195–210). Presses de Sciences Po.
Meadows, D. H., Randers, J., & Meadows, D. L. (2012). *Les limites à la croissance (dans un monde fini) – Le rapport Meadows, 30 ans après*. Rue de l'échiquier. (original edition 2004), French translation.
Semal, L., & Szuba, M. (2015). Théorie politique verte/Green political theory. In D. Bourg & A. Papaux (Eds.), *Dictionnaire de la pensée écologique* (pp. 988–990). PUF.
Villalba, B. (2010). L'écologie politique face au délai et à la contraction démocratique. *Ecologie & Politique, 40*, 95–113.
Zin, J. (2010). Qu'est-ce que l'écologie politique? *Ecologie & Politique, 40*, 41–49.

3. Governance

Biermann, F. (2007). "Earth system governance" as a crosscutting theme of global change research. *Global Environmental Change, 17*, 326–337.
Biermann, F. (2014). The Anthropocene: A governance perspective. *The Anthropocene Review, 1*, 57–61.
Lövbrand, E., Stripple, J., & Wiman, B. (2009). Earth system governmentality: Reflections on science in the Anthropocene. *Global Environmental Change, 19*(1), 7–13.
Nations Unies. (1992, June 3–14). *Déclaration de Rio sur l'environnement et le développement*. Sommet Planète Terre, Conférence des Nations Unies, Rio de Janeiro, Brazil. http://www.un.org/french/events/rio92/rio-fp.htm
Reid, W. V., Chen, D., Goldfarb, L., Hackmann, H., Lee, Y. T., Mokhele, K., Ostrom, E., Raivio, K., Rockström, J., Schellnhuber, H. J., & Whyte, A. (2010). Earth system science for global sustainability: Grand challenges. *Science, 330*, 916–917.
Sinaï, A. (Ed.). (2013a). *Penser la décroissance – Politiques de l'Anthropocène*. Presses de Sciences Po.
Sinaï, A. (2013b). Le destin des sociétés industrielles. In A. Sinaï (Ed.), *Penser la décroissance* (pp. 23–48). Presses de Science Po.
Statius, P. (2017). "La transition écologique et la démocratie: Quelques remarques philosophiques, politiques, et anthropologiques", *lapenseeecologique.com* (PUF), *1*(1), pp. 1–15.

4. Relationship to the Environment

Afeissa, H.-S. (2014). *La fin du monde et de l'humanité, Essai de généalogie du discours écologique*. PUF.

Atkinson, G., Dietz, S., & Neumayer, E. (Eds.). (2007). *Handbook of sustainable developpement*. Edward Elgar Publishing.

Berque, A. (2016a). La relation perceptive en mésologie: du cercle fonctionnel d'Uexküll à la trajection paysagère. *Revue du MAUSS, 47*, 87–104.

Berque, A. (2010). Logique des lieux de l'écoumène. *Communications, 87*, 17–26.

Berque, A. (2004). Ce qui fonde l'éthique environnementale. *Diogène, 207*, 3–14.

Berque, A. (2009a). Les travaux et les jours. Histoire naturelle et histoire humaine. *L'Espace géographique, 38*, 73–82.

Berque, A. (2015). Mythologie de l'urbain diffus. *Annales de géographie, 704*, 351–365.

Berque, A. (2016b). Perception de l'espace ou milieu perceptif? *L'Espace géographique, 45*, 168–181.

Berque, A. (2009b). *Ecoumène – Introduction à l'étude des milieux humains*. Belin.

Blanc, G., Demeulenaere, E., & Feuerhahn, W. (Eds.). (2017b). *Humanités environnementales*. Publications de la Sorbonne.

Cabanes, V. (2017). *Homo natura – En harmonie avec le vivant*. Buchet Chastel.

Chakrabarty, D. (2009). The climate of history: Four theses. *Critical Inquiry, 35*(Winter), 197–222.

Choné, A., Hajek, I., & Hamman, P. (Eds.). (2016). *Guide des humanités environnementales*. Septentrion.

Dunlap, R. E., & Michelson, W. (Eds.). (2002). *Handbook of environmental sociology*. Greenwood Press.

Fox, T., Pope, M., & Ellis, E. C. (2017). Engineering the Anthropocene: Scalable social networks and resilience building in human evolutionary timescales. *The Anthropocene Review, 4*(3), 199–215.

François. (2015). *Laudato si', Sur la sauvegarde de la maison commune*. Salvator. French translation.

Hess, G., & Bourg, D. (Eds.). (2016). *Science, conscience et environnement*. PUF.

Hornborg, H. (2014, September). Does the Anthropocene really imply the end of culture/nature and subject/object distinctions?, *Os mil nomes de Gaïa*, Rio de Janeiro, www.osmilnomesdegaia.eco.fr, pp. 1–16.

Latour, B. (2013). *Facing Gaia Six lectures on the political theology of nature*. Being the Gifford Lectures on Natural Religion Edinburgh, 18th–28th of February 2013. https://macaulay.cuny.edu/eportfolios/wakefield15/files/2015/01/LATOUR-GIFFORD-SIX-LECTURES_1.pdf, consulté le 25 October 2017.

Latour, B. (1991). *Nous n'avons jamais été modernes*. La Découverte.

Latour, B. (1999). *Politiques de la nature*. La Découverte.

Lockie, S., Sonnenfled, A., & Fischer, D. (Eds.). (2014). *Routledge international handbook of social and environnemental change*. Routledge.

Lovelock, J. E. (1979). *Gaia: A new look at life on Earth*. Oxford University Press.

Lovelock, J. E. (1988). *The ages of Gaia: A biography of our living on Earth*. W. W. Norton & Co.

Ollitrault, S. (2016). Engagement et trajectoires socio-spatiales à l'heure de la mondialisation: les bénévoles de Greenpeace et les agriculteurs biologiques. In D. C. Goirand, D. Garibay, & H. Combes (Eds.), *Les lieux de la colère – Occuper l'espace pour contester, de Madrid à Sanaa*. Karthala.

Ollitrault, S. (2001). Les écologistes français, des experts en action. *Revue française de science politique, 51*, 105–130.

Ollitrault, S. (2008). *Militer pour la planète, Sociologie des écologistes*. PUR.

Papaux, A. (2015). Homo faber. In D. Bourg & A. Papaux (Eds.), *Dictionnaire de la pensée écologique* (pp. 536–540). PUF.

Papaux, A., & Frigerio, V. (2015). Droit de l'homme et écologie. In D. Bourg & A. Papaux (Eds.), *Dictionnaire de la pensée écologique* (pp. 292–296). PUF.
Parizeau, M.-H. (2016). De l'Apocalypse à l'Anthropocène: parcours éthique des changements climatiques. *Revue de métaphysique et de morale, 89*, 23–38.
Porcelijn, B. (2018). *Notre empreinte cachée*. Seuil.
Raworth, K. (2017). *Doughnut economics: Seven ways to think like a 21st century economist*. Chelsea Green Publishing.
Rees, W. E. (1992). Ecological footprints and appropriated carrying capacity: What urban economics leaves out. *Environment and Urbanization, 4*, 121–130.
Scranton, R. (2013, November 10). Learning how to die in the Anthropocene. *The New York Times*.
Scranton, R. (2015). *Learning to die in the Anthropocene*. City Light Publisher.
Stiegler, B. (2015). Sortir de l'Anthropocène. *Multitudes, 60*, 137–146.

5. Relationship to Technology

(a). Technoscientific Progress

Gibson, D. G., Glass, J. I., Lartigue, C., Noskov, V. N., Chuang, R.-Y., Algire, M. A., Benders, G. A., Montague, M. G., Ma, L., Moodie, M. M., Merryman, C., Vashee, S., Krishnakumar, R., Assad-Garcia, N., Andrews-Pfannkoch, C., Denisova, E. A., Young, L., Qi, Z.-Q., Segall-Shapiro, T. H., Calvey, C. H., Parmar, P. P., Hutchison, C. A., III, Smith, H. O., & Venter, J. C. (2010). Creation of a bacterial cell controlled by a chemically synthesized genome. *Science, 329*, 52–56.
Haff, P. K. (2014). Humans and technology in the Anthropocene: Six rules. *The Anthropocene Review, 2*, 126–130.
Silver, D., Schrittwieser, J., Simonyan, K., Antonoglou, I., Huang, A., Guez, A., Hubert, T., Baker, L., Lai, M., Bolton, B., Chen, Y., Lillicrap, T., Hui, F., Sifre, L., van den Driessche, G., Graepel, T., & Hassabis, D. (2017). Mastering the game of go without human knowledge. *Nature, 550*, 354–359.

(b). Geo-Engineering and Earth System Stewardship

Angel, R. (2006). Feasibility of cooling the Earth with a cloud of small spacecraft near the inner Lagrange point (L1). *Proceedings of the National Academy of Sciences, USA, 103*(46), 17184–17189.
Breeze, P. (2008). Coping with carbon: A near-term strategy to limit carbon dioxide emissions from power stations. *Philosophical Transactions of the Royal Society, London, 366*, 3891–3900.
Budyko, M. I. (1977). *Climatic changes*. American Geophysical Union. (original edition 1974).
Crutzen, P. J. (2006). Albedo enhancement by stratospheric sulfur injections: A contribution to resolve a policy dilemma? *Climate Change, 77*, 211–219.
Latham, J., Rasch, P., Chen, C.-C., Kettles, L., Gadian, A., Gettelman, A., Morrison, H., Bower, K., & Choularton, T. (2008). Global temperature stabilization via controlled albedo enhancement of low-level maritime clouds. *Philosophical Transaction of the Royal Society, 366*, 3969–3987.
Lovelock, J. (2008). A geophysiologist's thoughts on geoengineering. *Philosophical Transaction of the Royal Society, London, 366*, 3883–3890.
Marchetti, C. (1977). On geoengineering and the CO_2 problem. *Climatic Change, 1*, 59–68.
Neyrat, F. (2014). Critique du géo-constructivisme. Anthropocène & géo-ingénierie. *Multitudes, 56*, 37–47.
Österblom, H., Jouffray, J.-B., Folke, C., & Rockström, J. (2017). Emergence of a global science-business initiative for ocean stewardship. *Proceedings of the National Academy of Sciences, 114*(34), 1–6.

Rasch, P. J., Crutzen, P. J., & Coleman, D. B. (2008a). Exploring the geoengineering of climate using stratospheric sulfate aerosols: The role of particle size. *Geophysical Research Letters, 35*, 1–6.
Rasch, P. J., Tilmes, S., Turco, R. P., Robock, A., Oman, L., Chen, C.-C., Stenchikov, G. L., & Garcia, R. R. (2008b). An overview of geoengineering of climate using stratospheric sulphate aerosols. *Philosophical Transactions of the Royal Society, London, 366*, 4007–4037.
Schneider, S. H. (2008). Geoengineering: Could we or should we make it work? *Philosophical Transaction of the Royal Society, 366*, 1–20.
Steffen, W., Persson, A., Deutsch, L., Zalasiewicz, J., Williams, M., Richardson, K., Crumley, C., Crutzen, P., Folke, C., Gordon, L., Molina, M., Ramanathan, V., Rockström, J., Scheffer, M., Schellnhuber, H. J., & Svedin, U. (2011b). The Anthropocene: From global change to planetary stewardship. *Ambio, 40*(7), 739–761.

(c). A Good Anthropocene

Bourg, D., Prouteau, F., Hétier, R., & Wallenhorst, N. (interview). (2019). *Bad Anthropocene* et anthropologie politique postprométhéenne (*forthcoming*). *Raisons politiques*, Presses de Science Po.
Ellis, E. C. Prouteau, F., Hétier, R., & Wallenhorst, N. (interview). (2019). *Good Anthropocene* et anthropologie politique prométhéenne (*forthcoming*). *Raisons politiques*, Presses de Science Po.
Ellis, E. C. (2011a). The planet of no return: Human resilience on an artificial Earth. *The Breakthrough Journal, 2*, 37–44.
Guillaume, B. (2015b). Géoingénierie. In D. Bourg & A. Papaux (Eds.), *Dictionnaire de la pensée écologique* (pp. 468–470). PUF.
Hamilton, C., (2016). The theodicy of the "Good Anthropocene". *Environmental Humanities, 7* (1):233–238. https://read.dukeupress.edu/environmental-humanities/article/7/1/233/8202/The-Theodicy-of-the-Good-Anthropocene

B. Authors Regularly Cited or Studied

I. Hannah Arendt

1. Primary Sources

Arendt, H. (2002). *Les origines du totalitarisme III. Le système totalitaire*. Seuil. (original edition 1948), French translation.
Arendt, H. (1983). *Condition de l'homme moderne*. Calmann-Lévy. (original edition 1958, French translation 1961), French translation.
Arendt, H. (1972a). *La crise de la culture*. Gallimard. (original edition 1961), French translation.
Arendt, H. (1972b). Préface – La brèche entre le passé et le futur. In H. Arendt (Ed.), *La crise de la culture* (pp. 11–27). Gallimard. (original edition 1961a), French translation.
Arendt, H. (1972c). Qu'est-ce que l'autorité? In H. Arendt (Ed.), *La crise de la culture* (pp. 121–185). Gallimard. (original edition 1961b), French translation.

Arendt, H. (1972d). La crise de l'éducation. In H. Arendt (Ed.), *La crise de la culture* (pp. 223–252). Gallimard. (original edition 1961c), French translation.
Arendt, H. (1972e). Qu'est-ce que la liberté? In H. Arendt (Ed.), *La crise de la culture* (pp. 186–222). Gallimard. (original edition 1961d), French translation.
Arendt, H. (1964). *De la révolution*. Gallimard. (original edition 1963), French translation.
Arendt, H. (1974). *Vies politiques*. Gallimard. (original edition 1970), French translation.
Arendt, H. (2016a). *Vita activa – oder Vom tätigen Leben*. Piper. (original German edition 1967).
Arendt, H. (2016b). *La vie de l'esprit*. PUF. (original edition 1978), French translation.
Arendt, H. (1991). *Juger – Sur la philosophie politique de Kant*. Seuil. (original edition 1982), French translation.
Arendt, H. (1995). *Qu'est-ce que la politique?* Seuil. (original edition 1993), French translation.
Arendt, H. (2005). *Journal de pensée, 1950–1973 – 2 tomes*. Seuil. French translation (original edition 2002).

Correspondence

Arendt, H., & Blumenfeld, K. (2012). *Correspondance 1933–1963*. Desclée de Brouwer. (original edition 1995), French translation.
Arendt, H., & Heidegger, M. (2001). *Lettres et autres documents 1925–1975*. Gallimard. (original edition 1998), French translation.
Arendt, H., & Blücher, H. (1999). *Correspondances – 1936–1968*. Calman-Lévy. French translation.

Correspondence in English

Arendt, H., & Jaspers, K. (1992). *Correspondence 1926–1969*. Harcourt Brace, Jovanovich.

2. Secondary Sources

Albanel, V. (2010). *Amour du monde – Christianisme et politique chez Hannah Arendt*. Cerf.
Foray, P. (2001). Hannah Arendt, l'éducation et la question du monde. *Le Télémaque, 19*, 79–101.
Frogneux, N. (1996). Hans Jonas développe-t-il une anthropologie arendtienne? *Revue philosophique de Louvain, Tome, 94*(4), 677–686.
Genel, K. (2006, January 11). Hannah Arendt et l'école de Francfort, deux critiques de la modernité. https://philolarge.hypotheses.org/files/2017/09/11-01-2006_genel.pdf
Goetz, B., & Younès, C. (2009). Hannah Arendt: Monde – Déserts – Oasis. In T. Paquot & C. Younès (Eds.), *Le territoire des philosophes* (pp. 29–46). La Découverte.
Iakovou, V. (2001). La critique entre Hannah Arendt et l'école de Francfort. *Tumultes, 2*(17–18), 259–278.
Muhlmann, G. (2001/2002). Pensée et non-pensée selon H. Arendt et T. W. Adorno. *Tumultes, 17–18*, 279–319.
Poché, F. (2003). *Penser avec Arendt et Lévinas*. Lyon.
Poizat, J.-C. (2009). *Assumer l'humanité, Hannah Arendt: la responsabilité face à la pluralité* de Gérôme Truc. *Le Philosophoire, 31*, 177–188.
Pommier, E. (2012). La responsabilité en discussion: Appel/Jonas. *Revue philosophique de la France et de l'étranger, 137*, 495–514.
Pommier, E. (2013a). Ethique et politique chez Hans Jonas et Hannah Arendt. *Revue de métaphysique et de morale, 78*, 271–286.
Pommier, E. (2013b). Le sens de la vie chez Hans Jonas. *Etudes, 418*, 485–495.
Taminiaux, J. (2007). La déconstruction arendtienne des vues politiques de Heidegger. *Cahiers Philosophiques, 111*, 16–30.
Tassin, E. (2017). *Le trésor perdu – Hannah Arendt, l'intelligence de l'action politique*. Klincksieck.
Revault d'Allonnes, M. (2011). Hannah Arendt penseur de la crise. *Etudes, 415*, 197–206.
Ricœur, P. (1961). Préface. In H. Arendt (Ed.), *Condition de l'homme moderne* (pp. I–XXVIII). Calmann-Lévy.

II. Christian Arnsperger

Arnsperger, C. (2001). Entre impartialité, horizons de sens et précarité existentielle: les fondements de l'éthique économique et sociale. In C. Arnsperger, C. Larrère, & J. Ladrière (Eds.), (pp. 19–67). *Trois essais sur l'éthique économique et sociale*, Editions Quae.

Arnsperger, C. (2002). Justice et économie. Latitudes d'égalisation et obstacles existentiels. *Revue de métaphysique et de morale, 33*, 7–26.

Arnsperger, C. (2003). Peut-il exister une critique sociale rawlsienne? *Mouvements, 27–28*, 165–170.

Arnsperger, C. (2004). L'inestimable valeur de l'autre: argent, altérité et socialité. In M. Drach (Ed.), *L'argent* (pp. 33–59). La découverte.

Arnsperger, C. (2005a). *Critique de l'existence capitaliste – Pour une éthique existentielle de l'économie*. Cerf.

Arnsperger, C. (2005b). Voie large ou voie étroite de l'économie politique? Une interrogation toujours pertinente. *Reflets et perspectives de la vie économique, 44*, 49–55.

Arnsperger, C. (2005c). La "moralisation du capitalisme": une approche du problème. L'expérimentalisme démocratique et la coévolution des normes. *Revue internationale des sciences sociales, 185*, 477–488.

Arnsperger, C. (2006a). Les sciences sociales face aux problèmes sociaux. In C. Arnsperger (Ed.), *L'économie c'est nous, pour un savoir citoyen* (pp. 33–42). Eres.

Arnsperger, C. (2006b). Qu'est-ce qu'être libre dans un système? In R. de Borchgrave (Ed.), *Le philosophe et le manager* (pp. 157–170). De Boeck.

Arnsperger, C. (2008). L'angoisse dans la financiarisation: les causes existentielles de la crise. *Finance & Bien Commun, 31–32*, 94–102.

Arnsperger, C. (2009). *Éthique de l'existence post-capitaliste – Pour un militantisme existentiel*. Cerf.

Arnsperger, C. (2010a). Changer d'existence économique: enjeux anthropologiques de la transition du capitalisme au post-capitalisme. *Revue d'éthique et de théologie morale, 258*, 23–50.

Arnsperger, C. (2010b). Monnaie, dette et croissance sans prospérité: portée et limites du 'tournant' jacksonnien. *Etopia, 8*, 109–116.

Arnsperger, C. (2011a). *L'homme économique et le sens de la vie*. Textuel.

Arnsperger, C. (2011b). Dépasser le capitalisme, mais par étapes. *Revue Projet, 324–325*, 73–81.

Arnsperger, C. (2011c, October 3). *Economie existentielle, critique sociale et plasticité anthropologique*. Texte accompagnant le séminaire "Anthropologie et Justice Sociale" au Centre Sèvre.

Arnsperger, C. (2013). Fonder l'économie écologique. *Revue d'éthique et de théologie morale, 276*, 93–120.

Arnsperger, C. (2016). Progrès et conscience, Eléments pour une anthropologie économique non réductionniste de la durabilité. In G. Hess & D. Bourg (Eds.), *Science, conscience et environnement, Penser le monde complexe* (pp. 179–206). PUF.

Arnsperger, C. (2019). Serons-nous enfin, un jour, indigènes? Permaculture et éducation des profondeurs. In N. Wallenhorst & J.-P. Pierron (Eds.), *Eduquer en Anthropocène*. Le Bord de l'eau.

Arnsperger, C., & Bourg, D. (2017). *Ecologie intégrale, Pour une société permacirculaire*. PUF.

Arnsperger, A., & Deibler, L. (2017, May). *Suffisamment, c'est mieux* (p. 43). Série "Réflexions", Banque alternative Suisse.

Arnsperger, C., & Parijs, P. (2003). *Éthique économique et sociale*. Éditions La Découverte.

III. Maurice Bellet

1. Primary Sources

Bellet, M. (1967). *La peur ou la foi*. Desclée de Brouwer.
Bellet, M. (1976). *Le lieu du combat*. Desclée de Brouwer.
Bellet, M. (1979). *Le Dieu pervers*. Desclée de Brouwer.
Bellet, M. (1988). *L'épreuve*. Desclée de Brouwer.
Bellet, M. (1990). *Dire ou la vérité improvisée*. Desclée de Brouwer.
Bellet, M. (1993). *La seconde Humanité. De l'impasse majeure de ce que nous appelons l'économie*. Desclée de Brouwer.
Bellet, M. (1994). *Incipit*. Desclée de Brouwer.
Bellet, M. (1996). *L'Europe "Au-delà" d'elle-même*. Desclée de Brouwer.
Bellet, M. (1997). *L'insurrection*. Desclée de Brouwer.
Bellet, M. (1998). *Le sauvage indigné*. Desclée de Brouwer.
Bellet, M. (2002). *La longue veille 1934–2002*. Desclée de Brouwer.
Bellet, M. (2004). *Le paradoxe infini*. Desclée de Brouwer.
Bellet, M. (2008). La contestation évangélique. *Etudes, 408*, 636–648.
Bellet, M. (2011). *Translation*. Bayard.
Bellet, M. (2013). *L'avenir du communisme*. Bayard.
Bellet, M., & Wallenhorst, N. (interviews). (2022). *Sapere. Saveurs et savoirs d'un monde qui nous parle*. Le Pommier.

2. Secondary Sources

Bertrand, D. (2005, February). Maurice Giuliani, Maurice Bellet. *Etudes*, 271–273.
Caillé, A. (2003). Maurice Bellet: Invitation. Plaidoyer pour la gratuité et l'abstinence. *Revue du MAUSS, 22*, 434.
Cupillard, D. (2001, April). Maurice Bellet: L'amour déchiré. *Etudes*, 565–566.
Euvé, F. (2013, December). Recension de Maurice Bellet, *L'avenir du communisme*, Paris, Bayard, 2013. *Etudes*, 713.
Lagacé-Leroy, D. (2003). *L'expérience du point-lieu chez Maurice Bellet – Etude d'un parcours discursif*. Doctoral thesis in Theology, supervised by Guy-Robert St-Arnaud, University of Montreal.
Le Corre, F. (2006, January). Maurice Bellet: La traversée de l'en-bas. *Etudes*, 135–136.
Lefebvre, P. (2005). Une conversion "hors les murs": *Les Allées du Luxembourg* de Maurice Bellet. In N. Brucker (Ed.), *La conversion. Expérience spirituelle, expression littéraire* (Vol. 8, pp. 573–582). Actes du colloque de Metz (5–7 June 2003).
Maldamé, J.-M. (1988). L'itinéraire de Maurice Bellet. Une phénoménologie de la création. *Revue Thomiste, 88*(2), 299–314.
Moussavi Chirazi, S. D. (1999). *L'évolution de l'écriture de Maurice Bellet*. Doctoral thesis in 20th-century, supervised by Philippe Lefebvre, Université de Nancy 2.
Moussavi Chirazi, S. D. (2007, July). Un style moderne pour le langage religieux: l'écriture de Maurice Bellet. *Archive of Scientifique Information Database*. http://www.sid.ir/en/VEWSSID/J_pdf/1046320100106.pdf, consulted online on 3 October 2013.
Wallenhorst, N. (2014, July). Recension de L'avenir du communisme de M. Bellet. *Esprit*, 142–143.

IV. Dominique Bourg

Bourg, D. (1996). *L'Homme artifice*. Gallimard.

Bourg, D. (2001). Le nouvel âge de l'écologie. *Le Débat, 113*, 92–105.

Bourg, D. (2009, December). L'impératif écologique. *Esprit*, 59–71.

Bourg, D. (2010a, July). L'éco-scepticisme et le refus des limites. *Etudes*, 29–40.

Bourg, D. (2010b). Un système qui ne peut répondre au défi environnemental. Interview with Pierre Le Hir, *Le Monde*, 31 October 2010.

Bourg, D. (2012a). Pour une anthropologie de la finitude. In J.-P. Pierron & M.-H. Parizeau (Eds.), *Repenser la nature: dialogue philosophique, Europe, Asie, Amériques*. Presses Universitaires de Laval.

Bourg, D. (2012b). Transition écologique, plutôt que développement durable. *Vraiment durable, 1*, 77–96.

Bourg, D. (2013a, July). Position. La guerre écologique. *Esprit*, 5–7.

Bourg, D. (2013b). Peut-on encore parler de crise écologique? *Revue d'éthique et de théologie morale, 276*(HS), 61–71.

Bourg, D. (2013d). Dommages transcendantaux. In D. Bourg, P.-B. Joly, & A. Kaufmann (Eds.), *Du risque à la menace, Penser la catastrophe* (pp. 109–126). PUF.

Bourg, D. (2017). Gouverner le long terme. *lapenseeecologique.com*, Points de vue, *1*(1), PUF, 1–9.

Bourg, D. (2018). *Une nouvelle Terre*. Desclée de Brouwer.

Bourg, D., Augagneur, F., Blondiaux, L., Cohendet, M.-A., Fourniau, J.-M., François, B., & Prieur, M. (2017). *Inventer la démocratie du XXIème siècle*. Les liens qui libèrent.

Bourg, D., & Bompard, J.-P. (2012, May). Comment verdir la démocratie? *Projet, 330*, 14–22.

Bourg, D., & Fragnière, A. (2014). *La pensée écologique – Une anthologie*. PUF.

Bourg, D., & Papaux, A. (2010a). Ecologie, 1980–2010: de l'exception française à la normalisation. *Le Débat, 160*, 94–114.

Bourg, D., & Papaux, A. (2010b). *Vers une société sobre et désirable*. PUF.

Bourg, D., & Papaux, A. (Eds.). (2015a). *Dictionnaire de la pensée écologique*. Paris.

Bourg, D., & Papaux, A. (2015b). Anthropologie de la finitude. In D. Bourg & A. Papaux (Eds.), *Dictionnaire de la pensée écologique* (pp. 44–46). PUF.

Bourg, D., & Papaux, A. (2015c). Pensée écologique. In D. Bourg & A. Papaux (Eds.), *Dictionnaire de la pensée écologique* (pp. 756–759). PUF.

Bourg, D., & Whiteside, K. (2010). *Vers une démocratie écologique. Le citoyen, le savant et le politique*. Seuil.

Bourg, D., & Whiteside, K. (2011). Ecologie, démocratie et représentation. *Le Débat, 164*, 145–153.

Bourg, D., & Whiteside, K. (2017). Ecologies politiques: essai de typologie. *lapenseeecologique.com, PUF, 1*(1), 1–26.

V. Jean-Pierre Boutinet

Boutinet, J.-P. (1990). *Anthropologie du projet*. PUF.

Boutinet, J.-P. (1998). *L'immaturité de la vie adulte*. PUF.

Boutinet, J.-P. (2004). *Vers une société des agendas*. PUF.

Boutinet, J.-P. (2006). Interview avec Jean-Pierre Boutinet. *Communication et organisation, 28*, 194–196.

Boutinet, J.-P. (2007). Vulnérabilité adulte et accompagnement de projet: un espace paradoxal à aménager. In J.-P. Boutinet, N. Denoyel, G. Pineau, & J.-Y. Robin (Eds.), *Penser l'accompagnement adulte* (pp. 27–49). PUF.
Boutinet, J.-P. (2010). *Grammaires des conduites à projet*. PUF.
Boutinet, J.-P., & Dominicé, P. (Eds.). (2009). *Où sont passés les adultes?* Téraèdre.
Boutinet, J.-P. (2012). Donner de la valeur au projet éducatif et pédagogique, mais à quelles conditions? *Colloque du CERFOP*, 23 May 2012.
Boutinet, J.-P., & Bréchet, J.-P. (2014). *Logiques de projet, logiques de profit*. Lyon.
Boutinet, J.-P., & Heslon, C. (2010). Le *Life designing* face aux aléas postmodernes du conseil en orientation. *L'orientation scolaire et professionnelle, 39*, 53–71.

VI. Jean-Philippe Pierron

Coutellec, L., & Pierron, J.-P. (2017). Penser une éthique alimentaire. *Revue française d'éthique appliquée, 4*, 19–24.
Pierron, J.-P. (2012). L'indignation. *Etudes, 416*, 57–66.
Pierron, J.-P. (2013a). Ressource naturelle ou service écologique gratuit? Ce que l'eau nous donne. *Revue du MAUSS, 42*, 271–280.
Pierron, J.-P. (2013b). Au-delà de l'anthropocentrisme: la nature comme partenaire. *Revue du MAUSS, 42*, 41–48.
Pierron, J.-P. (2014a, February). Insolite ou insolente idée de nature? *Etudes*, 55–65.
Pierron, J.-P. (2014b). *Mythopées*. Vrin.
Pierron, J.-P. (2016). L'écologie, lieu d'une renaissance spirituelle. *Revue Projet, 354*, 84–89.
Pierron, J.-P. (2017). Pour le meilleur et pour le pire? *Jusqu'à la mort accompagner la vie, 129*, 93–103.
Pierron, J.-P. (2018). *La poétique de l'eau – Pour une nouvelle écologie*. Editions François Bourin.
Pierron, J.-P. (2019). *Prendre soin de la nature et des humains – Médecine, travail, écologie*. Les Belles Lettres.
Wallenhorst, N., & Pierron, J.-P. (Eds.). (2019a). *Eduquer en Anthropocène*. Le Bord de l'eau.
Wallenhorst, N., & Pierron, J.-P. (2019b). L'éducation en Anthropocène. Une métamorphose pour pérenniser l'aventure humaine. In N. Wallenhorst & J.-P. Pierron (Eds.), *Eduquer en Anthropocène*. Le Bord de l'eau.

VII. Hartmut Rosa

Lübbe, H. (2009). The contraction of the present. In H. Rosa & W. Scheuerman (Eds.), *High-speed society. Social acceleration, power and modernity* (pp. 159–178). Pennsylvania State University.
Rosa, H. (2013). *Accélération. Une critique sociale du temps*. La découverte. (original edition 2010), French translation.
Rosa, H. (2014). *Aliénation et accélération – Vers une théorie critique de la modernité tardive*. La Découverte. (original 2012), French translation.
Rosa, H. (2018). *Résonance – Une sociologie de la relation au monde*. La découverte. (original edition 2016), French translation.
Rosa, H. (2016). *Resonanz, Eine Soziologie der Weltbeziehung*. Suhrkamp.
Rosa, H., & Wallenhorst, N. (interview). (2017a). Apprendre à écouter le monde. *Chemins de formation, 21*, 19–31.

Rosa, H., & Wallenhorst, N. (interview). (2017b). Apprendre ensemble dans la société de l'accélération. *Bildungsforschung, 2017*(1), 1–7.
Rosa, H., & Wallenhorst, N. (interview). (2022). *Accélérons la résonance!*. Le Pommier.

VIII. Bruno Villalba

Semal, L., & Villalba, B. (2013). Obsolescence de la durée. La politique peut-elle continuer à disqualifier le délai? In F.-D. Vivien, J. Lepart, & P. Marty (Eds.), *L'évaluation de la durabilité* (pp. 81–100). Editions Quæ.
Villalba, B. (2009). Conclusion. Une appropriation politisée mais peu politique. In B. Villalba (Ed.), *Appropriations du développement durable – Emergences, diffusions, traductions* (pp. 363–369). Presses Universitaires du Septentrion.
Villalba, B. (2015). Au fondement matériel de la démocratie. *Revue Projet, 344*, 56–63.
Villalba, B. (2016). Sobriété: ce que les pauvres ont à nous dire. *Revue Projet, 350*, 39–49.

IX. Political Scientists from the Rennes School

Baudouin, J. (2002a). *Les idées politiques contemporaines*. PUR.
Baudouin, J. (2002b). *Le mouvement catholique français à l'épreuve de la pluralité (une militance éclatée)*. PUR.
Baudouin, J., & Bruneteau, B. (Eds.). (2014). *Le totalitarisme – Un concept et ses usages*. PUR.
Bruneteau, B. (2010). *Le Totalitarisme. Origines d'un concept, genèse d'un débat. 1930–1942*. Cerf.
Bruneteau, B. (2014). *Les totalitarismes*. Armand Colin.
Bénéton, P. (2010). *Introduction à la politique*. PUF.
Lambert, F. (2001). *Introduction à l'histoire des idées politiques*. Armand Colin.
Lambert, F., & Lefranc, S. (2012). *50 fiches pour comprendre la Science politique*. Bréal. (4ème éd).
Lambert, F. (2012). Le 'silence' des intellectuels catholiques français. In F. Hourmant & A. Leclerc (Eds.), *Les intellectuels et le pouvoir*. PUR.
Lambert, F. (2014). Le 'totalitarisme': l'impensable dans la philosophie politique? In J. Baudouin & B. Bruneteau (Eds.), *Le totalitarisme* (pp. 69–81). PUR.
Portier, P. (2009). La modernité libérale à l'épreuve de la théorie politique. *Revue française de science politique, 59*, 813–820.
Portier, P. (2017). Jean Baudouin et la science politique – Trajectoire d'une hétérodoxie. In B. Bruneteau, G. Châton, & P. Portier (Eds.), *L'aventure démocratique – Cheminement en compagnie de Jean Baudouin* (pp. 13–28). PUR.

X. The Convivialists

Chapelle, G., & Decoust, M. (2015). *Le vivant comme modèle: la voie du biomimétisme*. Albin Michel.
Fistetti, F. (2016). Le convivialisme, 'contre-mouvement' du 21ème siècle. *Revue du MAUSS, 48*, 247–258.
Glémain, P. (2017). Penser le convivialisme en économie sociale contemporaine. *RECMA, 4*(346), 27–41.

Humbert, M. (Ed.). (2017). *Reconstruction de la société – Analyses convivialistes*. Presses Universitaires de Rennes.
Illich, I. (1973a). *La convivialité*. Seuil. French translation.
Illich, I. (1971a). *Une société sans école*. Seuil.
Illich, I. (1973b). *Tools for conviviality*. Harper and Row.
Servigne, P., & Stevens, P. (2016). *Comment tout peut s'effondrer*. Seuil.
Servigne, P., & Chapelle, G. (2017). *L'entraide – L'autre loi de la jungle*. Les Liens qui Libèrent.
Wallenhorst, N., Mellot, S., & Theviot, A. (2020). *Interconnectés? Numérique et convivialisme*. Le bord de l'eau. (*forthcoming*).

1. Alain Caillé

Caillé, A. (2008). *Théorie anti-utilitariste de l'action. Fragments d'une sociologie générale*. La Découverte.
Caillé, A. (2011). Au-delà du libéralisme, le convivialisme. *Revue Projet, 324–325*, 94–97.
Caillé, A. (2015). *Le convivialisme en dix questions. Un nouvel imaginaire politique*. Le Bord de l'Eau.
Caillé, A. (2016). *Les Convivialistes, Eléments d'une politique convivialiste*. Le Bord de l'eau.
Caillé, A. (2018). "Foreword" to Graeber, D. In *La démocratie aux marges*. Flammarion.
Caillé, A., & Chanial, P. (Eds.). (2014). *Du convivialisme comme volonté et comme espérance* (Vol. 43). Revue du MAUSS.
Caillé, A., Chanial, P., Dufoix, S., & Vandenberghe, F. (Eds.). (2018). *Des Sciences sociales à la Science sociale – Fondements antiutilitaristes*. Le Bord de l'Eau.

2. François Flahault

Flahault, F. (2005). Vers une nouvelle pensée sociale. *Revue du MAUSS, 26*, 377–382.
Flahault, F. (2008a). *Le crépuscule de Prométhée – Contribution à une histoire de la démesure humaine*. Mille et une nuit.
Flahault, F. (2008b). Comment l'homme peut-il être à la fois égoïste, bon et méchant? *Revue du MAUSS, 31*, 307–317.
Flahault, F. (2013a). Pour une conception renouvelée du bien commun. *Études, 418*, 773–783.
Flahault, F. (2013b). L'homme fait-il partie de la nature? *Revue du MAUSS, 42*, 125–128.
Flahault, F. (2018). *L'homme, une espèce déboussolée – Anthropologie générale à l'âge de l'écologie*. Fayard.

3. Edgard Morin

Morin, E. (1990). *Introduction à la pensée complexe*. Seuil.
Morin, E. (1999a). *Introduction à une politique de l'homme*. Seuil.
Morin, E. (1999b). *La tête bien faite*. Seuil.
Morin, E. (2000). *Les sept savoirs nécessaires à l'éducation du futur*. Seuil.
Morin, E. (2001). *L'identité humaine*. Seuil.
Morin, E. (2012). Krisis. In E. Morin (Ed.), *Journal 1962–1987* (pp. 1123–1180). Seuil.
Morin, E. (2014). *Enseigner à vivre*. Actes Sud/Play Bac.
Morin, E., & Lafay, D. (interview). (2017). *Le temps est venu de changement de civilisation*. L'Aube.

4. Corine Pelluchon

Pelluchon, C. (2015). *Les nourritures*. Seuil.
Pelluchon, C. (2017a). L'éthique des vertus: une condition pour opérer la transition environnementale. *lapenseeecologique.com, PUF, 1*(1), 1–18.
Pelluchon, C. (2017b). *Manifeste animaliste – Politiser la cause animale*. Alma.
Pelluchon, C. (2018). *Ethique de la considération*. Seuil.

5. Bernard Perret

Perret, B. (2010, January). Quand la valeur n'est plus monnayable. *Esprit*, 98–117.
Perret, B. (2012). L'urgence occultée de la crise climatique. *Etudes, 4162*, 151–161.
Perret, B. (2014a). Transition écologique ou choc de la finitude? *Revue du MAUSS, 43*, 35–40.
Perret, B. (2014b). De la difficulté de penser un avenir sous contrainte écologique. *Transversalités, 130*, 69–81.

XI. The Accelérationists

Avanessian, A. (2016). In L. de Sutter (Ed.), *Accélération !* (pp. 225–251). PUF. French translation.
Bastoni, A. (2015, October). Inventing the future. *Novara Media*.
Brassier, R. (2016). Accélérer la raison. In L. de Sutter (Ed.), *Accélération !* (pp. 157–181). PUF. French translation.
Citton, Y. (2016). Accélérer l'écologie. In L. de Sutter (Ed.), *Accélération !* (pp. 205–223). PUF.
Cuboniks, L. (2016). Accélérer le féminisme. In L. de Sutter (Ed.), *Accélération !* (pp. 253–273). PUF. French translation.
de Sutter, L. (Ed.). (2016a). *Accélération !* PUF.
de Sutter, L. (2016b). Introduction. In L. de Sutter (Ed.), *Accélération !* (pp. 7–25). PUF.
Negri, A. (2016). Accélérer la politique. In L. de Sutter (Ed.), *Accélération !* (pp. 49–67). PUF. French translation.
Seibt, S. (2019). L'accélérationnisme, l'autre idéologie invoquée par le terroriste de Christchurch. *France 24*, 19 March 2019, https://www.france24.com/fr/20190319-accelerationnisme-ideologie-christchurch-nouvelle-zelande-tireur-mosquees-terrorisme-extrem
Srnicek, N., & Williams, A. (2016). #Accelerate. Manifeste pour une politique accélérationniste. In L. de Sutter (Ed.), *Accélération !* (pp. 27–47). PUF. French translation.
Srnicek, N., & Williams, A. (2013). #ACCELERATE. Manifesto for an Accelerationist Politics. *Critical Legal Thinking*, 14 May 2013. http://criticallegalthinking.com/2013/05/14/accelerate-manifesto-for-an-accelerationist-politics
Srnicek, N., & Williams, A. (2015). *Inventing the futur – Postcapitalism and a World without work*. Verso.
Wallenhorst, N. (2018). L'accélération: opportunité ou inopportunité pour quels apprentissages? *Education permanente, 215*, 69–80.
Wallenhorst, N. (2016e). Vitesse et politique: accélérer ou décélérer? *Raisons Politiques (Presses de Sciences Po), 64*, 137–147.

XII. Andreas Weber

Weber, A. (2007). *Alles fühlt: Mensch, Natur und die Revolution der Lebenswissenschaften*. Berlin Verlag.

Weber, A. (2008). *Biokapital. Die Versöhnung von Ökonomie, Natur und Menschlichkeit*. Berlin Verlag.

Weber, A. (2016a). *Enlivenment. Eine Kultur des Lebens. Versuch einer Poetik für das Anthropozän*. Matthes und Seitz.

Weber, A. (2017). *Sein und Teilen – Eine Praxis schöpferischer Existenz*. Transcript Verlag.

Weber, A. (2016b). *The biology of Wonder – Aliveness, Feeling, and the Metamorphoses of Science*. New Society Publishers.

Weber, A., & Kurt, H. (2015). *Lebendigkeit sei! Für eine Politik des Lebens. Ein Manifest für das Anthropozän*. Klein Jasedow, Think Oya.

XIII. David Abram

Abram, D. (2013). *Comment la Terre s'est tue – Pour une écologie des sens*. Paris La Découverte. (original edition 1996), French translation.

XIV. The Early Critical Theoreticians

Primary Sources

Adorno, T. W. (2003). Eduquer après Auschwitz. In *Modèles critiques* (pp. 235 à 251, fr. fr). Payot. (original edition 1966).

Adorno, T. W. (2001). *Sur Walter Benjamin*. Gallimard. (original edition 1968), French translation.

Anders, G. (2012). *L'Obsolescence de l'homme – Sur la destruction de la vie à l'époque de la troisième révolution industrielle*. Editions Fario. (original edition 1980), French translation.

Horkheimer, M. (1974). *Théorie traditionnelle et théorie critique*. Gallimard. (original edition 1970), French translation.

Horkheimer, M., & Adorno, T. W. (1974). *La Dialectique de la Raison – Fragments philosophiques*. Gallimard. (original edition 1947), French translation.

Marcuse, H. (1963). *Eros et civilisation*. Editions de minuit. (original edition 1955), French translation.

Marcuse, H. (1968). *L'Homme unidimentionnel – Essai sur l'idéologie de la société industrielle avancée*. Editions de minuit. (original edition 1964), French translation.

Secondary Sources

Surya, M. (2003). *Lignes*, themed issue, "Theodor W. Adorno et Walter Benjamin", 11.

Assoun, P.-L. (2016). *L'Ecole de Francfort*. PUF. (first edition 1997).

Works by Karl Marx

Marx, K. (1996). *Manuscrits de 1844*. Flammarion. (original edition 1844, unpublished), French translation.
Marx, K. (1973). *Grundrisse: Foundations of the Critique of Political Economy*. Harmondsworth, Penguin. (original edition 1857–1858), English translation; French translation: *Manuscrits de 1857–1858 dits « Grundrisse »*, 2011, Paris, La Dispute.
Marx, K. (2014a). *Le capital*. PUF. (original edition 1867), French translation.
Marx, K. (2014b). *Thèses sur Feuerbach*. Syllepse. (or. 1888), French translation.

C. Citizenship

I. Citizenship

Balibar, E. (2001). *Nous, citoyens d'Europe, Les frontières, l'Etat, le peuple*. La Découverte.
Balibar, E. (2009). Insurrection et constitution: la citoyenneté ambiguë. In *Revue Mouvements, Pensée critiques* (pp. 9–28). La Découverte.
Boulbina, S. (2004). Penser au pluriel: l'homme universel et le citoyen du monde. In H. Vincent (Ed.), *Citoyens du monde: Enjeux, responsabilités, concepts* (pp. 143–157). L'Harmattan.
Foessel, M. (2004). Kant: du droit cosmopolitique à l'habitation du monde. In H. Vincent (Ed.), *Citoyens du monde: Enjeux, responsabilités, concepts* (pp. 19–31). L'Harmattan.
Foray, P. (2004). L'école et la citoyenneté du monde. In H. Vincent (Ed.), *Citoyens du monde, Enjeux, responsabilités, concepts* (pp. 179–193). L'harmattan.
Gaille, M. (2015). Désir de liberté, citoyenneté et démocratie. Retour sur la question de l'actualité politique de Machiavel. *Astérion, 13*. [consulté online le 8 June 2015 http://asterion.revues/org/2623].
Holston, J. (2009). *Insurgent citizenship, disjunctions of democracy and modernity in Brazil*. Princeton University Press.
Lamarre, J.-M. (2002a). La personne. *Le Télémaque, 21*, 19–28.
Lamarre, J.-M. (2002b). Fichte et l'éducation: devenir homme parmi les hommes. *Le Télémaque, 21*, 65–80.
Lamarre, J.-M. (2006). Seule l'altérité enseigne. *Le Télémaque, 29*, 69–78.
Lamarre, J.-M. (2012). L'éducation cosmopolitique: apprendre le propre, apprendre l'étranger. *Le Télémaque, 41*, 31–46.
Lamarre, J.-M. (2020). Citoyenneté mondiale et éducation cosmopolitique. In N. Wallenhorst & E. Mutabazi (Eds.), *D'une citoyenneté empêchée à une éducation citoyenne*. Le Bord de l'eau. (*forthcoming*).
Murard, N., & Tassin, E. (2006). La citoyenneté entre les frontières. *L'Homme et la société, 160–161*, 17–35.
Neveu, C. (2004). Les enjeux d'une approche anthropologique de la citoyenneté. *Revue européenne des migrations internationales, 20*(3), 89–101.
Payan, S. (2011). De l'autochtonie à la citoyenneté: vers une construction psychique de la citoyenneté. *Topique, 114*, 115–123.
Pélabay, J. (2011). Former le "bon citoyen" libéral. *Raisons politiques, 44*, 117–138.
Sanchez-Mazas, M., & Gély, R. (2005). Des appartenances aux identités: vers une citoyenneté politique européenne. *Connexions, 84*, 73–86.

Vincent, H. (Ed.). (2004a). *Citoyens du monde: Enjeux, responsabilités, concepts*. L'Harmattan.
Vincent, H. (2004b). Citoyen du monde: éléments de problématiques. In H. Vincent (under supervision), *Citoyens du monde: Enjeux, responsabilités, concepts* (pp. 7–16). L'Harmattan.
Wallenhorst, N. (2016c). Citoyenneté existentielle et reconfiguration du politique – Les pratiques écologiques de deux jeunes professionnels. *Revue des Sciences Sociales, 55*, 116–123.
Wallenhorst, N., & Mutabazi, E. (Eds.). (2020). *D'une citoyenneté empêchée à une éducation citoyenne*. Lormont.

II. Crisis in Politics

Batout, J., & Constantin, E. (2014). Croissance, crise et dépérissement de la politique. *Le Débat, 182*, 145–155.
Chauvel, L. (2007). L'âge de l'Assemblée (1946–2007). Soixante ans de renouvellement du corps législatif: bientôt la troisième génération. *La Vie des idées*, 22 October 2007. http://www.laviedesidees.fr/L-age-de-l-Assemblee-1946-2007,81.html
De Lapparent, O. (2017). La crise de la civilisation selon Raymond Aron à travers l'exemple européen. *Bulletin de l'Institut Pierre Renouvin, 45*(1), 177–183.
Donegani, J.-M. (2011). L'autocompréhension du catholicisme, entre critique et attestation. *Raisons Politiques, 4*, 5–14.
Donegani, J.-M. (2013). Crise de l'Occident, crise du christianisme, crise de la différence. *Recherches de Science Religieuse, 101*, 351–376.
Donegani, J.-M., & Sadoun, M. (1999). Des institutions en crise. *Le Débat, 106*, 182–191.
Ehrwein Nihan, C. (2013). Penser la crise ou plaidoyer pour une réflexion critique sur la crise de la crise à partir de l'œuvre de Hannah Arendt. *Revue d'éthique et de théologie morale, 276*, 43–60.
Rozès, S. (2005). Aux origines de la crise politique. *Le Débat, 134*, 4–18.
Touraine, A. (2010). *Après la crise*. Seuil.

III. Other Works of Political Theory

Aron, R. (1962). A propos de la théorie politique. *Revue française de science politique, 1*, 5–26.
Braud, P. (2002). *Sociologie politique*. LGDJ.
Delannoi, G. (1982). *Crise intellectuelle et tentatives de fondation d'une politique de l'homme: « Arguments », Edgar Morin, Jean-Paul Sartre*, Doctoral thesis, Institut d'études politiques de Paris.
Demonet, M.-L. (2005). Quelques avatars du mot 'politique' (XIVème-XVIIème siècles). *Langage et société, 113*, 33–61.
Donegani, J.-M. (2006). De l'anthropologie au politique. *Raisons politiques, 22*, 5–14.
Donegani, J.-M., & Sadoun, M. (2007). *Qu'est-ce que la politique?* Gallimard.
Graeber, D. (2018a). *Pour une anthropologie anarchiste*. Lux. (original edition 2004), French translation.
Graeber, D. (2018b). *La démocratie aux marges*. Flammarion. (original edition 2014), French translation.
Jacq, A., & Guespin-Michel, J. (2015). Science et démocratie: une articulation difficile mais nécessaire. *Ecologie & politique, 51*, 107–120.
Oakeshott, M. (1991). *Rationalism in Politics and Others Essays*. Liberty Fund.

Wilhelm Graf, F. (2002). Le politique dans la sphère intime, Protestantisme et culture en Allemagne au XIXème siècle. *Annales. Histoire, Sciences Sociales, 3.*, 57ème année, 773–787.

IV. Hospitality

Boudou, B. (2012a). Eléments pour une anthropologie politique de l'hospitalité. *Revue du MAUSS, 40*, 267–284.

Boudou, B. (2012b). La traversée du politique: Derrida et Ricœur entre pureté de la philosophie et tragique de l'action. *Raisons Politiques, 45*, 211–233.

Cornu, L. (2007). Confiance, étrangeté et hospitalité. *Diogène, 220*, 15–29.

Cornu, L. (2004). Espace du monde, lieux de citoyenneté, liens d'hospitalité. In H. Vincent (Ed.), *Citoyens du monde: Enjeux, responsabilités, concepts* (pp. 284–300). L'Harmattan.

Gotman, A. (1997). La question de l'hospitalité aujourd'hui. *Communications, 65*, 5–19.

Honneth, A. (2004). La Théorie de la reconnaissance: une esquisse. *Revue du MAUSS, 23*, 133–136. French translation.

Moreau, D. (2012). L'étrangeté de la formation de soi. *Le Télémaque, 41*, 115–132.

Poché, F. (2014). *Le temps des oubliés, refaire la démocratie*. Lyon.

Poché, F., & Piolot, F. (interviews). (2013). *A-t-on encore le droit d'être fragile?* Chronique Sociale.

Schérer, R. (1997). Cosmopolitisme et hospitalité. *Communications, 65*, 59–68.

Tassin, E. (2004). Le cosmopolitique à l'épreuve de l'étranger. In H. Vincent (Ed.), *Citoyens du monde: Enjeux, responsabilités, concepts* (pp. 269–282). L'Harmattan.

Theobald, C. (2008). Le christianisme comme style. Entrer dans une manière d'habiter le monde. *Revue d'éthique et de théologie morale, 251*(HS), 235–248.

V. Other Works on the Future

Badré, M., Bourg, D., Brunetière, J.-R., Gadrey, J., Granjean, A., Perret, A., & Villalba, B. (2011). Les droits de l'avenir. *Esprit, 3*, 205–235.

Dryzek, J. S., & Niemeyer, S. (2008). Discursive representation. *American Political Science Review, 102*(4), 481–493.

Dupuy, J.-P. (2009). Le Futur bifurque-t-il? Vers une nouvelle science du futur. In M. Grossetti, M. Bessin, & C. Bidart (Eds.), *Bifurcations* (pp. 373–386). La Découverte.

Dupuy, J.-P. (2012). On peut ruser avec le destin catastrophique. *Critique, 783–784*, 729–737.

Dupuy, J.-P. (2007). Some Pitfalls in Philosophical Foundations of Nanoethics. *Journal of Medicine and Philosophy, 32*, 237–261.

Illich, I. (1971b). *Libérer l'avenir*. Seuil. French translation.

Jonas, H. (1990). *Le principe responsabilité*. Cerf. French translation 1990 (or. 1979).

Jonas, H. (1998). *Pour une éthique du futur*. Payot & Rivages. French translation.

Read, R. *Guardians of the future: A constitutional case for representing and protecting future people*. http://www.greenhousethinktank.org/uploads/4/8/3/2/48324387/guardians_inside_fina l.pdf

Rosanvallon, P. (2010). Le souci du long terme. In D. Bourg & A. Papaux (Eds.), *Vers une société sobre et désirable* (pp. 151–162). PUF.

Saudan, A. (2008). Hans Jonas: la tentative ou la tentation d'un fondement ontologique du devoir. *Le Philosophoire, 30*, 111–125.

VI. Politics and Education

Ardoino, J. (1999). *Education et politique*. Anthropos. (original edition 1977).
Biesta, G. J. J. (2006). *Beyond learning. Democratic education for a human future*. Paradigm Publishers.
Hess, R. (1999). Préface. In J. Ardoino (Ed.), *Education et politique* (pp. 9–14). Anthropos.
Lavelle, P. (2007). Marino Pulliero, *Le désir d'authenticité Walter Benjamin et l'héritage de la* Bildung *allemande*. *Labyrinthe, 26*, 107–112.
Monjo, R. (2013). L'école peut-elle être, à la fois, libératrice et obligatoire? *Le Télémaque, 43*, 87–102.
Ricken, N. (2002). La philosophie de l'éducation de langue allemande. Rupture avec l'un: différence – pluralité – socialité. *Le Télémaque, 21*, 121–142.
Wallenhorst, N. (2016d). Politique et éducation en anthropocène. *Raisons Politiques (Presses de Sciences Po), 62*, 153–161.
Wallenhorst, N., Robin, J.-Y., & Boutinet, J.-P. (2016). L'émancipation éducative comme posture paradoxale. *Recherches et éducations, 16*, 155–171.
Wallenhorst, N., Poché, F., Robin, J.-Y., Heslon, C., Bergier, B., & Boutinet, J.-P. (2017). Il est grand temps de rallumer les étoiles. revue-*esprit.fr. Online.* http://www.esprit.presse.fr/actualites/nathanael-wallenhorst-bertrand-bergier-jean-pierre-boutinet-christian-heslon-fred-poche-jean-yves-robin/il-est-grand-temps-de-rallumer-les-etoiles-543

VII. Other Manifestos

Allard, L. (2007). A propos du *Manifeste cyborg*, d'*Ecce Homo* et de *La promesse des monstres* ou comment Haraway n'a jamais été posthumaniste. In D. Haraway (Ed.), *Manifeste cyborg: science, technologie et féminisme socialiste à la fin du 20ème siècle* (pp. 19–27). Exils éditeur. (original edition 1985), French translation.
Askenazy, P., Coutrot, T., Orléan, A., & Sterdyniak, H. (2010a). *Manifeste d'économistes atterrés*. Les liens qui libèrent.
Batho, D. (2019). *Ecologie intégrale – Le manifeste*. Les éditions du Rocher.
Bauwens, M., Kostakis, V., & Pazaitis, A. (2018) *The Commons Manifesto – Peer to Peer, A paraître*. http://www.p2plab.gr/en/archives/117, version 2018.
Gardey, D. (2007). Avant-propos. Deux ou trois choses que je dirais d'elle. In D. Haraway (Ed.), *Manifeste cyborg: science, technologie et féminisme socialiste à la fin du 20ème siècle* (pp. 9–16). Exils éditeur. (original edition 1985), French translation.
Haraway, D. (2007). *Manifeste cyborg: science, technologie et féminisme socialiste à la fin du 20ème siècle*. Exils éditeur. (original edition 1985), French translation.
Les convivialistes. (2013). *Manifeste convivialiste*. Le bord de l'eau.
Les économistes atterrés. (2015). *Nouveau Manifeste des économistes atterrés*. Les liens qui libèrent.
Notre affaire à tous. (2019). *Comment nous allons sauver le monde – Manifeste pour une justice climatique*. Massot éditions.

D. Education and Pedagogy

I. Educational Reflections

1. General or Specialised Contributions on Education

Brougère, G., & Ulmann, A.-L. (Eds.). (2009). *Apprendre de la vie quotidienne*. PUF.

Chalier, C. (2008). *Transmettre de génération en génération*. Buchet Castel.

Chartrin, T. (2015). *Apprendre à vivre, c'est se savoir mortel: récits "épiphaniques » d'autoformation existentielle*. Doctoral thesis in Educational Science, University of Nantes, defended in December 2015.

Circulaire no. 77-300 du 29 August 1977, Programmation et Coordination: bureau DGPC 9, Instructions pédagogiques, Article 525-0.

Dumont, L. (1991). *L'idéologie allemande*. Galimard.

Lapassade, G. (1997). *L'entrée dans la vie, Essai sur l'inachèvement de l'homme*. Anthropos. (original edition 1963).

Maigné, C. (2002). J. F. Herbart: Pédagogie humaniste et critique du sujet. *Le Télémaque, 21*, 51–64.

Masschelein, J. (2002). Education et humanisme? *Le Télémaque, 21*, 37–50.

Parkes, C. M. (1971). Psycho-social transitions: A field for Study. *Social Sciences and Medicine, 5*, 101–115.

Von Bonsdorff, P., & Grenaudier-Klijn, F. (2011). Esthétique et *Bildung*. *Diogène, 233–234*, 178–195.

Wallenhorst, N. (2013). *L'école en France et en Allemagne, Comparer des expériences scolaires*. Peter Lang.

Wallenhorst, N. (2020b). La valeur de l'expression du désaccord dans l'espace de la classe. In A. Wagener & L. Nicolas (Eds.), *La valeur du désaccord*. Les presses de la Sorbonne. (*forthcoming*).

2. Educational Sociology

Dubet, F. (1991). *Les lycéens*. Seuil.

Dubet, F., Cousin, O., & Guillemet, J.-P. (1991). Sociologie de l'expérience lycéenne. *Revue française de pédagogie, 94*, 5–12.

Dubet, F., & Martuccelli, D. (1996). *A l'école ! Sociologie de l'expérience scolaire*. Seuil.

Dubet, F., & Martuccelli, D. (1998). Sociologie de l'expérience scolaire. *L'orientation scolaire et professionnelle, 27*(2), 169–187.

3. Education for the Environment and Education in the Anthropocene

Brière, L., Sauvé, L., & Jickling, B. (2010–2011). Vivre ensemble sur Terre: un projet éducatif à dimension politique. *Education relative à l'environnement, 9*, 251–263.

Curnier, D. (2017) *Quel rôle pour l'école dans la transition écologique? Esquisse d'une sociologie politique, environnementale et prospective du curriculum prescrit*. Doctoral thesis in Environmental Sciences from the University of Lausanne, supervised by Dominique Bourg and Philippe Hertig.

Delorme, D. (2019). Écologie et éducation: l'exemple des écotopies. In N. Wallenhorst & J.-P. Pierron (Eds.), *Eduquer en Anthropocène*. Le Bord de l'eau.

Federau, A. (2019). Martiens et terriens: quelle anthropologie pour l'Anthropocène? In N. Wallenhorst & J.-P. Pierron (Eds.), *Eduquer en Anthropocène*. Le Bord de l'eau.

Garnier, A., Martinez, M.-L., Hétier, R., & Wallenhorst, N. (coord.). (2021). L'éducation en Anthropocène. *Recherches et éducation*.

Garric, G., & Miller, M. (2019, March 23). Le dérèglement climatique est trop peu enseigné, de l'école à l'université. *Le Monde*. https://www.lemonde.fr/planete/article/2019/03/23/le-climat-trop-peu-enseigne-de-l-ecole-a-l-universite_5440113_3244.html

Hétier, R. (2019). Apprendre à faire vivre en Anthropocène. In N. Wallenhorst & J.-P. Pierron (Eds.), *Eduquer en Anthropocène*. Le Bord de l'eau.

Hétier, R. (2021). *L'humanité contre l'Anthropocène. Lutter contre les effondrements*. PUF.

Hétier, R. (2022). *La liberté à corps perdu*. Le Pommier.

Hétier, R., & Wallenhorst, N. (coord.). (2021). L'éducation politique en Anthropocène. *Le Télémaque, 58*.

Hétier, R., & Wallenhorst, N. (2022). *Enseigner à l'époque de l'Anthropocène*. Le Bord de l'eau.

Jagodzinski, J. (Ed.). (2018). *Interrogating the Anthropocene. Ecology, aesthetics, pedagogy, and the future in question*. Palgrave Macmillan.

Lange, J.-M., & Kebaïli, S. (2019). Penser l'éducation au temps de l'anthropocène: conditions de possibilités d'une culture de l'engagement. *Education et Socialisation, 51*, 1–15.

Leinfelder, R. (2013). Assuming responsibility for the Anthropocene: Challenges and opportunities in education. In H. Trischler (Ed.), *Anthropocene – Envisioning the future of the age of humans, RCC perspectives* (Vol. 2, pp. 9–28). Rachel Carson Center.

Paulsen, M., Jagodzinski, J., & Hawke, S. M. (2022). *dir. Pedagogy in the Anthropocene. Re-wilding education for a New Earth*, Palgrave Macmillan.

Pineau, G. (dir). (1992). *De l'air. Essai sur l'éco-formation*. Païdéia/Sciences et culture.

Priyadharshini, E. (2021). *Pedagogies for the post-Anthropocene – Lessons from apocalypse, revolution and Utopia*. Springer.

Prouteau, F. (2021). *Odyssée pour une Terre habitable*. Le Pommier.

Sauvé, L. (2015). Education à l'environnement. In D. Bourg & A. Papaux (Eds.), *Dictionnaire de la pensée écologique* (pp. 376–379). PUF.

Sauvé, L. (2009). Le rapport entre éthique et politique: un enjeu pour l'éducation relative à l'environnement. *Education relative à l'environnement, 8*, 147–162.

Sigaut, O. (2010–2011). L'éducation à l'environnement, entre politique et politiques publiques. *Education relative à l'environnement, 9*, 59–75.

Taleb, M. (2016). Ecoformation. In A. Choné, I. Hajek, & P. Hamman (Eds.), *Guide des Humanités environnementales* (pp. 83–91). Presses universitaires du Septentrion.

Tannock, S. (2021). *Educating for radical social transformation in the climate crisis*. Palgrave MacMillan.

Wallace, M. F. G., Bazzul, J., Higgins, M., & Tolbert, S. (Eds.). (2022). *Reimagining Science Education in the Anthropocene*. Palgrave Macmillan.

Wallenhorst, N., Hétier, R., Pierron, J.-P., & Wulf, C. (Eds.). (2023). *Political education in the Anthropocene*. Springer.

Wallenhorst, N., Robin, J.-Y., & Boutinet, J.-P. (2019). L'émergence de l'Anthropocène, une révélation étonnante de la condition humaine? In N. Wallenhorst & J.-P. Pierron (Eds.), *Eduquer en Anthropocène*. Le Bord de l'eau.

Wallenhorst, N., & Wulf, C. (Eds.). (2022). *Humains. Un dictionnaire d'anthropologie prospective*. Vrin.

Wulf, C. (2019). La transformation à l'ère de l'Anthropocène – Mimesis, rituels, gestes. In N. Wallenhorst & J.-P. Pierron (Eds.), *Eduquer en Anthropocène*. Le Bord de l'eau.

Wulf, C. (2013). *Anthropology. A continental perspective*. University of Chicago Press.

Wulf, C., & Zirfas, J. (Eds.). (2014). *Handbuch Pädagogische Anthropologie*. Springer VS.

Wulf, C., & Zirfas, J. (Eds.). (2020, January 29). *Den Menschen neu denken.* Paragrana: Internationale Zeitschrift für Historische Anthropologie.
Wulf, C. (2022a). *Education as human knowledge in the Anthropocene. An Anthropological perspective*. Routledge.
Wulf, C. (2022b). *Human beings and their images. Imagination, mimesis and performativity.* Bloomsbury.

II. Pedagogical Practice

Avanzini, G. (1996). *L'éducation des adultes*. Anthropos.
Colin, L., & Le Grand, J.-L. (Eds.). (2008). *L'éducation tout au long de la vie*. Anthropos.
Delory-Momberger, C. (2003). *Biographie et éducation*. Anthropos.
Delory-Momberger, C. (2005). *Histoire de vie et recherche biographique en éducation*. Anthropos.
Delory-Momberger, C. (2004). *Les histoires de vie*. Anthropos.
Delory-Momberger, C. (2009). Postface, Etre adulte ou la condition impossible. In J.-P. Boutinet & P. Dominicé (Eds.), *Où sont passés les adultes?* Téraèdre.
Dominicé, P. (2007). *La formation biographique*. L'Harmattan.
Dominicé, P. (2016). *Au risque de se dire*. Téraèdre.
Hess, R., & Weigand, G. (2008). L'éducation tout au long de la vie: une théorie de l'expérience comme processus d'apprentissage et de connaissance. In L. Colin & J.-L. le Grand (Eds.), *L'éducation tout au long de la vie* (pp. 5–22). Anthropos.
Manifet, C. (2015). L'éducation des adultes dans les universités en France. Une mission hétérodoxe? *Cahiers de la recherche sur l'éducation et les savoirs, 14*, 267–294.
Pineau, G. (2003). Vers une anthropo-formation en deux temps trois mouvements. *Spirale, 31*, 35–46.
Pineau, G., Allard, D., Deroy, F., Fouilloux, G., Lerbet, G., Vidricaire, A., Villevieille, A., Zayed, J., & Jay-Rayon, J.-C. (1992). *De l'air, Essai sur l'écoformation*. Païdeia.
Robin, J.-Y. (2007). Ingénium de l'accompagnement. In J.-P. Boutinet, N. Denoyel, G. Pineau, & J.-Y. Robin (Eds.), *Penser l'accompagnement adulte* (pp. 243–249). PUF.
Robin, J.-Y. (2015). Adulte certes, mais postmoderne ! Le recours biographique. In A. Slowik & O. Czerniawska (Eds.), *Trajets de formation et approche biographique* (pp. 29–50). L'Harmattan.
Robin, J.-Y., & Wallenhorst, N. (2016). Une pédagogie de l'engagement face à la désorientation. *Revue Internationale de Pédagogie de l'Enseignement Supérieur, 32–33*, 1–12.
Wallenhorst, N. (2015). L'étudiant face à la désorientation. In M.-H. Jacques (Ed.), *Les transitions en contexte scolaire* (pp. 281–293). Rennes.

E. Other Elements for Political Anthropology

I. Other Philosophical Works

1. Primary Sources

Aristotle. (1995). *La politique, I., 2*. Vrin. (original edition circa -350), French translation.
Blanchot, M. (1980). *L'Ecriture du désastre*. Gallimard.

Deleuze, G. (2015). *Lettres et autres textes*. Les Editions de Minuit.
de Montaigne, M. (2002). *Les essais*. LGF.
Descartes, R. (1953). *Le discours de la méthode, Œuvre et lettres*. La Pléiade, Gallimard. (original edition 1637).
Descola, P. (2005). *Par-delà nature et culture*. Gallimard.
Ellul, J. (2008). *La technique ou l'enjeu du siècle*. Economica. (original edition 1954).
Kant, I. (1991). *Qu'est-ce que les Lumières?* Flammarion. (original 1784), French translation.
Kant, I. (2006). *Vers la paix perpétuelle*. Flammarion. (original 1795), French translation.
Kemp, P. (2013). *Le Prince*. Seuil.
Koselleck, R. (1995). Glühende Lava. Vielerlei Abschied vom Krieg. In H. L. Arnold, B. Sauzay, & R. von Thadden (Eds.), *Vom Vergessen, Vom Gedenken. Erinnerung und Erwartungen in Europa zum 8. Mais 1945* (pp. 19–25). Göttigen.
Lacoue-Labarthe, P. (1997). *La poésie comme expérience*. Christian Bourgeois.
Land, N. (2011). *Fanged Noumena: Collected writings 1987–2007*. Urbanomic.
Lévinas, E. (1991). *Entre nous – Essais sur le penser-à-l'autre*. Grasset.
Lévinas, E. (1972). *L'Humanisme de l'autre homme*. Fata Morgana.
Lévi-Strauss, C. (1973). *Anthropologie structurale II*. Plon. (original edition 1962).
Lyotard, J.-F. (1979). *La condition postmoderne*. Editions de minuit.
Merleau-Ponty, M. (1945). *Phénoménologie de la perception*. Gallimard.
Mounier, E. (2001). *Le personnalisme*. PUF. (original edition 1950).
Plato. (2002). *La République*. Flammarion. (written -385 / -370), French translation.
Plato. (1999). *Phédon*. Flammarion. (written -385 / -370), French translation.
Plato. (1997). *Protagoras, 312b*. Flammarion. French translation (written -399 / -390), French translation.
Poché, F. (2012). Après la dé-symbolisation. Quel avenir pour les quartiers populaires? *Cahiers de l'Atelier, 532*, 45–54.
Ricœur, P. (1990). *Soi-même comme un autre*. Seuil.
Rorty, R. (1990). *L'homme spéculaire*. Seuil. (original edition 1979), French translation.
Saint Augustin. (1993). *Les Confessions*. Flammarion. (original edition 401), French translation.
Serres, M. (2019). *Morales espiègles*. Le Pommier.
Theobald, C. (2007). *Le christianisme comme style – Une manière de faire de la théologie en postmodernité*. Cerf.
Tillich, P. (1999). *Le courage d'être*. Cerf. French translation.
Tonus, M. (2019). *Ouvrir l'espace du christianisme – Introduction à la pensée pionnière de Maurice Bellet*. Albin Michel.
Weber, M. (2004). *L'éthique protestante et l'esprit du capitalisme*. Gallimard. (original edition 1904–1905), French translation.
Weil, E. (1982). *Philosophie et réalité, Derniers essais et conférences*. Beauschene.
Zundel, M. (1990). *Recherches sur la Personne*. Desclée de Brouwer.

2. Secondary Sources

Benveniste, E. (1969). *Le vocabulaire des institutions indo-européennes, Tome II Pouvoir, Droit, Religion*. Les Editions de Minuit.
Collective Work. (1997). *Autour de Jacques Rancière*, *Critique*, no. 601–602.
Collective Work. (2006). *La philosophie déplacée. Autour de Jacques Rancière*. Actes du colloque de Cerisy, Bourg en Bresse, Horlieu.
Derrida, J. (1997). *Adieu (à Emmanuel Lévinas)*. Galilée.
Fiasse, G. (Ed.). (2008). *Paul Ricœur. De l'homme faillible à l'homme capable*. PUF.
Goethe, J. F. (1893). "Studie nach Spinoza" (1784–1785), in *Goethe-Jahrbuch*, Band 12 (1891), in *Seine Nachwerke in Anmerkungen*, I, 2, Abt., Band 11, 1. Teil, Weimar.
Guibal, F. (1995, November). Eric Weil. Le défi de la violence. *Etudes*, 495–504.

Jacquinod, B. (1988). Etude de vocabulaire grec: αὔξω et ἄωτος. *Revue des études anciennes, 90*(3–4), 315–323.
Kearney, R. (2001). L'autre et l'étranger: entre Derrida et Ricœur. In J.-F. Mattéi (Ed.), *Philosopher en français* (pp. 95–111). PUF.
Legendre, P. (1998). La totémisation de la société. Remarques sur les montages canoniques et la question du sujet. In A. van Debeek & K. van der Toorn (Eds.), *Canonization und Decanonization* (pp. 425–434). Brill.
Virgoulay, R. (2005). Chronique blondélienne quelques publications récentes (2001–2004). *Recherche de sciences religieuses, 93*, 65–76.

II. Other Economic Works

Askenazy, P., Coutrot, T., Orléan, A., & Sterdyniak, H. (2010b). *Manifeste d'économistes atterrés*. Les liens qui libèrent.
Delannoy, I. (2017). *L'économie symbiotique – Régénérer la planète, l'économie et la société*. Actes sud.
Economistes atterrés. (2015). *Nouveau manifeste des économistes atterrés*. Les liens qui libèrent.
Jackson, T. (2010). *Prospérité sans croissance: la transition vers une économie durable*. De Boeck-Etopia. French translation.
Latouche, S. (2010). La décroissance est-elle la solution de la crise? *Ecologie & Politique, 40*, 51–61.
Mathieu, P. L. (1969). *La pensée politique et économique de Teilhard de Chardin*. Seuil.
Peyrelevade, J. (2005). *Le capitalisme total*. Seuil.
Menger, P.-M. (2003). *Portrait de l'artiste en travailleur, Métamorphoses du capitalisme*. Seuil.

III. Sociology of the Contemporary Individual

Aubert, N. (Ed.). (2005). *L'individu hypermoderne*. Eres.
Baumann, Z. (2013). *La vie liquide*. Fayard. French translation.
Bergier, B. (1992). *Compagnons d'Emmaüs – Sociologie du quotidien communautaire*. Editions Ouvrières.
Bergier, B. (2014). *Comment vivre ensemble?* Chronique Sociale.
Bréchet, J.-P., & Prouteau, L. (2010). A la recherche de l'entrepreneur. Au-delà du modèle du choix rationnel: une figure de l'agir projectif. *Revue française de socio-économie, 6*, 109–130.
Cabanas, E., & Illouz, E. (2018). *Happycratie*. Premier Parallèle.
Charles, S. (2008). *Les temps hypermodernes*. Le livre de poche.
de Singly, F. (2003). *Les uns avec les autres. Quand l'individualisme crée du lien*. Armand Colin.
Dubet, F. (1994). *Sociologie de l'expérience*. Seuil.
Dubet, F. (2002). *Le déclin de l'institution*. Seuil.
Dubet, F., & Wieviorka, M. (Eds.). (1995). *Penser le sujet autour d'Alain Touraine*. Paris.
Ehrenberg, A. (1998). *La fatigue d'être soi. Dépression et société*. Odile Jacob.
Gauchet, M. (1985). *Le désenchantement du monde*. Gallimard.
Giddens, G. (1994). *Les conséquences de la modernité*. L'Harmattan. French translation.
Giddens, A. (2005). *La constitution de la société*. PUF. French translation.
Lahire, B. (1998). *L'homme pluriel*. Nathan.
Martuccelli, D. (2002). *Grammaires de l'individu*. Gallimard.
Schmitt, L. (2014). *La valse des égo*. Odile Jacob.

Szuba, M. (2015). Sociologie environnementale. In D. Bourg & A. Papaux (Eds.), *Dictionnaire de la pensée écologique* (pp. 942–946). PUF.
Touraine, A. (1984). *Le retour de l'acteur*. Fayard.
Touraine, A. (1997). *Pourrons-nous vivre ensemble? Egaux et différents*. Fayard.
Touraine, A. (2013). *La fin des sociétés*. Seuil.
Touraine, A. (2015). *Nous, sujets humains*. Seuil.
Touraine, A., & Khosrokhavar, F. (2000). *La recherche de soi: dialogue sur le sujet*. Fayard.

IV. Anthropological Analyses

1. Anthropological Reflections

Anders, G. (2002). *L'obsolescence de l'homme, Sur l'âme à l'époque de la deuxième révolution industrielle*. Ivrea. (original edition 1956), French translation.
Balibar, E. (2012). L'introuvable humanité du sujet moderne. L'universalité "civique-bourgeoise" et la question des différences anthropologiques. *L'Homme, 203–204*, 19–50.
Balibar, E., & Gebauer, G. (2012). L'anthropologie philosophique et l'anthropologie historique en débat. *Rue Descartes, 75*, 81–101.
Breuvart, J.-M. (2009). Eric Weil et la question de Dieu selon H. Bouillard. *Recherches de Science Religieuse, 97*, 185–193.
Clément, B. (Ed.). (2006). *L'Homme capable. Autour de Paul Ricœur, HS de la Revue du Collège International de Philosophie*. PUF.
Faber, M., Petersen, T., & Schiller, J. (2002). Homo oeconomicus and homo politicus in Ecological Economics. *Ecological Economics, 40*, 323–333.
Génard, J.-L. (1999). *La grammaire de la responsabilité*. Cerf.
Harari, Y. N. (2017). *Homo deus, une brève histoire de l'avenir*. Albin Michel. (original edition 2015), French translation.
Haudry, J. (2016). *Le feu dans la tradition indo-européenne*. Archè.
Leroi-Gourhan, A. (1992). *Le geste et la parole, I. Technique et langage*. Albin-Michel. (original edition 1964).
Nyborg, K. (2000). Homo Economicus and Homo Politicus: interpretation and aggregation of environmental values. *Journal of Economic Behavior & Organization, 42*, 305–322.
Prouteau, F. (2004). *Anthropologie et pédagogie de la vocation*. Doctoral thesis in Educational Sciences, University of Lyon 2, ANRT, Lille.
Prouteau, F. (2006). *Former, oui... mais dans quel sens?* L'Harmattan.
Siebenhüner, B. (2000). Homo sustinens als Menschenbild für eine nachhaltige Ökonomie. *sowi-online, 1*, 1–13. www.sowi-onlinejournal.de/nachhaltigkeit/siebenhuener.htm

2. Transhumanism

(a). Transhumanist Theories and Research

Alexandre, L. (2011). *La mort de la mort*. Lattès.
Barthélémy, P. (2018, December 23). Sergio Canavero, l'homme qui veut greffer des têtes. *Le Monde*.
Bostrom, N. (2003). Are you living in a computer simulation? *Philosophical Quarterly, 53*(211), 243–255.
Bostrom, N. (2002). *Anthropic Bias, Observation Selection Effects in Science and Philosophy*. Routledge.

Bostrom, N. (2014). *Superintelligence: Paths, Dangers, Strategies*. Oxford University Press.

Hottois, G., Missa, J.-N., & Perbal, L. (Eds.). (2015). *L'humain et ses préfixes – Une encyclopédie du transhumanisme et du posthumanisme*. Paris.

Hughes, J. (2004). *Citizen Cyborg: Why democratic societies must respond to the redesigned human of the future*. Basic Books.

Huxley, J. (1957). *New bottles for new wine*. Chatto and Windus.

Maestrutti, M. (2011). Humain, transhumain, posthumain. Représentations du corps entre incomplétude et amélioration. *Journal International de Bioéthique, 22*, 51–66.

More, M. (2003). *Principles of extropy*. http://editions-hache.com/essais/more/more1.html

More, M. (1993). Technological self-transformation: Expanding personal extropy. *Extropy, 10*(4/2), 15–24.

More, M. (2010). The overhuman in the Transhuman. *Journal of Evolution and Technology, 21*(1), 1–4.

Savulescu, J., & Bostrom, N. (2008). *Human enhancement*. Oxford University Press.

(b). Critiques of Transhumanism

Baetschi, B. (2018). Plus vite, plus haut, plus fort, plus… pour aller où? Transhumanisme et aspiration au bonheur. *ESKA Journal International de bioéthique et d'éthique des sciences, 29*(3), 170–188.

Blin, T. (2015). Et si on liquidait l'homme! Sur le bonheur transhumaniste qui nous menace. *Le Débat, 183*, 87–100.

Caire, A.-B. (2018). La cryogénisation. *ESKA Journal international de bioéthique et d'éthique des sciences, 29*(3), 54–70.

Capelle-Dumont, P. (2019). Technique, finitude et éthique de la décision. In P. Capelle-Dumont & C. Roche (Eds.), *Puissance technologique et éthique de la décision*. Parole et silence.

Damour, F. (2017). Le transhumanisme, une idée chrétienne devenue folle? *Etudes, 7*(July-August), 51–62.

De Lauzun, P. L'économie des GAFAM. In P. Capelle-Dumont & C. Roche (Eds.), *Puissance technologique et éthique de la décision*. Parole et silence.

De Rosnay, J. (2018). Transhumanisme ou hyperhumanisme? L'avenir de l'humanité. *ESKA Journal international de bioéthique et d'éthique des sciences, 29*(3), 233–240.

Dibi Kouadio, A. Le prochain dans un univers du "sans distance". In P. Capelle-Dumont & C. Roche (Eds.), *Puissance technologique et éthique de la décision*. Parole et silence.

Frippiat, L. (2011). L'amélioration technique de l'être humain: introduction aux différents courants du débat. *Journal International de Bioéthique, 22*, 33–50.

Giesen, K.-G. (2018). Le transhumanisme comme idéologie dominante de la quatrième révolution industrielle. *ESKA Journal international de bioéthique et d'éthique des sciences, 29*(3), 189–203.

Goffi, J.-Y. (2011). Nature humaine et amélioration de l'être humain à la lumière du programme transhumaniste. *Journal International de bioéthique, 22*, 18–22.

Guényveau, F.-R. (2017). *Un dissident*. Albin Michel.

Habermas, J. (2002). *L'avenir de la nature humaine. Vers un eugénisme libéral?* Gallimard. (original edition 2001), French translation.

Hétier, R. (2018). Augmenter le pouvoir de sentir à l'heure de l'homme augmenté. In N. Wallenhorst, D. Coatanéa, & F. Prouteau (Eds.), *Eduquer l'homme augmenté – Pour une société postprométhéenne* (pp. 119–131). Le Bord de l'eau.

Hottois, G. (2018). L'anthropologie philosophique technicienne du transhumanisme. *ESKA Journal international de bioéthique et d'éthique des sciences, 29*(3), 135–153.

Larrieu, P. (2018). La neuro-amélioration des sujets "sains": enjeux anthropologiques, sociologiques et juridiques. *ESKA Journal international de bioéthique et d'éthique des sciences, 29*(3), 71–91.

Le Dévédec, N., & Collin, J. (2018). Le médicament augmenté: l'usage du médicament dans les discours transhumanistes et ses significations sociales. *ESKA Journal international de bioéthique et d'éthique des sciences, 29*(3), 93–108.

Liautaud, J.-M. (2018). La Conscience contre la Condition. Penser l'éducation de l'homme augmenté avec Eric Weil. In N. Wallenhorst, D. Coatanéa, & F. Prouteau (Eds.), *Eduquer l'homme augmenté – Pour une société postprométhéenne* (pp. 57–77). Le Bord de l'eau.

Luzi, J. (2018). Le capitalisme transhumaniste et la mort. *ESKA Journal international de bioéthique et d'éthique des sciences, 29*(3), 33–62.

Magnin, T. La dynamique du vivant entre robustesse et vulnérabilité. In P. Capelle-Dumont & C. Roche (Eds.), *Puissance technologique et éthique de la décision*. Parole et silence.

Parizeau, M.-H. Des cycles d'innovations technologiques. In P. Capelle-Dumont & C. Roche (Eds.), *Puissance technologique et éthique de la décision*. Parole et silence.

Prouteau, F. (2018). Tous connectés, ou comment on devient un homme numérique. In N. Wallenhorst, D. Coatanéa, & F. Prouteau (Eds.), *Eduquer l'homme augmenté – Pour une société postprométhéenne* (pp. 15–34). Le Bord de l'eau.

Rialle, V. (2018). Robotique humanitaire *versus* robotique suicidaire: ou comment ré-enchanter la 'silver économie'. *ESKA Journal international de bioéthique et d'éthique des sciences, 29*(3), 17–25.

Robin, J.-Y. (2018). Conditions anthropologiques de l'émergence d'un homme sans limite. In N. Wallenhorst, D. Coatanéa, & F. Prouteau (Eds.), *Eduquer l'homme augmenté – Pour une société postprométhéenne* (pp. 79–95). Le Bord de l'eau.

Thiel, M.-J. (2015). L'homme augmenté aux limites de la condition humaine. *Revue d'éthique et de théologie morale, 286*, 141–161.

Vidal, C. (2019). *Nos cerveaux resteront-ils humains?* Le Pommier.

V. Other Epistemological Works

Coutellec, L. (Ed.). (2015a). *Le temps des sciences impliquées, Ecologie et politique*, no. 51.

Coutellec, L. (2015b). Pour une philosophie politique des sciences impliquées – Valeurs, finalités, pratiques. *Ecologie et politique, 51*, 15–25.

Deléage, J.-P., & Coutellec, L. (2015). L'écologie scientifique, une science impliquée? *Ecologie et politique, 51*, 55–64.

Newman, G., Wiggins, A., Crall, A., Graham, E., Newman, S., & Crowston, K. (2012). The future of citizen science: emerging technologies and shifting paradigms. *Frontiers in Ecology and the Environment, 10*(6), 298–304.

Oreskes, N. (2015). How earth science has become a social science. *Historical Social Research, 40*, 246–270.

References

Abram, D. (2013). *Comment la Terre s'est tue – Pour une écologie des sens*. La Découverte (ed. or. 1996), tr. fr.

Arendt, H. (1972a). *La crise de la culture*. Gallimard (ed. or. 1961), tr. fr.

Arendt, H. (1972b). « Qu'est-ce que l'autorité? ». In H. Arendt (Ed.), *La crise de la culture*. Gallimard (pp. 121–185) (ed. or. 1961b), tr. fr.

Arendt, H. (1972c). « La crise de l'éducation ». In H. Arendt (Ed.), *La crise de la culture*. Gallimard (pp. 223–252) (ed. or. 1961c), tr. fr.

Arendt, H. (1972). « Qu'est-ce que la liberté ? ». In H. Arendt, *La crise de la culture*. Gallimard (pp. 186–222) (ed. or. 1961d), tr. fr.
Arendt, H. (1974). *Vies politiques*. Gallimard (ed. or. 1970), tr. fr.
Arendt, H. (1983). *Condition de l'homme moderne*. Calmann-Lévy (ed. or. 1958, tr. fr. 1961), tr. fr.
Arendt, H. (1995). *Qu'est-ce que la politique?*. Seuil (ed. or. 1993), tr. fr.
Arendt, H. (2002). *Les origines du totalitarisme III. Le système totalitaire*. Seuil (ed. or. 1948), tr. fr.
Arendt, H. (2016). *La vie de l'esprit*. PUF (ed. or. 1978), tr. fr.
Arendt, H., & Heidegger, M. (2001). *Lettres et autres documents 1925–1975*. Gallimard (ed. or. 1998), tr. fr.
Barnosky, A. D., Matzke, N. Tomiya, S., Wogan, G. O. U., Swartz, B., Quentall, T. B., Marshall, C., McGuire, J. L., Lindsey, E. L., Maguire, K. C., Mersey, B., & Ferrer, E. A. (2011). « Has the Earth's sixth mass extinction already arrived? ». *Nature, 471*, 51–57.
Bellet, M. (1992). *Incipit*. Desclée de Brouwer.
Bourg, D. (2017a). « Gouverner le long terme », lapenseeecologique.com, Points de vue (Vol. 1(1), pp. 1–9). PUF.
Bourg, D., Augagneur, F., Blondiaux, L., Cohendet, M.-A., Fourniau, J.-M., François, B., & Prieur, M. (2017b). *Inventer la démocratie du XXIème siècle*. Les liens qui libèrent.
Caillé, A. (2016). *Les Convivialistes, Eléments d'une politique convivialiste*. Bord de l'eau.
Faber, M., Petersen, T., & Schiller, J. (2002). « Homo oeconomicus and homo politicus in Ecological Economics ». *Ecological Economics, 40*, 323–333.
Haraway, D. (2007). *Manifeste cyborg: science, technologie et féminisme socialiste à la fin du 20ème siècle*. Exils éditeur (ed. or. 1985), tr. fr.
Haraway, D. (2015). « Anthropocene, Capitalocene, Plantationocene, Chthulucene : making kin ». *Environmental Humanities, 6*, 159–165.
IUCN. *The IUCN red list of threatened species, Version 2017-2*. http://www.iucnredlist.org
Jonas, H. (1990). *Le principe responsabilité*. Cerf, tr. fr. 1990 (or. 1979).
Lamarre, J.-M. (2020). « Citoyenneté mondiale et éducation cosmopolitique ». In N. Wallenhorst & E. Mutabazi (dir.), *D'une citoyenneté empêchée à une éducation citoyenne*. Le Bord de l'eau.
Leroi-Gourhan, A. (1992). *Le geste et la parole, I. Technique et langage*. Albin-Michel (ed. or. 1964).
Lussault, M. (2013). *L'avènement du Monde – Essai sur l'habitation humaine de la Terre*. Seuil.
Malm, A., & Hornborg, A. (2014). « The geology of mankind ? A critique of the Anthropocene narrative », *The Anthropocene Review, 1*, 62–69.
Mounier, E. (2001). *Le personnalisme*. PUF (ed. or. 1950).
Poché, F. (2007). *Penser avec Derrida*. Chronique Sociale.
Poché, F. (2013). (entretien avec Piolot, F.), *A-t-on encore le droit d'être fragile ?* Chronique Sociale.
Rosa, H. (2014). *Aliénation et accélération – Vers une théorie critique de la modernité tardive*. La Découverte (or 2012), tr. fr.
Shantser, E. V. (1973). « The anthropogenic system (period) ». In *The Great Soviet Encyclopedia* (Vol. 2, pp. 139–144). Macmillan.
Touraine, A., & Khosrokhavar, F. (2000). *La recherche de soi: dialogue sur le sujet*. Fayard.
Villalba, B. (2021). *Les collapsologues et leurs ennemis*. Le Pommier.

Index

© The Editor(s) (if applicable) and The Author(s), under exclusive license to Springer Nature Switzerland AG 2023
N. Wallenhorst, *A Critical Theory for the Anthropocene*, Anthropocene – Humanities and Social Sciences, https://doi.org/10.1007/978-3-031-37738-9

MIX
Papier aus verantwortungsvollen Quellen
Paper from responsible sources
FSC® C105338

If you have any concerns about our products,
you can contact us on
ProductSafety@springernature.com

In case Publisher is established outside the EU,
the EU authorized representative is:
Springer Nature Customer Service Center GmbH
Europaplatz 3, 69115 Heidelberg, Germany

Printed by Libri Plureos GmbH
in Hamburg, Germany